THOMAS REID ON MATHEMATICS AND NATURAL PHILOSOPHY

THE EDINBURGH EDITION
OF THOMAS REID

General Editor
Knud Haakonssen

1 *Thomas Reid on the Animate Creation: Papers Relating to the Life Sciences* edited by Paul Wood (1995)

2 *An Inquiry into the Human Mind on the Principles of Common Sense* edited by Derek R. Brookes (1997)

3 *Essays on the Intellectual Powers of Man* edited by Derek R. Brookes and Knud Haakonssen (2002)

4 *The Correspondence of Thomas Reid* edited by Paul Wood (2002)

5 *Thomas Reid on Logic, Rhetoric and the Fine Arts: Papers on the Culture of the Mind* edited by Alexander Broadie (2005)

6 *Thomas Reid on Practical Ethics* edited by Knud Haakonssen (2007)

7 *Essays on the Active Powers of Man* edited by Knud Haakonssen and James A. Harris (2010)

8 *Thomas Reid on Society and Politics* edited by Knud Haakonssen and Paul Wood (2015)

9 *Thomas Reid on Mathematics and Natural Philosophy* edited by Paul Wood (2017)

10 *Thomas Reid and the University* edited by Alexander Broadie and Paul Wood

THOMAS REID ON MATHEMATICS AND NATURAL PHILOSOPHY

EDITED BY

Paul Wood

The Pennsylvania State University Press
University Park, Pennsylvania

First cloth edition published in 2017 in the United States by
The Pennsylvania State University Press

First published in the United Kingdom in 2017 by
Edinburgh University Press Ltd

Typeset in Times New Roman
by R. J. Footring Ltd, Derby, and
printed and bound in Great Britain by
CPI Group (UK) Ltd, Croydon CR0 4YY

Library of Congress Cataloging-in-Publication Data

Names: Reid, Thomas, 1710–1796, author. | Wood, Paul, 1952– , editor. | Reid, Thomas, 1710–1796. Works. Selections. 1996 ; v. 9.
Title: Thomas Reid on mathematics and natural philosophy / edited by Paul Wood.
Description: University Park, Pennsylvania : The Pennsylvania State University Press ; Edinburgh : Edinburgh University Press, [2017] | Series: The Edinburgh edition of Thomas Reid ; 9 | Includes bibliographical references and index.
Summary: "Reconstructs the Scottish Enlightenment philosopher Thomas Reid's lifelong engagement with the physical sciences and makes clear why these fields were central to his epistemology and moral and social philosophy"—Provided by publisher.
Identifiers: LCCN 2017032943 | ISBN 9780271078496 (cloth : alk. paper)
Subjects: LCSH: Mathematics—Early works to 1800. | Physical sciences—Early works to 1800. | Reid, Thomas, 1710–1796.
Classification: LCC QA35 .R45 2017 | DDC 500—dc23
LC record available at https://lccn.loc.gov/2017032943

ISBN 978-0-271-07849-6

The Pennsylvania State University Press is a member of the Association of American University Presses.

It is the policy of The Pennsylvania State University Press to use acid-free paper. Publications on uncoated stock satisfy the minimum requirements of American National Standard for Information Sciences—Permanence of Paper for Printed Library Material, ANSI Z39.48–1992.

Contents

Acknowledgements

I am grateful to the Sir Duncan Rice Library of the University of Aberdeen for permission to publish the manuscripts included in this volume and for permission to reproduce on p. viii the image of MS 2131/7/II/22, and to the Library of Birmingham for permission to quote from the letter from Patrick Wilson to James Watt discussed below. I am also greatly indebted to a number of friends and colleagues for their help during the preparation of this book: Geoffrey Cantor; Patrick Corbeil; Niccolò Guicciardini; Knud Haakonssen; Michael Silverthorne; Stephen Snobelen; the Reading Room Manager, Michelle Gait, and the staff of the Wolfson Reading Room, Special Collections Centre, University of Aberdeen; and Julie Gardham, Senior Assistant Librarian, Special Collections, Glasgow University Library. Stephen Snobelen deserves special mention for generously providing an English translation of the Latin text 'Scholium ad Propositionem 26 Liber 3 Principia Newtoni' transcribed below. In addition, I am especially grateful to Ralph Footring for his patience and for his meticulous work on this and the previous volume of the Edinburgh Edition of Thomas Reid. Lastly, my sincere thanks go to the Social Sciences and Humanities Research Council of Canada and to the Hugh Campbell and Marion Alice Small Fund for Scottish Studies at the University of Victoria for the funding which supported my editorial work. I am solely responsible for any errors of transcription, fact or interpretation that remain in this volume.

Victoria
June 2017

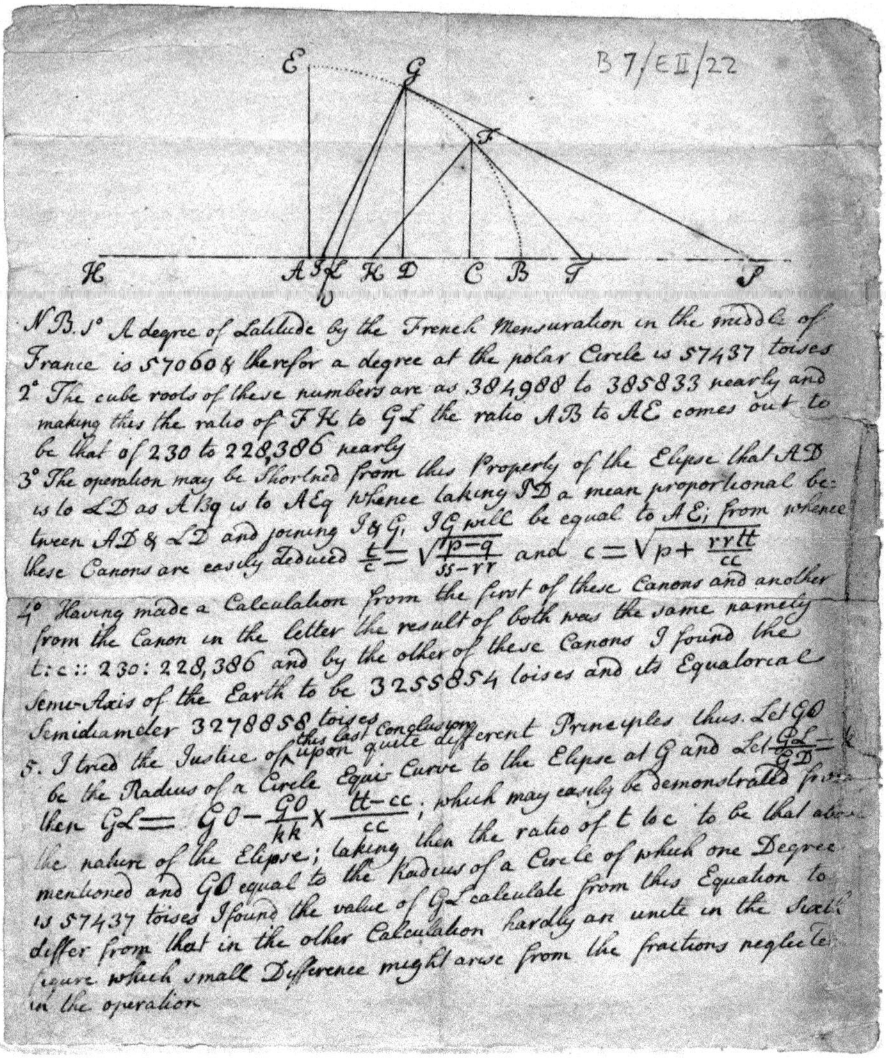

A folio from the group of Reid's manuscripts related to the controversy over the shape of the Earth, AUL 2131/7/II/22 – see p. 62. Image courtesy of the University of Aberdeen.

The Manuscripts and Editorial Principles

The manuscripts included in this volume are held in the Special Collections Centre in the Sir Duncan Rice Library, University of Aberdeen. The majority of the manuscripts transcribed here are found in the Birkwood Collection, MSS 2131/1–8, while the remainder are from the deposit of Reid's papers catalogued as MSS 3061/1–26. All of the manuscripts in these two collections have been digitised and are available through the Special Collections Centre website (https://www.abdn.ac.uk/historic/Thomas_Reid/index.shtml). The history of these manuscripts, as well as others housed in the Special Collections Centre, will be discussed in an appendix to volume 10 of the Edinburgh Edition of Thomas Reid.

Prior to Reid's death on 7 October 1796 he appears to have sorted at least some of his papers into thematic groups, for among the Birkwood Collection wrappers survive in his hand headed 'Politicks' and 'Nat Hist' (i.e. 'Natural History'). Immediately after he died, his papers were organised into bundles by his close colleague, the Glasgow Professor of Logic and Rhetoric George Jardine, who a few months later collaborated with his fellow professors Patrick Wilson and Archibald Arthur in choosing the titles from Reid's personal library that were gifted to the University in early 1797 by Reid's daughter Martha. Of the eleven known bundles assembled by Jardine, three consisted of manuscripts on mathematics and the natural sciences, including one made up exclusively of notes from the lectures on natural history and natural philosophy Reid gave as a regent at King's College, Aberdeen, in the period 1751 to 1764. A fourth bundle, which included papers Reid had read to the Glasgow Literary Society, contained material on mathematical and scientific subjects which eventually resurfaced in the public domain as part of the MS 3061 deposit. Apart from this particular bundle, the contents of the other ten bundles have, over time, lost their original arrangement. The Birkwood Collection as we now know it was given shape initially by A. T. W. Liddell in the 1950s, whose ordering of the surviving Reid manuscripts in the Collection was revised by David Fate Norton in the 1970s. Other materials in Reid's hand are to be found in the cache of Thomas Gordon's and Robert Eden Scott's papers discovered in 1982 and catalogued in the Special Collections Centre as MSS 3107/1–9, as well as in MSS 2341 and 2343. The former is a manuscript on book-keeping for farmers that Reid submitted to the Gordon's Mill Farming Club in 1761, while the latter is a lengthy and highly detailed discussion of chemical topics of uncertain date. The provenance of these two manuscripts is unknown; hence it is unclear if they were once among the original body of Reid's papers or whether one or both remained in Aberdeen when Reid departed for Glasgow in 1764.

The Birkwood Collection and MSS 3061 together contain approximately 640 separately catalogued items, of which a little over 40% deal with mathematical and scientific topics. The items in this group range from fair copies of Reid's 'An Essay on Quantity' and formal discourses he delivered to the Glasgow Literary Society to heavily worked over drafts of papers he read not only to the Literary Society but also to the Aberdeen Philosophical Society, and from sets of reading notes of varying lengths to random jottings and mathematical calculations. Given the sheer quantity and disparate nature of the surviving manuscripts on mathematics, medicine and the natural sciences, a number of editorial decisions had to be made about how best to illustrate Reid's work in these areas. I have already discussed some of these issues in volume 1 of the Edinburgh Edition of Thomas Reid, which contains an extensive selection of his manuscripts dealing with the life sciences. Although Reid's papers directly related to his teaching of mathematics and natural philosophy form a distinct and coherent group which merits separate treatment, in Part Five I have included one manuscript containing lecture notes on electricity because it provides the only systematic statement of his interpretation of electrical phenomena to be found in the Birkwood Collection. I have largely excluded sets of reading notes, except for those which document significant steps in the development of Reid's thought, as in Parts One, Five and Six.

Reid's voluminous mathematical manuscripts pose the most difficult editorial problems. As I will suggest in my Introduction, part of the distinctiveness of Reid's intellectual identity is based on the fact that he was as much a mathematical practitioner as he was an epistemologist during the course of his career. Yet the sheer diversity of his interests in pure and practical mathematics means that it is difficult to present in a readily comprehensible fashion the full range of manuscripts that illustrate this aspect of his identity. Moreover, it is equally difficult to pick out individual narrative threads in the array of papers dealing with the diverse mathematical topics that attracted Reid's attention. Nevertheless, his surviving manuscripts do tell a coherent and self-contained story about his engagement with one branch of mathematics that he studied for much of his adult life, namely Euclidian geometry. From among Reid's mathematical manuscripts, I have therefore included only those on the definitions and axioms in Euclid's *Elements* and, in particular, on Euclid's problematic treatment of straight and parallel lines. I do, however, provide an overview of Reid's work in various branches of mathematics in my Introduction in order to enable readers to consult online items of interest in the Birkwood Collection.

My aim in transcribing the manuscripts included in this volume has been to provide accurate and easily readable texts. Because most of the manuscripts contain numerous deletions, insertions, corrections and the like, the creation of such texts is not a straightforward process. I have therefore adopted the following editorial principles in order to achieve my editorial end:

1. I have made no attempt to normalise Reid's erratic spelling and punctuation, except where the spelling was clearly mistaken or its eccentricity too distracting.

2. In my transcriptions, words or characters which are missing because of damage to the manuscript or which are judged to have been inadvertently omitted by Reid are enclosed thus, '‹ ›'. Illegible letters or words are indicated thus, '‹?›'. On two occasions, Reid has included in quotations from the Greek symbols or characters that are no longer in use. I have replaced these early modern characters with modern equivalents and these equivalents are enclosed thus, '{ }'.

3. Any characters written as superscripts are printed on the regular line.

4. In Reid's manuscripts, he typically overlines for emphasis. The relevant words or passages are here printed in italics without editorial comment.

5. I have silently normalised Reid's contractions and abbreviations where no modern equivalent exists, or where they are not self-explanatory or readily pronounceable by the modern reader. Thus, on p. 4, l. 30, I have not expanded 'Sept' because it is still recognised as an abbreviation for 'September', but on p. 4, l. 16, I have normalised Reid's contraction 'El' because it does not obviously stand for 'Elements'. I have also silently expanded some of Reid's contractions in the interests of readability. For example, on p. 4, l. 12, I have normalised Reid's 'Elem.' to 'Elements' because it occurs in a reading context.

6. In his manuscripts on astronomy, Reid occasionally uses conventional astronomical symbols to refer to the Sun, the Moon and other bodies in the solar system. I have silently replaced these symbols with their English equivalents.

7. I have silently omitted repetitions of words or phrases, catchwords and any material deemed to have been mistakenly included the manuscripts.

8. Folio/page breaks are indicated by a vertical line, '|', in the text. In the page margin at the line in which Reid began a new folio or page the conventionally abbreviated manuscript number is printed and followed, after a comma, by an indication of the folio or page number in the original manuscript. In the case of unpaginated manuscripts, the folio number and the side – recto 'r' or verso 'v' – of the folio are given.

9. Variants in Reid's manuscripts are recorded in the Textual Notes. These notes are keyed to the relevant texts using page and line numbers (that is, '43/14' in the textual notes refers to p. 43, l. 14). In these notes, editorial comment is in italics and Reid's wordings are in regular typeface. Words or phrases repeated and left undeleted by Reid have not been recorded, nor have catchwords, nor those instances where Reid has changed an unfinished text by superimposing a letter or word over what he had originally written, or revised a phrase in the course of his initial writing. I have also not recorded those instances where Reid has merely gone back and corrected his spelling or grammar, or supplied a missing character, word or words. Variants are indicated in the following manner. In MS 2131/5/II/55 (p. 89, ll. 12–13), for example, Reid initially wrote 'their Reflexions or Refractions'. Reid's change is recorded in the textual notes thus:

89/12–13 their deviations form a right line] their Reflexions or Refractions

Where there are variants of variants I have usually followed this method of indicating Reid's changes, but in cases where this was not practicable I have explained the textual alterations in the notes.

Cancelled passages have been identified and recorded in the Textual Notes. Reid often failed to replace deleted material with a new word or phrase. On p. 34, ll. 9–10, for example, he first wrote 'Quantity, as Aristotle long since observed in his Categories is either Proper or Improper', and subsequently deleted 'in his Categories', so that in the state in which he left his manuscript only 'Quantity, as Aristotle long since observed is either Proper or Improper' remains. This change is recorded in the Textual Notes thus:

34/9 observed is either] observed in his Categories is either

Reid sometimes left his initial formulation of a passage uncancelled. For instance, on p. 35, ll. 5–6, Reid had initially written the phrase 'cannot consist', and then wrote over the line 'is not compatible', without cancelling his original formulation. This change is recorded in the Textual Notes thus:

35/5–6 is not compatible] cannot consist

In recording Reid's revisions, I have made a distinction between insertions and additions. Where Reid has indicated where to place additional material in the text (typically with a '^' symbol) I have used the annotation '*inserted*'. Where Reid has not indicated where to place additional material with a symbol, I have used the annotation '*added*'. Thus at 21/35, Reid has used a marker to indicate the insertion of 'falling upon a right line'. In this (and similar cases) I have recorded the revision thus:

21/35 falling upon a right line *inserted.*

But at 8/26, Reid has not specifically indicated where to place the additional wording 'at Pavia', although the context makes it clear where he intended the wording to go. In this (and other similar instances) I have recorded the revision thus:

21/35 at Pavia *added.*

In ambiguous cases I have specified the revision using the normal convention. Where Reid has written his insertion or addition in the margin of the page, I have noted the location of the insertion or addition in the Textual Notes.

10. In editing the published text of 'An Essay on Quantity' I have retained the original punctuation, spelling, capitalisation and italicisation. But I have silently normalised Reid's abbreviations where no modern equivalent exists. I have also silently deleted two misplaced commas and have corrected 'ingenuous' to 'ingenious' on p. 58, l. 35, based on the wording found in Reid's manuscripts.

11. The Editorial Notes preceding the Textual Notes contain translations of Latin and Greek passages, along with the details of papers and books Reid quotes from or refers to in his texts. The Editorial Notes are indicated in Reid's texts by asterisks '*' and are keyed to the texts using the same convention employed in the Textual Notes. Detailed commentary on the contents of the manuscripts has been confined to my Introduction.

Where known, life dates for all figures active prior to 1800 mentioned in the Introduction and the Editorial Notes are given in the Index of Persons and Titles.

Index of Manuscripts and Published Texts

Special Collections Centre, University of Aberdeen

MSS 2131: Birkwood Collection

MSS 3061

Printed Text

Introduction

1. Thomas Reid: Mathematician and Natural Philosopher

Thomas Reid was the offspring of a marriage linking two significant families in the north-east of Scotland. His father, the Rev. Lewis Reid, was a descendant of the Presbyterian clergyman James Reid, who in the late sixteenth century sired two notable sons: Thomas, who served as the Latin secretary to James VI and I, and Alexander, who eventually practised in London as a physician and surgeon. His mother, Margaret Gregory, was the eldest daughter of David Gregory, whose life as the Laird of Kinnairdie allowed him to study mathematics, medicine and natural philosophy and to correspond with leading men of science such as the French *physicien*, Edme Mariotte.[1] Reid's fascination with the genealogical details of the Gregory family strongly suggests that he identified himself as a member of the notable line of mathematicians and natural philosophers spawned by his maternal grandfather, David Gregory.[2] Reid's contemporaries also made this identification. For example, an anonymous biographer of his cousin, the physician John Gregory, asserted that Reid had 'inherit[ed] largely the mathematical genius of his ancestors'.[3] Yet Reid's most influential nineteenth-century biographer, Dugald Stewart, portrayed him as being, quintessentially, an anatomist of human nature. In downplaying his subject's mathematical and scientific predilections, Stewart obscured two important facts about Reid, namely that the catholicity of Reid's intellectual interests reflected the culture of the virtuosi in which he had been educated and that Reid was one of the most distinguished polymaths of the Enlightenment era.[4] Because of the hold on the historical imaginary exercised by

[1] For details of Reid's family background see A. Campbell Fraser, *Thomas Reid*, ch. 1, and Dugald Stewart, *Account of the Life and Writings of Thomas Reid* (1803), pp. 3–9.

[2] For evidence of Reid's fascination, see Reid to James Gregory, 24 August 1787 and late February 1788, in Reid, *Correspondence*, pp. 187–91, 196–7; Thomas Reid, 'Some Farther Particulars of the Family of the Gregorys and Andersons, Communicated by Dr Thomas Reid, Professor of Moral Philosophy in the University of Glasgow, a Nephew of the Late Dr David Gregory Savilian Professor at Oxford', in Charles Hutton, *A Mathematical and Philosophical Dictionary* (1795–96), vol. I, pp. 555–8. The original manuscript of this work survives in Aberdeen University Library (henceforth AUL), MS 3061/25.

[3] Anon., 'An Account of the Life and Writings of Dr John Gregory', in John Gregory, *The Works of the Late John Gregory, M. D.* (1788), vol. I, p. 26 note.

[4] On this issue, see Paul Wood, 'The Hagiography of Common Sense: Dugald Stewart's *Account of the Life and Writings of Thomas Reid*'; Paul Wood, 'Who Was Thomas Reid?' On Reid's

Stewart's biography throughout the nineteenth century and for much of the twentieth, scant attention was paid to Reid's scientific interests until the appearance of the initial volume of the Edinburgh Edition of Thomas Reid, which revealed for the first time the full extent of his work in the life sciences.[5] Reid's lifelong engagement with the natural sciences is also in evidence in his correspondence published in the fourth volume of the Edinburgh Edition.[6] The papers on mathematics and the physical sciences included in this volume of the Edinburgh Edition demonstrate that the image of Reid delineated by Stewart is one-dimensional. Moreover, the manuscripts transcribed in this volume also enable us to understand why virtually all of the tributes paid to Reid in his own day emphasised that he was an accomplished mathematician and natural philosopher as well as a noted analyst of the powers of the human mind. Stewart's partial view of his mentor's intellectual identity derives its plausibility from the fact that Reid's major publications, which made his name as a man of letters in the wider world and which still attract the attention of historians and philosophers, are largely concerned with the anatomy of our mental powers. In order to understand Reid's choices as to what to publish in the form of books and what to develop in other forms, however, we have to consider issues that go to the heart of how he saw himself as a thinker, or, to put it differently, what kind of Enlightenment figure he was. Such questions lie beyond the scope of the present Introduction, but will be central to my biography of Reid.[7]

We know almost nothing about Reid's formal education before he entered Marischal College, Aberdeen, in October 1722. Hence it is unclear whether he was given anything more than a superficial introduction to the rudiments of arithmetic, algebra or geometry at the parish school he most likely attended in Kincardine

indebtedness to the generation of virtuosi who fostered the Scottish Enlightenment, see Paul Wood, 'A Virtuoso Reader: Thomas Reid and the Practices of Reading in Eighteenth-Century Scotland'. The presence of a number of 'multi-competent intellects' in the Scottish Enlightenment is perceptively discussed in J. R. R. Christie, 'The Origins and Development of the Scottish Scientific Community, 1680–1760', pp. 126–7.

[5] Reid, *Animate Creation*. Although Reid's scientific and mathematical interests were noted by nineteenth-century commentators, they were subsequently ignored in the literature on Reid. The first writer in the twentieth century to take Reid's scientific pursuits seriously was L. L. Laudan; see L. L. Laudan, 'The *Vis viva* Controversy, a Post-mortem', and L. L. Laudan, 'Thomas Reid and the Newtonian Turn of British Methodological Thought'. The first systematic study of Reid's scientific and mathematical investigations was my unpublished PhD thesis, 'Thomas Reid, Natural Philosopher: A Study of Science and Philosophy in the Scottish Enlightenment'.

[6] Reid's letters to Joseph Black, Lord Kames, William Ogilvie, Richard Price, John Robison, Robert Simson, David Skene and John Stewart are included in Reid, *Correspondence*, and will be discussed in detail below.

[7] For a preliminary discussion of these issues, see Wood, 'Who Was Thomas Reid?'

O'Neil or at the grammar school in Aberdeen where he was sent to prepare for his studies at Marischal. Once he was a Marischal student, he may have briefly come under the tutelage of Colin Maclaurin, who had filled the College's vacant chair of mathematics in 1717 and had quickly established a reputation as one of Isaac Newton's most gifted mathematical disciples.[8] Maclaurin soon became frustrated by the limitations of Marischal's collection of scientific instruments and by the large number of students he was obliged to teach. Consequently, by 1720 he was looking for employment elsewhere. Throughout Reid's first session at the College, Maclaurin was absent without permission in France serving as tutor to Patrick Hume, the eldest son of Lord Polwarth. When Hume died in the autumn of 1724, Maclaurin unwillingly returned to Marischal in January1725 and, after classes were finished in late March or early April, he left Aberdeen in order to secure the Edinburgh chair of mathematics. Given Maclaurin's lengthy absence from the College, Reid could therefore have studied under him for only the first three or four months of 1725. Otherwise, he would have been taught by substitutes hired by the College to cover for the absentee professor or by the Marischal regent, Daniel Gordon, who filled in for Maclaurin in November and December 1724.

Few details survive regarding Maclaurin's teaching of mathematics during his stint at Marischal. In addition to giving public lectures, he gave a private course that was probably for more advanced students, and those who substituted for him may have followed his example. At the very least, the basics of arithmetic, the elements of geometry, algebra, spherical geometry, trigonometry, navigation and surveying likely featured in the curriculum. Maclaurin also probably introduced his pupils to Newton's method of fluxions and the fundamentals of mathematical astronomy. The details of what Reid's regent at Marischal, George Turnbull, covered in his natural philosophy classes are equally sketchy. Turnbull's graduation theses, published in 1723 and 1726, provide us with the most reliable evidence we have regarding the contents of his courses. His theses suggest that he summarised the projectile theory of light Newton expounded in the *Opticks* and the system of the world Newton outlined in the *Principia mathematica*. Turnbull echoed Newton and the early Newtonians in emphasising the power of the method of analysis and synthesis. He took the innovative step of quoting Newton's comment in the concluding paragraph of Query 31 of the *Opticks* that 'if natural Philosophy in all its Parts, by pursuing this Method, shall

[8] In what follows I summarise the account of Maclaurin's chequered career at Marischal College found in Paul Wood, *The Aberdeen Enlightenment: The Arts Curriculum in the Eighteenth Century*, pp. 14–19, and Betty Ponting, 'Mathematics at Aberdeen: Developments, Characters and Events, 1717–1860', pp. 162–5.

at length be perfected, the Bounds of Moral Philosophy will be also enlarged', to legitimise his recommendation of the use of Newton's method to transform moral philosophy into a genuinely empirical science. Lastly, Turnbull drew out the religious implications of 'the true Physiology [physics]' discovered by Newton. He apparently elaborated on Newton's claims in the *Opticks* and the *Principia* that the forces of gravity, cohesion and fermentation all proved that immaterial, active principles are continually at work in nature, and that the order of the solar system demonstrated that there must be a wise, benevolent and all-powerful creator who rules the universe through the laws governing the material and moral realms. It is likely that Turnbull spent little, if any, time in his classes on the mathematical intricacies of the Newtonian system, since there is no evidence to suggest that he had anything more than a rudimentary competence in mathematics. Moreover, he presumably illustrated at least some of his lectures with experimental demonstrations using the collection of scientific instruments acquired by the College.[9]

Reid graduated under Turnbull in April 1726 and soon afterwards entered the Divinity Hall in Aberdeen. The following year, his close friend and kinsman from Turnbull's class, John Stewart, was chosen to succeed Colin Maclaurin as the Marischal Professor of Mathematics.[10] According to Dugald Stewart, Reid recalled 'with much pleasure' that his 'predilection for mathematical pursuits, was confirmed and strengthened' by their friendship, and 'he recollected the ardour with which they both prosecuted these fascinating studies, and the lights which they imparted mutually to each other in their first perusal of the *Principia*, at a time when a knowledge of the Newtonian discoveries was to be acquired only in the writings of their illustrious author'.[11] Evidence for their study of the *Principia* and, in particular, Newton's theory of gravitation is contained in the two earliest surviving manuscripts in the Birkwood Collection. The first dates from 19 July 1729 and includes memoranda on the nature of mathematical quantity in relation to the operation of multiplication, and on the deviation from the perpendicular in the path of a falling body caused by the shape of the Earth. Significantly, Reid's comments on the multiplication of like quantities register the fact that he had read closely his late uncle David Gregory's *The Elements of*

[9] George Turnbull, *Education for Life: Correspondence and Writings on Religion and Practical Philosophy*, pp. 43–74; Isaac Newton, *Opticks: Or, a Treatise of the Reflections, Refractions, Inflections and Colours of Light* (1730), p. 405. At Marischal Turnbull was closely associated with Colin Maclaurin and may well have relied on Maclaurin's help in understanding the elements of Newton's natural philosophy.

[10] Wood, *Aberdeen Enlightenment*, p. 19; Ponting, 'Mathematics at Aberdeen', pp. 165–6.

[11] Stewart, *Account*, pp. 12–13.

Astronomy, Physical and Geometrical.[12] The second dates from 6 October 1729 and is a set of detailed reading notes on the first edition of Newton's *Principia* (1687). As Niccolò Guicciardini has observed, this manuscript documents Reid's critical engagement with 'the most advanced parts of the *Principia*' and attests to the fact that 'Reid was reading the *Principia* in depth, with a competence comparable to David Gregory's'.[13] That is, by the time Reid reached the age of nineteen, his mathematical expertise had matched that displayed earlier by his uncle, who before his death in 1708 had served as both the Professor of Mathematics at Edinburgh and the Savilian Professor of Astronomy at Oxford. To gauge accurately Reid's abilities as a mathematician, we should remember that Newton said of Gregory that '[he] is respected [as] the greatest Mathematician in Scotland & that deservedly so far as my knowledge reaches. For I take him to be an Ornament to his Country'.[14]

Reid was licensed as a preacher and probationer by the presbytery of Kincardine O'Neil on 22 September 1731. He failed to receive a call from a parish, however, and hence was not yet ordained as a minister in the Church of Scotland. After occupying a series of temporary posts in the Kirk, his family connections gained him the position of librarian at Marischal College in July 1733. If we can credit Reid's brother-in-law, the Rev. John Rose, Reid studied Newton's *Principia* and Locke's *Essay concerning Human Understanding* (1690) in earnest following his return to Marischal.[15] The brief period in which Reid was employed as the librarian at his alma mater is poorly documented; consequently, we know little about his activities apart from the fact that he was a member of a 'Philosophical Club' that probably met at Marischal in early 1736. Although John Stewart was most likely a member as well, the discussions of the group seem to have focused primarily on topics in metaphysics and moral philosophy rather than mathematics or the natural sciences. But Reid's notes from the meetings of the Club indicate that they did consider topics central to the metaphysical foundations of Newton's system of the world, such as whether God is the direct cause of the phenomena of nature. It may be that he made his reading notes from

[12] AUL, MS 2131/5/II/1. Reid cryptically refers to propositions xxi and xxii in Book I of David Gregory's *The Elements of Astronomy, Physical and Geometrical* (1715), vol. I, pp. 42–4. Gregory's book had first appeared in Latin in 1702; a second edition of the English translation was published in 1726.

[13] AUL, MS 2131/7/III/15; Niccolò Guicciardini, 'Thomas Reid's Mathematical Manuscripts: A Survey', p. 79.

[14] Newton to Arthur Charlett, 27 July 1691, in Isaac Newton, *The Correspondence of Isaac Newton*, vol. III, p. 155. Gregory became the Savilian Professor of Astronomy at Oxford in 1691 thanks to Newton's patronage.

[15] Campbell Fraser, *Thomas Reid*, p. 27.

the Leibniz–Clarke correspondence while preparing for the conversation the members had about this issue.[16]

Reid's dissatisfaction with his salary and the management of the bequest made in 1624 by his ancestral namesake to pay the salary of the 'Bibliothecar' prompted him to take a leave of absence in the summer of 1736. During his leave, he travelled to London to consult with his uncle, the physician George Reid, about taking legal action against the Aberdeen Town Council.[17] On the trip, Reid was accompanied by John Stewart, whose father Robert was in a position to mediate the simmering dispute between Reid's family and the Council because he had served three terms as Lord Provost of Aberdeen. According to a brief biographical sketch which may have been written for Dugald Stewart's use by Reid's kinsman the Rev. James Leslie, in 1736 Reid '& Mr. Stewart went to England, where they continued for some Months; they visited the Universities of Oxford & Cambridge & formed Acquaintance with some of the most distinguished Literary Men of that time'.[18] Reid later recalled that in Oxford the two of them visited their relation, the Canon of Christ Church, David Gregory, who may have introduced them to men of science such as the Savilian Professor of Astronomy, James Bradley.[19] While in Cambridge, they conversed with the famous blind mathematician Nicholas Saunderson and met the redoubtable Master of Trinity College, Richard Bentley, who was one of the principal architects of the Newtonian ascendancy in the University during the early decades of the eighteenth century.[20] At Trinity, they may also have encountered Bentley's protégé, the Plumian Professor of Astronomy, Robert Smith, for Stewart's name appears in the list of subscribers to Smith's *A Compleat System of Opticks* (1738).[21] Their visit to Trinity College was not without controversy, however, for the essayist and biographer John Ramsay of Ochtertyre states that Reid offended some of the Fellows of the College by

[16] 'Minutes of a Philosophical Club 1736', AUL, MS 2131/6/I/17, fol. 1r. For Reid's notes from the Leibniz–Clarke correspondence, see AUL, MS 2131/3/II/7. Given the handwriting, these notes most likely date from the 1730s.

[17] On this episode, see Wood, *The Life of Thomas Reid*.

[18] AUL, MS 2814/1/50, fol. 1v; on the identity of the possible author of the sketch see Stewart, *Account*, pp. 219–20.

[19] Reid to James Gregory, 24 August 1787, in Reid, *Correspondence*, p. 190. We shall see in Section 4 that Reid greatly admired the work of Bradley.

[20] Reid mentions one of their conversations with Saunderson in Reid, *Inquiry*, p. 79.

[21] Robert Smith, *A Compleat System of Opticks in Four Books, viz. a Popular, a Mathematical, a Mechanical and a Philosophical Treatise* (1738), vol. I, unpaginated. An unusually large number of Scots subscribed to Smith's work, which was partly because Colin Maclaurin collected subscriptions; see Maclaurin to Robert Smith, 28 November 1738, in Colin Maclaurin, *The Collected Letters of Colin Maclaurin*, pp. 306–7, where Maclaurin mentions Stewart's subscription.

explaining an inscription on a sundial that was thought to be indecipherable.[22] In London, their stay was partly taken up with business but, thanks to David Gregory, they gained entrée to learned circles in the metropolis through Gregory's contact Martin Folkes, who was a Fellow of both the Royal Society and the Society of Antiquaries.[23] Moreover, the pair attended a meeting of the Royal Society on 1 July 1736, sponsored by a physician originally from the north-east of Scotland, Alexander Stuart. At the meeting, they listened to a report on the work of the French naturalist René Antoine Ferchault de Réaumur presented by Mark Catesby, and witnessed demonstrations of electrical experiments performed by the Society's Curator of Experiments, J. T. Desaguliers.[24]

After consulting with George Reid in London, Reid and his family evidently decided against proceeding with a lawsuit against Marischal College over the management of the Bibliothecar's fund. Reid returned to his post and continued to oversee the College library until 1737, when he was finally ordained as a minister in the Church of Scotland in the rural parish of New Machar, ten miles north-west of Old Aberdeen.[25] Despite the move to the country, he remained in close contact with John Stewart, not least because Stewart's father was a ruling elder of Reid's new parish. Stewart had earlier been recruited to join the network of observational astronomers in Scotland that Colin Maclaurin gradually built up prior to the founding of the Philosophical Society of Edinburgh in 1737, and, by 1748 at the latest, Reid was collaborating with Stewart in making observations of celestial phenomena.[26] Moreover, the two friends also shared interests in physical astronomy and, in particular, the debate between Cartesian and Newtonian natural philosophers over the figure of the Earth that dominated the European Republic of Letters during the 1720s and 1730s. We shall return to this dispute in Section 4.

[22] Stewart, *Account*, p. 13; Alexander Allardyce (ed.), *Scotland and Scotsmen in the Eighteenth Century: From the MSS. of John Ramsay, Esq. of Ochtertyre*, vol. I, p. 476 note. Bentley's pivotal role in the institutionalisation of Newtonian natural philosophy at Cambridge is documented in John Gascoigne, *Cambridge in the Age of the Enlightenment: Science, Religion and Politics from the Restoration to the French Revolution*.

[23] Stewart, *Account*, p. 13.

[24] Royal Society of London, 'Journal Book (copy)', xv (1735–36), pp. 363–4. Reid and Stewart may have been introduced to Alexander Stuart by Reid's uncle George, who had overlapped with Stuart as a student at Marischal College. Desaguliers played a prominent role in the popularisation of Newtonian natural philosophy in Britain during the first half of the eighteenth century. On Desaguliers, see especially Larry Stewart, *The Rise of Public Science: Rhetoric, Technology and Natural Philosophy in Newtonian Britain, 1660–1750*.

[25] On the controversy surrounding Reid's appointment at New Machar see Stewart, *Account*, pp. 14–15, and Campbell Fraser, *Thomas Reid*, pp. 30–1. I deal with the controversy surrounding Reid's appointment at New Machar in my forthcoming *Life of Reid*.

[26] See below, p. cxiii.

A second notable scientific controversy of the period that Reid monitored in the late 1730s likewise pitted Newton and his followers against continental critics of the Newtonian system. Whereas Newton had affirmed in the *Principia* that the correct measure of the force of a moving body is mv, Leibniz had asserted in 1686 that the measure was, in fact, mv^2, which for him referred to the quantity of *vis viva* or living force in a body in motion.[27] But it was not until the exchanges between Leibniz and Newton's spokesman Samuel Clarke were published in 1717 that the two radically different views came into direct conflict. The controversy between the followers of Leibniz and Newton began in earnest soon thereafter, with the publication of experimental evidence supporting Leibniz's concept of *vis viva* by the Italian mathematician and natural philosopher Giovanni Poleni in 1718 and by the Leiden Professor of Astronomy and Mathematics Willem Jacob 's Gravesande in 1722.[28] In the 1720s Poleni and 's Gravesande, as well as other continental proponents of *vis viva* such as Jean Bernoulli and Jacob Hermann, were attacked by the Newtonians Henry Pemberton (whose paper included an anonymous postscript by Newton), J. T. Desaguliers, John Eames and Samuel Clarke for having 'raise[d] a *Dust* of Opposition against Sir *Isaac Newton*'s Philosophy' by defending the '*Absurdity*' that 'the *Force* of any *Body in Motion*, is proportional, not to its *Velocity*, but to the *Square of its Velocity*'.[29] During the 1720s, Bernoulli was also embroiled in a controversy at the Académie royale des sciences in Paris over the legitimacy of Leibniz's notion of *vis viva*, although in the French context his main opponents were not committed Newtonians but rather academicians loyal to either Descartes or Malebranche.[30] Reid's admission that in the 1740s he 'knew so little of the History of the Controversy about the force of moving Bodies as to think that the British Mathematicians onely, opposed the Notion of Leibnitz, and that all the foreign Mathematicians adopted it' (p. 50) indicates that his familiarity with the voluminous literature generated by the *vis viva* dispute was limited. Nevertheless, his notes on the Leibniz–Clarke correspondence show

[27] In what follows, my account of the *vis viva* controversy is based on: Thomas L. Hankins, 'Eighteenth-Century Attempts to Resolve the *Vis viva* Controversy'; Carolyn Iltis, 'The Leibnizian–Newtonian Debates: Natural Philosophy and Social Psychology'; Laudan, 'The *Vis viva* Controversy, a Post-mortem'; and Mary Terrall, '*Vis viva* Revisited'.

[28] Giovanni Poleni, *De castellis per quae derivantur fluviorum aquae habentibus latera convergentia liber: Quo etiam continentur nova experimenta ad aquas fluentes, & ad percussionis vires pertinentia* (1718); W. J. 's Gravesande, 'Essai d'une nouvelle théorie du choc des corps fondée sur l'expérience'.

[29] Samuel Clarke, 'A Letter from the Rev. Dr Samuel Clarke to Mr Benjamin Hoadly, F.R.S. Occasion'd by the Present Controversy Among Mathematicians, concerning the Proportion of Velocity and Force in Bodies in Motion', p. 382.

[30] Terrall, '*Vis viva* Revisited', pp. 191–9.

that by the mid-1730s he was scrutinising the arguments of the two interlocutors regarding the concept of *vis viva*. What appears to be the earliest surviving statement of the ideas central to his 'An Essay on Quantity' indicates that, at the very least, he clearly understood the basic issues at stake even if he had not read widely in the relevant literature.[31] As we will see in Section 3, Reid subsequently devoted a considerable amount of time and intellectual energy to refining his analysis of the *vis viva* debate while caring for his parishioners at New Machar.

In addition to mulling over the nature of quantity after relocating to the manse at New Machar, Reid considered other foundational issues in mathematics, as well as specific mathematical topics. Even though there is no evidence in Reid's extant manuscripts regarding his initial reading of David Hume's *A Treatise of Human Nature* (1739–40), Hume's contention that 'geometry is founded upon ideas that are not exact, and axioms that are not precisely true' in Book I of the *Treatise* may have prompted Reid to reflect on the soundness of the basic concepts of Euclidean geometry.[32] Moreover, a group of undated manuscripts dealing with Newton's fluxional calculus suggests that in the late 1730s or early 1740s Reid, like other Scottish mathematicians such as Colin Maclaurin and John Stewart, gave considerable thought to the critique of Newton's method of fluxions advanced in George Berkeley's *The Analyst* (1734).[33] In order to understand Reid's response to Berkeley, we must first consider Berkeley's critique and the counter-arguments of Maclaurin and Stewart.

According to Berkeley, the concepts upon which the fluxional and differential forms of the calculus were founded were effectively meaningless because basic terms such as 'moments', 'velocities' of points, lines and planes, and 'infinitesimals' had no precise empirical referents. Furthermore, he contended that both Newton's and Leibniz's methods were logically flawed insofar as their respective versions of the calculus contained internal contradictions. Hence he claimed that the calculations of Newton, Leibniz and other 'modern analysts' lacked the logical and evidential rigour of classical geometry, a feature which for him undermined the pretensions of his target, namely the unnamed 'infidel' mathematician who contrasted the demonstrative and self-evident character of

[31] AUL, MS 2131/3/II/7, fols 1r–2v, and MS 2131/5/I/20, fol. 1v. A further manuscript (AUL, MS 2131/5/II/8), which appears to be written in an early hand, also refers to the *vis viva* debate.

[32] Reid, *Intellectual Powers*, p. 419; we will return to this point in Section 2, pp. lxxxvi–lxxxvii.

[33] For Reid's manuscripts, see AUL, MSS 2131/5/II/6, 7, 9–11. The handwriting in these manuscripts suggests that they date from the 1730s or 1740s. Moreover, on the verso of MS 2131/5/II/7 there is an incomplete draft of a letter that Reid intended to send to David Gregory in Oxford. In the letter Reid mentions his trip to London in 1736. This particular manuscript thus dates from late 1736 or 1737 at the earliest.

mathematics with the supposedly unintelligible and self-contradictory mysteries at the heart of religion.[34]

The publication of *The Analyst* sparked a hostile response from members of the Newtonian camp in Britain, who took exception to Berkeley's attack on the basic principles and method of the fluxional calculus and especially to his insinuation that some of Newton's disciples were not good Christians.[35] Although Colin Maclaurin intended to answer Berkeley immediately after *The Analyst* appeared, he soon changed his mind. In November 1734 he wrote to James Stirling saying that 'upon more consideration I did not think it best to write an Ansuer to Dean Berkeley but to write a treatise of fluxions which might ansuer the purpose and be useful to my scholars'.[36] By 1737 he had composed and printed 'the greatest Part of the first Book' of his *A Treatise of Fluxions*, but it was not until 1742 that the complete work issued from the press.[37] In the *Treatise*, Maclaurin countered three specific claims made by Berkeley in *The Analyst*: (i) that Newton's method of fluxions lacked the rigour and evidence of the geometry of the ancients; (ii) that the fluxional calculus involved the use of infinitesimals; and (iii) that Newton's notion of velocities had no empirical content.

In the *Treatise*, Maclaurin recast what he called 'the Elements' of Newton's method by 'deduc[ing] those Elements after the Manner of the Antients, from a few unexceptionable Principles, by Demonstrations of the strictest Form', and presented the fluxional calculus as an extension of geometrical techniques found in the writings of Archimedes. In doing so, he emphasised the continuities between classical geometry and Newton's method of fluxions in order to make the point against Berkeley that, properly understood, the fluxional calculus was just as rigorous and epistemically well founded as the geometry of the Greeks.[38] Secondly, for Maclaurin the question of the use of infinitesimals was a complex

[34] George Berkeley, *The Analyst: Or, a Discourse Addressed to an Infidel Mathematician* (1734), in *The Works of George Berkeley, Bishop of Cloyne*, vol. IV, especially pp. 65–9, 71–4. On Berkeley's criticisms of the fluxional and differential calculus, see Douglas M. Jesseph, *Berkeley's Philosophy of Mathematics*, ch. 6.

[35] The response to *The Analyst* is surveyed in Jesseph, *Berkeley's Philosophy of Mathematics*, ch. 7, and Niccolò Guicciardini, *The Development of Newtonian Calculus in Britain, 1700–1800*, ch. 3.

[36] Maclaurin to Stirling, 16 November 1734, in Maclaurin, *The Collected Letters*, p. 250.

[37] Colin Maclaurin, *A Treatise of Fluxions. In Two Books* (1742), vol. I, p. iii. The fact that Maclaurin had completed Book I of the *Treatise* by 1737 raises the possibility that he may have sent a manuscript or printed copy of the text to John Stewart, who would undoubtedly have passed the copy on to Reid.

[38] Maclaurin, *Treatise of Fluxions*, vol. I, pp. i–ii; Maclaurin presents the method of fluxions as being rooted in the geometry of Archimedes in the lengthy introduction to the *Treatise of Fluxions*, vol. I, pp. 1–50.

one since he had himself employed them in his early writings.[39] However, in order to obviate Berkeley's objections to the method of fluxions, Maclaurin was obliged to reconsider the issue of whether the use of infinitesimals was compatible with the standards of rigour and intelligibility set by the geometers of classical antiquity. He now came to the conclusion that although magnitude can be increased or decreased without limit, this did not mean that mathematicians can reason about infinitely large or small magnitudes 'with that perspicuity that is required in geometry'. He insisted that even though 'there is something marvellous in the doctrine of infinites, that is apt to please and transport us; and that the method of infinitesimals has been prosecuted of late with an acuteness and subtlety not to be parallelled in any other science', he was nevertheless convinced that

> geometry is best established on clear and plain principles; and these speculations are ever obnoxious to some difficulties. If the greatest accuracy has been always required in this science, in reasoning concerning finite quantities, we apprehend that Geometricians cannot be too scrupulous in admitting or treating of infinites, of which our ideas are so imperfect. Philosophy probably will always have its mysteries. But these are to be avoided in geometry: and we ought to guard against abating from its strictness and evidence the rather, that an absurd philosophy is the natural product of a vitiated geometry.

Consequently, Maclaurin relied on the ancient method of exhaustion rather than the modern method of infinitesimals in order to place the fluxional calculus on as sound an evidential and conceptual footing as classical geometry.[40] Thirdly, to counter Berkeley's claim that the notion of velocity upon which the method of fluxions was founded was empirically vacuous, Maclaurin argued that velocity was a 'power' that can be measured in terms of the relations between space, time and motion. And given that motion can be measured with reference to space and time, whose 'limited parts' we can 'clearly conceive', Maclaurin showed that the cluster of concepts related to velocity that were fundamental to the fluxional calculus were not only firmly grounded in experience but also definable in rigorous axiomatic form.[41]

John Stewart's response to Berkeley forms a small but nonetheless significant part of the extensive editorial commentary that accompanies his English translation of Newton's two Latin texts outlining the method of fluxions, 'Tractatus de quadratura curvarum' (1704) and 'De analysi per æquationes numero terminorum

[39] Guicciardini, *Development of Newtonian Calculus in Britain*, p. 48. On Maclaurin's reply to Berkeley, see also Jesseph, *Berkeley's Philosophy of Mathematics*, pp. 279–84.

[40] Maclaurin, *Treatise of Fluxions*, vol. 1, pp. 41–2, 47.

[41] Maclaurin, *Treatise of Fluxions*, vol. 1, pp. 51–9.

infinitas' (1711). In his commentary Stewart covers much the same ground as Maclaurin, although his estimate of the capacities of his intended audience of 'young Geometrician[s]' may account for the fact that the level of his discussion is not as advanced as that found in Maclaurin's *Treatise*.[42] Like Maclaurin, he distanced Newton's method of fluxions from 'the specious Analysis of the Moderns', and insisted that even though Newton's method was entirely new it nevertheless built on the achievements of the ancients.[43] And in emphasising that Newton's demonstrations were just as rigorous as those found in the writings of Euclid and Archimedes, he implicitly followed Maclaurin in rejecting Berkeley's assertion that Newton's analytical techniques lacked the rigour and evidence of classical geometry. As for the issue of infinitesimals, Stewart echoed Maclaurin in denying that the fluxional calculus necessarily involved the use of infinitesimals. In Stewart's commentary on Newton's 'Treatise of the Quadrature of Curves', he demonstrated how the method of fluxions could be formulated without relying on either infinitesimals or the concept of prime and ultimate ratios (which Berkeley had also criticised in *The Analyst*). He returned to Berkeley's accusation that infinitesimals were an integral part of the fluxional calculus in his commentary on Newton's 'Analysis by Equations of an Infinite Number of Terms'. Here he maintained that Newton's method of fluxions, as well as his 'Method of *prime and ultimate Ratios*' rested firmly on the classical method of exhaustion initially developed by Euclid and subsequently refined by Archimedes.[44] In doing so, he adopted essentially the same strategy as Maclaurin had employed in the *Treatise* to answer Berkeley. He was, however, apparently somewhat less concerned than Maclaurin with Berkeley's charge that the basic concepts of the fluxional calculus were ill defined if not meaningless. There is an element of bluster in Stewart's

[42] John Stewart, *Sir Isaac Newton's Two Treatises of the Quadrature of Curves, and Analysis by Equations of an Infinite Number of Terms, Explained* (1745), p. 344. The different levels of sophistication evident in Maclaurin's *Treatise of Fluxions* and Stewart's work to some extent register their respective talents as mathematicians. That said, Maclaurin was also able to teach more advanced courses on fluxions at Edinburgh than Stewart was able to give at Marischal College. For a description of Maclaurin's teaching, see Anon., 'A Short Account of the University of Edinburgh, the Present Professors in It, and the Several Parts of Learning Taught by Them', p. 372; on Stewart, see Wood, *Aberdeen Enlightenment*, pp. 19–23.

[43] Stewart, *Sir Isaac Newton's Two Treatises*, p. 33; see also p. 344, where Stewart emphasises the continuity between the mathematical work of the ancients and that of Newton.

[44] Stewart, *Sir Isaac Newton's Two Treatises*, pp. 36–47, 344–60. In his account of what he called 'the Rise and Progress of the Contemplation of Infinites and the Quadrature of Curves' (p. 344), Stewart highlighted the work on infinites of James Gregory, the brother of David Gregory of Kinnairdie. Given that Stewart was distantly related to the Gregory clan, it is difficult not to see an element of family pride in his historical sketch, especially since Maclaurin said little about Gregory as a precursor to Newton.

remark that 'the great Dust which has been raised of late about the Whole of this Doctrine [of prime and ultimate ratios], must be owing to Weakness, or some worse Principle'. But he did take seriously Berkeley's contention that the notion of there being different orders of fluxions was akin to a mystery of religious faith, and he resorted to defining fluxions much the same way that Maclaurin had done, in terms of time, space and motion, to rebut Berkeley's objection. With this task completed to his satisfaction, Stewart was led to proclaim that the different orders of fluxions 'have a real Foundation in Nature; and may be distinctly conceived', which he presumably thought was true also of Newton's method of fluxions more generally.[45]

Although Stewart's introduction to the fluxional calculus appeared in 1745, the records of the Society for the Encouragement of Learning (which sponsored the publication of the book) tell us that something like the final version of the complete text was submitted to the Society in February 1742.[46] The chronology of the composition and publication of Maclaurin's *Treatise* and Stewart's textbook thus suggests that some of Reid's manuscripts on fluxions could date from as early as 1737. But it is more likely that they were written in the early 1740s, around the time when the *Treatise* was published and Stewart was finalising his translations of Newton's mathematical tracts and his editorial apparatus.[47] What are probably the earliest of this group of manuscripts deal with specific details of infinite series and the direct method of fluxions, whereas those that appear to be written at a slightly later date contain broader discussions of the basics of the fluxional calculus.[48] What is striking about the later manuscripts is that they lay down the fundamental elements of Newton's fluxions in sequences of postulates, axioms, definitions, lemmas and propositions. In presenting the method of fluxions in the format of Euclid's *Elements* and other works on geometry written in classical antiquity, Reid can be seen as implicitly claiming that, Berkeley's criticisms notwithstanding, the fluxional calculus possessed the same degree of evidence and

[45] Stewart, *Sir Isaac Newton's Two Treatises*, pp. 40, 67.

[46] 'Memoirs of the Society for the Incouragement of Learning taken from the Register of their Meetings, and Minute Books of the Committee', British Library, Add MS 6185, p. 125. At a meeting of the Society held in February 1745, it was noted that Stewart had added 'a large appendix' to the final text (p. 150). The sponsorship of Stewart's book was apparently brokered by Alexander Stuart, who had been a founding member of the Society in 1735.

[47] It should be noted that Reid appended sixteen 'Queries With respect to Infinite Series' to the earliest surviving version of his 'An Essay on Quantity'. It may be, therefore, that he was discussing the mathematical issues raised by infinite series and the method of fluxions with John Stewart *circa* 1737. See AUL, MS 2131/5/I/20, fol. 2r–v.

[48] AUL, MSS 2131/5/II/6–7 resemble manuscripts written in the 1730s, whereas MSS 2131/5/II/9–11 appear to have been written at a slightly later date.

rigour as classical geometry. Furthermore, as Niccoló Guicciardini has observed, the debate sparked by *The Analyst* serves as the subtext for these manuscripts, insofar as the axioms Reid used to define the term 'fluxion' are similar to those found in Maclaurin's reformulation of the method of fluxions in the *Treatise.*[49] Given the similarities, Reid's manuscripts imply that his formulation of the fluxional calculus was likewise intended as an answer to the objections raised by Berkeley in *The Analyst.*[50]

Reid's axioms also register his contemporaneous musings on the nature of quantity, which may likewise have been partly stimulated by *The Analyst* controversy. For in the earliest surviving version, from *circa* 1737, of what ultimately became 'An Essay on Quantity', fluxions are included among his examples of quantities that must be measured in terms of the 'primary and direct objects of the Mathematical Sciences', namely 'Duration extension and Number' (p. 32). And in a subsequent undated draft of the 'Essay', Reid follows Aristotle in distinguishing between 'proper' and 'improper' quantities, with the former encompassing extension, number and duration, and the latter being 'Measured by other Quantities, but not by Quantities of the same kind'. Fluxions are here listed among the various kinds of improper quantities considered in the mathematical sciences, of which he says that they 'Must all be measured by Space Duration or Number or by Something which is Measured by these'. Consequently, he defined fluxions as 'Velocitys of Increase or Decrease', and such velocities were to be measured by 'the length which would be passed over in a given time'.[51] Reid's formulation of the fluxional calculus in the early 1740s, therefore, can be seen as being guided by his belief that 'Improper Quantitys ought to be defined and the measure of them taken into the Definition otherwise there can be no clear reasoning about them' (p. 37).[52]

The extent of Reid's mathematical enquiries during his years at New Machar was not, however, limited solely to his reflections on the soundness of the conceptual foundations of Newton's fluxional calculus. One of his surviving manuscripts documents his close study of Newton's 'Treatise of the Quadrature of Curves' and his attempt to assimilate and master the advanced techniques

[49] Guicciardini, 'Thomas Reid's Mathematical Manuscripts', pp. 76–7. Compare AUL, MS 2131/5/II/9, fol. 1r–v, with Maclaurin, *Treatise of Fluxions*, vol. 1, p. 59.

[50] That said, Reid did not touch on the problematic concept of infinitesimals in his manuscripts. Reid later indicated to Lord Kames that Berkeley's criticisms of the Newtonian calculus were misguided and implied that Berkeley did not completely understand the mathematical issues involved; see Reid to Lord Kames, [October 1782], in Reid, *Correspondence*, p. 158.

[51] Compare Stewart, *Sir Isaac Newton's Two Treatises*, p. 63, and Guicciardini, 'Thomas Reid's Mathematical Manuscripts', p. 77.

[52] Guicciardini, 'Thomas Reid's Mathematical Manuscripts', pp. 76–7.

of integration involved in the fifth proposition of the 'Treatise'.[53] The various deletions and additions to the text suggest that the manuscript embodies the results of a prolonged period of work. But the last folio of the manuscript, dated June 1743, shows that Reid was also carefully studying Newton's *Universal Arithmetick* in addition to the 'Treatise'. This folio contains a cryptic reference to 'Newtons Algebra Pag 42 &c', which is a citation of the English translation of Newton's *Arithmetica universalis* published by Joseph Raphson in 1720.[54] Reid was thus honing his skills in advanced algebraic analysis in the early 1740s as well as schooling himself in the technicalities of the direct and inverse methods of fluxions. Moreover, a second manuscript, which partly dates from 1748, headed 'Some properties of Numbers', apparently distills what Reid had discovered up to that point about the arithmetical relationships between numbers produced by the operations of multiplication and division. The different later forms of Reid's handwriting in this manuscript point to the fact that number theory was a branch of arithmetic that continued to fascinate him throughout the rest of his life.[55]

Unfortunately, there is almost no trace of Reid's reading of other works in mathematics while he lived at New Machar, apart from a brief set of comments dating from November 1750 on the opening pages of Colin Maclaurin's posthumously published *Treatise of Algebra*. In his notes, Reid queries two claims made by Maclaurin. First, he had reservations about Maclaurin's assertion that 'other Symbols [are] admitted into Algebra beside the Letters and Numbers which represent the Magnitude of Quantities' because 'there are many Affections and Properties of Quantities, and Operations to be performed upon them, that are necessarily to be considered'. Secondly, he questioned Maclaurin's view that 'Quantity is what is made up of Parts, or is capable of being greater or less'.[56] While he acknowledged that Maclaurin had 'justly observed' that 'the Mathematical Sciences contemplate not onely the Magnitude but the Figures of Extended Quantitys, not onely the Velocity but Direction of Motion &c', he objected that

[53] AUL, MS 2131/5/II/54. As Niccolò Guicciardini points out, the inverse method of fluxions dealt with in this manuscript is also discussed in AUL, MS 2131/5/I/20; Guicciardini, 'Thomas Reid's Mathematical Manuscripts', p. 75.

[54] AUL, MS 2131/5/II/54. Folios 1r–4v of the manuscript discuss proposition 5 of Newton's 'Treatise', while fol. 5r–v is dated June 1743 and is headed 'Queries concerning Affected Equations & the Orders of Analytical Problems'. On this manuscript, see also Guicciardini, 'Thomas Reid's Mathematical Manuscripts', p. 75. Reid's reference on fol. 5r is to Isaac Newton's *Universal Arithmetick: Or, a Treatise of Arithmetical Composition and Resolution* (1720), p. 42. Newton's Latin text was first edited and published by William Whiston in 1707.

[55] AUL, MS 2131/7/III/8. Reid has paginated this manuscript and pages 1–3 date from 1748. The writing in pages 5–7 is in a variety of Reid's later hands.

[56] Colin Maclaurin, *A Treatise of Algebra, in Three Parts* (1748), p. 3.

'this is not the Reason … that other Symbols besides th[o]se which represent the Magnitude of Quantities are admitted into Algebra'. Rather, he argued that 'there is no Character used in Algebra to denote [the] Affections of Extension or Motion but by Means of Magnitudes that measure them. The attributes of Particular Quantities can never enter into Algebra which Considers Quantity abstractedly'.[57] That is, Maclaurin's explanation was faulty because it overlooked the fact that the 'Affections and Properties of Quantities' considered in algebra were entirely abstract, highly generalised and not, as Reid read Maclaurin as implying, specific examples of quantitative relationships. As Reid put the point in his 'An Essay on Quantity', algebra 'treats of Quantity in general, or of those Relations and Properties which are common to all Kinds of Quantity'. In his view, Maclaurin had failed to recognise this defining characteristic of algebra.[58] Moreover, he dismissed Maclaurin's definition of quantity as 'a bad Definition'. According to Reid, 'some improper Quantities are not made up of Parts as Velocity' and 'Many things are capable of being greater or less which yet are not Quantities as Anger Wisdom &c'.[59] Here too Maclaurin was mistaken for reasons that Reid had already considered in his 'An Essay on Quantity', notably that quantity ought to be defined as 'What may be measured', in order to avoid the abuses of mathematical reasoning that were entailed by Maclaurin's more capacious definition. Given Reid's criticisms in these notes, we can see why he later wrote to Lord Kames saying that 'I have no high Opinion of Mclaurens metaphysical accuracy, though he was great in Mathematicks'.[60]

Reid's career took an unexpected turn when, on 25 October 1751, he was elected to replace the recently deceased Alexander Rait as a regent at King's College, Aberdeen. That Reid was chosen to succeed Rait tells us something about how his talents were perceived by the masters at King's, insofar as Rait had earlier served as the nominal Professor of Mathematics at the College, and had been teaching mathematics privately as well as maintaining the College's

[57] AUL, MS 2131/3/I/15, fol. 1r.

[58] See below, p. 51. Reid elsewhere defined algebra as 'that part of Mathematics which treats of Quantity in General or of the Relations and Proportions common to all kinds of Quantity' (AUL, MS 2131/5/II/31, fol. 5r); compare 'A View of Algebra in the Order in which it is to be taught', AUL, MS 2131/7/I/1, fol. 1r, where Reid repeats his criticism of Maclaurin and states that 'Algebra considers Quantity in General. Geometry treats particularly of th[o]se Quantitys which have figure'. On fol. 1v of this manuscript, Reid makes the distinction in a slightly different way, arguing that algebra considers magnitude in general, whereas geometry studies figured magnitudes.

[59] AUL, MS 2131/3/I/15, fol. 1r.

[60] Compare below, p. 50; Reid to Lord Kames, 13 June 1782, in Reid, *Correspondence*, p. 153.

collection of mathematical and scientific instruments.[61] When the session began shortly after his election, Reid took over his predecessor's class of students who were then entering the third year of their four-year course. Consequently, in the sessions for 1751–52 and 1752–53 he had to prepare lectures on ethics, politics, 'general' and 'special' physics (natural history and natural philosophy) and, at the very least, arithmetic, algebra and Euclidean geometry. Even though there appear to be no surviving lecture notes from his first cycle of lectures written either by himself or by one of his students, some evidence regarding his teaching is contained in the graduation oration he delivered at King's College on 9 April 1753. In this oration, Reid outlined his view of the true aim and proper method of 'philosophy', that is, the subject taught by the regents at King's College.[62] While acknowledging that many ancient and modern figures had distinguished themselves individually as philosophers, he affirmed that philosophy in general had made little progress because there was no consensus about the laws regulating philosophical practice. What is striking about his assessment of the confused and impoverished state of philosophy in his day is that he held up mathematics as the model that philosophers ought to emulate, on the ground that from classical antiquity onwards mathematicians had collectively agreed on the methods and standards of proof appropriate to their discipline. The pursuit of mathematics had led to discoveries 'befitting the human mind' and had 'brought forth [the] most abundant fruits' thanks to 'the most renowned Newton' and a succession of mathematicians both ancient and modern. By contrast, philosophy continued to be riven by disputes between rival sects and schools and had barely moved beyond 'the state of infancy or at any rate of childhood'.[63] Despite this gloomy prognosis, Reid thought that there had been signs of progress in the three main branches of philosophy (namely moral philosophy, physics and logic). He maintained that the progress that had been made resulted from the fact that philosophers

[61] On Rait and the chequered history of the chair of mathematics at King's, see Wood, *Aberdeen Enlightenment*, p. 29. Reid also owed his appointment to his family connections; see Roger L. Emerson, *Professors, Patronage and Politics: The Aberdeen Universities in the Eighteenth Century*, pp. 69, 74–5. Although most of the teaching at King's was done by the regents, the College also had specialist professors in civil law, medicine, divinity, Greek and Hebrew, as well as a 'Humanist' who taught Latin and related subjects. Some of these chairs were sinecures.

[62] In the regenting system, each regent was responsible for teaching all of the subjects covered in the three-year philosophy curriculum, whereas in the professorial system of fixed chairs each professor would teach a single subject such as moral philosophy. King's College finally switched from regenting to the professorial system in 1800, the last of the Scottish universities to do so. Edinburgh was the first to switch to fixed chairs, in 1707, and was followed by Glasgow (1727), St Andrews (1747) and Marischal College (1753).

[63] Thomas Reid, *Philosophical Orations*, oration I, paras 5–6; the *Orations* are reproduced in Reid, *University*.

had finally eschewed groundless conjectures and recognised that observation, experiment and the methods of analysis and synthesis constituted the true path to knowledge.[64] For Reid, this crucial methodological shift was primarily due to the influence of Sir Francis Bacon, who was the pivotal figure in Reid's reading of the history of philosophy. For Bacon had not only renovated the study of natural history but had also proclaimed that 'natural philosophy is not a child of the human mind but a fair and legitimate interpretation of the works of nature itself or of God' and had shown that new discoveries could not be made using the tools of Aristotelian syllogistic logic. And it was the Lord Chancellor, Reid said, who had 'led [Newton] by the hand' and taught him the correct rules of reasoning in natural philosophy.[65]

Reid's tribute to 'the very great genius of Newton' strongly suggests that his first set of lectures on natural philosophy was largely taken up with expounding the essentials of the Newtonian system of the world and explaining Newton's methodological dicta. Presumably he covered much the same range of subjects that was typically taught in courses on natural philosophy in the period (mechanics, astronomy, pneumatics, hydrostatics, magnetism, electricity, optics), although his oration indicates that he may have spent a good deal of his classroom time on those sciences most closely associated with Newton, namely astronomy and optics. His synopsis of the rise of 'true physical astronomy' implies that in his classes devoted to astronomy he mentioned the most notable telescopic observations made by leading seventeenth-century astronomers such as Galileo, the laws of planetary motion discovered by Kepler and Newton, and the work subsequently done by Edmond Halley, James Bradley and the French expeditions to Lapland and Peru that confirmed the truth of Newton's account of the physical principles governing the solar system. In optics it would seem that he discussed the basics of Newton's theory of light and colours and perhaps performed some of the key experiments upon which the theory was based. And whatever he had to say about such subjects as electricity and magnetism was likely to have been derived from Newton's ideas, since Reid commented in his oration that 'whatever ground has been gained in natural philosophy by the followers of this great man

[64] Because the distinction between 'special' and 'general' physics continued to structure the teaching of the natural sciences in the first half of the eighteenth century, Reid would have taken 'physics' to encompass both natural history and natural philosophy. However, the scope of 'physics' also came to be identified with that of natural philosophy in the period.

[65] Reid, *Philosophical Orations*, oration I, paras 10, 14, 18–20. Reid's depiction of Newton as the methodological disciple of Bacon is indebted to his own regent, George Turnbull; see Turnbull, 'On the Association of Natural Science with Moral Philosophy', in Turnbull, *Education for Life*, p. 49.

[Newton] ... is in general due to the questions and conjectures proposed by him with a modesty that is unique and worthy of a philosopher'.[66]

Shortly after he delivered his first graduation oration, Reid and his colleagues at King's College embarked on a thorough reform of various aspects of College life, including the curriculum.[67] As well as extending the length of the academic session and altering the rules governing student discipline, the Principal and masters broke with the legacy of scholasticism by adding new subjects to the *cursus philosophicus* and by reordering the sequence in which the different branches of philosophy were taught. Reid was apparently the main architect of the 1753 reforms, and it was thus largely due to his influence that more class time was to be devoted to the teaching of mathematics and the natural sciences than had previously been the case.[68] Whereas in the first half of the eighteenth century students at King's were given a rudimentary introduction to mathematics in their final year, they were now to be taught 'a Course of Mathematics, both speculative and practical' during their second and third years at the College.[69] In addition, their second year was given over to lectures on Greek, along with 'all the Branches of Natural History, and ... the study of Geography, and Civil History', while their third year was to be taken up with 'a Course of Natural and Experimental Philosophy' that would complement their studies in mathematics.[70] In the reformed curriculum, natural history and natural philosophy were thus recognised as distinct and separate branches of the sciences, and with this recognition went a significant increase in the overall number of lectures allotted to the study of nature than had earlier been the norm. The change in terminology from 'general' and 'special' physics to 'natural history' and 'natural and experimental philosophy' is, therefore, historically significant. For it registers the determination of Reid and his fellow faculty members to eliminate the last vestiges of Aristotelian scholasticism from the curriculum and to modernise the *cursus philosophicus* by

[66] Reid, *Philosophical Orations*, oration I, paras 16–17. In Reid's survey of recent work confirming Newton's system of the world he also refers to an extensively annotated three-volume edition of Newton's *Principia* published by the Minim friars Thomas Le Seur and François Jacquier in collaboration with Jean Louis Calandrini, who was a Professor of Mathematics at the Academy in Geneva; see Niccolò Guicciardini, 'Editing Newton in Geneva and Rome: The Annotated Edition of the *Principia* by Calandrini, Le Seur and Jacquier'.

[67] On the 1753 reforms at King's, see Wood, *Aberdeen Enlightenment*, ch. 3.

[68] Wood, *Aberdeen Enlightenment*, p. 67. A template for the curriculum reforms at King's can be found in Reid's manuscript entitled 'Scheme of a Course of Philosophy', AUL, MS 2131/8/V/1; for a transcription of this manuscript, see Reid, *University*.

[69] On the distinction between speculative and practical mathematics, see Ephraim Chambers, *Cyclopædia: Or, an Universal Dictionary of Arts and Sciences* (1728), vol. II, p. 509.

[70] Anon., *Abstract of Some Statutes and Orders of King's College in Old Aberdeen. M.DCC.LIII. With Additions M.DCC.LIV* (1754), p. 13. For this pamphlet, see Reid, *University*.

grounding their students more thoroughly in the sciences of nature. But having emphasised the importance of natural knowledge in the revised curriculum, Reid and his colleagues were now faced with the pressing need to improve the material resources available at the College that would facilitate the study of mathematics and the natural sciences. They therefore launched a public appeal to 'their *Alumni*, in different Parts of the World, and others, who wish well to this University, and the Improvement of natural Knowledge'. They asked for help in establishing a natural history museum, acquiring 'the best Instruments for Surveying, Mensuration, Navigation, Astronomy and Optics' and in constructing 'a Laboratory for chymical Experiments and Operations, and a Room for anatomical Dissections'.[71] There was little response to their appeal, however, which meant that both teaching and research at King's were compromised by a lack of facilities, instruments and funding throughout the second half of the eighteenth century.

Following the formal adoption of the reform proposals in August 1753, Reid began teaching the newly expanded course of mathematics in the sessions for 1753–54 and 1754–55. Although no complete set of notes either in Reid's hand or that of a student survive, we can nevertheless reconstruct the sequence and range of topics that he covered with reasonable accuracy. He apparently began the first year of his revamped course with an introductory lecture in which, *inter alia*, he defined mathematics as 'a Science which instructs us in the Relations and Proportions of things Measurable' and informed his students that 'the Object of it is Quantity[,] which is wrong[ly] defined by many to be whatever is capable of More or less & ought rather to be defined [as] Whatever is capable of being Measured or what is capable of being doubled trebled halved &c'.[72] After making some general remarks on the relations between geometry, algebra and arithmetic, he turned to the basics of plane geometry using the first three books of Euclid's *Elements*. Once this initial segment of the course was complete, he moved on to lectures on algebra 'as far as Quadratick Equations', ratios as expounded in Book VI of Euclid, the application of algebra to geometry with illustrations drawn from Book II of Euclid, plane trigonometry, logarithms, mensuration and surveying.[73] For textbooks, it would seem that he briefly used David Gregory's

[71] Anon., *Abstract*, pp. 20–1.

[72] AUL, MS 2131/5/II/31, fol. 5v. Reid's remarks on the nature of quantity and his definition of the science of algebra echo his 'An Essay on Quantity'; compare below, p. 50.

[73] See AUL, MSS 2131/5/I/12 (logarithms), 2131/5/I/21 (trigonometry), 2131/5/II/13 (quadratic equations), 2131/5/II/18 (surveying), 2131/5/II/20 (quadratic equations), 2131/5/II/25 (geometry), 2131/5/II/35, fol. 1r (trigonometry), 2131/5/II/37 (trigonometry), 2131/7/I/1 (algebra), 2131/7/III/12 (surveying) and 2131/8/V/1, fol. 1v (geometry). Given that Reid taught his students mensuration and surveying we can understand why he persuaded his colleagues to appeal for funds to purchase the mathematical instruments required to take accurate measurements and surveys. Reid clearly assumed

edition of Euclid's *Elements* until the appearance of Robert Simson's new edition in 1756, Colin Maclaurin's *Treatise of Algebra* and Maclaurin's edition of David Gregory's *Treatise of Practical Geometry*.[74] In the advanced course given to the third-year students, Reid lectured on various branches of pure mathematics, including spherical geometry, conic sections, higher-order equations, the properties of higher-order curves, 'the Principles of Fluxions' and 'The Method of Exhaustions of Indivisibles & of Prime & Ultimate Ratios'. He also taught more practical subjects such as spherical trigonometry, dialling, navigation, the mathematical principles of astronomy, the use of globes and astronomical instruments and, if there was sufficient time, the principles of perspective.[75] In addition to the textbooks he recommended in the second year, he apparently referred his students to Colin Maclaurin's *Geometria organica* on the properties of curved lines and it is possible that he suggested that his advanced pupils consult Maclaurin's *Treatise of Fluxions* as well.[76] In its general outline and specific contents, Reid's syllabus was much the same as John Stewart's at Marischal College, although his course was more compressed than his friend's, for Stewart had the luxury of teaching over three years rather than two. Nevertheless, Reid taught the most ambitious course of mathematics offered at King's in the mid-eighteenth century, and it was not equalled in scope or sophistication until the 1790s, when Dugald Stewart's protégé, Robert Eden Scott, began to teach at the College.[77] Reid was thus able

that his students were already numerate. This was, however, not always the case; see the comments in Reid to Archibald Dunbar, 4 September 1755, in Reid, *Correspondence*, p. 10.

[74] AUL, MSS 2131/3/I/12–13 and 2131/7/I/1, fol. 1r. Reid's friend John Stewart likewise used Gregory's *Treatise* as a textbook; see Wood, *Aberdeen Enlightenment*, p. 229. The works in question are: Euclid, ΕΥΚΛΕΙΔΟΥ ΤΑ ΣΩΖΟΜΕΝΑ. *Euclidis quae supersunt omnia*, edited by David Gregory (1703); David Gregory, *A Treatise of Practical Geometry* (1745); Robert Simson, *The Elements of Euclid, viz. the First Six Books, Together with the Eleventh and Twelfth* (1756); Maclaurin, *Treatise of Algebra*. Although Maclaurin is not mentioned in the text of Gregory's *Treatise*, his involvement in the publication of the work is noted in Patrick Murdoch's 'An Account of the Life and Writings of the Author', which prefaces Maclaurin's *An Account of Sir Isaac Newton's Philosophical Discoveries, in Four Books* (1748), p. xix.

[75] AUL, MS 2131/8/V/1, fols 1v–2r; for more details on what he probably taught in the 'higher parts of Algebra', see AUL, MS 2131/7/I/1, fol. 1r–v. See also AUL, MSS 2131/5/I/30 (spherical trigonometry), 2131/5/II/31 (conic sections), 2131/5/II/34 (properties of ellipses), 2131/5/II/36 (spherical trigonometry) and 2131/8/V/1, fol. 1v (spherical trigonometry). On the 'higher parts of Algebra' and fluxions, see AUL, MSS 2131/5/I/6–8, 2131/5/I/10–11, 2131/5/II/3, 2131/5/II/32–32a, and 2131/5/II/50. On perspective, see AUL, MS 2131/5/I/18. On the construction of sundials, see AUL, MS 2131/7/VIII/11.

[76] For an allusion to Maclaurin's *Geometria organica: Sive descriptio linearum curvarum universalis* (1720), see AUL, MS 2131/8/V/1, fol. 2r; a note taken from Maclaurin's *Treatise of Fluxions* survives in AUL, MS 2131/6/V/23, fol. 1v.

[77] On John Stewart's course at Marischal, see Alexander Gerard, *Plan of Education in the Marischal College and University of Aberdeen, with the Reasons of It* (1755), pp. 32–3; Wood, *Aberdeen Enlightenment*, pp. 19–23, 87, 89–90.

to draw on his considerable expertise as a mathematician in the classroom, and he undoubtedly gave those students capable of grasping the material a firm grounding in the mathematical sciences.

In addition to having considerable biographical and local institutional significance, Reid's lectures on mathematics are of broader historical importance because they indicate that George Davie's interpretation of the cultural meaning and philosophical orientation of Scottish mathematics during the eighteenth century requires thorough revision. Davie has hitherto been the most influential commentator on the teaching of mathematics in Enlightenment Scotland, and it was he who coined the phrase 'mathematical Hellenism' in order to capture the distinctive features of the mathematical style that held sway in the Scottish universities from the 1710s until the late 1830s.[78] Davie emphasised that mathematics was taught in Scotland not as a 'mechanical knack' but as an integral part of a liberal education with strong connections to philosophy and the study of the classics.[79] Consequently, beginning with the distinguished Glasgow Professor of Mathematics, Robert Simson, and Simson's pupil Colin Maclaurin (whom Davie cast as the co-founders of the Scottish 'school' of mathematics), the lectures of Scottish professors differed from those given by their counterparts in England and elsewhere not only in terms of their overall pedagogical aims, but also in terms of their course contents.[80] According to Davie, the Scots typically introduced their students to the philosophical issues posed by the basic concepts of Euclidean geometry, algebraic analysis and Newtonian fluxions and, in addition, provided their students with a sophisticated overview of the historical evolution of the various branches of mathematics.[81] Moreover, Davie claimed that the Scots inherited

[78] George Davie, *The Democratic Intellect: Scotland and Her Universities in the Nineteenth Century*, chs 5–7; the phrase 'mathematical Hellenism' appears on p. 112. Davie's interpretation was subsequently taken up in Richard Olson, 'Scottish Philosophy and Mathematics, 1750–1830', and Richard Olson, *Scottish Philosophy and British Physics, 1750–1880: A Study in the Foundations of the Victorian Scientific Style*.

[79] The phrase 'mechanical knack' comes from the Edinburgh mathematician and natural philosopher John Leslie, as quoted in Davie, *Democratic Intellect*, p. 108.

[80] Although Davie sees Simson and Maclaurin as the founders of the Scottish 'school', he implies that the 'earliest Gregories' were also a part of the tradition of 'mathematical Hellenism', although he provides no evidence to substantiate his characterisation of their work; Davie, *Democratic Intellect*, p. 131.

[81] Even though Davie insists that the Scots dealt with such philosophical issues in their teaching, he does note that 'the philosophical tradition was, apparently, of a very elementary level and was kept going by means of a very few citations, which would often be reconsidered'; Davie, *Democratic Intellect*, p. 132. The quoted passage undermines the main thrust of his argument, insofar as it reduces the philosophical dimension of Scottish mathematical teaching to the repetition of a few commonplaces.

from Isaac Newton and Edmond Halley a preference for the purity and conceptual clarity of ancient Greek geometry, which meant that the analytical techniques and modes of proof found in the writings of the Greek geometers were regarded as the exemplars of rigorous mathematical reasoning. And because algebra and the form of modern mathematical analysis cultivated by continental mathematicians did not conform of the ideals set by classical geometry, Davie contended that the Scots downgraded the study of algebra in the curriculum and largely ignored the advances made in mathematics on the continent.

Niccolò Guicciardini, Judith Grabiner and Helena Pycior have, however, shown that key components of Davie's account of eighteenth-century Scottish mathematics are historically inaccurate.[82] As they have pointed out, Davie ignored the fact that the Scottish mathematical community in the period was deeply divided over the relative merits of geometry and algebra. Rather than collapse the work of Robert Simson and Colin Maclaurin together into a homogeneous Scottish 'school' of mathematics as Davie has done, Guicciardini and Grabiner have highlighted the technical differences between the mathematical thinking of the two Scots, and have demonstrated that Maclaurin's approach to shoring up the conceptual foundations of Newton's fluxional method was at odds with the geometrical programme of Simson and his disciple, the Edinburgh Professor of Mathematics, Matthew Stewart. Guicciardini has identified a group of mathematicians he calls the 'analytical fluxionists', who were not willing to sacrifice the power of the 'new analysis' developed in the seventeenth century by Newton and other mathematicians on the altar of classical geometry, and makes the telling point that Reid's extant mathematical manuscripts reveal that he is best understood as an 'analytical fluxionist'.[83] Furthermore, the details of Reid's mathematics course do not correspond to what Davie would lead us to expect regarding his teaching.[84] There is no evidence to suggest that in the classroom he dwelt on the foundational issues which otherwise preoccupied him in the privacy of his study or in his conversations with colleagues such as Stewart. Nor is there any indication that he lectured on the history of mathematics in any systematic

[82] Judith V. Grabiner, 'Maclaurin and Newton: The Newtonian Style and the Authority of Mathematics'; Judith V. Grabiner, 'Was Newton's Calculus a Dead End? The Continental Influence of Maclaurin's *Treatise of Fluxions*'; Guicciardini, *Development of Newtonian Calculus in Britain*; Guicciardini, 'Thomas Reid's Mathematical Manuscripts'; Helena M. Pycior, *Symbols, Impossible Numbers and Geometric Entanglements: British Algebra through the Commentaries on Newton's* Universal Arithmetick.

[83] Guicciardini, *Development of Newtonian Calculus in Britain*, ch. 6; Guicciardini, 'Thomas Reid's mathematical manuscripts', pp. 73–5.

[84] The same can be said about the teaching of mathematics at Aberdeen more generally in the eighteenth century; see Wood, *Aberdeen Enlightenment*, p. 91.

way, or remarked on the problems associated with the editing of ancient mathematical texts. And while some of what appear to be his lecture notes are in Latin, we should not conclude from this fact that he presented mathematics as a cultural offshoot of the classics, as Davie would have it. Instead, Reid undoubtedly used Latin for sound pedagogical reasons, in that he wanted to ensure that his second- and third-year pupils became proficient in the use of the language and thereby to allay the anxieties expressed in the wake of the reform of the curriculum at King's that students were not being given sufficient instruction in Latin and Greek.[85] More importantly, Reid neither maintained nor told his students that the epistemic status of geometry was in any way superior to that of algebra. Whereas Davie claimed that in the curricula of the Scottish universities the 'new algebra took very much a second place to the old geometry, and its introduction to the curriculum was delayed till late in the student's career',[86] we have seen that Reid taught algebra alongside geometry in both years of his course and that he probably devoted more classroom time to algebra than to geometry. He also did not share Simson's antipathy towards algebra. In the manuscripts related to his mathematics lectures, Reid made two suggestions that Simson would not have countenanced, namely that Book II of Euclid's *Elements* might 'be demonstrated in the Algebraic way' and that 'in the order of Nature' the elements of algebra preceded those of geometry, since 'Algebra considers Quantity in General' whereas 'Geometry treats particularly of those Quantitys which have figure'.[87] When compared with the actual record of the teaching practices of figures such as Reid, Davie's interpretation of the place of mathematics in the Scottish universities during the eighteenth century is not corroborated by the available evidence.[88]

Reid structured the second leg of his mathematics course so that it would complement his lectures on natural philosophy for his third-year class. We know far more about the contents of his classes on natural philosophy than we do about those he gave on the other subjects included in the revised *cursus philosophicus* because a reasonably complete set of student notes from his course survives and these notes enable us to make sense of the inchoate mass of papers in his own hand that are related to his lectures.[89] The extant archival material shows that the

[85] Anon., *Abstract*, pp. 18–19.

[86] Davie, *Democratic Intellect*, p. 110.

[87] AUL, MSS 2131/8/V/1, fol. 1v, and 2131/7/I/1, fol. 1r.

[88] As Knud Haakonssen has pointed out to me, the available evidence can also be seen as implying that Davie has exaggerated the contrast between the teaching of mathematics in the Scottish and English universities. If anything, Davie's argument needs to be heavily qualified in light of Guicciardini's work on Scottish 'analytical fluxionists' such as Maclaurin and Reid referred to above.

[89] Anon., 'Natural Philosophy 1758', AUL, MS K 160. The notes contain twenty-one pages of preliminary material, 115 on mechanics, forty-three on astronomy, six on corpuscular attraction and

scope and structure of his course were essentially the same as for the course given at Marischal College in the period.[90] That is, he began with a preliminary series of lectures in which he defined the basic terms used in natural philosophy, stated Newton's rules of philosophising and laid down what he called the 'axioms of physics', along with the laws of motion.[91] Then he proceeded to consider mechanics, observational and physical astronomy, cohesion, magnetism, electricity, hydrostatics and pneumatics, and concluded with a lengthy treatment of optics, in which he covered physical optics, catoptrics and vision.[92] The organisation of

cohesion, four on magnetism, four on electricity, thirteen on hydrostatics, fourteen on pneumatics and sixty-eight on optics. Not all of the sections are complete and a number of pages have been left blank. In addition to this set of student notes, there are a number of manuscripts in Reid's hand related to his course. There are various course outlines, which he may have drafted to submit to the Principal of the College for approval following the procedure enacted by the Parliamentary Commission on the Scottish universities in 1695; see Robert Sangster Rait, *The Universities of Aberdeen: A History*, p. 174. For the outlines see AUL, MSS 2131/6/V/10, 2131/6/V/31, fol. 3r–v, and 2131/8/V/2. The latter manuscript dates from the 1760–61 session – the names of the students that Reid has jotted down can be correlated with the relevant class list; see Peter John Anderson, *Officers and Graduates of University and King's College Aberdeen, 1495–1860*, pp. 242–3. For further manuscripts related to his teaching of natural philosophy see AUL, MSS 2131/2/I/5, 2131/6/V/23 and 2131/8/V/6 (introductory material); 21316/V/7–9, 2131/6/V/15–15a, 2131/6/V/25–33 and 2131/7/II/14 (mechanics); 2131/6/I/23, 2131/6/V/18, 2131/7/III/2 and 2131/8/VI/2 (optics); 2131/8/V/4–5 (hydrostatics); 2131/6/V/24 (pneumatics); 2131/6/V/11 (electricity and magnetism); and 2131/8/V/3, which contains fragments from different sections of his course. For manuscripts associated with his lectures on astronomy, see below, p. cxvii and note 354.

90 Gerard, *Plan of Education*, p. 32; Wood, *Aberdeen Enlightenment*, p. 101. After the reform of the curriculum which took place at Marischal College in 1753, William Duncan became the first Professor of Natural Philosophy. At his death in 1760, he was succeeded by George Skene, who was a close associate of Reid; see Peter John Anderson, *Fasti Academiae Mariscallanae Aberdonensis: Selections from the Records of the Marischal College and University, 1593–1860*, vol. II, pp. 45, 46.

91 Significantly, Reid had apparently not formulated his *vera causa* reading of Newton's first rule of philosophising by the time that he delivered his natural philosophy lectures in 1757–58; on this point, see below, p. lxiii.

92 Although a section devoted specifically to music is not included in the surviving volume of student notes, Reid proposed to teach the subject in his 'Scheme of a Course of Philosophy'; AUL, MS 2131/8/V/1, fol. 1v. He might have dealt with music in the context of his mechanics lectures on the elasticity of chords. The extant student notes contain some material related to music, but the section of the notes in which the subject features is clearly incomplete (Anon., 'Natural Philosophy 1758', pp. 116–19). Manuscripts on musical theory which could be Reid's own lecture notes survive in AUL, MSS 2131/6/V/16–16a, 2131/7/III/4 and 2131/7/V/25. For a highly technical set of notes taken from Robert Smith's *Harmonics, or the Philosophy of Musical Sounds* (1749), see AUL, MS 2131/6/V/17. Reid thus lectured on, and was primarily interested in, the cluster of sciences that Thomas Kuhn anachronistically referred to as the 'classical' or 'mathematical' sciences, namely astronomy, geometrical optics, statics, mathematics and harmonics or music. According to Kuhn, in classical antiquity these sciences acquired their own separate and independent identities based on technical knowledge specific to each of them; Thomas S. Kuhn, 'Mathematical Versus Experimental Traditions in the Development of Physical Science', especially pp. 35–41. In fact, as J. L. Heilbron

his lectures was thus relatively conventional, and followed the general format of the standard textbooks of the period.[93] The contents of his course were also largely drawn from these textbooks, although his lectures on mechanics, astronomy and optics were less derivative because he here incorporated material based on the work he had carried out in these sciences prior to his appointment at King's.[94] Moreover, his optics lectures deviated from the norms of the period, insofar as he included a considerable amount of detail on the theory of vision, even though this topic played an increasingly vestigial role in the mainstream of the study of optics during the eighteenth century.[95] Lastly, Reid's own enthusiasm for both mathematics and natural philosophy occasionally got the better of him in the classroom, because we find him warning a student's father in September 1755 that rather than start on 'the Philosophy of the Mind, Logic, Morals, and Politics' as he was scheduled to do in the session for 1755–56, he would be spending 'one hour in the day, for about two months, in the beginning of the session ... employed

has pointed out, from Aristotle until the turn of the nineteenth century this cluster of sciences constituted what was called 'mixed mathematics'; see J. L. Heilbron, 'A Mathematicians' Mutiny, With Morals', pp. 107–10. For an influential definition of 'mixed mathematics' available to Reid see Chambers, *Cyclopædia*, vol. II, p. 509.

[93] We have seen above that Reid was familiar with J. T. Desaguliers' *A Course of Experimental Philosophy* (1734–44); see also the undated note from this work in which Reid states that he found four errors in a set of Desaguliers' calculations: AUL, MS 2131/5/I/2. The other popular natural philosophy textbooks available to Reid were: Willem Jacob 's Gravesande, *Mathematical Elements of Natural Philosophy Confirm'd by Experiments, or an Introduction to Sir Isaac Newton's Philosophy*, second edition (1721–26); Willem Jacob 's Gravesande, *An Explanation of the Newtonian Philosophy, in Lectures Read to the Youth of the University of Leyden* (1735); Richard Helsham, *A Course of Lectures in Natural Philosophy* (1739); John Keill, *An Introduction to Natural Philosophy: Or, Philosophical Lectures Read in the University of Oxford, Anno Dom. 1700* (1720); Pieter van Musschenbroek, *The Elements of Natural Philosophy* (1744); and John Rowning, *A Compendious System of Natural Philosophy* (1753).

[94] Other works that Reid drew on in his lectures include: George Berkeley, *An Essay Towards a New Theory of Vision* (1709); Robert Boyle, *Essays of the Strange Subtilty, Determinate Nature [and] Great Efficacy of Effluviums* (1673); James Bradley, 'A Letter from the Reverend Mr James Bradley Savilian Professor of Astronomy at Oxford, and F.R.S. to Dr Edmond Halley Astronom. Reg. &c. Giving an Account of a New Discovered Motion of the Fix'd Stars'; James Jurin, 'An Essay upon Distinct and Indistinct Vision'; James Keill, *The Anatomy of the Humane Body Abridg'd: Or, a Short and Full View of All the Parts of the Body*, second edition (1703); Benjamin Martin, *Philosophia Britannica: Or, a New and Comprehensive System of the Newtonian Philosophy, Astronomy and Geography* (1747); Bernard Nieuwentijdt, *The Religious Philosopher: Or, the Right Use of Contemplating the Works of the Creator* [1715] (1718–19); William Porterfield, 'An Essay concerning the Motions of Our Eyes: Of Their External Motions'; William Porterfield, 'An Essay concerning the Motions of Our Eyes: Of Their Internal Motions'.

[95] On the changing definition of the science of optics during the eighteenth century, see Geoffrey Cantor, *Optics after Newton: Theories of Light in Britain and Ireland, 1704–1840*, p. 20; on Reid's optics lectures, see below, pp. clxxxviii–cxli.

upon Optics and some branches of Mathematicks, which I could not overtake last session'.[96]

Beyond the walls of King's College, Reid discussed various topics related to agricultural improvement with fellow members of the Gordon's Mill Farming Club. The Farming Club (*fl.* 1758–64) brought together Reid, the Principal and four other faculty members from King's with noted local agricultural improvers, including Reid's patron, Sir Archibald Grant of Monymusk, Reid's friend John Douglas of Fechil and Robert Barclay of Urie.[97] Although he was a founding member of the Club, he rarely attended meetings, for reasons which remain unclear. Nevertheless, he was consulted about such matters as the use of lime as a fertiliser, the distillation of potatoes and the most effective method of harnessing draught animals to carts and ploughs.[98] Fertilising with lime was problematic in Aberdeenshire in the eighteenth century because limestone was in short supply, costly to transport inland and expensive to burn.[99] The minutes of the meeting at which the 'conversation having fallen in with respect to lime as a manure' imply that Reid was known to have an opinion on the subject, which may well have been based on his expertise in natural history and chemistry rather than rooted in any practical experience. Similarly, his knowledge of the basic principles of mechanics was likely the reason why he was asked about how best to harness draught animals.[100] From jottings in his manuscripts dating from early 1758, however, we know that he was growing his own potatoes in Old Aberdeen and that he was also making simple qualitative experiments on the chemical properties of the tubers.[101] Insofar as potatoes were only just beginning to be widely grown in Scotland in the 1750s, he was in the vanguard of those in the middling and lower ranks of society who began to cultivate potatoes for food and his experimentation is redolent of the improving spirit that informed the activities of the Farming

[96] Reid to Archibald Dunbar, 4 September 1755, in Reid, *Correspondence*, p. 10.

[97] 'Minute Book of the Farming Club at Gordon's Mill 1758', AUL, MS 49; J. H. Smith, *The Gordon's Mill Farming Club, 1758–1764*. Reid was present at the inaugural meeting of the Club, held on 14 December 1758. The group remained active until January 1765.

[98] 'Minute Book of the Farming Club at Gordon's Mill 1758', pp. 26, 38 and 231.

[99] Smith, *The Gordon's Mill Farming Club*, p. 54.

[100] 'Minute Book of the Farming Club at Gordon's Mill 1758', p. 26. Reid may have discussed the use of lime with his colleagues at King's, John Gregory and Thomas Gordon, who were also members of the Farming Club. Gregory wrote about liming in his undated manuscript treatise 'Reflexions on the Principles of Agriculture', AUL, MS 2206/7/18, pp. 38–44. In Reid's natural philosophy lectures, he surveyed the practical application of mechanical principles to such problems as the design of wheel carriages and the construction of roads; see Anon., 'Natural Philosophy 1758', pp. 74–9, and AUL, MS 2131/6/V/31, fol. 3r.

[101] AUL MSS 2131/5/II/2, fol. 1r, and below, p. 127.

Club.[102] But Reid's most substantial contribution to the proceedings of the Club was his outline of a method of book-keeping designed specifically for the use of farmers, which he presented at a meeting held on 13 September 1760.[103] His scheme reflects the desire of the Club members to systematise and rationalise all aspects of farming life in order to increase efficiency, productivity and profitability, but it also speaks to his firm belief that the principles of mathematics and natural philosophy ought to be applied for the public good. We will return to this point below.

A second enlightened coterie to which Reid belonged while he was teaching at King's was the Aberdeen Philosophical Society or Wise Club (*fl.* 1758–73), which held its inaugural meeting on 12 January 1758.[104] Like the Farming Club, he was a founding member of the Philosophical Society. But while he was only minimally involved in the affairs of the Gordon's Mill group, he was a leading light in the Wise Club prior to his departure for Glasgow in the summer of 1764.[105] As one of the founders and most active members of the Aberdeen Philosophical Society, he did much to shape the Society's aspirations and proceedings. His Baconian commitment to moral and material improvement through the application of human knowledge as well as his understanding of the aims and scope of philosophy are expressed in the rules governing the meetings of the Society adopted at its inception, notably in Rule 17, which stipulated that 'the Subject of the Discourses and Questions' handled by the members of the Club 'shall be Philosophical'. The Rule then defined 'Philosophical Matters' as encompassing 'Every Principle of Science which may be deduced by Just and Lawfull Induction from the Phænomena either of the human Mind or of the material World' and 'All

[102] As T. C. Smout notes, potatoes had been grown as a food crop in the gardens of the Scottish nobility since the late seventeenth century; T. C. Smout, *A History of the Scottish People, 1560–1830*, pp. 251–2. Reid may also have been prompted to grow his own potatoes by the pressing need to feed a growing family on a limited income. To paraphrase the old proverb, necessity was in his case most likely the mother of improvement.

[103] 'Minute Book of the Farming Club at Gordon's Mill 1758', p. 140. Reid had been asked about a 'plan of book-keeping proper for a farmer' following a meeting held on 16 April 1760 (p. 122). A copy of Reid's paper survives; see 'A Short System of Book-keeping for the Farmer drawn up at the desire of the Farmers Club at Gordon Milne Aberdeen A° 1761', AUL, MS 2341. This manuscript is not in Reid's handwriting, but the text incorporates corrections in his hand. For a transcription of this manuscript see Walter R. Humphries, 'The Philosopher, the Farmer and Commercial Education'.

[104] The minute books of the Society are transcribed and analysed in H. Lewis Ulman (ed.), *The Minutes of the Aberdeen Philosophical Society, 1758–1773* (1990).

[105] Reid served as the first secretary of Wise Club (1758–60), acted on various occasions as the President (who was responsible for chairing meetings) and attended 101 out of a possible 122 meetings (82.8%). In the period 1758 to July 1764, Reid was the most regular attender, followed by George Campbell, John Gregory, David Skene and John Stewart; see Wood, 'Thomas Reid, Natural Philosopher', p. 87.

Observations & Experiments that may furnish Materials for such Induction; The Examination of False Schemes of Philosophy & false Methods of Philosophizing; The Subserviency of Philosophy to Arts, the Principles they borrow from it and the Means of carrying them to their Perfection'.[106] Because the scope of 'philosophy' was so broadly defined, the conversations within the Society prompted by the questions proposed for discussion ranged widely over a number of topical issues. Of these questions, in the period Reid was a member of the Wise Club, the highest proportion (seventeen or 25%) dealt with political, social and economic matters, whereas a much smaller number addressed natural philosophical topics (six or 9%).[107] And of the formal discourses delivered by each member in rotation, only two (3%) were devoted to the physical sciences.[108]

While Reid did not himself propose any of the six questions related to the physical sciences, he participated in the collective consideration of topics in astronomy and chemistry that were of interest to his colleagues. In particular, he took part in the conversation that occurred on 12 April 1758, which was prompted by the question submitted by the Rev. Robert Traill, 'What are the proper Methods of determining the Suns Paralax by the Transit of Venus over his Disk in 1761?'[109] The transits of Venus that took place in 1761 and 1769 were among the most eagerly anticipated celestial events to occur in the eighteenth century, because in 1716 Edmond Halley had pointed out that accurate observations of the transits would allow astronomers to establish the distance of the Sun from the Earth and hence to calculate the exact size of the solar system. As well as predicting the dates and times when the transits of Venus would occur, Halley specified various locations across the globe where the most useful observations could be made and implied that what was needed was the international collaboration of men of science, supported by the major colonial powers of Britain, France and the United Provinces.[110] No record of the specifics of the conversation that took place on 12 April survive, but the fact that it took place is significant insofar as the Wise

[106] Ulman, *Minutes of the Aberdeen Philosophical Society*, p. 78. Rule 17 should be read in the context of Reid's discussion of the laws of philosophising in his 1756 graduation oration; see Reid, *Philosophical Orations*, oration II.

[107] For Reid's contributions to the Society dealing with social, economic and political issues see Reid, *Society and Politics*.

[108] Wood, 'Thomas Reid, Natural Philosopher', pp. 87–8.

[109] Ulman, *Minutes of the Aberdeen Philosophical Society*, pp. 83, 189.

[110] Edmond Halley, 'Methodus singularis quâ Solis parallaxis sive distantia à Terra, ope Veneris intra Solem conspiciendæ, tuto determinari poterit'. See further below, pp. cxviii–cxxii. On the collaborative efforts to observe the transits of Venus in the 1760s, see Harry Woolf, *The Transits of Venus: A Study of Eighteenth-Century Science*. The work carried out by the Wise Club figures in neither Woolf's study nor in the more recent popular account, Andrea Wulf, *Chasing Venus: The Race to Measure the Heavens*.

Club was one of the first groups in Europe to consider Halley's scheme and to prepare for the transit of Venus in 1761. As we shall see, Reid subsequently read a discourse to the Club summarising the results of the observations that he and others made of the transit, which occurred, as predicted, on 6 June 1761. The following year, he gave the only discourse on mathematics presented to the Wise Club during the period he was a member. This was his 'voluntary discourse at the laying down his office as President', delivered in January 1762, which no longer survives but which the Society's minutes inform us was on 'Euclid's definitions & axioms'.[111] Reid's discourse summarised at least six years' reflection on the foundations of Euclidean geometry; what can be inferred about the contents of his discourse will be discussed more fully in Section 2 below.

Regrettably, little evidence survives regarding Reid's reading of books and papers on mathematics and natural philosophy while he taught at King's College. While it is reasonable to assume that he read widely during the time he lived in Old Aberdeen, not least because he was preparing lectures on the array of subjects covered in the *cursus philosophicus*, only four sets of reading notes from the period are extant. The first, which is undated, records Reid's reading of Robert Simson's edition of Euclid's *Elements*, which appeared in Latin and English versions in 1756; we will return to this set in Section 2.[112] The other three all deal with writings on the mathematical sciences of optics, mechanics and astronomy. In March 1757, Reid read the second volume of the *Essays and Observations, Physical and Literary*, published the previous year by the Edinburgh Philosophical Society. This particular volume contained, *inter alia*, notable essays by Matthew Stewart on a simplified solution to what is known as 'Kepler's problem' regarding the mathematical analysis of planetary motion, William Cullen on the cold produced by evaporation and Joseph Black's seminal paper 'Experiments upon Magnesia alba, Quick-lime, and Some Other Alkaline Substances', which Reid came to regard as a model of inductive reasoning.[113] But the paper which initially attracted his attention was Thomas Melvill's 'Observations on Light and Colour', and he made notes on what he thought were the salient points in the first six sections of Melvill's paper. His comments will be analysed in Section 5. Then, in September of that year, he copied out brief passages from a work that defended

[111] Ulman, *Minutes of the Aberdeen Philosophical Society*, p. 107.

[112] AUL, MS 2131/3/I/13 (see pp. 10–12); on Simson's edition of Euclid see Philip Gaskell, *A Bibliography of the Foulis Press*, pp. 314–15.

[113] AUL, MS 2131/3/I/10; Matthew Stewart, 'A Solution of Kepler's Problem'; William Cullen, 'Of the Cold Produced by Evaporating Fluids, and of Some Other Means of Producing Cold'; Joseph Black, 'Experiments upon Magnesia alba, Quick-lime and Some Other Alkaline Substances'; Reid, *Logic, Rhetoric and the Fine Arts*, p. 183.

Leibniz's account of the force of moving bodies, Jacob Hermann's *Phoronomia, sive de viribus et motibus corporum solidorum et fluidorum libri duo* (1716). Reid, however, seems not to have been interested in this aspect of the book for he concentrated on Hermann's method of calculating centres of gravity. To these notes, he added material derived from William Emerson's *The Principles of Mechanics* (1754) and Leonhard Euler's *Mechanica sive motus scientia analytice exposita* (1736), as well as jottings on the composition of forces.[114] The only other mathematical and scientific text that we know he read while he was in Old Aberdeen is Edmond Halley's posthumous work, *Tabulæ astronomicæ* (1749). At some unspecified date in 1761, Reid studied Halley's astronomical tables with care and took meticulous notes from, and commented upon, Halley's tables of the motions of the Moon, the five inferior and superior planets, the satellites of Jupiter and Saturn, and of the orbits of comets.[115] The scrupulousness with which these notes were made reinforces the impression given by his detailed reading notes from Maupertuis and Pierre Bouguer, namely that Reid was both an experienced and a highly competent observational astronomer who had the practical, mathematical and theoretical skills necessary to engage critically with the works that served to advance the science of astronomy during the course of the eighteenth century.[116] And some of the data that he took from Halley's tables indicate that Reid not only collaborated with his Wise Club colleagues in making telescopic observations but also worked independently as an observer of the heavens while he lived in Old Aberdeen.[117]

After months of political manoeuvring by both his supporters and opponents, Thomas Reid was elected to succeed Adam Smith as the Glasgow Professor of Moral Philosophy at a meeting of the University Senate held on 22 May 1764.[118] Four days later he accepted the appointment in a letter to the Lord Advocate of Scotland and Rector of the University, Thomas Miller, and subsequently travelled

[114] AUL, MS 2131/3/I/8; see also Gucciardini, 'Thomas Reid's Mathematical Manuscripts', p. 73, and Niccolò Guicciardini, *Reading the Principia: The Debate on Newton's Mathematical Methods for Natural Philosophy from 1687 to 1736*, pp. 205–16, for an illuminating discussion of Hermann's *Phoronomia*.

[115] AUL, MS 2131/3/I/21; Reid's notes occupy both sides of four folios and are thus one of the longest sets of his reading notes to have survived.

[116] Reid's reading notes from Maupertuis and Bouguer are discussed below, p. cxiv–cxv.

[117] See also Reid's comments in the *Inquiry* on having injured his right eye while making telescopic observations in May 1761, which imply that he had been working independently as an observational astronomer for some time; Reid, *Inquiry*, p. 131.

[118] This hotly contested election will be discussed in detail in my forthcoming *The Life of Thomas Reid* but, for now, see Paul Wood, '"The Fittest Man in the Kingdom": Thomas Reid and the Glasgow Chair of Moral Philosophy', pp. 289–91.

to Glasgow, where he was formally admitted to his new post on 11 June.[119] By the end of September, Reid and his family were ensconced in a recently built house in the Drygate which, he reported to his old friend Andrew Skene back in Aberdeen, had 'the best Air and the finest prospect in Glasgow'.[120] Reid's letter to Skene shows that he was excited by his new surroundings, not least because Glasgow offered him material and cultural resources that were markedly superior to those available in Aberdeen. Even though he appears not to have been to Glasgow before his visit in June 1764, he received intelligence about the University in 1763 from his protégé William Ogilvie, who sent him an account of Joseph Black's theory of latent heat and who also seems to have encouraged an exchange of letters between Reid and Robert Simson in early 1764.[121] Traces of his own initial excitement linger in a letter he sent to Andrew Skene's son, David , in 1766, in which he encouraged the younger Skene to put himself forward as a candidate to succeed Black as the Professor of the Practice of Medicine at Glasgow. Reid informed his friend that 'there is a great Spirit of Enquiry here[.] among the young people Literary Merit is much regarded. and I conceive the opportunities a man has of improving himself are much greater than at Aberdeen'.[122]

Reid's enthusiasm was well founded. By the 1760s the University of Glasgow had been to some extent eclipsed academically by the Town's College in Edinburgh, largely because of the meteoric rise of the Edinburgh medical school. Nevertheless, Glasgow was a vibrant centre for the cultivation of natural knowledge.[123] The doyen of Scottish mathematicians, Robert Simson, had retired from teaching in 1761 but continued to be active until his death in 1768. Black's pioneering work on latent and specific heats was taken up by a circle of younger chemists whose researches turned Glasgow into one of the most important sites in Europe for the study of the science of heat in the second half of the eighteenth century.[124] The

[119] Reid to [Thomas Miller], 26 May 1764, in Reid, *Correspondence*, p. 34; 'Minutes of the University Meetings, 1763–1768', Glasgow University Archive Services, MS 22643, p. 28.

[120] Reid to Andrew Skene, 14 November 1764, in Reid, *Correspondence*, p. 37.

[121] Reid to [William Ogilvie], [1763], and Reid to [Robert Simson], [1764], in Reid, *Correspondence*, pp. 23–6, 32–4; both of these letters will be discussed further below. One other possible source for news about Glasgow was Reid's colleague in the Wise Club, Robert Traill, who became the Glasgow Professor of Divinity in 1761.

[122] Reid to David Skene, 18 April [1766], in Reid, *Correspondence*, p. 51. Reid did, however, point out in this letter that the chair in Glasgow would prove to be an advantageous stepping stone to a position in the Edinburgh medical school.

[123] On science and medicine in eighteenth-century Glasgow, see Roger L. Emerson and Paul Wood, 'Science and Enlightenment in Glasgow, 1690–1802'.

[124] Black's circle in Glasgow included James Watt, John Robison, William Irvine and William Trail, who moved to Aberdeen to become the Marischal Professor of Mathematics in 1766. What distinguished this group was its quantitative approach to the science of heat; see John Robison to

University also boasted an extensive cabinet of scientific instruments, models of water wheels and other technological devices, and collections of mineral and fossil specimens, thanks to a succession of Professors of Natural Philosophy that included Robert Dick I and II, and Reid's new colleague John Anderson. A considerable institutional advantage that the University of Glasgow enjoyed over its Scottish academic rivals was its astronomical observatory. A public subscription to fund the erection of an observatory had been launched in 1754, with Dick II and Robert Simson spearheading efforts to obtain the necessary observational hardware. While Dick II and Simson met with some success, it was the arrival of a notable collection of instruments gifted to the University by the wealthy merchant and amateur astronomer Alexander Macfarlane in 1756 that provided the impetus for the formal laying of the foundation stone of the observatory, on a site adjacent to the College buildings, in August 1757. By 1760 the building was finished and the astronomical instruments were installed.[125] The primary function of the now completed observatory (which was named the Macfarlane Observatory in honour of the University's alumnus and benefactor) was to support the work of the Regius Professor of Practical Astronomy and Observer in the College, Alexander Wilson, whose post was created largely due to the machinations of the third duke of Argyll.[126] Significantly, Reid mentioned the observatory as one of the most noteworthy features of the University when he first wrote to Andrew Skene in late autumn 1764 and, as we see shall see, Reid became a close friend and collaborator with Wilson and Wilson's son and successor, Patrick.[127]

Despite the fact that Reid's colleagues were deeply divided over his appointment as the Professor of Moral Philosophy, Reid quickly established a personal rapport with the men of science affiliated with the College. Recognised as the University's mathematical instrument maker in 1756 and given a workshop in the College buildings on the High Street, James Watt was mulling over the improvement of steam engines when Reid arrived in Glasgow and hit upon his revolutionary idea for a separate condenser engine in early 1765, just as Reid was finishing his first session of teaching. It is not clear if Reid was allowed

James Watt, [October 1800], in Eric Robinson and Douglas McKie (eds), *Partners in Science: Letters of James Watt and Joseph Black*, pp. 359–60.

[125] On the building of the Macfarlane Observatory, see Emerson and Wood, 'Science and Enlightenment in Glasgow', pp. 105–6; David Murray, *Memories of the Old College of Glasgow: Some Chapters in the History of the University*, pp. 260–3; James Coutts, *A History of the University of Glasgow: From Its Foundation in 1451 to 1909*, pp. 229–30. For a contemporary report on the laying of the foundation, see Anon., 'Affairs in Scotland', *Scots Magazine* 19 (1757), p. 431.

[126] Roger L. Emerson, *Academic Patronage in the Scottish Enlightenment: Glasgow, Edinburgh and St Andrews Universities*, pp. 134–40.

[127] Reid to Andrew Skene, 14 November 1764, in Reid, *Correspondence*, p. 37.

to see one of the earliest examples of Watt's new design, but he apparently did examine Newcomen engines built by Watt that incorporated changes to increase their efficiency. In December 1765 he reported to David Skene that 'Mr Watt has made two small improvements of the Steam Engine' and noted that Watt was 'just now employed in setting up An Engine for the Caron Company with these Improvements'. In the same letter, he also gave Skene detailed instructions for assembling the 'Perspective machine' that he had recently purchased from Watt on Skene's behalf and was sending to Aberdeen. While Skene had apparently wanted one of the devices to draw with, Reid pointed out that it could also be used to copy documents such as maps.[128] Presumably, Reid had visited Watt's shop in the nearby Trongate to buy the perspective machine, and it may be that his conversations with Watt usually took place at the shop.[129] Although they apparently did not correspond with one another after Watt left Glasgow for Birmingham in 1774, they would have received intelligence about each other through mutual friends such as Patrick Wilson, who wrote to Watt a few months after Reid's death in 1796 and mentioned 'our late truly excellent and venerable Doctor Reid'.[130]

The man who unintentionally set Watt on the path to the invention of a steam engine with a separate condenser, John Anderson, was Reid's staunch ally in the acrimonious battles that were fought at the Faculty and Senate meetings of the University during the 1760s and 1770s. Anderson had initially been appointed as the Glasgow Professor of Oriental Languages in 1754, before switching to the natural philosophy chair in 1757.This switch has been regarded as a sign that Anderson was merely a talented, if somewhat irascible, pedagogue who was at best a dilettante who lacked a discernible intellectual identity. Such interpretations of Anderson's career are, however, highly problematic, for they overlook the fact that Anderson, like Reid, had inherited the Baconian outlook of the generation of virtuosi who had helped to create the Scottish Enlightenment. When viewed as a virtuoso, Anderson emerges as a dedicated man of science who cultivated natural knowledge for the practical benefit of humankind. Consequently, we should see Anderson as an accomplished natural philosopher, even though his scientific style was far less mathematically oriented than that of his friend Thomas Melvill, or

[128] Reid to David Skene, 20 December [1765], in Reid, *Correspondence*, p. 41–3. On Watt's perspective machine, see Richard L. Hills, *James Watt, Volume I: His Time in Scotland, 1736–1774*, pp. 109–11.

[129] Watt's shop in the Trongate, which opened in 1763, was run in partnership with John Craig and with funding from Joseph Black; D. J. Bryden, 'James Watt, Merchant: The Glasgow Years, 1754–1774', pp. 10–12.

[130] Patrick Wilson to James Watt, 29 May 1797, James Watt Papers, Birmingham Archives and Heritage.

that of Reid.[131] Only glimpses of Anderson and Reid's conversations and collaborative investigations remain. As we shall see in Section 4, the pair joined forces with Alexander Wilson in 1769 to observe the second transit of Venus to occur in the eighteenth century. The three of them also entertained the natural philosopher and virtuoso Benjamin Franklin when he visited Glasgow in 1771, and, in the company of Patrick Wilson, they travelled to Schielhallion in Perthshire in the summer of 1774 to converse with the Astronomer Royal, Nevil Maskelyne, who was carrying out an experiment to test Newton's claim in the *Principia* that the gravitational attraction of a mountain causes the deflection of a plumb line from the vertical.[132]

Anderson temporarily relocated to Dumbarton Castle in the late summer and early autumn of 1782, where he performed experiments on cannon and shot, took samples from the volcanic rock formation where the Castle is situated, monitored 'the Heat of the Loch or Spring, of the Rock, of the River [Clyde] &c' and began writing 'a Plan for improving Artillery &c' as well as 'an Essay on Moisture'. While he was there, he was visited by various British and European men of science and a number of his Glasgow colleagues, including Alexander and Patrick Wilson, as well as Reid. Reid stayed with Anderson for four days and, because of the inclement weather, they were forced to stay indoors for the whole period. In his diary, Anderson noted that the two of them spent their time reading unidentified books by Lord Monboddo and Joseph Priestley, a pamphlet on the contentious topic of ecclesiastical patronage and a recent issue of what was most likely the *Monthly Review*. Presumably their reading was interspersed with conversation, which probably touched on Anderson's gunnery experiments and his other scientific activities at the Castle.[133] However, this may have been one of

[131] On Anderson's intellectual identity, see Paul Wood, '"Jolly Jack Phosphorous" in the Venice of the North; or, Who Was John Anderson?' Anderson's natural philosophy is analysed in David B. Wilson, *Seeking Nature's Logic: Natural Philosophy in the Scottish Enlightenment*, ch. 5.

[132] J. Bennet Nolan, *Benjamin Franklin in Scotland and Ireland, 1759 and 1771*, p. 191; Nevil Maskelyne, 'An Account of Observations Made on the Mountain Schehallien for Finding Its Attraction', pp. 524–5.

[133] 'Dumbarton Castle', John Anderson Papers, University of Strathclyde Archives and Special Collections, MS 33. Anderson included his measurements of temperature in his 1782 discourse, 'Of the Moisture in Houses that are situated on prominent Rocks', MS 40, fols 8–9. His gunnery experiments are recorded in his 'Journal of Experiments', MS 8. Although Anderson had been reading Monboddo's *Antient Metaphysics* while he was at Dumbarton Castle, his reference to '3d Volumes' in the relevant passage makes no sense because the third volume of *Antient Metaphysics* did not appear until 1784. Priestley published a number of works in 1782, notably second editions of his *Institutes of Natural and Revealed Religion* and his *Disquisitions Relating to Matter and Spirit*. In the former he attacked the common sense philosophy of Reid. In the latter he presented a version of materialism that Reid attacked in the 1780s, notably in the essay 'Some Observations on the Modern

the last occasions when the two men were able to enjoy one another's company unreservedly because, from 1783 onwards, Anderson was involved in a series of vitriolic disputes with his colleagues and in 1784 began to lobby for the establishment of a Royal Commission to investigate the management of the University. Anderson's increasingly disruptive behaviour alienated former friends such as Reid and he became an isolated figure at the College as a consequence.[134]

In Glasgow Reid continued his work on astronomy and natural philosophy in collaboration with Alexander Wilson and his son Patrick, who succeeded his father as the Professor of Practical Astronomy in 1784 when Wilson senior officially resigned from the chair.[135] After graduating from the University of St Andrews in 1733, the elder Wilson was encouraged to master the craft of making thermometers by the physician George Martine, who wrote extensively on thermometry and the science of heat.[136] Later, Wilson moved to London, where a chance visit to a type foundry inspired him with an idea of how to improve the process of type founding. When he returned to St Andrews in 1739, he entered into a business partnership to open a type foundry and transformed himself into a highly skilled manufacturer of type. He subsequently moved the manufactory to Camlachie, which was then to the east of Glasgow. Such was the quality of type produced by Wilson and his sons that they began to supply the type for the Foulis Press. Wilson senior may have introduced Reid to the Foulis brothers, whose books Reid had purchased before moving to Glasgow. Reid was also apparently aware of the quality of the thermometers that Wilson and his sons made, for he supplied them to David Skene back in Aberdeen.[137] And in a letter to Skene from July 1765, Reid stated that 'I never considered Dollonds Telescopes till I came here' and mentioned that he had examined a prism made of 'brazil peeble', along

System of Materialism'; see Reid, *Animate Creation*, pp. 173–217. The pamphlet on patronage was most likely Anon., *The Case of Patronage Stated, According to the Laws, Civil and Ecclesiastical, of the Realm of Scotland* (1782). Anderson's reference indicates that it may have been written by the Glasgow Professor of Ecclesiastical History, Hugh Macleod.

[134] Coutts, *History of the University of Glasgow*, pp. 283–93.

[135] See Patrick Wilson's biography of his father, 'Biographical Account of Alexander Wilson, M.D. Late Professor of Practical Astronomy in Glasgow', p. 16; on Alexander and Patrick Wilson, see the recent account of their activities in David Clarke, *Reflections on the Astronomy of Glasgow: A Story of Some Five Hundred Years*, ch. 4.

[136] George Martine, *Essays Medical and Philosophical* (1740); Wilson's thermometers are praised for their 'perfection' and 'exactness' on p. 206. The second edition of this work carried the title *Essays and Observations on the Construction and Graduation of Thermometers, and on the Heating and Cooling of Bodies* (1772). In this version of the work the medical essays contained in the first edition were dropped.

[137] Reid to David Skene, 20 December [1765] and 23 March 1766, and Reid to Andrew Skene, 8 May 1766, 15 July 1766 and 17 December 1766, in Reid, *Correspondence*, pp. 41, 47, 53, 54–5, 56.

with a micrometer manufactured by the renowned London instrument maker John Dollond, which was 'fitted to a Reflecting Telescope of about 18 inches by which one may take the Apparent Diameter of the Sun or of any Planet within a second of a degree'.[138] His remarks suggest that he was discussing Dollond's achromatic telescopes and micrometers, as well as Dollond's recently published experiments on the refrangibility of light, with Wilson senior at the Macfarlane Observatory. By 1769 at the latest, we know that the Dollond telescope was in use at the observatory, because it is mentioned in Alexander Wilson's paper summarising the observations made in Glasgow of the transit of Venus which occurred on 3 June of that year.[139]

Like John Anderson, Alexander Wilson was an active researcher in the field of meteorology. Wilson's investigations were rooted in those of his mentor at St Andrews, George Martine, who devoted a section of his *Essays Medical and Philosophical* to a discussion of the heat and cold of the air and climate.[140] Wilson evidently launched his own research on heat and cold while running the type foundry at Camlachie, and it was in 1748 or 1749 that he met the young Glasgow divinity student and budding natural philosopher Thomas Melvill. During the summers of 1749 and 1750 Wilson and Melvill took measurements of the temperature of the Earth's atmosphere by attaching thermometers to kites, which the pair flew in a field adjacent to Wilson's home.[141] As a Glasgow professor, Wilson continued to be interested in monitoring the heat and cold of the air, not least because he recognised that atmospheric conditions affected the reliability of his astronomical observations. Evidence for his interest comes in a paper published in the *Philosophical Transactions* in which he reported the thermometer readings taken at the Macfarlane Observatory when a severe cold snap hit Glasgow at the beginning of January 1768.[142] Patrick Wilson inherited his father's fascination with the nature of cold, as can be seen in his description of the freezing temperatures experienced in Glasgow in January 1780. The frigid weather led the younger Wilson to perform a series of observations and experiments that suggested that neither evaporation nor chemical solution are the cause of cold. In making his observations and performing the variety of simple experiments he described,

[138] Reid to David Skene, 13 July 1765, in Reid, *Correspondence*, p. 40. The achromatic telescope and microscope made by Dollond were acquired by the University in 1754.

[139] Alexander Wilson, 'Observations of the Transit of Venus Over the Sun, Contained in a Letter to the Reverend Nevil Maskelyne, Astronomer Royal', p. 333.

[140] Martine, *Essays Medical and Philosophical*, pp. 293–306.

[141] Wilson, 'Biographical Account of Alexander Wilson', pp. 5–7.

[142] Alexander Wilson, 'An Account of the Remarkable Cold Observed at Glasgow, in the Month of January, 1768'.

Wilson stated that he was endeavouring to test the theories of Joseph Black, William Irvine and Adair Crawford regarding the production of cold; implicitly, he was also addressing ideas initially put forward by William Cullen in 1752.[143] Although there is no evidence that Reid collaborated actively with either of the Wilsons in their researches on cold, we know that they apprised him of this facet of their enquiries, for Patrick Wilson mentions having shown Reid the effects of a hoar frost upon an iron railing in the College buildings.[144] However, as we shall see in Section 5, Reid was far more closely involved in Patrick Wilson's investigation of the aberration of light, which was a topic they first discussed in 1770. It was thus their shared interest in the mathematical sciences of astronomy and optics, rather than in meteorology or chemistry, which cemented a friendship which lasted until Reid's death in 1796.

Successive lecturers in chemistry at Glasgow provided Reid with varying degrees of companionship and intellectual stimulus. According to Dugald Stewart, 'in Dr Black, to whose fortunate genius a new world of science had just opened, Reid acknowledged an instructor and a guide; and met a simplicity of manners congenial to his own'.[145] Although Stewart's characterisation of the relationship between Reid and Black is misleading insofar as he implies that Black was the first to introduce his new colleague to the science of chemistry, the available evidence indicates that the two men struck up a cordial, if not overly close, relationship after Reid's arrival at the College.[146] As noted above, in 1763 Reid's ex-student William Ogilvie reported on Black's lectures and, in reply, Reid enthused about Black's theory of latent heat. The pressures of preparing new lectures prevented him from attending Black's classes in the academic session for 1764–65, but he managed to do so in the winter of 1765–66, when he again praised Black's theory, despite being mildly critical of the rest of the course.[147] Once Black left Glasgow in 1766, they seem to have lost direct contact with one

[143] Patrick Wilson, 'An Account of a Most Extraordinary Degree of Cold at Glasgow in January Last; Together with Some New Experiments and Observations on the Comparative Temperature of Hoar-Frost and the Air Near to It, Made at the Macfarlane Observatory Belonging to the College'; Cullen, 'Of the Cold Produced by Evaporating Fluids'.

[144] Patrick Wilson, 'An Account of a Most Extraordinary Degree of Cold at Glasgow', p. 471. In this paper Wilson mentions that he was assisted in making his observations and experiments by John Anderson, William Irvine and Adair Crawford.

[145] Stewart, *Account*, p. 47.

[146] Their relationship had a rocky start because Black was one of those who opposed Reid's appointment; see Wood, '"The Fittest Man in the Kingdom"', p. 289. We know that Black served as a physician for the Reid family on at least two occasions; see Reid to Andrew Skene, 14 November 1764 and 15 July 1766, in Reid, *Correspondence*, pp. 37, 55.

[147] See above, p. xlviii; Reid to [William Ogilvie], 1763, and Reid to David Skene, 20 December [1765], in Reid, *Correspondence*, pp. 26, 44.

another, apart from an exchange of letters in January 1773 regarding the chemical analysis of the mineral waters at Peterhead.[148]

One of Black's protégés in Glasgow, John Robison, had the unenviable task of replacing his mentor as the University's Lecturer in Chemistry. Although Robison confessed to his friend James Watt in 1783 that he was 'a very unpopular teacher' as an Edinburgh professor, at the beginning of his academic career in Glasgow he met with some success in the classroom.[149] Mid-way through Robison's first session of teaching chemistry, Reid wrote to David Skene that 'the Lecturer in Chemistry has general Approbation. He chiefly follows Dr Black and Stahl. There is a book of Stahls called three hundred Experiments which he greatly admires, and very often quotes'.[150] But beyond this reference to Robison's course, we know nothing about their relationship during the brief period in which they overlapped at Glasgow. After Robison travelled to the court of Catherine the Great at St Petersburg in 1770 as secretary to Admiral Sir Charles Knowles, the two ex-colleagues apparently lost touch until the 1780s, when they renewed their friendship and corresponded about Robison's work in optics.[151] We will return to their correspondence and discuss their conversations on scientific subjects in Section 5. Robison's main rival for the chemistry lectureship in Glasgow in 1766 had been another ex-student of Black's and a fellow member of Black's inner circle, William Irvine. Irvine was bitterly disappointed to lose the election to Robison, whose candidacy Black had endorsed. However, Reid and other faculty members intervened because Irvine had established his credentials as a competent teacher.[152] At their instigation, a lectureship in materia medica was created for Irvine and, in addition, he took over the teaching of chemistry when Robison left Glasgow in 1769. Apart from passing references in Reid's correspondence

[148] Reid to Joseph Black, 17 January 1773, in Reid, *Correspondence*, pp. 74–5. It may be, however, that Reid met with Black on one of his periodic visits to Edinburgh.

[149] Robison to James Watt, 22 October 1783, in Robinson and McKie, *Partners in Science*, p. 130; Paul Wood, 'Science, the Universities and the Public Sphere in Eighteenth-Century Scotland', p. 103. During his time in Glasgow Robison also did some substitute teaching for John Anderson; see Robison to Watt, [25 February 1800], in Robinson and McKie, *Partners in Science*, p. 337.

[150] Reid to David Skene, 25 February 1767, in Reid, *Correspondence*, p. 58. The book by the German chemist and physician Georg Ernst Stahl that Robison used was *Experimenta, observationes, animadversiones, CCC numero, chymicae et physicae* (1731). On Robison's use of Stahl and on his lectures on chemistry more generally, see Wilson, *Seeking Nature's Logic*, pp. 150–7.

[151] Reid to [John Robison], [April 1788–90], in Reid, *Correspondence*, pp. 197–200.

[152] Shortly after Irvine's death in 1787, an anonymous biographer claimed that although Irvine had 'at first resolved to quit his native city [Glasgow] ... some members of the University were sorry to lose one whose bright dawn promised a splendid noon. The venerable Dr Reid, in particular, interested himself so warmly, that a new lectureship was instituted, with a salary equal to that annexed to the chemical chair'; Anon., 'Biographical Account of Dr William Irvine', p. 456.

indicating that he and Irvine were regular companions, almost nothing is known about their relationship in the years following the events of 1766.[153]

When Irvine died in July 1787, he was succeeded as the lecturer in chemistry by Thomas Charles Hope, who went on to become the Glasgow Professor of Medicine (1791–95) and, ultimately, Black's replacement in Edinburgh. Hope seems to have befriended the ageing Reid and, as an early exponent of Antoine-Laurent Lavoisier's revolutionary system of chemistry, he may have been the catalyst for Reid's study of the English translations of key chemical texts by Antoine-François de Fourcroy and Lavoisier, in 1789 and 1790.[154] Hope's appointment as the College's Professor of Medicine allowed the physician Robert Cleghorn to switch from the position of lecturer on materia medica to the chemistry lectureship in 1791. Like Reid, Cleghorn was a member of the phalanx of Foxite Whigs at the University and, apart from sharing similar political views, Cleghorn acted as Reid's personal physician during his last illness and wrote the first biography of Reid to be published after his death. Presumably the two of them conversed about chemical topics as well as the French Revolution in the 1790s because Cleghorn states in his tribute to his late colleague that Reid 'studied the late improvements in Chemistry, he observed the great political events which have happened, and contemplated those with which the time seems pregnant, with the keen interest of one just entering on life'.[155]

Evidence for Reid's contacts with Glasgow's mathematicians is limited. Dugald Stewart's assertion that Reid's 'early passion for the mathematical sciences was revived by the conversation of Simson' and Simson's close associate, the Glasgow Professor of Greek, James Moor, is problematic for two reasons: first, his statement cannot now be corroborated and, secondly, he misleadingly implies that Reid's scientific interests had waned while he was a regent at King's College, Aberdeen.[156] Moreover, there is no hint that Simson's successor, James

[153] Emerson, *Academic Patronage in the Scottish Enlightenment*, pp. 159–60; Alexander Duncan, *Memorials of the Faculty of Physicians and Surgeons of Glasgow, 1599–1850: With a Sketch of the Rise and Progress of the Glasgow Medical School and of the Medical Profession in the West of Scotland*, pp. 131–2; Reid to Andrew Skene, 15 July 1766, Reid to David Skene, 25 February 1767 and 14 September 1767, in Reid, *Correspondence*, pp. 55, 57, 58, 61.

[154] See below, pp. clxxxviii–cxci.

[155] [Robert Cleghorn], *Sketch of the Character of the Late Thomas Reid, D.D. Professor of Moral Philosophy in the University of Glasgow; with Observations on the Danger of Political Innovation, from a Discourse Delivered on 28th. Nov. 1794 by Dr Reid, Before the Literary Society in Glasgow College* (1796), p. 5. For Cleghorn's biography, see Reid, *University*. For evidence that Cleghorn acted as Reid's physician, see Stewart, *Account*, pp. 178–9. On the group of Foxite Whigs at the University, see Bob Harris, *The Scottish People and the French Revolution*, p. 25.

[156] Stewart, *Account*, p. 49. Like Simson, Moor was an accomplished mathematician, and the two men were regular companions.

Williamson, ever discussed mathematical matters with Reid, although the two of them collaborated in observing the transit of Venus in 1769.[157] Yet Reid seems to have had some knowledge of the mathematical community at the University, because when he was asked by the Aberdeen Town Council to act as an 'Examinator' of the candidates for the vacant chair of mathematics at Marischal College in 1766 he wrote to Andrew Skene that 'there is one Candidate for your Profession of Mathematicks to go from this College and if your Colledge get a better Man or a better Mathematician they will be very lucky'.[158] Reid's fellow examiners apparently agreed. After a rigorous and lengthy series of tests the Glasgow graduate, William Trail, was appointed as the Marischal Professor of Mathematics in August 1766.[159] Only James Millar, the son of the distinguished Glasgow Regius Professor of Civil Law John Millar, seems to have shared in Reid's mathematical pursuits. Their discussions prompted Millar to affirm that even though Reid had expended a considerable amount of mental energy in refurbishing the science of the mind in order to refute Hume, 'it is well known to Dr. Reid's literary acquaintance, that these exertions have not diminished the original bent of his genius, nor blunted the edge of his inclination for mathematical researches; which, at a very advanced age, he still continues to prosecute with a youthful attachment, and with unremitting assiduity'.[160]

After moving to Glasgow, Reid's friendship with Henry Home, Lord Kames, flourished. Whereas his earliest surviving letters to Kames are politely deferential and bear the hallmarks of a client cultivating a prospective patron, his letters to Kames dating from the 1770s and early 1780s are far more candid.[161] Like the virtuosi who helped create the Scottish Enlightenment, Kames was a man of encyclopaedic interests who published books on antiquarian topics, agricultural improvement, philosophical history, legal theory and moral philosophy, but who

[157] Wilson, 'Observations of the Transit of Venus Over the Sun', pp. 333–4. In 1774 Reid borrowed a copy of Alexander Gordon's *Itinerarium septentrionale* (1726) from the University library on behalf on Williamson, so it may be that the two men were on reasonably friendly terms; see 'Professors Receipt Book, 1770–[1789]', Glasgow University Library (henceforth GUL), Spec Coll MS Lib (uncatalogued) in the listings for 1774.

[158] Aberdeen Town Council to Reid, 30 April 1766, and Reid to Andrew Skene, 8 May 1766, in Reid, *Correspondence*, pp. 52, 54. Reid served as an examiner along with the professors of mathematics at Edinburgh and St Andrews. The invitation from the Town Council speaks to his perceived standing as a mathematician, as well as his family connection with Marischal College.

[159] Ponting, 'Mathematics at Aberdeen', pp. 166–7. Trail later published a life of Robert Simson, under whom he had studied at Glasgow; see William Trail, *Account of the Life and Writings of Robert Simson, M.D. Late Professor of Mathematics in the University of Glasgow* (1812).

[160] Hutton, *Mathematical and Philosophical Dictionary*, vol. I, p. 558.

[161] On the relationship between Reid and Kames, see especially Allardyce, *Scotland and Scotsmen in the Eighteenth Century*, vol. I, p. 474–5.

also penned essays on such topics as evaporation.[162] When one of his papers, 'Of the Laws of Motion', was published in 1754, it caused a considerable stir within the Philosophical Society of Edinburgh, which the Society's joint secretary, David Hume, did his best to contain. Kames' ascription of a 'power of motion' to matter, his claim that 'Chemistry discovers various powers in matter of the most active kind', and his critique of Newton's analysis of motion and gravitation elicited a polemical reply from the Edinburgh Professor of Natural Philosophy, John Stewart. Stewart systematically refuted Kames' arguments regarding the fundamentals of Newton's natural philosophy. In doing so, he suggested that his opponent's philosophical views were no different from those of heterodox thinkers such as Buffon, John Turberville Needham, Thomas Hobbes, John Toland, Anthony Collins, Spinoza and Kames' kinsman David Hume, who had all questioned in their different ways the passivity of matter. As an orthodox Newtonian, Stewart read his adversary's challenge to the matter–spirit distinction drawn by Newton and his disciples as implying that Kames was advancing a version of materialism which could only lead to irreligious consequences.[163]

After Reid and Kames were introduced to one another in 1762, the latter sent Reid an 'Essay on moving Forces', which may have been a reformulation of some of the ideas contained in Kames' controversial essay of 1754. Writing in late December 1762, Reid promised to 'send you my remarks upon it sometime in March' but warned that 'I fear they will be against you'.[164] In 1780–82 Kames revisited his earlier treatment of motion and his critique of the foundations of Newtonian physics, partly in relation to a discourse on Newton's theory of gravitation that he read to the Philosophical Society of Edinburgh in 1781 but also with the intention of publishing a revised and expanded version of his earlier

[162] The standard biography of Kames remains Ian Simpson Ross, *Lord Kames and the Scotland of His Day*.

[163] Henry Home, Lord Kames, 'Of the Laws of Motion', pp. 7, 9, 16–60; John Stewart, 'Some Remarks on the Laws of Motion, and the Inertia of Matter', pp. 72, 117, 130, 139. Stewart's questioning of Kames' religious orthodoxy and especially his claim that Kames' *Essays on the Principles of Morality and Natural Religion* (1751) was a 'useful commentary' on David Hume's philosophical writings (p. 117, note) were particularly damaging. When the Kames–Stewart exchange was published in 1754, Kames and Hume were facing the possibility of formal censure by the General Assembly of the Church of Scotland for their philosophical doctrines; on the fracas within the General Assembly, see Richard B. Sher, *Church and University in the Scottish Enlightenment: The Moderate Literati of Edinburgh*, pp. 65–74. For Hume's response to Stewart's provocation see Hume to [John Stewart], [February 1754], in David Hume, *The Letters of David Hume*, vol. I, pp. 185–8.

[164] Reid to Lord Kames, 29 December 1762, in Reid, *Correspondence*, p. 20; Ross, *Lord Kames and the Scotland of His Day*, p. 360.

essays.[165] On this occasion, Reid adopted the role of the defender of Newtonian orthodoxy previously played by John Stewart.[166] Although Kames' letters to Reid are no longer extant, the first in the sequence of Reid's letters indicates that Kames initiated their exchanges on 15 April 1780. Reid's initial letter also implies that Kames intended to supplement his earlier material with a survey of the history of responses to Zeno of Elea's paradoxes regarding motion, which evidently included a quote from an unidentified passage in Newton's works which may have related to Newton's treatment of infinites.[167] Over the course of the next two and a half years, the pair debated a number of issues addressed in Kames' 'Of the Laws of Motion', including the status and meaning of the definitions and axioms prefacing the *Principia*, the physical causes of natural phenomena such as the weight of the air and the force of gravity, the principles of hydrostatics and the final causes of the laws of motion.[168] Like Stewart before him, Reid rejected Kames' arguments outright, although he was far less obsessed with exposing the irreligious implications of Kames' attribution of activity to matter than Stewart had been. Nevertheless, Reid's letters to Kames illustrate distinctive features of his conception of the proper aims, scope and method of natural philosophy and his interpretation of the true meaning of Newton's scientific writings.

Three themes in the correspondence are especially relevant here. First, although I. Bernard Cohen and other historians of science following his lead have argued that Newton's *Principia* and *Opticks* embodied two different styles of scientific enquiry that spawned divergent traditions within eighteenth-century

[165] Kames presented a discourse on the cause of gravity to the Philosophical Society of Edinburgh on 1 February 1781; for two slightly different versions of his discourse see National Library of Scotland, MS Acc 10,073/4. Reid did his best to dissuade Kames from publishing his 'Essays Upon the Laws of Motion &c'; see Reid to Lord Kames, [October 1782], in Reid, *Correspondence*, pp. 157–9. It may be that Reid succeeded in doing so, but Kames was also seriously ill in the late autumn of 1782 and died on 27 December 1782; see Ross, *Lord Kames and the Scotland of His Day*, pp. 369–71. Reid alludes to Kames' ill health in Reid to Lord Kames, 11 November 1782, in Reid, *Correspondence*, p. 159.

[166] Reid explicitly identifies himself as a Newtonian at various points in the course of their exchanges; see especially Reid to Lord Kames, 13 June 1782, in Reid, *Correspondence*, p. 153, where he refers to 'we Newtonians'.

[167] Reid to Lord Kames, 23 April 1780, in Reid, *Correspondence*, pp. 121–4. Kames had also mentioned John Keill's refutation of Zeno, which Reid also notes; see John Keill, *An Introduction to Natural Philosophy*, pp. 69–71. Keill's resolution of Zeno's paradox invokes the nature of infinite series. Kames' 'Of the Laws of Motion' begins with a brief review of definitions of motion, and it would seem that in 1780 Kames intended to expand this part of the essay.

[168] Apparently some of Reid's letters to Kames have also not survived because there is a gap in the correspondence between February 1781 and June 1782. Otherwise, the two men exchanged letters on a regular basis. They presumably also discussed Kames' 'Essays' when Reid visited his friend and patron at the latter's family estate at Blair Drummond.

natural philosophy, it is striking that Reid emphasised the structural and methodological similarities between the two works.[169] According to Reid the *Principia* and the *Opticks* were both modelled on Euclid's *Elements*, insofar as Newton began each of them by laying down the definitions and axioms which provided the conceptual and empirical foundations for the arguments that unfolded deductively through the remainder of each text. He acknowledged, however, that the nature of the axioms employed in the *Principia* and the *Opticks* differed significantly from those in the *Elements*, for in his view Newton's axioms were not self-evident truths but rather empirical generalisations which rested on 'what had before been discovered' in the sciences of motion and optics.[170] In calling attention to the axiomatic structure common to the two books, Reid was echoing the point he made in his King's College graduation oration for 1756 that Newton had been the first natural philosopher to employ the axiomatic method used by mathematicians and to erect a system of physics on axioms encapsulating matters of fact and principles of common sense. Consequently, on Reid's reading of the recent history of natural philosophy, Newton had 'raised a fabric in [mechanics and optics] which is not liable to be shaken by doubtful disputation, but stands immoveable upon the basis of self-evident principles', while in the eighteenth century Newton's 'fabric' had been enriched 'by the accession of new discoveries' but was no longer 'subject to revolutions'.[171] Furthermore, Reid affirmed in his oration that *all* of the major branches of learning could be individually reconstructed on an axiomatic basis, stating that 'there are axioms and phenomena in ethics and politics no less than in physics on which every sound argument in these sciences rests'.[172] We see, therefore, that the format of Euclid's *Elements*, as well as Newton's *Principia* and *Opticks*, were part of the inspiration for Reid's view that human knowledge ultimately rests on principles of common sense that are akin to the axioms of ancient geometry and the mathematical sciences.[173]

[169] I. Bernard Cohen, *Franklin and Newton: An Inquiry into Speculative Newtonian Experimental Science and Franklin's Work in Electricity as an Example Thereof.*

[170] Reid to Lord Kames, 19 May 1780, in Reid, *Correspondence*, p. 125; compare the slightly more elaborate formulation in Reid, *Intellectual Powers*, pp. 455–6, 457.

[171] Reid, *Intellectual Powers*, p. 457; compare Reid, *Philosophical Orations*, oration IV, para. 4, where he states that astronomy, mechanics, hydrostatics, optics and chemistry had all been placed on firm foundations.

[172] Reid, *Philosophical Orations*, oration II, para. 20. Reid's claim in this oration that Newton's rules of philosophising expressed principles of common sense was repeated in the *Inquiry*; see Reid, *Inquiry*, p. 12. In Reid's lectures at King's College, he adopted an axiomatic presentation of his material; see the preliminary material in the set of student notes from his lectures on natural philosophy, AUL, MS K 160, pp. 1–21, and Reid, *Society and Politics*, pp. 23–32.

[173] Reid, *Intellectual Powers*, pp. 457–9.

The second significant theme in Reid's letters to Kames is the distinction that he repeatedly draws between the domains of physics and metaphysics, which he employed to counter Kames' assertion that the phenomena of nature imply that matter is active. Reid maintained that Kames and his judicial colleague, Lord Monboddo, had both 'mixed too much Metaphysicks with Physicks' in their criticisms of the Newtonian system because they had failed to recognise that the investigation of efficient and final causes has no place in natural philosophy.[174] He insisted that the 'bussiness' of natural philosophy is 'to collect by just Induction the Laws that are less general, and from these the more general as far as we can go' on the basis of 'particular facts in the material World' established through observation and experiment. Hence he argued that when natural philosophers refer to the 'causes' of natural phenomena they mean nothing more than 'the law of Nature, of which that Phenomenon is an instance or a necessary consequence'.[175] For him, the scope of natural philosophy was thus restricted to the investigation of 'particular facts' and the 'general rules' or laws governing those facts, whereas the consideration of efficient and final causes was the proper province of metaphysics, wherein metaphysicians employ 'abstract Reasoning' rather than empirical and inductive methods in their contemplation of such causes.[176]

While Reid believed that 'final Causes … stare us in the face wherever we cast our Eyes', he was far less optimistic regarding our capacity to discover the efficient causes of gravity and other natural phenomena. Beyond accepting that 'the Deity is the first Efficient Cause of all Nature', he believed that it was beyond the power of human reason to discover 'how far [God] operates in Nature immediately, [or] how far by the Ministry of Subordinate Efficient Causes, to which he has given Power adequate to the task committed to them'. Consequently, he regarded the very different explanations of the phenomena of nature given by Malebranche, Leibniz, Ralph Cudworth, Monboddo, Joseph Priestley and Kames as 'conjectures onely, about Matters where we have not Evidence', and he observed that none of these rival philosophical systems could 'be either proved

[174] Reid to Lord Kames, [October 1782], in Reid, *Correspondence*, p. 158. Reid had in mind Monboddo's *Antient Metaphysics*; the first two volumes of the work appeared in 1779 and 1782. Reid would have been especially interested in, and critical of, Monboddo's 'Dissertation on the Principles of the Newtonian Philosophy', published in James Burnett, Lord Monboddo, *Antient Metaphysics: Or, the Science of Universals* (1779–99), vol. I, pp. 497–555.

[175] Reid to Lord Kames, 16 December 1780, in Reid, *Correspondence*, pp. 142–3; compare Reid to Kames, 19 May 1780 and [October 1782], in Reid, *Correspondence*, pp. 127, 158. Reid later elaborated on his discussion of the different meanings of the word 'cause' in Reid, *Active Powers*, pp. 33–8.

[176] Reid to Lord Kames, [October 1782], in Reid, *Correspondence*, p. 159.

or refuted from the principles of natural Philosophy'.[177] Even some of Reid's fellow Newtonians came in for implicit censure. Against those who had taken up Newton's postulate of an ethereal medium to account for gravity and a range of other phenomena, he emphasised that Newton himself had viewed his ether speculations as entirely conjectural and that 'it [the ether] was no part of his System; and to hold it as such is to do him injustice'.[178] In order to read Newton correctly, therefore, Reid stipulated that one had to recognise that his system of physics was expounded in the 'Propositions and Corollaries' of the *Principia* and *Opticks*. 'In Scholia and Queries', on the other hand, Newton 'gives a Range to his thoughts, & sometimes enters into the Regions of Natural Theology & Metaphysicks'. And even though Reid deemed the 'Queries & Conjectures' in these texts to be 'valuable', he nevertheless warned that Newton 'never intended that they should be taken for granted, but made the Subject of Inquiry'.[179]

The third notable theme in Reid's letters to Kames concerns the use and abuse of conjectures, hypotheses and queries in natural philosophy.[180] Reid's remarks in this sequence of letters register the evolution of his critique of hypothetical reasoning over the span of some thirty years. The earliest concrete evidence we have for his view of the use of hypotheses and conjectures is to be found in his lectures on natural philosophy, his graduation orations and his manuscripts related to his teaching at King's College. In his natural philosophy lectures, he warned his students against relying on hypotheses and pointed to Fermat's and Leibniz's derivation of Snell's law governing the refraction of light as a cautionary tale illustrating the methodological lesson that 'the most ingenious men when they trust to Hypothesis … have only the chance of going wrong in a more ingenious

[177] Reid to Lord Kames, 16 December 1780, in Reid, *Correspondence*, pp. 143–5; on final causes, see also Reid to Kames, [autumn 1782], in Reid, *Correspondence*, pp. 154–7, where Reid cautions against searching for the final causes of necessary truths.

[178] Reid to Kames, 31 October 1780, in Reid, *Correspondence*, p. 137; see also Reid to Kames, [October 1782], in Reid, *Correspondence*, p. 159. For Newton's ether speculations see Isaac Newton, 'General Scholium', in *The Principia: Mathematical Principles of Natural Philosophy* [1687], translated by I. Bernard Cohen, Anne Whitman and Julia Budenz, pp. 943–4; the 'Advertisement' and Queries 17–24 added in 1717 to Newton, *Opticks*, pp. cxxiii, 347–54; and Newton to Robert Boyle, 28 February 1678/79, in Isaac Newton, *Isaac Newton's Papers and Letters on Natural Philosophy and Related Documents*, pp. 250–3. Newton's letter to Boyle was first published in Thomas Birch, *The Life of the Honourable Robert Boyle* (1741), pp. 234–47. One text that was instrumental in reviving the popularity of Newton's ether speculations in the mid-eighteenth century was Bryan Robinson, *A Dissertation on the Æther of Sir Isaac Newton* (1743). On this revival, see Robert E. Schofield, *Mechanism and Materialism: British Natural Philosophy in an Age of Reason*, especially part II.

[179] Reid to Lord Kames, 16 December 1780, in Reid, *Correspondence*, p. 147.

[180] On Reid's view of the use of conjectures and hypotheses, compare Shannon Dea, 'Thomas Reid's Rigourised Anti-hypotheticalism'.

way'.[181] After the curriculum reforms of 1753 Reid also touched on the use of hypotheses in his pneumatology lectures, for one of his surviving course outlines includes the heading, 'The Danger & Mischief of Hypotheses'.[182] The most detailed examination of the abuse of hypotheses, however, was included in Reid's graduation orations for 1759 and 1762. In these orations he attacked the theory of ideas as being a groundless hypothesis because the proponents of the theory had failed to demonstrate that ideas actually exist and to prove that such entities were capable of explaining the phenomena of memory or sensory perception. Yet even though his critique of the theory of ideas involved the two criteria central to his *vera causa* interpretation of Newton's first rule of philosophising (that the assumed cause be shown to exist and proved to be sufficient to explain the phenomena), it is significant that Reid neither appealed to Newton's first rule nor explicitly formulated his objections to the theory of ideas in terms of the *vera causa* principle.[183]

Published in 1764, Reid's *Inquiry* was likewise taken up with refuting the theory of ideas, as well as offering an alternative approach to the study of our mental powers. Drawing heavily on Newton's methodological dicta, Reid again argued that the theory of ideas was an unwarranted hypothesis which had led philosophers such as Berkeley and Hume to question undeniable facts about human nature, including 'a fixed belief of an external material world … which is neither got by reasoning nor education, and … which we cannot shake off'. Paraphrasing Newton's fourth rule of philosophising, he declared that such facts 'are phænomena of human nature, from which we may justly argue against any hypothesis … But to argue from a hypothesis [the theory of ideas] against facts, is contrary to the rules of true philosophy'.[184] In condemning hypotheses in the *Inquiry*, he also drew on Bacon and, in particular, Bacon's distinction between the anticipation and interpretation of nature. Proclaiming that 'a just interpretation of nature is the only sound and orthodox philosophy', and that we must study God's works 'with attention and humility, without daring to add any thing of ours to what they declare', Reid dismissed 'conjectures and theories' as being 'the creatures of men [which] will always be found very unlike the creatures of

[181] AUL, MS K 160, pp. 8, 262–3. At the beginning of his natural philosophy course Reid discussed Newton's four rules of philosophising in the second edition of the *Principia* but apparently said little about Newton's rejection of hypotheses in the fourth rule.

[182] AUL, MS 2131/8/V/1, fol. 2r.

[183] Reid, *Philosophical Orations*, orations III and IV. Newton's first rule of philosophising reads: 'No more causes of natural things should be admitted than are both true and sufficient to explain their phenomena'; Newton, *The Principia*, p. 794. Newton grounded this rule on the simplicity of nature, and his use of it was very different from Reid's.

[184] Reid, *Inquiry*, p. 76; compare Newton, *The Principia*, p. 796.

God'. Consequently, for him not only the theory of ideas but also 'all our curious theories of the formation of the earth, of the generation of animals, [and] of the origin of natural and moral evil, so far as they go beyond a just induction from facts', were to be dismissed as 'vanity and folly, no less than the vortices of Des Cartes or the Archæus of Paracelsus'.[185]

Yet for all of his criticisms of the abuse of hypotheses, as a careful student of Newton's writings he nevertheless recognised that hypotheses and conjectures have a legitimate, but circumscribed, use in philosophy, provided that they are clearly distinguished from empirically proven propositions. According to Reid, 'the great Newton first gave an example to philosophers, which always ought to be, but rarely hath been followed, by distinguishing his conjectures from his conclusions, and putting the former by themselves, in the modest form of queries'.[186] Following Newton's precedent, he included the section 'Squinting considered hypothetically' in the *Inquiry*, wherein he framed a series of queries regarding squinting that were intended to shed light on the laws governing human vision.[187] But even though much of what he later wrote about hypotheses and conjectures can be found already in the *Inquiry*, there is no indication in the text that by the time it was published he had arrived at his distinctive, and subsequently influential, *vera causa* interpretation of Newton's first rule of philosophising.

In revising and expanding his pneumatology lectures after his move to Glasgow, Reid initially hit on something like his mature interpretation of Newton's first rule in 1765 and, by 1768–69, he explicated the rule in the terms he later published in his *Essays on the Intellectual Powers of Man*. Having stated Newton's first rule for his students in his pneumatology lectures for 1768–69, Reid commented:

> This is a Golden Rule in Philosophy, by which we may always distinguish what is sound and solid … from what is hollow and vain. If a Philosopher therefore pretends to tell us the Cause of any Natural Effect, whether relating to Matter or to Mind; Let us first consider whether there is sufficient Evidence that the Cause he assigns really exists. If we find sufficient Evidence of its Existence we are next to consider, whether the Effect it is brought to explain, necessarily follows from it. If the Cause he Assigns has these two properties it is to be admitted as the true Cause; Otherwise it is not.[188]

185 Reid, *Inquiry*, p. 12; in this section of his introduction Reid also refers to Newton's rules of philosophising as 'maxims of common sense'.

186 Reid, *Inquiry*, p. 163.

187 Reid, *Inquiry*, pp. 140–8.

188 AUL, MS 2131/4/II/2, insert, p. 5; for his formulation in 1765 see AUL, MS 2131/4/II/1, p. 20. Compare Reid, *Intellectual Powers*, p. 51.

Reid's *vera causa* reading of Newton's first rule, as well as his distinction between the illegitimate utilisation of hypotheses and the proper, circumspect use of queries, surface in his exchange with Lord Kames. Writing to Kames in October 1780, Reid was highly critical of his friend's explanation of gravity in terms of 'an inherent Power, or Tendency' in the particles of matter 'to move towards every other particle', and deemed this theory 'a mere Hypothesis without any Evidence'. He also accused Kames of invoking an occult quality to account for the cause of gravity, and pointedly contrasted Kames' posture with that of a Newtonian natural philosopher. Unlike Kames, he averred that in response to the question 'what is the Cause of Gravitation of Matter?' the Newtonian 'confesses his Ignorance, & says I do not know, and I never trust to Hypotheses & Conjectures about the Works of God, being perswaded that they are more like to be false than to be true'.[189]

Stung by these remarks, Kames accused Reid of 'valuing [him] self upon [his] Ignorance of the Cause of Gravity' and maintained that 'never to trust to Hypotheses & Conjectures, about the works of God, & being perswaded that they are more like to be false than true, is a discouraging Doctrine and damps the Spirit of Inquiry'.[190] Reid countered that he regarded the confession of ignorance as 'a Sign, not of Pride, but of Humility, & of that Candor which becomes a Philosopher'. Moreover, he insisted that his censure of conjectures and hypotheses reflected the teachings of Bacon and Newton and served as 'the very Key to Natural Philosophy, & the Touchstone by which every thing that is Legitimate & Solid in that Science is to be distinguished from what is Spurious & Hollow'. And while he recognised that conjectures and hypotheses are of heuristic value if they lead us to make observational and experimental tests of their validity, he emphasised that we should not confuse theories used in this way with established facts or empirically proven laws. For Reid, such confusion had to be avoided and he believed that we could do so by evaluating explanatory causal hypotheses or conjectures in terms of the two criteria specified in his *vera causa* interpretation of Newton's first rule of philosophising. He put the point to Kames thus:

> A Cause that is conjectured ought to be such, that if it really does exist, it will produce the Effect. If it have not this Quality it hardly deserves the name even of a Conjecture. Supposing it to have this Quality, the Question remains,

[189] Reid to Lord Kames, 31 October 1780, in Reid, *Correspondence*, pp. 137–9; the Newtonian's reply echoes the rhetoric of the passages in Reid's *Inquiry* quoted above on pp. lxiii–lxiv.

[190] Reid to Lord Kames, 16 December 1780, in Reid, *Correspondence*, pp. 139–40; Reid was here quoting from Kames' reply to his letter of 31 October.

> Whether does it exist or not? And this being a Question of Fact, is to be tried by positive Evidence.[191]

Descartes' theory of vortices, according to Reid, signally failed to meet these criteria. While he granted that a vortex of subtle matter was sufficient to explain the orbit of a planet around the Sun, he asserted that the existence of vortices had never been established. By contrast, Newton's theory of universal gravitation satisfied Reid's criteria because Newton had demonstrated in the third book of the *Principia* that gravity exists as a property of bodies and that this property is sufficient to explain the orbits of the planets and comets about the Sun, the motions of our Moon and the moons of Jupiter and Saturn, the tides and the behaviour of falling bodies here on Earth.[192]

In Reid's view, therefore, Kames' criticisms of Newton were seriously misplaced. For not only had Newton scrupulously observed the distinction between physics and metaphysics and carefully distinguished between empirically demonstrated truths and conjectures intended to stimulate further observational and experimental investigation, he had also provided the strongest possible proof of his theory of gravitation. The methodological lessons to be drawn from Newton's works were clear to Reid. True Newtonians adhered to the method of reasoning spelled out in Newton's rules of philosophising and other passages in their master's writings. Hence they eschewed conjectures and hypotheses and restricted themselves to discovering general laws on the basis of observation, experiment, induction and rigorous quantitative reasoning. And, while they might propose queries as heuristic tools to promote future enquiries, they recognised that hypotheses and conjectures so used were in no way to be considered established truths in natural philosophy. As natural philosophers, they also avoided making claims about efficient and final causes. Consequently, they recognised that Newton's ether was not to be taken as the efficient cause of gravitation or of the optical and electrical phenomena catalogued in Newton's works. They also understood that when Newton spoke of forces of attraction and repulsion he was not referring to efficient causes but rather to powers like gravity, whose action could be described mathematically. Reid's letters to Kames were thus as much about defining Newton's scientific legacy as they were about disagreements over specific elements of Newtonian natural philosophy.[193]

[191] Reid to Lord Kames, 16 December 1780, in Reid, *Correspondence*, p. 140.

[192] Reid to Lord Kames, 16 December 1780, in Reid, *Correspondence*, pp. 140–2.

[193] The same can be said of Reid's defence of Newton addressed to the Bampton lecturer, Edward Tatham, and of his response to Joseph Priestley; see Reid to Tatham, October 1791, in Reid, *Correspondence*, pp. 224–7, and my introduction to Reid, *Animate Creation*, pp. 30–47.

In addition to the mathematical and scientific activities stimulated by Reid's friends and colleagues surveyed above, his surviving manuscripts and other sources show that he pursued other lines of enquiry. In Glasgow, Reid benefited from the conversations and debates held in the College Literary Society, better known as the Glasgow Literary Society (*fl.* 1752–*circa* 1802).[194] As in the Aberdeen Philosophical Society, the membership of the Literary Society was dominated numerically by University men; they were joined by local physicians, lawyers, merchants, clergymen and members of the landed classes, as well as notables from elsewhere, such as David Hume. Because of the intimate connection between the University and the Literary Society, the Society's institutional life followed the rhythms of the academic session and, by the time Reid was elected a member in November 1764, meetings typically began once the session was underway in early November and finished around the end of classes, in mid-May.[195] While the Literary Society flourished, it was one of the major sites for the discussion of scientific, medical, agricultural and technological matters in Glasgow. As in the Wise Club, members of the Society were obliged to give a formal discourse annually on a topic of their choosing and, in addition, they were expected to propose questions for debate. According to Roger Emerson's statistics, most of the 308 discourses listed in the records of the Society fall into the category of languages and literature (sixty-three or 21%). But the next highest number were devoted to natural philosophy (thirty-nine or 13%), followed at some distance by medicine (twenty-four or 8%), natural history (ten or 3%), technological subjects (seven or 2%), mathematics (four or 1%) and agriculture (three or 1%). The topical spread of the questions proposed for the consideration of the members was, however, very different. Of the 183 questions canvassed by the Society, only eight (4%) addressed the natural sciences or issues related to agriculture.[196] In the period 1764 to 1796, Reid is known to have delivered twenty discourses and introduced eleven questions. Of these presentations, he presented five discourses on materialism and scientific metaphysics, one on muscular motion and one on Euclid's *Elements*, whereas none of the questions he posed for discussion dealt with mathematical or scientific subjects.[197] His paper on Euclid, which he probably read in the period

[194] The best account of the Glasgow Literary Society is found in Roger L. Emerson, *Neglected Scots: Eighteenth Century Glaswegians and Women*, ch. 2.

[195] For Reid's election on 9 November 1764 see 'Laws of the Literary Society in Glasgow College', GUL, MS Murray 505, p. 10.

[196] Emerson, *Neglected Scots*, pp. 29–36, 128–31. The total number of all of the discourses and questions is unknown because the Society minutes for 1752–55, 1759–61, 1765, 1771–73 and 1780–94 are either incomplete or no longer survive.

[197] Reid's extant discourses on materialism and scientific metaphysics, as well as his essay on muscular motion, are reproduced in Reid, *Animate Creation*, pp. 103–24, 164–8, 217–40. For a list

circa 1790–93, is transcribed below (pp. 15–31) and will be analysed in Section 2. Reid also participated in the discussion of some notable discourses on the physical sciences and mathematics, including an essay on sunspots by Alexander Wilson that was awarded a gold medal by the Royal Academy of Copenhagen in 1771, papers by Patrick Wilson on optics, a series of discourses on chemical topics by William Irvine and various presentations on meteorology by Irvine, John Walker and John Anderson.[198] However, Reid's participation in the conversations and discussions of the Society was undoubtedly hampered by his increasing deafness, which began to trouble him in the late 1780s.[199]

Because of the paucity of evidence, it is difficult to trace Reid's reading of works on mathematics and natural philosophy prior to 1764. His reading of books and journals after his move to Glasgow is, by contrast, much better documented, not least because the texts he borrowed from the University library are recorded in the surviving lists of books issued to professors that cover the years Reid spent in the 'northern Venice'.[200] As I have shown elsewhere, of the 459 titles that he took out of the library in the period 1767–89, twenty-eight (6%) were on natural philosophy, twenty-three (5%) on mathematics and six (1%) on chemistry.[201] He also frequently borrowed learned journals, especially the major periodicals dealing with the natural sciences such as the Royal Society's *Philosophical Transactions*, the *Transactions* of the American Philosophical Society and the publications of the academies in Paris, Berlin and St Petersburg.[202] His reading in

of Reid's discourses and questions see Wood, 'Thomas Reid, Natural Philosopher', appendix III.

[198] Emerson, *Neglected Scots*, pp. 86, 97, 100, 101, 102, 103, 105, 112, 113. Alexander Wilson's interpretation of sunspots was published in Alexander Wilson, 'Observations on the Solar Spots'. Irvine's discourses were later collected together in William Irvine and William Irvine, Jr, *Essays, Chiefly on Chemical Subjects* (1805).

[199] Reid to John Robison, [April 1788–90], and Reid to Dugald Stewart, [May 1792], in Reid, *Correspondence*, pp. 200, 230. See also Stewart, *Account*, p. 177, where he notes that during a visit to Edinburgh in the summer of 1796 Reid's 'deafness prevented him from taking any share in general conversation, [although] he was still able to enjoy the company of a friend'.

[200] Reid's book borrowing is documented in 'Professors Receipt Book, 1765–1770', GUL, Spec Coll MS Lib (uncatalogued), and 'Professors Receipt Book, 1770–[1789]', referred to above in note 157. The phrase 'northern Venice' is John Galt's; see John Galt, *The Last of the Lairds: Or, the Life and Opinions of Malachi Mailings Esq. of Auldbiggings* (1826), p. 64.

[201] Wood, 'A Virtuoso Reader', pp. 55–9. The category containing the largest number of titles borrowed was theology and church history (sixty-two or 14%), followed by moral philosophy and history (forty-two or 9% each), polite literature (thirty-nine or 9%), miscellaneous works (thirty-eight or 8%) and then natural philosophy. Taken together, the titles Reid borrowed on natural philosophy, natural history, mathematics, medicine and chemistry amounted to ninety-six (roughly 20%).

[202] In consulting the transactions of these academies, Reid acquainted himself with the work of leading men of science on the continent. For example, when he borrowed the second volume of the *Commentarii Academiae scientiarum imperialis Petropolitanae* in 1789 he was able to read

natural philosophy ranged over works by major figures of the Scientific Revolution, such as Robert Boyle, Pierre Gassendi, Robert Hooke and Johannes Kepler, various editions of Newton's writings, standard commentaries on Newton's system by Colin Maclaurin and Henry Pemberton, as well as a miscellany of more recent texts by contemporary men of science such as Neville Maskelyne and John Elliott. He also borrowed books by prominent savants on the continent such as Bernard Le Bovier de Fontenelle, Leonhard Euler, Jean-André Deluc and Nicolas-Louis de La Caille.[203] And even though he was not a Fellow of the Royal Society of London, he took an interest in the Society's affairs and read one of Sir John Hill's attacks on the Society, as well as an account of the dispute that arose in 1783–84 between the supporters of the President, Joseph Banks, and a group of mathematicians led by Maskelyne and Samuel Horsley. Banks' opponents were scandalised by the Society's treatment of the Professor of Mathematics at the Royal Military Academy in Woolwich, Charles Hutton, whose resignation as the Society's Foreign Secretary had been engineered by Banks. The dispute precipitated a brief pamphlet war between the two camps, which drew public attention to the crisis, but, after an exchange of polemical salvos from both factions, Banks saw off his challengers and consolidated his control over the proceedings of the Society.[204] As for his limited reading of works on chemistry, the six titles he borrowed from the Library while he was in Glasgow encompassed both standard works by eminent chemists of the early modern period as well as texts closely related to the revolutionary transformation of the science that took place in the

important papers on mathematics by Daniel Bernoulli, Leonhard Euler and Jacob Hermann. I shall return to the significance of his reading of these journals below, pp. cxxxii–cxxxiii, cxciii.

[203] Over the years, Reid borrowed a number of individual volumes from Fontenelle's *Oeuvres*, using the Amsterdam edition published in 1764 that survives in Special Collections, Glasgow University Library. He also took out: Leonhard Euler, *Opuscula varii argumenti* (1746); Jean-André Deluc, *Récherches sur les modifications de l'atmosphere* (1772); and Nicholas-Louis de La Caille, *The Elements of Astronomy, Deduced from Observations; and Demonstrated upon the Mathematical Principles of the Newtonian Philosophy: With Practical Rules Whereby the Principal Phenomena Are Determined* (1750). Reid's interest in meteorology and thermometry that he shared with the Wilsons and William Irvine probably prompted his reading of Deluc.

[204] Sir John Hill, *A Review of the Works of the Royal Society of London; Containing Animadversions on such of the Papers as Deserve Particular Observation*, second edition (1780); Anon., *An Authentic Narrative of the Dissensions and Debates in the Royal Society* (1784). On the dispute over Charles Hutton's resignation, see John Gascoigne, *Joseph Banks and the English Enlightenment: Useful Knowledge and Polite Culture*, pp. 10–13, and Heilbron, 'A Mathematicians' Mutiny, With Morals', pp. 81–91. Personal connections may at least partly explain Reid's interest in these two works. Hill had been an associate of Reid's contact in London, Alexander Stuart; see Kevin J. Fraser, 'John Hill and the Royal Society in the Eighteenth Century', p. 44. Given that Reid later contributed to Hutton's *Mathematical and Philosophical Dictionary*, the two men apparently knew of one another, with John Anderson being a possible intermediary.

closing decades of the eighteenth century. This group of books will be discussed in greater detail in Section 7.

Reid's borrowings on mathematics form a miscellaneous group. As well as different editions of Apollonius and Archimedes, he consulted John Napier of Merchiston's seminal work on logarithms in versions printed in Edinburgh and Lyon, Johann Castillon's annotated edition of Newton's *Arithmetica universalis*, James Stirling's exposition of Newton's treatment of infinite series in Stirling's *Methodus differentialis*, James Dodson's *The Mathematical Repository* (a collection which dealt with properties of numbers), practical manuals on book-keeping and the calculation of annuities, and works on algebra by John Kersey the elder and Thomas Simpson.[205] The last title was borrowed – with the permission of the University Senate – from the collection of mathematical books bequeathed to the College by his erstwhile colleague Robert Simson. Reid was evidently fascinated with Simson's bequest, for over the years he made regular use of books in the collection and, in 1782, he borrowed the 'Mss Press Catalogue of Dr Simsons Books'. In addition to Thomas Simpson's *A Treatise of Algebra*, Reid also borrowed from the Simson collection copies of Sir Henry Savile's lectures on Book I of Euclid's *Elements* published as *Praelectiones tresdecim in principium Elementorum Euclidis* (1621), John Wallis' *Opera mathematica* (1693–99), the Marquis de l'Hôpital's *Analyse des infiniment petits, pour l'intelligence des lignes courbes* (1696), Joseph Raphson's account of Newton's development of the fluxional method in *Historia fluxionum* (1715), Willem Jacob 's Gravesande's *Matheoseos universalis elementa* (1727), the Marquise du Châtelet's 1759 French translation of Newton's *Principia* (which contained an extensive editorial commentary), Jean Étienne Montucla's *Histoire des mathématiques* (1758) and an unidentified work or set of works by a distant ancestor on the Gregory side of his family, Alexander Anderson, who settled in Paris at the turn of the seventeenth century and became a close associate of the algebraist François Viète. But not all of the books he took out from the Simson collection were on mathematics, for he also used Simson's copies of Arthur Collier's *Clavis universalis: Or, a New Inquiry After Truth* (1713) and James Jurin's pamphlet, *A Reply to Mr Robins's Remarks on the Essay upon Distinct and Indistinct Vision Published at the End of Dr Smith's Compleat System of Opticks* (1739).[206]

[205] For material related to his reading in 1786 of editions of the works of Archimedes published by David Rivault de Fleurance (Paris, 1615) and Isaac Barrow (London, 1675) see AUL, MS 2131/5/II/31, fol. 4r–v; see also the manuscript dated 1786 entitled 'Of the Helix of Archimedes', AUL, MS 2131/5/II/39.

[206] 'Professors Receipt Book, 1770–[1789]'; 'Minutes of Senate Meetings, 1771–1787', Glasgow University Archive Services, SEN 1/1/1, pp. 7, 107, 139, 294; 'Minutes of Senate Meetings,

Reid's surviving reading notes from scientific titles other than those borrowed from the University library are taken primarily from works on natural philosophy and mathematics. Of the sixteen sets of notes found among Reid's papers, ten come under the broad heading of natural philosophy, five derive from texts on mathematics and one deals with a book related to chemistry, namely Axel Fredrik Cronstedt's *An Essay Towards a System of Mineralogy* (translated 1770). What apparently interested Reid most about the *Essay* was the translator's description of a portable chemical laboratory used for the analysis of mineral samples; he made a list for himself of the instruments contained in the 'pocket-laboratory' and seemingly ignored the rest of the book.[207] The focus of his notes from Cronstedt implies that he periodically engaged in the art of chemical analysis, and further evidence for his practical experience as an analyst comes in a letter to Joseph Black from 1773, in which he reported that he had collaborated with the Marischal Professor of Natural Philosophy, Dr George Skene, in making 'experiments' on the medicinal waters at Pannanich Wells in Deeside.[208] Reid's reading in natural philosophy was almost exclusively devoted to books and pamphlets on electricity, optics and vision, and astronomy, although in June 1788 he also took notes from Ruđer Josip Bošković's annotated edition of Benedetto Stay's philosophical poem on Newtonian natural philosophy, *Philosophiae recentioris* (1755–60).[209] Individual items from this group of notes will be discussed further below.

As for his reading in mathematics, all but one set of notes date from the late 1780s, the exception being Reid's undated comments on the principles of navigation inspired by the remarks on the projection of the sphere found in William West's *Mathematics* (1762).[210] In 1786–87 Reid spent some time familiarising himself with the background to the development of logarithms by Napier of Merchiston, for in 1786 he took a detailed set of notes from the historical introduction to Charles Hutton's *Mathematical Tables: Containing Common, Hyperbolic and Logistic Logarithms* (1785) and, in May 1787, he read and criticised the earl

1787–1802', Glasgow University Archive Services, SEN 1/1/2, p. 174. Reid's reading notes from Collier's *Clavis universalis*, dated 21 February 1771, survive in AUL, MS 2131/3/II/10, fol. 1r. In 1785 Reid made a note on a method of extracting square and cube roots found in 's Gravesande's *Matheoseos universalis elementa*; see AUL, MS 2131/5/II/45, p. 3. A brief undated note taken from Montucla's *Histoire des mathématiques* is included in AUL, MS 2131/5/I/26, fol. 1v. On Reid's borrowings from the Simson collection, see also 'Rules &c Relating to Dr R Simsons Collection of Books 1768', GUL, MS Simson Ea5–b.1, unpaginated, which records at least some of the loans made in the eighteenth century. Reid is one of the professors listed. I am grateful to Julie Gardham for this reference.

207 AUL, MS 2131/3/I/14, fol. 2r.

208 Reid to Joseph Black, 17 January 1773, in Reid, *Correspondence*, p. 75.

209 Reid's notes from Stay are transcribed in Reid, *Animate Creation*, pp. 171–3.

210 AUL, MS 2131/3/III/10.

of Buchan (David Steuart Erskine) and Walter Minto's *An Account of the Life, Writings and Inventions of John Napier, of Merchiston* (1787), which had just been republished.[211] In July of the following year, he elaborated on 'a very good Rule for simple Alligation' taken from George Berkeley's *Arithmetica absque algebra aut Euclide demonstrata* (1707) and, in February 1789, he read Charles Hutton's *A Treatise on Mensuration, Both in Theory and Practice* (1770).[212] Although he wrote of Hutton's preface that 'one would be apt to think … that [the author] had neither a distinct Notion of the Quadrature of the Circle, nor of the Nature of non quadrate Numbers', and proceeded to correct Hutton's mathematical errors, he nevertheless read other parts of the book with greater satisfaction because he concluded his summary with the favourable observation that 'the Book is a very compleat Treatise of Mensuration both in Theory and in Practice; & gives an historical Account of the Methods by which Artificers, who work by Measure, measure their Work'.[213] Lastly, there is a miscellaneous list of titles apparently dating from the mid-1780s which includes Richard Kirwan's *Elements of Mineralogy* (1784) along with the *Philosophical Transactions* for 1783 and the *Mémoires* of the Berlin Academy for 1781. But it is unclear whether Reid had in fact read these works or was intending to borrow them from the University library.[214] Presumably Reid's colleagues also lent him books on scientific and mathematical subjects, which may have been the source for some of the reading notes surveyed above, and some of the notes may have been taken from books in his personal library, which we know contained titles on natural philosophy and mathematics. Unfortunately, little evidence survives regarding the circulation of books among Reid's associates or the contents of his own library.[215]

Reid engaged in a wide range of enquiries in mathematics and natural philosophy in Glasgow. In addition to his researches in Euclidean geometry, astronomy, optics, electricity and chemistry discussed below, Reid continued to reflect on various topics in mathematics and the physical sciences. As noted above, although Dugald Stewart implied that there had been a hiatus in Reid's

[211] AUL, MS 2131/3/I/1.

[212] AUL, MSS 2131/3/I/17, pp. 6–7 (Hutton), and 2131/5/II/51, fol. 2r–v (Berkeley); compare George Berkeley, *Arithmetica absque algebra aut Euclide demonstrata* [1707], in *The Works of George Berkeley Bishop of Cloyne*, vol. IV, p. 194. Alligation is defined as 'one of the rules in arithmetic, by which are resolved questions which relate to the compounding or mixing together of divers simples or ingredients, being so called from *alligare*, to tie or connect together'; Hutton, *Mathematical and Philosophical Dictionary*, vol. 1, p. 98.

[213] AUL, MS 2131/3/I/17, pp. 6–7.

[214] AUL, MS 2131/3/I/11, fol. 2r.

[215] On these issues, see Wood, 'A Virtuoso Reader', pp. 48–53. For a provisional list of books known to have been in Reid's personal library, see Wood, 'Who Was Thomas Reid?', pp. 47–8.

mathematical pursuits in the period 1751 to 1764, there is no evidence that this was the case.[216] If anything, there is a remarkable degree of continuity in Reid's scientific and mathematical work over the course of his career, which Stewart failed to capture. Moreover, while Stewart remarked on the extent to which the investigation of mathematical subjects dominated Reid's retirement, he framed his discussion in terms of a 'revival' of interests that had preoccupied Reid in his youth. According to Stewart:

> Among the various occupations with which [Reid] … enlivened his retirement, the mathematical pursuits of his earlier years held a distinguished place. He delighted to converse about them with his friends; and often exercised his skill in the investigation of particular problems. His knowledge of ancient geometry had not probably been, at any time, very extensive; but he had cultivated diligently those parts of mathematical science which are subservient to the study of Sir Isaac Newton's Works. He had a predilection, more particularly, for researches requiring the aid of arithmetical calculation, in the practice of which he possessed uncommon expertness and address. I think, I have sometimes observed in him a slight and amiable vanity connected with this accomplishment.[217]

Stewart proceeded to suggest that part of the reason why Reid exercised his talent for mathematics was that he hoped to maintain 'the state of his intellectual faculties' by working on difficult mathematical calculations and that he chose 'detached problems' in which 'all the *data* are brought at once under the eye, and where a connected train of thinking is not to be carried on from day to day' as a way of combating his failing mental powers. But even though Stewart claimed to have witnessed Reid engaged in this kind of 'recreation', a close examination of Reid's surviving mathematical papers dating from his years in Glasgow tells a different story.[218]

Stewart's assertion that Reid's familiarity with the writings of Greek geometers was limited is not borne out by the available evidence. Reid's dogged attempt to solve the problem of Euclid's treatment of parallel lines, as well his repeated reading of Robert Simson's reconstruction of Apollonius of Perga's work on plane loci and his study of various editions of Apollonius and Archimedes all indicate that he was reasonably well versed in the literature of the ancient geometers,

[216] Stewart, *Account*, p. 49.

[217] Stewart, *Account*, pp. 170–1.

[218] Stewart, *Account*, pp. 172–3. Most of the mathematical papers discussed in what follows are undated. I have dated these papers to the period 1764–96 on the basis of Reid's handwriting and other clues.

even if he was not steeped in their texts like those schooled by Simson, such as Stewart's father Matthew.[219] What may be one of Reid's earliest mathematical manuscripts from his Glasgow period attests to his engagement not only with the geometrical legacy of classical antiquity but also more recent work in the field, for it contains the solution to a straightforward problem in geometry along with a note entitled 'Properties of the Cassinian Curve', which deals with the properties of plane curves now known as Cassinian ovals that were initially analysed by the Italian-born mathematician and astronomer Giovanni Domenico Cassini.[220] Other manuscripts addressing the legacy of the ancients include treatments of geometric loci, a discussion of the forty-seventh proposition of Book I of Euclid's *Elements*, and jottings on the geometric properties of cylinders, cones and spheres along with the Archimedean helix based on Archimedes' *On the Sphere and Cylinder* and other works.[221] As for his engagement with early modern texts, Robert Simson and Matthew Stewart are the only two contemporary geometers to figure in Reid's manuscripts, apart from the Italian Jesuit Giovanni Girolamo Saccheri, to whom we will return in Section 2. When teaching conic sections at King's College, Aberdeen, Reid relied heavily on either the first or second edition of Robert Simson's *Sectionum conicarum libri V* (1735, 1750) for his survey of the geometry of circles, ellipses, parabolas and hyperbolas. In later additions to what appear to be notes for his lectures on hyperbolas, Reid returned to the topic of conic sections and again drew on Simson to consider the geometry of parabolas, including their quadrature.[222] Reid seems to have been less indebted in the mathematical writings of Matthew Stewart, although a loose sheet survives on which he recorded a theorem taken from Stewart on the quadrature of the circle.[223] In addition, a group of five manuscripts related to those on geometry deal with aspects of trigonometry, including basic topics such as the logistical curve and versed sines, as well as a theorem from Roger Cotes' *Harmonia mensurarum* (1722) which Reid used in attempting to clarify his thoughts on ascertaining the 'natural measure' of ratios.[224]

If we accept that the number of surviving manuscripts provides an approximate guide to the importance of a topic for Reid, then the fact that there are seventeen

[219] Dugald Stewart's assessment can thus be seen as reflecting the obsession with ancient geometry that characterised the work of Simson, Matthew Stewart and their associates.

[220] AUL, MS 2131/3/III/17, fols 1v–2r. Reid's jottings are written on the blank sides of a letter from the Rev. James Oswald of Methven from 16 October 1766; see Reid, *Correspondence*, p. 56.

[221] AUL, MSS 2131/5/II/30, 2131/5/II/31, fol. 4r–v, 2131/5/II/39, 2131/5/II/45, p. 4 (dated 8 March 1787), and 2131/5/II/48.

[222] AUL, MS 2131/5/II/31, fols 3r, 5v–8v.

[223] AUL, MS 2131/7/III/1, fol. 2r–v.

[224] AUL, MSS 2131/3/III/2, 2131/5/I/26, fol. 1r, 2131/5/II/23, 2131/5/II/45 and 2131/5/II/52.

extant papers on number theory written after 1764 indicates that he devoted a considerable amount of time and thought to this field of mathematics during his Glasgow years. Niccoló Guicciardini has described this group of manuscripts as 'rather elementary exercises', yet they nevertheless document a line of enquiry that Reid pursued from at least the late 1740s until the 1790s, with Reid's essay 'Of the Relation between the Series of odd Numbers, & the Products and Powers of whole Numbers', dated 23 February 1792, standing as the culmination of his investigation of the properties of numbers.[225] In what may be the first of the Glasgow sequence of manuscripts, Reid records a theorem given to him by Dugald Stewart regarding square numbers, and there follow manuscripts on the multiples of prime numbers, prime and composite numbers, perfect numbers and figurate numbers, with Reid summarising his work on the last topic in a paper entitled 'Of Figurate Numbers', dating from June 1790.[226] His exploration of number theory was grounded in some of the standard sources of the period: he refers to Books VII and IX of Euclid's *Elements*, Theon of Smyrna, François Viète, Claude Gaspard Bachet de Méziriac's edition of Diophantus' *Arithmetica*, Pierre de Fermat and John Wallis.[227] But he also made use of less expected sources insofar as he cites Newton's 'Methodus differentialis' (1711) and an anonymous review of the second and third volumes of the *Histoire et mémoires de l'Académie royale des sciences, inscriptions et belles-lettres de Toulouse* (1788) in the *Monthly Review* for April 1790, which mentions a property of prime numbers that the reviewer claimed was 'not unknown to mathematicians'. Reid disagreed, noting that 'I should be glad to know where it is demonstrated'.[228] Even though Reid may not

[225] Guicciardini, 'Thomas Reid's Mathematical Manuscripts', p. 74. AUL, MS 2131/2/I/6; for an undated draft of this essay see AUL, MS 2131/2/I/2. See also 2131/5/II/40 for a related draft essay entitled 'Of the Series of the Products of Numbers'. A section of AUL, MS 2131/2/I/2, headed 'Of prime and composite Numbers' (pp. 6–8), overlaps with AUL, MS 2131/5/I/28.

[226] AUL, MSS 2131/3/I/17, 2131/5/I/23, 2131/5/I/28, 2131/5/II/16, 2131/5/II/17, 2131/5/II/32, 2131/5/II/33, 2131/5/II/38, 2131/5/II/42–4, 2131/7/II/1, 2131/7/II/5 and 2131/7/VIII/4, fol. 1v. The manuscript containing the theorem from Dugald Stewart (2131/7/II/1) could conceivably date from as early as 1771–72, when he studied at Glasgow under Reid. Stewart then substituted for his father and taught mathematics until 1775, when he became the Edinburgh Professor of Mathematics. He switched to the chair of moral philosophy in 1785.

[227] AUL, MSS 2131/2/I/2, p. 5 (Fermat and Bachet), 2131/3/I/17 (Euclid, Fermat, Bachet, Viète and Wallis), 2131/5/I/23 (Viète), 2131/5/II/44 (Euclid) and 2131/7/II/5 (Theon of Smyrna and Euclid). From references in AUL, 2131/3/I/17, fol. 1r and numbered pages 19–20, it is clear that Reid had consulted Bachet's edition of Diophantus published in 1670, which also included a commentary and letters by Fermat. A copy of this work survives in the Simson collection.

[228] For Reid's use of Newton's 'Methodus differentialis' in the context of a discussion of pyramidal numbers see AUL, MS 2131/5/II/32, pp. 5, 8. On the property of prime numbers, compare *Monthly Review* 80 (1790), p. 496, and AUL, MS 21312/I/2, p. 7. Reid also implicitly cites the review in AUL, MS 2131/5/II/40, p. 3. Niccolò Guicciardini has pointed out that Reid was also indebted to

have made any significant advances in number theory, he evidently thought that he had made a genuine, if limited, contribution to the field, for in the extant draft of his essay 'Of the Relation between the Series of odd Numbers, & the Powers and Products of whole Numbers', he says of a relationship 'between the Series of odd Numbers & the cubical Numbers' stated in one of his corollaries that it was 'first discovered by Fermat, & is observed by Bachet in his Commentary on Diophantus, but I think both have deduced it from different Principles'.[229] Thus it seems that while Reid recognised that he had not made any profound discoveries about the properties of numbers, he was also confident that he had managed to come up with alternative proofs of what was already known. Moreover, the fact that in the early 1790s he drafted papers written in axiomatic form headed 'Of Prime and Composite Numbers' and 'Of Figurate Numbers' in addition to his essay from 1792 indicates that he intended to distil the work on the properties of numbers that he had carried out since the 1740s and to present his results in a systematic fashion.[230] For reasons that remain unclear, however, Reid's intention was only partly fulfilled.

By far the largest number of Reid's mathematical manuscripts written after 1764 (twenty-four) deal with different facets of algebra. As mentioned above, in April 1766 Reid was invited by the Aberdeen Town Council to examine the candidates for the vacant chair of mathematics at Marischal College. In order to 'rub up [his] Mathematicks' for the 'Tryal' the candidates were obliged to undergo, Reid compiled a lengthy manuscript dated 27 May 1766 on the 'History of Arithmetick and Algebra'.[231] The writings of John Wallis seem to have been Reid's primary source for the details of his historical survey and, as Guicciardini has noted, Reid followed Wallis in viewing the history of algebra through an Anglocentric lens.[232] For while he recognised that the French mathematician François Viète was the founder of modern algebra, he maintained that Viète's work had then been taken up by a succession of Englishmen who had greatly advanced the art. According to Reid, William Oughtred 'adopted Vieta's Notation

Newton's *Arithmetica universalis* and Thomas Simpson's *Algebra*; Guicciardini, 'Thomas Reid's Mathematical Manuscripts', p. 75.

229 AUL, MS 2131/2/I/2, p. 5.

230 'Of Prime & Composite Numbers', AUL, MS 2131/5/I/28; this manuscript is undated but is written the late hand found in manuscripts dating from the 1790s. 'Of Figurate Numbers', AUL, MS 2131/5/II/38, dated June 1790.

231 Aberdeen Town Council to Thomas Reid, 30 April 1766, and Reid to Andrew Skene, 8 May 1766, in Reid, *Correspondence*, pp. 52, 54; AUL, MS 2131/7/I/2.

232 Reid cites Wallis' *Mathesis universalis; seu opus arithmeticum* (1657) and *A Treatise of Algebra, Both Historical and Practical* (1685) in AUL, MS 2131/7/I/2, p. 6; Guicciardini, 'Thomas Reid's Mathematical Manuscripts', p. 84, note 28.

and improved upon it', while Thomas Harriot not only further refined and clarified the notation used in algebra but also made 'the most important' discovery 'in Algebra since it was introduced into Europe', namely 'the Composition of affected Equations from simple ones'. Oughtred's contributions spawned those of Isaac Barrow and Wallis, and the writings of these three mathematicians paved the way for Newton, who 'was enabled to make great discoveries in Mathematicks before his beard was grown'. Reid regarded Descartes, on the other hand, as something of plagiarist who 'followed Harriot ... without mentioning him'. And while others would have celebrated Descartes' accomplishments as a mathematician, Reid was less charitable in his assessment: 'The onely thing belonging to pure Algebra which Des Cartes added to Harriots Inventions was the reduction of all Biquadratick Equations to Cubicks'.[233]

When considered together, Reid's manuscripts on algebra form a more miscellaneous group than those on number theory, with no single overarching theme or issue linking the papers. Four of the manuscripts contain solutions to straightforward algebraic problems, while others address basic topics such as the solution of indeterminate problems, the extraction of roots of equations, interpolation, arithmetic progressions and trinomial, cubic and biquadratic equations. Reid also dealt with more advanced subjects in two manuscripts, one on the summing of infinite series and the other on Cotes' theorem.[234] There is a further set of four papers which overlap with mathematical astronomy, insofar as they are devoted to finding an equation to facilitate the calculation of the mathematical relationship between an indiction in the Julian calendar and the solar and lunar cycles.[235] In terms of dating, most of the papers on number theory are clustered in the period 1790 to 1792; by comparison, Reid's dated manuscripts on algebra range from his 'History of Arithmetick and Algebra' of May 1766 (mentioned above) to a brief note on an algebraic problem jotted down in November 1794.[236] We also know that Reid continued to work on algebra until shortly before his death. Dugald Stewart notes that in the summer of 1796 Robert Cleghorn reported to James Gregory that the recently appointed Glasgow Professor of Natural Philosophy, James Brown, 'found [Reid] occupied in the solution of an algebraical problem of considerable difficulty, in which, after the labour of a day or two, he at last

233 AUL, MS 2131/7/I/2, pp. 9–[13].

234 'Of Cotes's Theorem', AUL, MS 2131/5/I/15, and 'Of the Summing of infinite Series's', AUL, MS 2131/5/II/50. See also AUL, MS 2131/5/II/27 for a fragment on summing a series.

235 AUL, MSS 2131/5/I/5, 2131/7/II/16, 2131/7/VIII/1 and 2131/7/VIII/4.

236 For the note from 1794, see AUL, MS 2131/7/III/9, fol. 2v. Two further manuscripts are dated: AUL, MS 2131/5/I/5 is headed 'May 31 1776', while part of AUL, MS 2131/5/II/45 was written in 1785.

succeeded'.[237] Algebra was thus an abiding interest during Reid's Glasgow years, even if it did not at times preoccupy him as did number theory.

In his manuscripts on algebra, Reid refers to only a few standard texts, largely by English authors. In addition to the works by John Wallis mentioned above, he made use of Book VII of Euclid's *Elements*, one of the many editions of *Sherwin's Mathematical Tables*, the first edition of Thomas Simpson's *A Treatise of Algebra* (1745), Roger Cotes' *Harmonia mensurarum*, Colin Maclaurin's *A Treatise of Algebra*, the first volume of James Dodson's *The Mathematical Repository* and Newton's 'Methodus differentialis', along with the lemma 5 in Book III of the *Principia*.[238] Only two non-British mathematicians figure in the set of manuscripts on algebra: the Dutchmen Willem Jacob 's Gravesande and Frans van Schooten, whose 'De cubicarum equationem resolutione' is referred to by Reid in an entry written in a late hand added to a paper on cubic equations.[239] His familiarity with the writings of continental algebraists was thus extremely limited. Moreover, one manuscript indicates that he kept up with articles on mathematics that appeared in popular journals, for there is a reference to an essay on the extraction of roots by the St Andrews mathematician Nicolas Vilant, published in the *Scots Magazine* for March 1759.[240]

In addition to the manuscripts on geometry, number theory and algebra, three items dating from Reid's years in Glasgow deal with unrelated topics. Although Reid briefly mentioned the calculus of chances in 'An Essay on Quantity', the extent to which he studied probability theory during the course of his life remains unclear.[241] However, a short essay entitled 'Of our reasoning concerning Chance' survives, which appears to have been written after 1764 and which serves as an addendum to a long summary of the mathematical principles involved in the

[237] Stewart, *Account*, p. 178.

[238] AUL, MS 2131/5/I/8, fol. 1r (Newton's *Principia*), 2131/5/I/14, fols 1v and 4v (Maclaurin and Sherwin), 2131/5/I/15 (Cotes), 2131/5/II/26, fol. 1r (Euclid), 2131/5/II/32a, p. 5 (Newton's 'Methodus differentialis') and 2131/7/III/9, fol. 3r (Simpson). Reid's page reference to Simpson's *Treatise of Algebra* provides the clue that he used the first edition. Although we have seen that Reid was critical of Maclaurin, in AUL, MS 2131/5/I/14 he sketched the extraction of the root of a binomial equation and then commented, 'See this more elegantly handled by Mr McLaurin' (fol. 1v).

[239] As I have pointed out above in note 206, Reid copied material from 's Gravesande's *Matheoseos universalis elementa* in 1785; see AUL, MS 2131/5/II/45, p. 3. For the reference to van Schooten, see AUL, MS 2131/5/I/10, p. [9]. Frans Van Schooten's 'De cubicarum equationem resolutione' appeared as an appendix to his *De organica conicarum sectionum in plano descriptione, tractatus* (1646).

[240] AUL MS 2131/5/I/14, fols 1v and 3r; Nicolas Vilant, 'An Easy Method of Extracting the Cubic Roots of Binomials, whether Possible or Impossible'. Reid queried the usefulness of Vilant's method.

[241] See below, p. 52.

calculus of chances based on Abraham de Moivre's *The Doctrine of Chances* (1718).[242] Reid's involvement in the practical applications of mathematics is also in evidence: one manuscript deals with surveying while another includes a lengthy consideration of the construction of a sundial.[243] What is perhaps surprising is that there are no papers on fluxions akin to those he produced in the 1740s and 1750s, even though he was still engaged in various lines of enquiry in mathematics and the natural sciences that involved the use of fluxional techniques. Nevertheless, he retained his earlier expertise in the fluxional calculus, as can be seen in the fact that he identified mistakes in the equations contained in John Robison's highly technical paper on the refraction of light that Robison had sent to Reid for comment in the late 1780s.[244] Dugald Stewart also records an anecdote that attests to Reid's retention of his competence in higher mathematics. Stewart recalled that during Reid's last visit to Edinburgh, in the summer of 1796, he demonstrated his mathematical prowess:

> Mr Playfair and myself were both witnesses to the acuteness which he [Reid] displayed on one occasion, in detecting a mistake, by no means obvious, in a manuscript of his kinsman David Gregory, on the subject of *Prime and Ultimate Ratios*.[245]

But even though Reid retained his facility for grasping difficult mathematical problems, he evidently had little or no interest in exploring or expanding on what was already known about the fluxional calculus or other fields of advanced mathematics. Lastly, one manuscript illustrates Reid's deep knowledge of the mathematical infrastructure of Newton's *Principia* and, indirectly, his acceptance of Newton's reformulation of his analytical version of the fluxional calculus in geometrical terms.[246] As Niccoló Guicciardini has pointed out, the posthumous

[242] AUL MS 2131/5/II/49. See also Reid's reading notes from an unidentified work by Philip Doddridge on the calculus of chances in AUL, MS 2131/3/I/22.

[243] AUL, MSS 2131/5/II/18 and 2131/5/II/45, pp. 4–8.

[244] Reid to [John Robison], [April 1788–90], in Reid, *Correspondence*, p. 197. Significantly, Reid also questioned some of Robison's derivations (p. 199).

[245] Stewart, *Account*, p. 177. Stewart's reference to a manuscript written by David Gregory on first and ultimate ratios is puzzling, since no such manuscript survives. It is likely that the manuscript to which Stewart refers was a copy of Gregory's 'Notae in Isaaci Newtoni Principia Philosophiæ'. The copy of Gregory's 'Notae' that is now in Aberdeen was once owned by John Robison, and it may be that Reid had studied this copy; see David Gregory, 'Notae in Isaaci Newtoni Principia Philosophiæ', AUL, MS 465. Although the volume has been rebound, Robison's bookplate is affixed to the inside front cover. Reid was familiar with the 'Notae' because he refers to Gregory's extended commentary on Newton's *Principia* in 'Some Farther Particulars of the Family of the Gregorys and Andersons', in Hutton, *A Mathematical and Philosophical Dictionary*, vol. I, p. 558.

[246] On the mathematical style of the *Principia*, see Niccolò Guicciardini, 'Dot-Age: Newton's Mathematical Legacy in the Eighteenth Century', pp. 228–30.

collection of miscellaneous mathematical writings by Robert Simson published in 1776 contains an essay in which Simson questioned the veracity of lemma 1 of Book I, section one, of the *Principia* in order to show that Newton's use of first and ultimate ratios was less rigorous than the method of exhaustion employed by ancient geometers.[247] Reid took issue with Simson in an undated paper headed 'Of Lemma 1 Newt[on's] Princip[ia]', and argued that even though the wording of the lemma was open to misunderstanding (and had indeed been misunderstood by commentators) Newton's use of the lemma indicated that he had a clear grasp of the theory of limits, contrary to what Simson had contended.[248] Moreover, other manuscripts attest to Reid's familiarity with Newton's mathematical writings and techniques and these papers place him even more firmly in the tradition of Newtonian mathematics.[249]

What, then, do Reid's manuscripts on mathematics dating from his Glasgow period tell us about his mathematical pursuits, the style of mathematics that he cultivated and his identity as a mathematician? First, the manuscripts show that Dugald Stewart's claim that Reid was primarily occupied with solving individual mathematical problems primarily in order to keep his faculties sharp following his retirement from teaching in 1780 needs to be heavily qualified. While there are manuscripts containing discrete algebraic problems that Reid readily worked through, and there is one section of a manuscript which outlines what he labelled 'An Arithmetical Trick', a number of his manuscripts contain handwriting from different periods of his life, which implies that he revisited various issues, having given them further consideration.[250] This feature of his manuscripts is especially clear in some of his papers on number theory, which were slowly built up in differ-

[247] Guicciardini, 'Thomas Reid's Mathematical Manuscripts', pp. 77–8; Robert Simson, 'De limitibus quantitatum et rationum, fragmentum', in *Opera quaedam reliqua* (1776); each tract is separately paginated. Newton's lemma reads: 'Quantities, and also ratios of quantities, which in any finite time constantly tend to equality, and which before the end of that time approach so close to one another that their difference is less than any given quantity, become ultimately equal'; Newton, *The Principia*, p. 433.

[248] AUL, MS 2131/3/I/5. In this manuscript, Reid refers to the commentary on the lemma found in Anon., *Excerpta quædam e Newtonii Principiis philosophiæ naturalis, cum notis variorum* (1765), pp. 16–17.

[249] See AUL, MSS 2131/5/I/8, where Reid uses Newton's method of interpolation; 2131/5/I/12 and 2131/5/II/38, p. 3, where he employs Newton's binomial theorem; 2131/5/I/17, where he refers to Newton's *Arithmetica universalis* (fol. 2r); and 2131/5/II/32 and 32a, for references to Newton's 'Methodus differentialis'. In AUL, MS 2131/5/I/17, fol. 1v, Reid also refers to a mathematical detail in the account of the priority dispute between Newton and Leibniz over the discovery of the calculus found in John Collins, *Commercium epistolicum D. Johannis Collins, et aliorum, de analysi promota, jussu Societatis Regiæ in lucem editum* (1722), p. 182.

[250] For examples of discrete algebraic problems, see AUL, MSS 2131/5/I/2, 2131/5/I/4 and 2131/5/II/51. Reid's 'Arithmetical Trick' is found in AUL, MS 2131/5/I/26, fol. 2r–v.

ent chronological layers over extended periods of time.[251] Hence much of Reid's mathematical work gradually unfolded during his retirement, and was not largely episodic as Stewart would have us believe. Secondly, as I have already indicated, there was a good deal of continuity in Reid's mathematical investigations. Again, this continuity is particularly evident in his work on number theory, but it can also be seen in a manuscript on cubic equations which includes pages written in the 1750s to which Reid subsequently added material in the 1780s or 1790s.[252] And, as we shall see in Section 2, his efforts to reformulate Euclid's problematic treatment of parallel lines, which began in the 1750s if not earlier, continued after he left Aberdeen and culminated in the discourse he read to the Glasgow Literary Society in the period *circa* 1790–93.[253] Stewart's suggestion that Reid's retirement saw a 'revival' of his interest in mathematics is thus highly misleading because Reid was actively engaged in a variety of mathematical enquiries at every stage of his adult life and, in some instances, the work that he did late in his career built on, and occasionally went beyond, his earlier research.

Thirdly, the mathematical manuscripts written in Glasgow enrich our understanding of Reid's style as a mathematician and his place in the mathematical community of his day. I have suggested above that we should follow Niccoló Guicciardini in seeing Reid's mathematical style as being akin to that of the group of 'analytical fluxionists' who flourished in Britain during the second half of the eighteenth century. Reid's papers on mathematics from his Glasgow period confirm that he was a Newtonian mathematician who took his inspiration from the form of 'new analysis' found in some of Newton's published mathematical writings, although his papers also show that he had mastered and used the geometrical techniques associated with what Newton called the 'synthetic method of fluxions' found in the *Principia*.[254] The centrality of algebra in Reid's work on mathematics speaks to his cultivation of the tools employed by the analytical fluxionists, and what we know of his reading habits indicates that he had a working knowledge of the texts written by mathematicians active in the first half of the eighteenth century, such as Colin Maclaurin, James Stirling and Roger Cotes, who built on Newton's 'new analysis'. Moreover, Reid was also familiar with more recent works by analytical fluxionists such as Thomas Simpson. His mathematical style was, therefore, unmistakably that of a Newtonian mathematician schooled

[251] See especially AUL, MS 2131/3/I/17.

[252] AUL, MS 2131/5/I/10. Pages 1–6 are written in a hand that dates to the 1750s, while pages 7–[9] are written in a much later hand, characteristic of the 1780s and 1790s.

[253] On the dating of Reid's discourse, see below, pp. 164–5, editorial note 15/17.

[254] On Newton's distinction between the 'analytical' and 'synthetic' methods of fluxions, see Guicciardini, 'Dot-Age', p. 226.

in the analytical methods initially developed by Newton and eagerly adopted by his acolytes in the British mathematical community. Consequently, Reid's understanding of Newton's mathematical achievements differed sharply from that of Robert Simson, insofar as Simson's mathematical researches were rooted in Newton's engagement with the legacy of ancient geometry. After making his fundamental mathematical discoveries in the 1660s, Newton became increasingly enamoured of the geometry of the ancients and came to regard their work as more rigorous than the analysis of the moderns. This aspect of Newton's thought served as Simson's starting point, rather than the 'new analysis' that gave rise to the analytical style that Simson rejected. Reid's rebuttal of Simson's criticisms of lemma 1, Book I of the *Principia* thus symbolises the divergence of two Newtonian styles of mathematics, namely the analytical style adopted by Reid and the geometrical style promoted by Simson.[255] But the differences between the two men should not be overemphasised, for Reid greatly admired his Glasgow colleague and used Simson's writings as a guide to the geometry of the ancients.[256]

A fourth point concerns the question of whether or not progress had been made in the various branches of mathematics. Whereas Robert Simson believed that the 'new analysis' of the moderns lacked the rigour of classical geometry and hence represented a retrogressive step in the history of mathematics, Reid seemingly endorsed the position of Colin Maclaurin and John Stewart, that, notwithstanding the continuities between Newton's method of fluxions and the geometry of the ancients, Newton's discoveries heralded a new era in the annals of mathematics.[257] As we have seen, Reid's reading of the history of arithmetic and algebra also celebrated the superiority of the moderns over the ancients. In one of his Glasgow manuscripts he likewise affirmed that the invention of logarithms constituted another progressive step beyond the ancients:

> Is not the Doctrine of Logarithms an improvement of the doctrine of the Ancients concerning Ratios, somewhat similar to the improvement of the Arithmetick of th[e] Ancients by decimal Fractions. The Ancients had Names

[255] Reid's willingness to employ algebraic reasoning to solve geometrical problems also illustrates the stylistic differences between the two camps; see AUL, MS 2131/7/VIII/1, 4r–v, for an example of Reid's application of algebra to geometry.

[256] For evidence of Reid's admiration of Simson see Reid to [William Ogilvie], [1763], in Reid, *Correspondence*, p. 23.

[257] Maclaurin, *Treatise of Fluxions*, vol. I, pp. i–ii; Stewart, *Sir Isaac Newton's Two Treatises*, p. 33. Stewart here distinguishes between Newton's 'entirely new' method and what he calls 'the specious Analysis of the Moderns'. For a suggestive discussion of Simson's view of the relationship between algebra and geometry, see Trail, *Account of the Life and Writings of Robert Simson*, pp. 63–70.

> for certain determinate Ratios and Symbols to express them but not for the numberless intermediate ones. Logarithms furnish us with a way of expressing intelligbly all possible Ratios; even when one of the Terms of the Ratio is with respect to the other irrational; and that either accurately or nearer the truth than by an assignable Difference[.][258]

While Newton himself may not have subscribed to the Enlightenment doctrine of progress, Reid and the majority of his fellow Newtonian mathematicians saw the seventeenth and eighteenth centuries as a period in which Newton and others had made significant advances in the major branches of mathematics. Yet it must be said that Reid's view of the progress that had been made by the moderns was an insular one, for he was not widely read in the writings of continental mathematicians and he seems to have been largely unaware of the contributions of prominent contemporaries in Europe such as Jean le Rond d'Alembert or the various members of the Bernoulli family. His horizons as a mathematician were set by Newton and the analytical fluxionists; consequently, much of the most important work on mathematics carried out on the continent during the course of the eighteenth century was for him *terra incognita*.[259]

A fifth and final point to be made is that Reid's identity as a mathematician was not simply or exclusively that of a Newtonian analytical fluxionist. Over the past two decades, historians have come to recognise that the revolutionary transformation of the natural sciences that took place in Europe during the sixteenth and seventeenth centuries was bound up with the emergence of a new social role, namely that of the mathematical practitioner.[260] The earliest manifestations of this role can be found among mathematicians working in late-fifteenth-century Italy. During the course of the sixteenth century, however, other individuals who can be identified as mathematical practitioners began to appear in the German-speaking

[258] AUL, MS 2131/5/I/26, fol. 1r. Reid elsewhere suggested that progress had been made in geometry, writing that: 'Euclid's Elements ... exhibit a system of geometry which deserves the name of a science; and though great additions have been made by Appollonius, Archimedes, Pappus, and others among the ancients, and still greater by the moderns; yet what was laid down in Euclid's Elements was never set aside. It remains at the firm foundation of all future superstructures in that science'; Reid, *Intellectual Powers*, p. 62. Reid thus believed that progress in the sciences was cumulative.

[259] See also Guicciardini, 'Thomas Reid's Mathematical Manuscripts', p. 73.

[260] In what follows I draw on: Mario Biagioli, 'The Social Status of Italian Mathematicians, 1450–1600'; Stephen Johnston, 'Mathematical Practitioners and Instruments in Elizabethan England'; Katherine Hill, '"Juglers or Schollers?": Negotiating the Role of a Mathematical Practitioner'; John Henry, *The Scientific Revolution and the Origins of Modern Science*, pp. 14–30. The classic studies of mathematical practitioners in England (and to some extent in Scotland) are E. G. R. Taylor's *The Mathematical Practitioners of Tudor and Stuart England, 1485–1716*, and her *The Mathematical Practitioners of Hanoverian England, 1714–1840*.

regions of Europe, the Low Countries and, eventually, England. Although the social origins of the mathematical practitioners were diverse and the range of activities in which they engaged varied, the members of this group typically maintained that the study of arithmetic, geometry and, from the seventeenth century onwards, algebra served as the basis for the mastery of the various branches of 'mixed' mathematics, such as astronomy and optics, as well as applied fields such as navigation, surveying, mensuration, gunnery, engineering and book-keeping.[261] Some mathematical practitioners held university positions while others taught privately, found employment with civic corporations or states, or secured the patronage necessary to join princely or royal courts. Because many mathematical practitioners relied on the marketplace for their economic survival, they also made and sold maps and navigational charts, constructed globes and mathematical instruments, and published books, usually written in the vernacular, explaining the use of the hardware they manufactured. Underwriting these activities was the belief expressed so memorably by Galileo that the book of nature 'is written in the language of mathematics'.[262] In the sixteenth and early seventeenth centuries the mathematical practitioners rejected the Aristotelianism of the schools and looked to alternative philosophical systems, such as Platonism, in order to provide a metaphysical rationale for the usefulness of mathematics. On the continent the Scientific Revolution was advanced by mathematical practitioners such as Simon Stevin, Niccolò Tartaglia and Galileo Galilei. In Scotland their counterparts Napier of Merchiston, James Corss, George Sinclair and members of the Gregory family all furthered the rise of the new science through their writings, teaching or, in the case of Sinclair, their public lecturing.[263]

An understanding of the role of the mathematical practitioner enables us to make sense of the pattern of Reid's activities and interests as a mathematician. Reid clearly fits the mould of the mathematical practitioner in his emphasis on the practical applications of 'pure' mathematics. His conviction that knowledge ought to be applied for human benefit shaped the reform of the curriculum that he and his colleagues effected at King's College, Aberdeen, in 1753. We have seen that his teaching subsequently encompassed both 'speculative' and 'practical' mathematics, for he covered the basics of arithmetic, algebra and geometry, as well as dialling, navigation, the use of mathematical and astronomical instruments, the principles of perspective, mensuration and surveying.

[261] For an eighteenth-century definition of mixed mathematics, see Ephraim Chambers' *Cyclopædia*, as cited above in note 92.

[262] Galileo Galilei, *The Assayer* (1623), *Discoveries and Opinions of Galileo*, p. 238.

[263] On the significance of this group, see Paul Wood, 'Science in the Scottish Enlightenment', pp. 95–7.

And, following the precedent of earlier mathematical practitioners, he provided instruction in book-keeping through the scheme he designed for farmers that he submitted to the Gordon's Mill Farming Club. After moving to Glasgow he continued to take an active interest in practical mathematics, as can be seen in various manuscripts dealing with topics such as navigation, surveying and the construction of sundials. Moreover, a letter to Richard Price dating from *circa* 1772–73 provides an eloquent expression of his belief that mathematicians (and men of science more generally) had an obligation to apply their knowledge in order to protect the public interest. Writing to Price, Reid recalled that 'about forty years ago there was a Phrenzy in the Nation about mechanical Projects' which had caused the financial ruin of those misled by 'Projectors'. According to Reid, the mathematical practitioner and natural philosopher J. T. Desaguliers 'had no small Merit' in combating the spread of this 'Disease', and said that Desaguliers did so by teaching 'Men clearly upon principles of Science the utmost Effects that the Mechanical Powers can produce'. Reid then went on to praise Price for having likewise used his mathematical skills to counter 'the present Epidemical Disease of trusting to visionary Projects of Reversionary Annuities' and to warn against the future dangers posed by the alarming level of the national debt. Regarding the burgeoning amount of the debt, he observed that

> I have been long perswaded that the Administration may receive great aid in this Matter from those who have entered most deeply into the Science of Numbers. It is much to the Honour of [this] Science, & ought to draw more Respect to it than it commonly meets with in this Age of Dissipation that even the most abstract parts of it are found of so great utility in the affairs of Life.[264]

Reid here expressed his credo that the fundamentals of mathematics and natural philosophy ought to be used for the benefit of humankind. In stating his credo, he echoed the teachings not only of Bacon but also of the mathematical practitioners who, from the sixteenth century onwards, had applied the principles of 'speculative' mathematics to the solution of practical problems in fields such as gunnery, fortification and engineering.[265] Even though Reid may not have built and marketed his own mathematical instruments or acted as a consultant,

[264] Reid to Price, [1772–73], in Reid, *Correspondence*, p. 64. Desaguliers' efforts to define the public role of the natural philosopher are brilliantly described in Stewart, *Rise of Public Science*.

[265] Compare Reid's first law of philosophising, as stated in his graduation oration for 1756: 'it is the end of philosophy to enhance the fortune of humankind and to enhance humankind's power over things'; Reid, *Philosophical Orations*, oration II, para. 6. For a discussion of Reid's use of mathematics in political arithmetic, see the editorial introduction to Reid, *Society and Politics*, pp. xxviii–xxx, lxix–lxx, lxxv–lxxvi.

he nevertheless exhibited virtually all of the characteristic traits of a mathematical practitioner, given that he had taught 'speculative', 'mixed' and 'practical' mathematics, and pursued a broad range of enquiries in both pure and applied mathematics throughout his career.[266] It is to the detailed discussion of Reid's work on Euclidean geometry and the mixed mathematical sciences of mechanics, astronomy and optics, along with his investigations in the largely experimental sciences of electricity and chemistry that we now turn.

2. Euclidean Geometry

Unfortunately, we lack any direct evidence regarding the beginnings of Reid's serious reflection on the conceptual foundations of Euclidean geometry. Circumstantial evidence, however, suggests that he began to consider the logical cogency of Euclid's system at a relatively early stage in his career and that at some point in the 1740s or 1750s his interest in the *Elements* dovetailed with his desire to defuse the sceptical threat posed by David Hume's *A Treatise of Human Nature*. Writing in the early 1760s to William Ogilvie, Reid confessed that he was 'ashamed' to admit 'how much time I consumed long ago upon this Axiom [i.e. Euclid's axiom concerning parallel lines], in order to find Mathematical Evidence for what common sense does not permit any man to doubt'.[267] Reid's worries about straight and parallel lines, which we can begin to document with some precision in the 1750s, thus seem to have originated well before his reading of Robert Simson's English translation of Euclid's *Elements* (*circa* 1756–57) or his development of the non-Euclidean 'geometry of visibles' in the *Inquiry*.[268] His comments also indicate that epistemological considerations at least partly motivated his attempted emendation of the definitions and axioms of the *Elements*, insofar as Euclid's 'eleventh' axiom on parallel lines did not meet the

[266] The construction and sale of instruments arguably became less central to the role of the mathematical practitioner beginning in the latter part of the seventeenth century because of the rise of specialist instrument makers. Even though Reid may not have made his own mathematical or scientific instruments, we have seen above (pp. lii–liii) that he was fascinated with the practicalities of instrument making. See also a brief undated set of queries regarding the construction of James Watt's 'Perspective machine' (see above, p. l) that survives in AUL, MS 2131/7/II/5, fol. 2r. His letters to Andrew and David Skene written shortly after moving to Glasgow also record his efforts to obtain thermometers, as well as chemical furnaces for his Aberdeen friends. In addition to the letters cited above in note 137, see Reid to David Skene, 25 February 1767, in Reid, *Correspondence*, pp. 43, 47, 57.

[267] Reid to [William Ogilvie], [1763], in Reid, *Correspondence*, p. 23. On this letter see below, note 276.

[268] Reid, *Inquiry*, pp. 103–12.

standard of evidence characteristic of the mathematical sciences.[269] Moreover, a revealing passage in Reid's *Essays on the Intellectual Powers of Man* implies that his work on the foundations of Euclidean geometry began in response to Hume's sceptical critique of the basic concepts routinely employed by mathematicians. In the *Intellectual Powers*, Reid argued that the 'clear and accurate notions which geometry presents to us of a point, a right line, an angle, a square, a circle, of ratios direct and inverse, and others of that kind' are not ideas of sense but rather the product of analysis and judgement. Having made this claim, Reid rounded on his adversary, writing that,

> had Mr Hume attended duly to this, it ought to have prevented a very bold attempt, which he has prosecuted through fourteen pages of his Treatise of Human Nature, to prove that geometry is founded upon ideas that are not exact, and axioms that are not precisely true.[270]

In the section of the *Treatise* dealing with our ideas of space and time to which Reid here refers, Hume had questioned whether we can have exact conceptions of basic mathematical relations such as that of equality, or of elementary geometrical objects such as straight and curved lines.[271] His remarks may well have been the original catalyst for Reid's attempt to discover a more rigorous treatment of straight and parallel lines than that found in Euclid's *Elements*. But lacking any direct evidence to support this hypothesis, it is safer to suggest that, at the very least, Hume's scepticism regarding the absolute certainty of geometry (and mathematics more generally) served to increase Reid's anxieties about the logical flaws in Euclid's definitions and axioms.

The earliest direct evidence we have for Reid's critical engagement with the conceptual basis of Euclid's *Elements* dates from June 1756, that is, shortly after he had finished teaching the first cycle of the revised curriculum at King's College, Aberdeen. In his 'Observations on the Elements of Euclid' (below, pp. 3–4), he followed the example of earlier geometers by working through the definitions and axioms prefacing the *Elements*, along with some of the propositions demonstrated in Book I. And, like other commentators on Euclid, he also remarked on the

[269] As Reid himself notes (pp. 7, 23), in some editions of Euclid's *Elements* the axiom is the eleventh and in others it is numbered the twelfth. The axiom is also known as the fifth postulate. In Thomas L. Heath's modern translation, the axiom/postulate reads: 'That, if a straight line falling on two straight lines make[s] the interior angles on the same side less than two right angles, the two straight lines, if produced indefinitely, meet on that side on which are the angles less than the two right angles'; Euclid, *The Thirteen Books of Euclid's Elements*, trans. Thomas L. Heath, vol. I, p. 155. See also below, p. 160, editorial note 7/23.

[270] Reid, *Intellectual Powers*, p. 419.

[271] David Hume, *A Treatise of Human Nature*, pp. 33–40 (1.2.4).

unsatisfactory nature of the definition of ratios in Book V.[272] Reid's long-standing interest in mathematics, as well as his practical experience in teaching geometry, had clearly made him acutely aware of some of the logical problems found in the *Elements*, and his worries about Euclid's discussion of parallel lines emerge in his extended discussion of the relevant axioms and propositions in Book I. But even though he recognised that the problematic 'eleventh' axiom upon which Euclid's demonstrations involving parallel lines rest might have to be jettisoned, he seems not to have been unduly concerned, for he merely observed that 'Since therefore there seems to be a Necessity of making some Alteration in order to Reconcile the Doctrine of Parallel Lines to Geometrical Accuracy I conceive it may be most easily done thus by altering the Def[inition] of Parallel lines' (p. 3).[273]

The publication in late 1756 of Robert Simson's edition of Euclid's *Elements* in both Latin and English versions, however, seems to have underlined for Reid just how problematic the whole issue of parallel lines really was.[274] In an undated set of notes from Simson's edition (below, pp. 10–12) Reid queried the Glasgow geometer's demonstration of the corollary to proposition 11, Book I of the *Elements*, on the ground that the corollary had already been presupposed in earlier propositions; this prompted the comment that 'there are several things assumed in Euclids Demonstrations which are neither laid down as Axioms by his Editors nor demonstrated'. He proceeded to identify two such assumptions made in the first three propositions of Book I, and supplied a demonstration for a presupposition of the fourth, namely that 'two right lines cannot have a common segment'. He then remarked: 'It may be observed that the Simplest properties of Right lines cannot be demonstrated from Euclid[']s Definition of a Right line which indeed is too vague to found a Demonstration upon' (p. 11). Reid was, therefore, well aware of

[272] See, for example, Edmund Scarburgh, *The English Euclide, Being the First Six Elements of Geometry, Translated out of the Greek, with Annotations and Useful Supplements* (1705), pp. 178–83; Edmund Stone, *Euclid's Elements of Geometry, the First Six, the Eleventh and Twelfth Books; Translated into English, from Dr Gregory's Edition, with Notes and Additions* (1752), pp. 205–6. For a robust defence of Euclid's definition, see Isaac Barrow, *The Usefulness of Mathematical Learning Explained and Demonstrated: Being Mathematical Lectures Read in the Publick Schools at the University of Cambridge* (1734), pp. 383–440. Euclid's treatment of ratios in Book V of the *Elements* generated considerable controversy in the early modern period; a brief but useful overview is found in the editorial introduction to Girolamo Saccheri, *Euclid Vindicated from Every Blemish* [1733], pp. 17–24.

[273] In Edmund Stone's translation, axiom 11 reads: 'If a right line falling upon two right lines does make the internal angles on the same side less than two right angles, those right lines, being infinitely produced, do meet on that side where the angles are less than two right angles'; Stone, *Euclid's Elements*, pp. 5–6.

[274] On the publication of Simson's Latin and English texts, see Gaskell, *Bibliography of the Foulis Press*, pp. 206–7.

the difficulties plaguing Euclid's treatment of right lines, and his correspondence from the early 1760s indicates that he spent a good deal of time trying to shore up this aspect of the *Elements* in order to resolve the related problem of parallel lines.

Drafts of two fascinating letters dating from the period 1762 to 1764 provide a glimpse of the direction Reid's work on Euclidean geometry took in the late 1750s. The drafts also contain valuable clues regarding the contents of a manuscript which no longer survives, namely the text of Reid's 'voluntary' discourse, on 'Euclid's definitions and axioms', which he delivered before the Aberdeen Philosophical Society on 26 January 1762.[275] The first letter was most likely written during the winter of 1762–63 and was probably addressed to his ex-pupil William Ogilvie, who was apparently acting as a tutor in Glasgow.[276] Significantly, Reid here paid tribute to Robert Simson as 'the father of the Mathematicians now alive' and praised Simson's edition of Euclid, which he was then using to teach his students the elements of geometry. Given the Glasgow professor's pre-eminence as a geometer, Reid said that 'it mortifies me not a little' that Simson held that the '11 or 12 Axiom upon which so great a part of [Euclid's] System hangs is neither selfevident nor does [it] admit of a demonstration in a strict sense', for this implied that there was 'a defect in the Elements ... which is not to be attributed to [Euclid's editor] Theon but to the Science itself'. He confessed that he had 'long ago' devoted a considerable amount of time trying to demonstrate the validity of the axiom and that after he had 'laboured in vain', he 'quite despaired' to find that 'Dr Simpson [*sic*] was of [the] Opinion that [the axiom] could not be strictly demonstrated'. He then proceeded to summarise his thoughts regarding the problem of parallel lines and said that he 'would be very glad to know Dr Simpsons opinion of it'.[277]

[275] Ulman, *Minutes of the Aberdeen Philosophical Society*, p. 107. Reid volunteered to read a discourse because he had just finished his term as President of the Society. His offer implies that he had already written the text.

[276] Reid to [William Ogilvie], [1763], in Reid, *Correspondence*, pp. 23–6. The date of this letter can be inferred from Reid's comment that 'I am just now teaching Euclids Elements' (p. 23). He began a new cycle of teaching in 1762–63 and would therefore have been teaching Euclidean geometry in that session. William Ogilvie studied under Reid at King's College and graduated in 1759; see Anderson, *Officers and Graduates of University and King's College Aberdeen*, p. 241. In the winter of 1761–62 he was unable to substitute for the King's regent, Alexander Burnet, because he was apparently acting as a tutor to a 'Mr Graeme'. Nevertheless, the Principal and Masters of King's promised to appoint Ogilvie as a regent when the next vacancy occurred; the Masters of King's College to William Ogilvie, 25 November 1761, in Reid, *Correspondence*, pp. 17–18. During the winter of 1762–63 Ogilvie was in Glasgow, and attended some of the same classes as David Steuart Erskine, the future eleventh earl of Buchan; see the 'Extracts from the Diaries and Letter-Books of the 11th Earl of Buchan. No. 9. The Story of His Life, 1764', GUL, MS Murray 502/65.

[277] Reid to [William Ogilvie], [1763], in Reid, *Correspondence*, p. 23.

According to Reid, the properties of all of the geometrical figures described in the *Elements* could be derived from their respective definitions, apart from one. Not surprisingly, the exception was 'right lines', since, as he had already suggested in his reading notes on Simson's edition of Euclid (p. 11), the definition of a straight line in the *Elements* 'gives no mathematical conception of the thing defined' and hence could not be used as the basis for proper geometrical demonstrations.[278] Following Simson's example of ascribing all of the flaws in the *Elements* to copyists and editors, he attributed the faulty definition of straight lines to 'Theon or some interpolator', arguing that Euclid was too accomplished a mathematician to offer a tautology for a true definition, and that the definition of a straight line in the *Elements* must have been a spurious addition because it was not referred to elsewhere in the text. Moreover, he contended that a better definition of straight lines was to be found in the demonstration of proposition 1, Book XI, namely that a 'right line cannot meet another in more points than one; otherwise the right lines will coincide'.[279] On the basis of this definition he claimed that it was possible to demonstrate three of the four axioms he enumerated regarding straight lines, but admitted that he had been unable to derive the fourth axiom, dealing with parallel lines, remarking that 'Alas! I have spent much labour in vain to deduce that Axiom by mathematical Reasoning from this Definition. And after what Dr Simpson has said I almost despair of it'.[280]

Reflecting on his own failure to demonstrate the problematic axiom, Reid suggested that there were two options open to geometers. The first was to resort to a set of self-evident axioms that could be used to deduce the various properties of straight lines. He offered one possible candidate from which the axiom concerning parallel lines could be demonstrated: 'That if two points of a right line be equally distant from another right line in the same plane the intermediate points will be at the same distance from that right line'.[281] He was not, however, inclined to adopt this strategy for he believed that 'the Axioms in Mathematicks seem … to be onely a kind of *succedaneum* [substitute] to supply the want of proper definitions'. Furthermore, even though he was prepared to grant that the basic properties and relations of genuinely simple objects of thought such as quantity could only be

[278] Reid to [William Ogilvie], [1763], in Reid, *Correspondence*, p. 24.

[279] Stone, *Euclid's Elements*, p. 315. Simson omitted this wording from his edition of the *Elements* because he believed that the wording was 'an addition by some unskilful hand; for this is to be demonstrated, not assumed'; Simson, *The Elements of Euclid* (1756), p. 415. It is now accepted that the wording may have been an interpolation of Euclid's editor, Theon.

[280] Reid to [William Ogilvie], [1763], in Reid, *Correspondence*, p. 25; compare Simson, *The Elements of Euclid*, pp. 360–2.

[281] Reid to [William Ogilvie], [1763], in Reid, *Correspondence*, p. 25.

elucidated using axioms, he apparently doubted that straight lines were in fact simple objects akin to quantity. Consequently, Reid held that geometers should continue to search for a proper definition of a straight line because the discovery of such a definition 'would most effectually wipe of[f] that reproach from the Elements of their being in a great measure founded upon an Axiom that is neither self evident nor can be strictly demonstrated'.[282]

William Ogilvie apparently acceded to Reid's request to forward his reflections on Euclid to Robert Simson, for Reid and Simson subsequently corresponded and exchanged presentation copies of books. In a draft of a letter to Simson dating from March or April 1764, Reid thanked his recipient for a letter dated 26 August 1763 and for having invited him to 'send ... my observations upon your Edition of the data of Euclid'. Simson's letter presumably accompanied a copy of the second edition of his English translation of the *Elements*, published in 1762, which also included Simson's recension of Euclid's *Data*. Reid, in turn, had asked his Edinburgh publisher, Alexander Kincaid, to send Simson a copy of his *Inquiry*, which had appeared in January or early February 1764. Reid apologised to Simson for not having had the time required to read through the text of the *Data* with care because of the combined pressures of teaching and preparing the *Inquiry* for the press, and said that he would not be able to scrutinise Simson's text 'till our Session is up'.[283] But Reid did not wish to disoblige Simson and instead offered a few observations on the changes that his correspondent had made to the second edition of the *Elements*. Reid was clearly still exercised by the cluster of problems posed by Euclid's treatment of straight and parallel lines, for he repeats much of the argument of his letter to Ogilvie. What is most striking about his letter to Simson is that his comments strongly suggest that his own work on the contentious definition of right lines had reached an impasse and that this phase of his investigations, which had begun *circa* 1756, was effectively at an end.[284] Taken together, the draft letters to Ogilvie and Simson show that in the period 1756 to 1764 Reid devoted a considerable amount of time and intellectual energy

[282] Reid to [William Ogilvie], [1763], in Reid, *Correspondence*, 24, 25–6; the wording of the reproach comes from Simson, *The Elements of Euclid*, p. 360.

[283] Reid to [Robert Simson], [1764], in Reid, *Correspondence*, p. 32; Robert Simson, *The Elements of Euclid, viz. the First Six Books, Together with the Eleventh and Twelfth*, second edition (1762). For the publication date of Reid's *Inquiry*, see Reid, *Correspondence*, p. 276, note 32/31. When Reid drafted his letter to Simson, he presumably knew that he had been suggested as a successor to Simson's colleague, Adam Smith, but he chose to say nothing about the matter.

[284] Reid to [Robert Simson], [1764], in Reid, *Correspondence*, pp. 32–4. In this letter, Reid restates the criticism he had initially made in his reading notes on the first edition of Simson's translation of the *Elements* that the corollary demonstrated in proposition 11, Book I, was presupposed in proposition 4; see below, p. 10.

to the critical analysis of the foundations of Euclidean geometry, and that he tried unsuccessfully to emend the structure of the *Elements* in order to eliminate the logical flaws in the definitions, axioms and propositions dealing with straight and parallel lines. Moreover, it is reasonable to assume that Reid's 1762 discourse on 'Euclid's definitions and axioms' elaborated on the main themes of these letters, and that in his presentation to the Wise Club he discussed in greater detail his views on the logical properties of axioms and definitions. Moreover, he may have outlined his attempted derivation of the properties of straight and parallel lines from the definition of a straight line he took from proposition 1, Book XI of the *Elements*.

After Reid's move to Glasgow in the summer of 1764, his time was almost entirely occupied by revising his lectures and negotiating the acrimonious politics of the University. Nevertheless, we have seen that he kept up many of his earlier mathematical and scientific interests, with the conceptual foundations of Euclidean geometry figuring prominently in the range of subjects to which he devoted his spare moments. In the period 1764 to 1770, Reid continued to reflect on the issue of how best to recast Euclid's treatment of straight and parallel lines and developed his thoughts in a paper dealing with the topic. Evidence for this new phase of his work on Euclid comes in an intriguing memorandum dated 28 May 1770, in which he states that

> I resolve for the future to give up the Consideration of this Subject; having spent more time & thought in attempting to prove the simple properties of Streight lines from some one definition or Axiom than I can own without shame. (p. 13)

Reid had apparently revisited the problems that he had left unsolved in 1764 and drafted an 'Essay' on straight and parallel lines in which he seemingly again attempted to demonstrate formally Euclid's 'eleventh' axiom. What may be a draft or a fragment of the 'Essay' containing 'Proposition 6' is extant among his papers (pp. 12–13), but the complete 'Essay' has evidently been lost. By the time he came to write his memorandum in May 1770, however, he clearly regarded his 'Essay' as being unsatisfactory because he now recorded what he called his 'last thoughts upon this subject' (p. 13), which reveal that he had changed his mind about some key issues that he had left unresolved in 1764. Whereas previously he had doubted that a straight line was a simple object of thought, he here affirmed that 'a Streight line is an Object too Simple to admit of a proper mathematical Definition', and maintained that 'one or more Axioms' had to be laid down so that 'the Theory of Streight lines may be delivered Mathematically' (p. 13). The shift in his thinking is also illustrated by the fact that he referred to what he had earlier proposed as the definition of a straight line as his first axiom, and he granted that another axiom was required in order to derive all of the properties of straight and parallel lines. He then enumerated five potential candidates before the text breaks off and a further entry begins 'Sept[embe]r 13 1770 I find a Tract upon this Subject Intitled Euclides ab omni Nævo vindicatus…' (p. 15).

Having come across Girolamo Saccheri's *Euclides ab omni nævo vindicatus* (1733) among the extensive collection of books bequeathed to the University by Robert Simson, Reid subsequently made detailed reading notes from the work, which document his reaction to Saccheri's attempt to prove the validity of Euclid's 'eleventh' axiom (pp. 4–7). Although he was impressed by much of what he examined, and approved of the initial stages of Saccheri's reworking of Euclid, he believed that, in the end, the Italian geometer had failed to repair the flaws in the *Elements*, observing that Saccheri had been 'led into a large field and has recourse to reasonings about infinitesimals. Some of them … I think not Just' (p. 7). If Reid had turned to Saccheri hoping to find a way out of the impasse he had reached in his own research on straight and parallel lines, he must have been sorely disappointed to see that the Italian had also come unstuck in attempting to demonstrate the validity of the 'eleventh' axiom. And it appears that at this juncture Reid temporarily abandoned his work on the foundations of Euclidean geometry.

Reid was nothing if not tenacious, however, and he returned to his labours on the problem of straight and parallel lines in the period *circa* 1790 to 1793, when, as already mentioned, he delivered a discourse on Euclid's *Elements* before the Glasgow Literary Society.[285] As one might expect, his discourse covered much the same ground as his earlier manuscripts on the subject. But his discourse differed from his previous writings on Euclid (apart from his Wise Club paper, which has not survived) in two important respects. First, the fact that it was a text written for oral delivery conditioned the presentation of his material insofar as his annotations to the manuscript indicate that he confined himself to a verbal exposition of his ideas and that he did not resort to geometrical constructions in order to illustrate his argument (p. 232, editorial notes 30/18, 31/22; see also pp. 10, 28). Secondly, he was far more discursive in this late essay. For example, he provided a historical overview of previous attempts to resolve the issues surrounding Euclid's 'eleventh' axiom in which he discussed the writings of Ptolemy, Proclus, Christopher Clavius, the obscure English mathematician Thomas Oliver, Sir Henry Savile, John Wallis, David Gregory, Girolamo Saccheri and, last but by no means least, Robert Simson (pp. 7–8, 23–7). Thus, even though the contents of his discourse were to some extent determined by the conventions

[285] On the dating of Reid's discourse, see below, pp. 164–5, editorial note 15/17.

of an oral presentation, the text also gives the impression that he had finally abandoned any hope of being able to progress beyond what previous geometers had achieved. And it seems that he found some consolation in knowing that far greater mathematicians than himself had likewise failed to repair the foundations of Euclidean geometry.

Although the bulk of Reid's presentation recapitulated themes and arguments found in his earlier manuscripts, three features of his discourse deserve comment. First, Reid seems to have again expended considerable intellectual effort in composing his discourse. Drafts for different sections of the text survive (pp. 7–10), and these drafts show that he made a number of significant alterations. The most important of these changes is the note read to the meeting after concluding his discourse. In this note, he informed his colleagues that he had misled them about Simson's views because he had not collated the first edition of Simson's Euclid with later ones, which contained a revised version of Simson's editorial note on the 'eleventh' axiom and proposition 29, Book I (p. 29).[286] Thus while there is a good deal of conceptual overlap between his Glasgow Literary Society discourse and his earlier papers on Euclidean geometry, it should be emphasised that he was not content merely to rehash older material. The changes that Reid made to his discourse show that he persisted in trying to discover new solutions to the problem of straight and parallel lines and that he was no less inquisitive in his later years than he had been in his youth. Furthermore, the fact that one of his manuscripts juxtaposes drafts of passages in his Glasgow discourse with reading notes from Robert Simson and Girolamo Saccheri illustrates the point made in Section 1 regarding the chronological layering found in some of Reid's mathematical papers.[287] And, as I have suggested, the mixture of different temporal strata in these papers calls into question Dugald Stewart's suggestion that in retirement Reid revived his youthful interest in mathematics in order to exercise his mental faculties. Reid's manuscripts on Euclidean geometry document not only the continuity in his mathematical interests but also the evolution of his thinking about the conceptual foundations of Euclid's *Elements* over a span of roughly forty years. Reid clearly took the problem of straight and parallel lines

[286] Reid's memory failed him because he had written to Simson in early 1764 regarding the revised note in the second edition of Simson's English translation of the *Elements*. See Reid to [Robert Simson], [1764], in Reid, *Correspondence*, p. 33, and Simson, *The Elements of Euclid*, second edition, pp. 294–9. The revised version of the note that first appeared in 1762 is reproduced in the fifth edition of Simson's Euclid, to which Reid refers; see below, p. 29, and Robert Simson, *The Elements of Euclid, viz. the First Six Books, Together with the Eleventh and Twelfth*, fifth edition (1775), pp. 302–7.

[287] See pp. lxxx–lxxxi.

seriously over an extended period, and expended a considerable amount of mental energy searching for a solution. While Stewart may have viewed Reid's work on mathematics as an intellectual diversion, Reid himself undoubtedly regarded his mathematical researches, and especially those devoted to Euclid's *Elements*, in a different light. For Reid's historical survey of attempts to remedy the faulty 'eleventh' axiom suggests that he saw himself as standing in a long line of distinguished geometers who had likewise endeavoured to shore up the foundations of Euclidean geometry by reworking Euclid's flawed treatment of straight and parallel lines. Even though he may not have considered himself their equal, he nevertheless seems to have thought that he had a genuine if modest contribution to make to the perpetuation of the tradition of commentary on the *Elements* that originated in classical antiquity (p. 10).

Secondly, by the early 1790s Reid seems to have been more concerned than he had hitherto been to distinguish his approach to the study of Euclid's *Elements* from that of his deceased colleague, Robert Simson. On the one hand, Reid paid tribute to the 'ingenious Dr Simson' for having 'observed more of the defects of our Copies [of the *Elements*], and made more Judicious Amendments than all the Criticks that were before him; by which one of the most perfect Works in its kind is made more perfect'. Nevertheless, he made it clear that his interest in the text of the *Elements* was logical rather than exclusively editorial. He developed a point that he had hinted at some thirty years before and maintained that although Euclid's *Elements* was a model of 'Order Perspicuity and Elegance' which had 'justly been held in admiration by Mathematicians of the best Taste and Judgment in every Age since they were wrote', this did not mean that the work was beyond reproach. 'Euclid & his Editors' were, after all, 'Men as we are [and] No work of human Genius is above Criticism'. For his part, he maintained that it did 'more honour to him [Euclid] as well as to ourselves & to human Nature' to engage in the critical scrutiny of the *Elements* instead of accepting their truth based on 'implicit Admiration & blind Submission'. While he applauded Simson's efforts to improve on previous editions of the *Elements*, he distanced himself from his late colleague's exaggerated respect for Euclid's merits. And while he recognised that Simson had good grounds to attribute 'all the imperfections' in the *Elements* to Theon and other editors, he emphasised that he was primarily interested in 'the Work' itself 'rather than the Author or Editors'. Reid therefore proposed 'after Dr Simsons Example, to make some Observations which he hath not taken notice of, without concerning my self whether the things I censure are owing to Euclid or to his Editors' (pp. 15–16). Defined in these terms, Reid's approach registers his preoccupation with the question of whether Euclid's *Elements*, and in particular the 'eleventh' axiom, conformed to the standard of evidence demanded in the mathematical sciences. His discourse thus provides a further illustration of how his attempt to enumerate the different types of evidence found in mathematics and the other branches of human learning shaped his efforts to clarify and strengthen the foundations of geometry.[288]

Thirdly, Reid's geometrical investigations were related to his pedagogical concerns, for the clarification of Euclid's definitions and axioms facilitated both teaching and learning. Reid believed that Euclid had various pedagogical considerations in mind when composing the *Elements*, and such considerations came into play in the solution to the problem of the 'eleventh' axiom which Reid ultimately proposed. For as he indicated, the issue was not the *truth* of the axiom but rather 'whether it ha[s] that degree of selfevidence which intitles it to be assumed without proof as an Axiom'. Yet the answer to this question was far from straightforward because, according to Reid, 'the Sphere of self-evidence, must enlarge in proportion to the penetration of the Mind that judges', which implied that 'beginners in these studies' and more experienced mathematicians would assess the degree of evidence of the axiom differently (pp. 23, 27, 28).[289]

In Reid's Glasgow discourse, he outlined his method for deducing the properties of parallel lines without invoking the contested 'eleventh' axiom. His strategy involved replacing Euclid's definition of parallel lines with one that he claimed was more satisfactory, namely 'A right line is said to be parallel to a right line when being in the same plane it is in every point equally distant from it' (p. 27). From this definition, he said that he was able to infer the crucial propositions concerning parallel lines in the *Elements*, and that he was further able to demonstrate the 'eleventh' axiom, which, he said, 'is necessary in other parts of Geometry and is thought not to be selfevident' (p. 28). His demonstrations, however, all depended upon a corollary he drew in the course of proving the basic properties of straight lines. Unfortunately, he came to think that this corollary was no more self-evident than the contested 'eleventh' axiom, and he concluded that

[288] Reid maintained that 'the Observing & distinguishing' of the 'Various kinds of Evidence' was 'one of the Most important parts of that Logic which is really useful in Life'; Reid, *On Logic, Rhetoric and the Fine Arts*, p. 185. Among the mathematicians Reid knew, his friend John Stewart was similarly interested in the classification of the different kinds of evidence. Stewart read a number of discourses before the Aberdeen Philosophical Society on the theory of evidence, including two devoted to mathematical evidence; see Ulman, *Minutes of the Aberdeen Philosophical Society*, p. 238. So too was Stewart's successor, William Trail; see Ulman, *Minutes of the Aberdeen Philosophical Society*, p. 238.

[289] The fact that Reid never doubted the truth of the 'eleventh' axiom raises an important historiographical question regarding the sense in which his 'geometry of visibles' can be said to constitute the discovery of a system of non-Euclidean geometry. Arguably, it is anachronistic to interpret Reid as a precursor of later figures who developed non-Euclidean systems such as Karl Friedrich Gauss, but it is not clear how we are to contextualise his 'geometry of visibles'. On this issue, see especially Giovanni B. Grandi, 'Thomas Reid's Geometry of Visibles and the Parallel Postulate'.

'the common Sense of Men who are accustomed to judge in such Matters is the onely tribunal to which we can appeal for the resolution of this Doubt' (p. 30). In the end, he had to admit failure once again. But he was also now pulled in two different directions regarding the self-evidence of Euclid's axiom. He seemingly accepted that, for teaching purposes, Euclid's text required emendation in order to preserve its logical structure, yet he nevertheless excused Euclid's seemingly arbitrary introduction of the 'eleventh' axiom on the ground that what appeared self-evident to Euclid was not so to lesser mathematicians. Either way, a satisfactory resolution of the problems he had grappled with for over forty years still eluded him, and he finally seems to have resigned himself to defeat.

Reid's efforts were not entirely in vain, however. The distinguished Scottish mathematician John Playfair praised his work on Euclid's definitions and axioms. Earlier in his career, Playfair met Reid when he competed unsuccessfully for the mathematics chair at Marischal College, Aberdeen, in August 1766. He then served as a parish minister and briefly as a private tutor before being appointed in 1785 as the joint Professor of Mathematics at Edinburgh alongside Adam Ferguson. After Reid had delivered his discourse to the Glasgow Literary Society, Playfair seems to have received a report of it from Reid's colleague, the chemist Thomas Charles Hope. Playfair then asked to see the essay and Reid complied with his request. When he sent the discourse to the Edinburgh mathematician, Reid appended a dismissive covering note:

> If the Author had thought this Discourse worthy to be shewn to Mathematicians, he would have transcribed it fair, & put in their proper places the parts that are disjoyned; but as Dr Hope informs him that Mr. Playfair desires to see it, he will easily perceive that it is not worth that trouble.[290]

Playfair took a different view. In a note on the definition of a straight line given in his *Elements of Geometry* (1795), he records that he had been 'favoured by Dr Reid of Glasgow with the perusal of a MS. containing many excellent observations on the first Book of Euclid, such as might be expected from a philosopher distinguished for the accuracy as well as the extent of his knowledge'.[291] Among the mathematicians then working in Scotland, Playfair was in the best position to appreciate Reid's struggles with Euclid's definitions and axioms, for he too sought to establish Euclid's treatment of straight and parallel lines on a sound logical and

[290] Below, p. 232, textual note 31/22. Reid's comments imply that he had circulated some of his manuscripts in the past; on this point see Guicciardini, 'Thomas Reid's Mathematical Manuscripts', pp. 71–2.

[291] John Playfair, *Elements of Geometry; Containing the First Six Books of Euclid, with Two Books on the Geometry of Solids* (1795), pp. 351–2.

conceptual footing. It is therefore fitting that he should have paid tribute to his older colleague in print. Moreover, it may be that the Edinburgh mathematician also submitted some of his own papers on geometry to Reid for comment because there is a fragment surviving among Reid's manuscripts, but not in his hand, containing a demonstration concerning parallel lines, which Playfair could conceivably have sent to Reid.[292]

3. 'An Essay on Quantity'

Reid's first publication, 'An Essay on Quantity', probably began to take shape once he had settled into the routine of his new life at New Machar following his controversial ordination as parish minister in May 1737. According to Dugald Stewart, Reid was preoccupied with 'a careful examination of the laws of external perception, and of the other principles which form the groundwork of human knowledge' while living in the manse. We have seen in Section 1, however, that Reid's investigation of the foundations of human knowledge in this period was by no means his sole or even his main intellectual pursuit.[293] Moreover, even though Reid's papers provide only a partial snapshot of his activities, the manuscripts related to 'An Essay on Quantity' suggest that instead of being focused on the science of the mind, as Stewart would have it, he devoted much of his free time at New Machar to the consideration of the range of problems related to the 'Essay'. Numerically speaking, these manuscripts are one of the largest groups to survive from the period, with the earliest of them apparently dating from the mid- to late 1730s. The last in the sequence, which is a fair copy whose wording is close but not identical to the published version of the 'Essay', probably dates from 1748. Some of these manuscripts are very heavily worked over, which implies that Reid spent a considerable amount of time writing and revising the 'Essay' before sending it to the Rev. Henry Miles in Tooting.[294]

[292] AUL, MS 2131/5/I/29. The demonstration rests on a variant of the axiom which has come to be known as 'Playfair's axiom', although Sir Thomas Heath points out that it was first stated by Proclus; see Playfair, *Elements of Geometry*, p. 7, and Euclid, *Thirteen Books of Euclid's Elements*, vol. I, p. 220. As we have seen above, p. lxxix, Dugald Stewart noted that Reid and Playfair conversed about mathematical matters in the summer of 1796. The evidence thus suggests that Reid and Playfair were periodically in contact with one another in the years leading up to Reid's death in 1796.

[293] Stewart, *Account*, p. 18. Stewart also claimed here that Reid's 'chief relaxations were gardening and botany' and he says nothing about Reid's interests in mathematics and natural philosophy.

[294] Henry Miles communicated the paper to the President of the Royal Society, Martin Folkes, who in turn had it read at a meeting of the Society held on 3 November 1748. Miles was an English Dissenting minister and an accomplished man of science. He was elected a Fellow of the Royal Society in 1743. He is said to have been awarded an honorary DD by one of the Aberdeen colleges in 1744, but he is not listed amongst the recipients of honorary degrees of either King's or Marischal

The orthography and paper size of what appears to be the earliest surviving draft, entitled 'Concerning the Object of Mathematicks' (pp. 32–3), resemble those of manuscripts dating from the mid-1730s.[295] It is therefore reasonable to assume that this draft is roughly contemporary with these manuscripts, although it is impossible to say with any precision when Reid initially set down his ideas. The main points of the published 'Essay' are already present in this preliminary sketch: the rejection of the definition of quantity as 'what ever is capable of More or less' (p. 32); the contention that 'Duration extension and Number Seem to be the primary and direct objects of the Mathematical Sciences' (p. 32); the distinction between proper and improper quantities (pp. 32–3); the assertion that 'Mathematical Evidence is an Evidence Sui Generis' (p. 33); the claim that 'Tastes Smells, heat cold beauty pleasure pain all of the affections and Appetites of the mind, Probability, Wisdom folly &c &c' cannot be assigned quantitative measures (p. 32); the dismissal of Francis Hutcheson's attempt to formulate a moral calculus (pp. 32–3); and the recognition that the dispute between the followers of Newton and Leibniz over the force of moving bodies could only be resolved once the contending parties agreed on a definition of the measure of that force (p. 33). But the format of the sketch differs from subsequent versions, insofar as Reid initially adopted an explicitly geometrical method of presentation and divided his text into a definition and four corollaries, whereas in later drafts he switched to the essay genre and eventually incorporated three corollaries within the body of the published 'Essay'.

The contents of 'Concerning the Object of Mathematicks' indicate that the themes central to 'An Essay on Quantity' were rooted in the mathematical, scientific and philosophical issues that preoccupied Reid in the mid-1730s. In this early formulation of his ideas, we can see more clearly than in later versions of the 'Essay' that his reflections on the nature of quantity were initially prompted as much by his work in mathematics as they were by the *vis viva* controversy or by Francis Hutcheson's 'Attempt to introduce a *Mathematical Calculation* in Subjects of *Morality*'.[296] As noted above, Reid's comments on fluxions in 'Concerning

College; Philip Furneaux, *A Sermon Occasioned by the Death of the Reverend Henry Miles, D.D. and F.R.S.* (1763), p. 34. We do not know why Reid sent a copy of the 'Essay' to Miles in the first instance. It may be that David Fordyce put Reid into contact with Miles, because Fordyce corresponded with Miles and other leading Dissenters. We have seen above, p. xxiii, that Reid had met Folkes in London in 1736.

[295] AUL, MS 2131/5/I/20. Compare 'Minutes of a Philosophical Club', dated 1736, and a manuscript that may be related to the proceedings of the Club dated 22 December 1736–25 January [1737], AUL, MSS 2131/6/I/17 and 7/V/6.

[296] The quoted phrase appears on one version of the title page of the first edition of Francis Hutcheson's *An Inquiry into the Original of Our Ideas of Beauty and Virtue*, published in 1725.

the Object of Mathematicks' indicate that Berkeley's attack on Newton's fluxional calculus in *The Analyst* may well have led Reid to engage in a rigorous analysis of the foundational concept of quantity in order to answer Berkeley's critique of the method of fluxions. His concluding remark in the manuscript on the definition of number (p. 33) also points to his concern with a second concept fundamental to mathematics, and his mention of infinite series links together his discussion of quantity with the sixteen 'Queries with Respect to Infinite Series' that follow on the recto and verso of the second folio of the manuscript.[297] Moreover, Reid's references in these 'Queries' to Newton's tracts 'Tractatus de quadratura curvarum' and 'De analysi per aequationes numero terminorum infinitas' strongly suggest that key ideas expressed in 'Concerning the Object of Mathematicks' were closely tied to his efforts to master the mathematical techniques contained in Newton's writings.[298] Hence it seems that, at this stage, his thinking was largely, but not entirely, driven by his engagement with Newton's mathematical legacy, and that the *vis viva* debate as well as Hutcheson's application of mathematics to morality played a less significant role in the genesis of his argument regarding the nature of quantity than the published 'Essay' would imply.

The grounding in Aristotelian logic that Reid acquired as a student at Marischal College, along with one key aspect of his philosophical enquiries in the 1730s, also surface in 'Concerning the Object of Mathematicks'. Although Reid's regent at Marischal, George Turnbull, was highly critical of Aristotelian scholasticism, Turnbull was undoubtedly obliged to teach the rudiments of Aristotle's *Organon*. It is therefore likely that Reid was introduced to Aristotle's *Categories* while he was Turnbull's student. By the 1730s he would thus have been familiar with Aristotle's influential discussion of quantity in the *Categories*, which, as 'Concerning the Object of Mathematicks' shows, structured his own conception of the scope of quantitative analysis. For example, his claim that 'Duration extension and Number Seem to be the primary and direct objects of the Mathematical Sciences' (p. 32) is based on Aristotle's classification of the different kinds of continuous and discrete quantity. And in distinguishing between 'proper' and 'improper' quantities (p. 33), he reformulates Aristotle's contention that number, extension, time and place 'are called quantities strictly', whereas other types of quantity which are measured by at least one of these basic quantities are only

[297] AUL, MS 2131/5/I/20, fol. 2r–v. Reid's 'Queries' have been omitted from my transcription below.

[298] English translations of these tracts along with extensive commentary later appeared in John Stewart's *Sir Isaac Newton's Two Treatises*. It is therefore reasonable to assume that Reid studied these two works in collaboration with his friend Stewart.

quantities 'derivatively'.[299] Furthermore, Turnbull's lectures on logic may have been the inspiration for Reid's lifelong interest in identifying the different kinds of evidence appropriate to the various sciences, which Reid said was 'one of the Most important parts of that Logic which is really usefull in Life'.[300] Building on Aristotle and various seventeenth-century thinkers, John Locke fashioned a response to scepticism that hinged on establishing 'the *Bounds* between Opinion and Knowledge' and on ascertaining 'the Grounds and Degrees of Belief, Opinion, and Assent' to propositions whose truth was merely probable.[301] Locke's categorisation of the varying levels of probability and certainty associated with the sciences in turn contributed to the reorientation of the study of logic in the early eighteenth century. Turnbull was a notable exponent of this new 'logic of ideas', later writing that

> the nature and degrees of moral, probable, or historical evidence, tho' left out of what is commonly called logic, or but very superficially treated of in it, is, if not the most essential part of a science that merits to be called the art of reasoning, at least a very useful science in itself…[302]

Given that Turnbull was teaching material from Locke's *Essay* in his logic lectures at Marischal, he probably introduced Reid to this element of Locke's philosophical legacy. Moreover, circumstantial evidence suggests that Reid may have followed up on what Turnbull had taught him about the different categories of evidence after he became librarian at Marischal College in 1733.[303] It is thus plausible to see Reid's fourth corollary – which states that 'Mathematical Evidence is an Evidence Sui Generis not competent to any Proposition that does not express a relation of Quantities which may be Measured by lines or Numbers'

[299] Aristotle, *Categories*, 4b.20–5b.10, in Aristotle, *The Complete Works of Aristotle: The Revised Oxford Translation* (1984). For an undated set of notes from the *Categories*, which *inter alia* records Aristotle's distinctions between discrete and continuous, and proper and improper quantities, see AUL, MS 2131/3/II/9, esp. fol. 1r.

[300] See above, p. xcvi, note 288, as well as Thomas Reid, 'A System of Logic, Taught at Aberdeen 1763', Edinburgh University Library, MS Dk.3.2, pp. 38–58; Reid, *On Logic, Rhetoric and the Fine Arts*, pp. 185–7; and Reid, *Intellectual Powers*, pp. 228–33, 467–512, 555–62.

[301] John Locke, *An Essay concerning Human Understanding* (1690), pp. 43–4 (I.i.2 and I.i.3). For the classic study of Locke's treatment of probable belief, see Henry G. Van Leeuwen, *The Problem of Certainty in English Thought, 1630–1690*, ch. 5, along with Lorraine Daston, *Classical Probability in the Enlightenment*, esp. ch. 4, and Barbara J. Shapiro, *Probability and Certainty in Seventeenth-Century England: A Study of the Relationships between Natural Science, Religion, History, Law and Literature*, ch. 2.

[302] George Turnbull, *Observations upon Liberal Education, in All Its Branches*, p. 349. On the rise of the logic of ideas, see especially Wilbur Samuel Howell, *Eighteenth-Century British Logic and Rhetoric*, ch. 5.

[303] Wood, *Aberdeen Enlightenment*, p. 45; Campbell Fraser, *Thomas Reid*, p. 27.

(p. 33) – as emerging out of this Lockean context. And the fourth corollary further suggests that in the late 1730s he was giving thought to the distinctive characteristics of mathematical evidence as well as the other forms of evidence found across the natural and human sciences.

In what is probably the next written version of his material, Reid reframed his argument by giving the manuscript a revised title, 'Essay Concerning the Object of Mathematicks occasioned by reading a piece of Mr Hutchesons wherein Virtue is Measured by simple & Compound Ratios'. What was previously a series of reflections cast in geometrical form is now an 'essay', which suggests that, after writing the earlier paper, Reid decided to refine and to elaborate on his analysis of quantity.[304] The altered title and new format correspond to a change in focus, for in this second manuscript he explicitly deploys his definition of quantity in terms of the Aristotelian distinction between proper and improper quantities as a conceptual antidote to those who 'apply Mathematical Reasoning to subjects that do not admit of it' (p. 34). That he chose to target the moral calculus found in the first three editions of Francis Hutcheson's *An Inquiry into the Original of Our Ideas of Beauty and Virtue* speaks to the fact that during the 1730s Reid immersed himself in Hutcheson's moral philosophy, which served as a reference point for his own theorising about human nature and morality.[305] Reid was not, however, a slavish follower of Hutcheson, as the argument of the drafts and the published 'Essay' demonstrate. Moreover, in targeting Hutcheson's moral calculus, Reid was also questioning the validity of applying mathematical reasoning to morals more generally and, in doing so, rejecting an important strand of scientism in the Enlightenment.[306]

Inspired by the mathematical methods developed by natural philosophers during the Scientific Revolution of the sixteenth and seventeenth centuries, in 1690 the Halle professor Christian Thomasius formulated quantitative measures of the passions in order to investigate the varieties of human character and behaviour.[307] Shortly thereafter, in order to counter moral scepticism, the third

[304] The change of title came in the process of writing or was made as an afterthought. The title of this second manuscript was originally the same as that of the earlier sketch, which Reid then emended. The size of the letters in the word 'Essay' is smaller than those in the words 'Concerning the Object of Mathematicks', which suggests that Reid added the word 'Essay'. The remainder of the title, 'occasioned by reading a piece of Mr Hutchesons wherein Virtue is Measured by simple & Compound Ratios', is an interlinear addition; see textual notes 34/1 and 34/2–3 below (p. 233).

[305] Reid's engagement with Hutcheson's writings is detailed in my forthcoming *The Life of Thomas Reid*.

[306] See Paul Wood, 'Science, Philosophy and the Mind', esp. pp. 814–17.

[307] Robert J. Richards, 'Christian Wolff's Prolegomena to Empirical and Rational Psychology: Translation and Commentary', p. 229, note 19.

earl of Shaftesbury, in his *Characteristicks of Men, Manners, Opinions, Times* (1711), invoked the certainty of mathematics when he described his account of 'our Obligation to VIRTUE' as a '*Moral Arithmetick*' that 'may be said to have an Evidence as great as that which is found in Numbers, or Mathematicks'.[308] While Shaftesbury did not specify quantitative measures of virtue or vice in his *Characteristicks*, the young Colin Maclaurin used Newton's method of fluxions to calculate the attraction of the human mind to good and evil in an essay written in 1714, 'De viribus mentium bonipetis'. Hutcheson may have known about this essay, since both he and Maclaurin overlapped as students at the University of Glasgow.[309] Yet Reid was probably unaware of Maclaurin's sophisticated moral calculus, even though he may have been Maclaurin's pupil at Marischal College. On the other hand, it is conceivable that he was familiar with the rudimentary application of mathematics to morals in *An Enquiry into the Original of Moral Virtue* published in 1733 by the controversial St Andrews professor Archibald Campbell, who had been a contemporary of both Maclaurin and Hutcheson at Glasgow.[310] In criticising Hutcheson, Reid was thus taking issue with a broader intellectual phenomenon of the Enlightenment, namely the attempt to quantify moral reasoning in order to transform the study of morality into a science akin to natural philosophy. For Reid, the moral order, unlike the material world, was not susceptible to quantification and, furthermore, the evidence and ground of mathematics and natural philosophy were categorically different from those of morality. Moral philosophy was therefore methodologically distinct from natural philosophy and, in insisting on this point, Reid implicitly challenged the view espoused by his teacher George Turnbull that moral and natural philosophy shared a common method.[311]

Reid's newly retitled 'Essay Concerning the Object of Mathematicks' was also explicitly framed as a defence of Newtonian natural philosophy, with the Aristotelian distinction between proper and improper quantity playing a pivotal

[308] Anthony Ashley Cooper, third earl of Shaftesbury, 'An Inquiry concerning Virtue and Merit', in *Characteristicks of Men, Manners, Opinions, Times* (1711), vol. II, p, 99. Reid would almost certainly have been familiar with this passage.

[309] On Maclaurin's essay and the possible connection with Hutcheson see Grabiner, 'Maclaurin and Newton', pp. 145–6.

[310] Archibald Campbell, *An Enquiry into the Original of Moral Virtue* (1733), pp. 272–82. Campbell studied at Glasgow from *circa* 1712 to 1718.

[311] See, for example, George Turnbull's 1723 graduation thesis, 'On the Association of Natural Science with Moral Philosophy', in Turnbull, *Education for Life*, pp. 49–50. Reid later reaffirmed the methodological differences between moral and natural philosophy in the context of his dispute with Joseph Priestley; see Reid, 'Some Observations on the Modern System of Materialism', in Reid, *Animate Creation*, pp. 185–6.

role in his intervention in the *vis viva* controversy. Reid's interpretation of the exchanges between the rival Newtonian and Leibnizian camps is both original and insightful because no other contributor to the dispute traced the disagreement over the proper measure of the force of a moving body back to the conceptual foundations of the mathematical sciences. Of all the interlocutors in the debate, only Reid argued that the disagreement was rooted in the nature of improper quantities, such as the quantity of motion or the various forces postulated by Newton, which could be subjected to mathematical analysis only once a measure in terms of 'Space Duration or Number' was defined (p. 34). Like Reid, Willem Jacob 's Gravesande and J. T. Desaguliers claimed that the dispute revolved around competing definitions and measures of the force of moving bodies, but neither of these friendly adversaries argued that it was possible to formulate different definitions or measures, due to the distinctive nature of quantity. And while 's Gravesande and Desaguliers indicated that the dispute had reached an impasse because the choice between definitions was an arbitrary one, Reid contended that Newton's measure of mv was to be preferred to Leibniz's measure of mv^2, for two reasons. First, he said that 'the Simple ratio of the velocity [mv] will do as well for the Measure of the force' and, secondly, he held that whereas Leibniz's measure led to irresolvable 'Paradoxes' Newton's measure was consistent 'with the common Notion of Force' (p. 36). Moreover, Reid probably came to the conclusion that the controversy hinged on a definition independently of 's Gravesande and Desaguliers. It is unlikely that he knew of the paper published by 's Gravesande in 1729 in which the Dutchman suggested that the dispute was a verbal one. And it appears that while he was working on the successive drafts of the 'Essay' he was unaware of Desaguliers' contention that the *vis viva* controversy 'was only a Dispute about Words; the contending Parties meaning different Things by the Word FORCE'.[312] In this manuscript, therefore, we see an original and philosophically acute contribution to the debate between the followers of Newton and Leibniz in the making.

The manuscript of the 'Essay Concerning the Object of Mathematicks' shows that Reid struggled to express his ideas to his satisfaction, insofar as the draft contains numerous insertions and deletions, as well as whole paragraphs crossed out.[313] This is also the case with a further version of the 'Essay' entitled, 'An Essay concerning the Object of Mathematicks occasioned by reading a piece wherein Virtue is Measured by Simple and Compound Ratios', which has not

[312] On 's Gravesande, see Hankins, 'Eighteenth-Century Attempts to Resolve the *Vis viva* Controversy', pp. 287–8; Desaguliers, *A Course of Experimental Philosophy*, vol. II, p. vi.

[313] See the textual notes below, pp. 233–5.

been transcribed below because of the highly confused (and perhaps incomplete) state of the text.[314] This manuscript dates from *circa* July 1748, for on the recto of the second folio Reid has recorded his observation taken at the manse of New Machar of an eclipse of the Sun that occurred on 14 July that year (see p. 63). Like the previous undated draft of the 'Essay', this version is heavily worked over, with a number of lengthy passages cancelled. Moreover, the individual paragraphs are out of sequence and Reid has used a series of lower-case letters to indicate how to assemble the complete text he envisaged from the disjointed segments of prose found in the manuscript. If we reconstruct the text following his scheme, it emerges that this version is a revised and expanded variant of the previous draft.[315] Reid was thus still struggling with the wording and contents of his text as late as mid-July 1748, which implies that the fair copy of the 'Essay', which dates from *circa* October 1748 (pp. 38–50), was the product of a brief period of intensive work refining both his ideas and his prose.

The fair-copy and published versions of the 'Essay' present a succinct analysis of the defining characteristics of quantity and explore the conceptual foundations not only of mathematics but also of the mathematical approach to the study of nature championed in Newton's *Principia.* Whereas 'Concerning the Object of Mathematicks' was narrowly focused on foundational problems specific to mathematics, the 'Essay' features a broader discussion of the implications of the nature of quantity for our understanding of the scope and logical structure of the mathematical sciences which registers Reid's competences as both a mathematician and a Newtonian natural philosopher. As in earlier drafts, Reid adopts the common definition of quantity as that which can be measured, but makes a new point that it follows from this definition that whatever has quantity has parts.[316] These parts 'bear Proportion to one another, and to the Whole', which meant that quantities can be added, subtracted, multiplied and divided (pp. 39, 51). Consequently, mathematical quantities can 'bear any Proportion to another Quantity

[314] AUL, MS 2131/2/I/1; there are, in fact, two slightly different titles contained in this manuscript. One of them occurs in a deleted section of the draft.

[315] It may be that Reid intended to incorporate material from his previous draft into his reconstituted text. There is no paragraph marked 'h' in AUL, MS 2131/2/I/1 corresponding to the insertion point marked with a small letter 'h' found at the foot of fol. 3r. There is, however, a paragraph marked 'h' on fol. 2v of AUL, MS 2131/5/I/22; see p. 235, textual note 36/17.

[316] On the most common definition of quantity in the eighteenth century, see Pycior, *Symbols, Impossible Numbers and Geometric Entanglements*, p. 4. In asserting that quantity is made up of parts, Reid disagreed not only with Aristotle but also with, *inter alia*, Isaac Barrow, who argued for the primacy of geometry over arithmetic and algebra on the ground that quantity is continuous; see Aristotle, *Categories*, 4b.20–5a.37, and Barrow, *The Usefulness of Mathematical Learning Explained and Demonstrated*, pp. 20, 30.

of the same Kind, that one Line or Number can bear to another' and it was this feature that he regarded as the essential property of quantity (pp. 39, 51). It was this characteristic, he said, which was the basis of the 'Accuracy and Certainty' of mathematics insofar as quantitative concepts can be given rigorous definitions, and he argued that geometry most clearly illustrated the distinctiveness of quantitative reasoning (pp. 39, 51). Yet Reid did not maintain that the greater clarity of geometrical reasoning implied that geometry took precedence over arithmetic and algebra, as a number of prominent mathematicians such as Isaac Barrow had done. Rather, in seeing quantity as discrete and in characterising quantity in terms of arithmetical relationships, he was implicitly following John Wallis and others in suggesting that arithmetic and algebra, not geometry, were fundamental to the mathematical sciences.[317]

Even though Reid disagreed with Aristotle's view that there are two forms of quantity, namely discrete and continuous, in the final versions of the 'Essay' he nevertheless invoked the Stagyrite's distinction between proper and improper quantities, which he reformulated at length. In the *Categories* Aristotle had stated that 'number, … language, … lines, surfaces, bodies, … time and place' were all proper quantities in terms of which improper quantities were defined.[318] By contrast, Reid affirmed that 'Extension, Duration, Number & Proportion' were proper quantities which could all be 'measured by [their] own kind', whereas improper quantities such as velocity 'cannot be measured by [their] own Kind' and require 'a Measure in some Proper Quantity that is related to it' (pp. 39–40, 51–2). Thus, velocity had to be defined in terms of the space a body moves through in a given time so that we could reason about it mathematically. And the same held true of the other improper quantities enumerated by Reid that were fundamental to Newtonian natural philosophy, namely 'the Quantity of Motion, Density, Elasticity, the Vis Insita & Impressa, the various kinds of centripetal Forces & [the] different Orders of Fluxions' (pp. 40, 52). Consequently the application of mathematics to the investigation of nature required natural philosophers to specify appropriate measures for improper quantities, and he maintained that Newton had shown how best to do so in the Definitions which prefaced Book I of the *Principia* (pp. 40, 52–3).

Reid, however, added a crucial rider to his account of improper quantity that had important implications for his critique of moralists such as Francis Hutcheson, who attempted to apply mathematical reasoning to moral subjects. Rather than

[317] On the primacy of algebra 'in the order of Nature', see above, p. xl, and, on Wallis, see Pycior, *Symbols, Impossible Numbers and Geometric Entanglements*, pp. 121–5.

[318] Aristotle, *Categories*, 4b.20–5b.10.

allow that quantitative measures can be defined for virtually any property, he now stated that genuine improper quantities had to satisfy two conditions. First, he stipulated that they must 'admit of Degrees, of greater and less'. Secondly, he insisted they must be 'associated with or Related to something that hath proper Quantity so as that when one is increased the other is increased, when one is diminished the other is also diminished; [and] every Degree of the one must have a determinate Magnitude or Quantity of the other, corresponding to it' (pp. 41, 53). Consequently, while he acknowledged that 'there are many things capable of more and less' such as tastes, smells, beauty, pleasure and 'All the Affections and Appetites of the Mind', he denied that they were measurable in terms of proper quantities and he appealed to the evidence of language to illustrate his point. Citing the examples of pleasure and pain, he observed that if it were truly possible to quantify them, ''tis not to be doubted, but we should have had as many Names for their various Degrees, as we have Measures for Length or Capacity'. Furthermore, he noted that there was no commonly recognised standard by which pains or pleasures could be measured, unlike established measures of distance (pp. 38, 50–1). This meant that, for him, Hutcheson's moral calculus was a philosophical *jeux d'esprit*, which, he insisted, only rang 'Changes upon Words, and [made] a Shew of Mathematical Reasoning, without advancing one Step in real Knowledge' (pp. 43, 55).

The appeal to language also featured in Reid's proposed resolution of the *vis viva* dispute, which he had reformulated by the time he came to write the fair copy of the 'Essay'. According to Reid, members of both contending parties had confused the issues at stake because they had mistaken a definition, namely that of the force of a moving body, for a proposition susceptible to mathematical demonstration or experimental proof. Because the force of a moving body was an improper quantity, it required a definition that specified a measure, which is what Newton and Leibniz had done in using mv and mv^2 respectively. But Reid did not think that the two definitions were equally acceptable, for he contended that it was Newton's definition, rather than Leibniz's, that ought to be preferred, on the grounds that it was 'not onely clear and Simple, but [also] agree[d] best with the Use of the Word Force in common Language' (pp. 44, 56). Reid's case for the superiority of Newton's definition was thus now an entirely linguistic one, whereas he had earlier charged that Leibniz's measure of mv^2 led to 'Paradoxes' and 'Hypotheses' (p. 36) and hence ought to be rejected.[319]

[319] Compare AUL, MS 2131/2/I/1, fol. 3v, where Reid likewise complains of the 'Strange Paradoxes' and 'as Strange Hypotheses' implied by Leibniz's measure. Reid's arguments based on linguistic usage may have owed something to his teacher George Turnbull, who likewise appealed

As well as differences in accidentals and numerous variations in wording, Reid's fair copy of the 'Essay' differs from the published version in that it contains significant additional material elaborating on what eventually appeared in print. First, there is an insertion written on a separate loose sheet that serves as a coda to the final paragraph, in which Reid describes a disagreement in the science of harmonics over the correct mathematical ratio for expressing an octave that parallels the dispute over *vis viva* (pp. 47–8). Secondly, there is a passage of just over a folio in length in which he maintained that a truce ought to be declared in the war of words between the Newtonian and Leibnizian camps. To heal the 'Schism [that] hath been made in the Mathematicall World', he proposed that a 'Treaty' should be drawn up, which would include six articles that he thought would serve to lessen the hostilities between the combatants (p. 49). It is, however, difficult to know how to read this passage. He may have meant it sincerely or the passage may have been a somewhat laboured exercise in Scriblerian satire. If the passage was in fact a manifestation of Reid's pawky sense of humour, it appears that the Fellows of the Royal Society failed to see the joke because this lengthy conclusion of Reid's 'Reflections' on the *vis viva* dispute was omitted from the published 'Essay'.[320] Significantly, the manuscript of the fair copy also contains an undated postscript giving Reid's retrospective judgement on his first published work. In that postscript, written in the form of his handwriting characteristic of the last decade of his life, he admits that when 'this Essay was wrote in 1748' his lack of familiarity with some of the contributions to the *vis viva* controversy led him to the mistaken conclusion that 'the British Mathematicians onely, opposed the Notion of Leibniz [mv^2], and that all the foreign Mathematicians adopted it' (p. 50). Reid's postscript, however, raises the question of how many of the polemics generated by the dispute he had actually read over the course of his career, because his comment on the nationalities affiliated with the two camps of disputants paraphrases a passage in the work by Desaguliers that he cites.[321]

That said, Reid's interest in the *vis viva* debate extended beyond the writing of the 'Essay'. When Leibniz first advanced mv^2 as the measure of the force of moving bodies in 1686, he framed his argument in terms of the analysis of the motions of falling bodies. But in the opening decades of the eighteenth century his followers Jacob Hermann, Giovanni Poleni, Willem Jacob 's Gravesande, Pieter van Musschenbroek and Jean Bernoulli expanded the experimental basis for Leibniz's doctrine, with 's Gravesande, van Musschenbroek and Bernoulli citing

to the common use of language; see George Turnbull, *The Principles of Moral and Christian Philosophy* (1740), vol. II, p. 671.

320 An equivalent passage appears in AUL, MS 2131/2/I/1, fol. 2v, but not in earlier drafts.

321 Desaguliers, *A Course of Experimental Philosophy*, vol. II, pp. v, 50.

experiments on springs which they asserted provided empirical confirmation that mv^2 was the correct measure of this force. Moreover, Poleni performed experiments in which he dropped brass balls from different heights onto either wax or clay and then used the depth of the impressions made by the balls to calculate the force of the impact. According to Poleni, his calculations demonstrated that Leibniz's theory was correct.[322] The experimental results of the Leibnizians were in turn carefully scrutinised by Newton's followers, who came up with variations on the experiments using springs and falling bodies to counter the empirical findings of the supporters of *vis viva*, most notably in papers by James Jurin and Henry Pemberton.[323] A pair of undated manuscripts register Reid's knowledge of the experiments appealed to by the opposing camps in the dispute. One is a brief fragment which deals with establishing the accelerative force of springs in order to ascertain the forces involved in projectile motion and, significantly, to interpret the 'Pitts made in Clay by falling bodies'. The other, which appears to be incomplete, describes an experiment designed to measure the forces generated by a spring in communicating motion to two bodies of unequal mass. Implicitly, Reid distinguished between his estimate of the force and those of the Leibnizians. In a corollary, he writes: 'If we should suppose with the followers of Leibnitz that the Spring gives equal Quantities of force to both the Bodies it would necessarily follow that the Force of Bodies was as the Square of the Velocity multiplied into the Mass'. Unfortunately, the text breaks off at this point, and there is no hint as to how his argument was going to proceed.[324]

A third manuscript, which may have been written in the 1740s, consists of two separate entries and documents Reid's reflections on the conceptual and theoretical issues at stake in the *vis viva* controversy. The first entry echoes the 'Essay' in suggesting that empirical considerations cannot decide disagreements over different measures of force, while the second sketches an ingenious translation

322 Poleni's account of these experiments is translated in J. T. Desaguliers, 'Animadversions Upon Some Experiments Relating to the Force of Moving Bodies; with Two New Experiments on the Same Subject', pp. 285–6. Experiments akin to those performed by Poleni were also made by 's Gravesande; on their experiments, see Iltis, 'The Leibnizian–Newtonian Debates', pp. 355–61.

323 James Jurin, 'An Inquiry into the Measure of the Force of Bodies in Motion: With a Proposal of an *Experimentum crucis*, to Decide the Controversy About It', which describes experiments with springs; and Henry Pemberton, 'A Letter to Dr Mead, Coll. Med. Lond. & Soc. Reg. S. concerning an Experiment, Whereby It Has Been Attempted to Shew the Falsity of the Common Opinion, in Relation to the Force of Bodies in Motion', which attacks Poleni and includes an anonymous postscript written by Newton.

324 AUL, MSS 2131/6/V/33c and 33d. Larry Laudan implies that these manuscripts were written in the 1740s but provides no evidence for this; Laudan, 'The *Vis viva* Controversy, a Post-mortem', pp. 138–9. They may date from the 1740s, but they may also have been written in the 1750s, when Reid was working on his lectures on natural philosophy.

of Newton's laws of motion and their corollaries into 'the Leibnitzian Language [which] makes the Square of the Velocity the Measure of Force'.[325] Using wording similar to that found in the 'Essay', he affirms that 'Sir Isaac Newton seems to have been extreamly happy in his Definitions or Mathematical Conceptions as he Somewhere calls them'. Furthermore, he says that Newton's measures of the various forces defined at the beginning of the *Principia* 'serve admirably to [express] Concisely and Elegantly both the Laws of Motion discovered by those Mathematicians that went before Him Galileo Wrenn. Wallis Hugens & Mariot. and likewise those more Intricate ones discovered by himself'.[326] He then proceeds to reformulate Newton's laws in Leibnizian terms but pauses at the third corollary, which states: 'The quantity of motion, which is determined by adding the motions made in one direction and subtracting the motions made in the opposite direction, is not changed by the action of bodies on one another'.[327] According to Reid, this corollary cannot be translated into 'the Leibnitzian Language':

> This third Corollary cannot as far as I see be expressed without joyning the Ratios of the quantity of Matter & the Velocity simply. So that if we do not make these Ratios joyned to be [a] Measure of the Quantity of Motion we must even Make a fictitious Quantity of them and give them a Name[.] for the Laws of Motion plainly require that they should be joined together[.][328]

For him, this point was decisive because 'the Laws of Motion laid down as Axioms by Newton … undoubtedly hold in Experience'.[329] When writing this manuscript at least, Reid believed that Leibniz's measure of the force of moving bodies was incompatible with Newton's laws of motion and was thus to be rejected because it was inconsistent with the experimental data confirming the truth of those laws.[330] In this manuscript he also gives the clearest indication of the 'Paradoxes' Leibniz's measure of mv^2 led to, for he notes that Leibniz's account of the collision of bodies implied that 'Unequal Forces when directly opposed may

[325] AUL, MS 2131/5/II/18, fol. 1r. The handwriting in this manuscript is similar to that found in other manuscripts dating from the 1740s.

[326] AUL, MS 2131/5/II/8, fol. 1r; compare Newton, *The Principia*, pp. 416–30. Newton refers to all of the figures mentioned by Reid.

[327] Newton, *The Principia*, p. 420.

[328] AUL, MS 2131/5/II/8, fol. 1v.

[329] AUL, MS 2131/5/II/8, fol. 1r.

[330] On fols 1v–2r of AUL, MS 2131/5/II/8, Reid goes on to discuss the collision of elastic and inelastic bodies. The analysis of such collisions was hotly disputed by the rival Newtonian and Leibnizian camps.

mutually destroy one another that of Equal Forces one may Overcome the Other & that a less Force may overcome a greater'.[331]

Following Reid's move to King's College, Aberdeen in 1751, his interest in the *vis viva* debate apparently waned. He said little or nothing about the dispute in his natural philosophy lectures, and we have seen in Section 1 that when he consulted Jacob Hermann's *Phoronomia*, he seems to have read the work for the analysis of centres of gravity that it contained rather than Hermann's experimental defence of Leibniz's concept of *vis viva*.[332] He did, however, briefly revisit the *vis viva* controversy after he retired from teaching in Glasgow in the late spring of 1780.[333] Reading notes taken from van Musschenbroek's *Elements of Natural Philosophy* in 1781 are entirely devoted to recording the Dutchman's experimental demonstrations that were designed to show that mv^2 was the correct measure of the force of moving bodies.[334] Reid did not comment on van Musschenbroek's interpretation of his experiments, which might be taken as a sign that his notes were taken simply for information regarding van Musschenbroek's case for *vis viva*. And if this was Reid's rationale, it implies that his comment that 'When [the] Essay was wrote in 1748 I knew … little of the History of the Controversy about the force of moving Bodies' (p. 50) was more than just a self-deprecating remark.

4. Astronomy

The starting point for Reid's work in the fields of observational, physical and mathematical astronomy was Newton's system of the world. Newton's legacy in the science of astronomy was exceptionally rich, for astronomers in the Enlightenment not only grappled with the conceptual and mathematical details of his theory of gravitation and his celestial dynamics, but also with the observational and theoretical complexities of his theory of the Moon and the basics of his theory

[331] AUL, MS 2131/5/II/8, fol. 1v.

[332] In his lectures Reid is recorded as having stated without further elaboration that 'the quantity of motion in a Body is measured by the quantity of matter & velocity conjunctly'; Anon., 'Natural Philosophy 1758', p. 16.

[333] Although Reid apparently lost interest in the *vis viva* dispute in the 1750s, he did refer to it in his Glasgow lectures on the culture of the mind when he discussed what he called 'vicious' definitions. This type of definition, he said, was one in which 'the Definition is borrowed from some hypothesis about the thing defined' and he cited 'Disputes about the force of Moving Bodies[, and] about first and last Ratios of Variable Quantities' as examples; see Reid, *On Logic, Rhetoric and the Fine Arts*, p. 163.

[334] AUL, MS 2131/7/II/12; compare van Musschenbroek, *Elements of Natural Philosophy*, vol. I, pp. 76–88.

of comets. Moreover, Newton's ideas were quickly taken up in Britain by established astronomers such as Edmond Halley, as well as by Newton's early disciples David Gregory and William Whiston, who produced textbooks of astronomy that presented the science as being founded upon the principles enunciated in Books I and III of the *Principia.* Reid's introduction to the study of astronomy occurred at the point when Newtonian natural philosophy became the basis for the teaching of the sciences in the Scottish universities. At Marischal College, Newton's theories were given a guarded reception in the 1690s but by 1710 the regents were largely converts to Newtonianism. And with the appointment of a new cohort of regents in the wake of the Jacobite rebellion of 1715, Newton's theories of light and colour and of universal gravitation acquired hegemonic status in lectures on natural philosophy. The election of Colin Maclaurin to the Marischal Chair of Mathematics in 1717 effectively turned Aberdeen into the most northerly outpost of the Newtonian camp in Britain. If Maclaurin did teach Reid mathematics, he may well have introduced his young student to the elements of mathematical astronomy and hence to the basics of Newton's celestial dynamics. Reid's regent George Turnbull undoubtedly provided his pupils with a grounding in the elements of Newtonian physical astronomy, given that Turnbull affirmed in his 1723 graduation thesis that Newton had 'elaborated an Astronomy which was complete in every respect'.[335]

We have seen in Section 1 that by July 1729 Reid had scrutinised David Gregory's *The Elements of Astronomy*. In particular, he had apparently paid close attention to Gregory's exposition of Newton's analysis of planetary motion and to a problem that was to figure prominently in Reid's scientific researches in the ensuing decades, namely the question of the shape of the Earth. We have also seen that a set of reading notes dating from October 1729 reveal that he was working through the mathematical complexities of the *Principia.* As well as illustrating his prodigious gifts as a mathematician, these notes demonstrate that while he was a divinity student, he was already grappling with the technicalities of Newton's celestial dynamics, along with topics covered in Book III, such as the motions of comets, the theory of the Moon and the precession of the equinoxes.[336] A third set of notes, written in an early form of Reid's handwriting, are taken from William Whiston's *Prælectiones astronomicæ* (1707). These notes document not only his critical response to the second Newtonian textbook in the field (the first

[335] Christine M. Shepherd, 'Newtonianism in Scottish Universities in the Seventeenth Century', p. 78; Turnbull, 'On the Association of Natural Science with Moral Philosophy', p. 50.

[336] AUL, MS 2131/7/III/15. Although Reid was using a copy of the first edition of the *Principia* in taking his notes, he indicates in this manuscript that he was aware of textual changes in subsequent editions of the work; see fols 2r, 3v.

being David Gregory's Latin edition of the *Elements*, published in 1702) but also Reid's attempt to develop competence in the use of astronomical tables, which any aspiring astronomer had to acquire (pp. 60–1). Astronomy, and especially Newton's system of the world, thus appear to have been the subjects to which he was initially attracted while he was a student at Marischal College.

During the 1730s and 1740s, Reid honed his observational skills. In 1737 it may be that he collaborated with his friend John Stewart and others in recording the observations of an annular eclipse of the Sun; their data were subsequently transmitted by Stewart to Colin Maclaurin in Edinburgh and later published in the *Philosophical Transactions*.[337] Seven years later, Reid tracked the Great Comet of 1744 (C/1743 X1), which was first noted in the skies over Europe in December 1743 and which was still visible in March 1744. It is unclear, however, whether he observed the comet using the naked eye or with a telescope, and whether he viewed the comet purely out of personal interest or watched it in conjunction with Stewart and other astronomers.[338] We do know that Reid and Stewart worked together in July 1748 when a further annular eclipse of the Sun was observed by a group of Scottish men of science led by Alexander Monro *primus* in Edinburgh. Stewart forwarded Reid's data on this solar eclipse to Monro, who then sent a register of the observations made in Scotland to the Royal Society in London.[339] It was probably also in this period that he first encountered James Bradley's pioneering paper of 1728 on the apparent parallax of the fixed stars caused by the

[337] Maclaurin was instrumental in obtaining the gift of a telescope to John Stewart from the nobleman and man of science James Douglas in 1732; Colin Maclaurin to George Graham, 15 February 1731/2, in Maclaurin, *The Collected Letters*, pp. 243–4. The gifting of the telescope implies that by 1732 at the latest Stewart was a member of Maclaurin's circle of contacts in Scotland and that Maclaurin wanted to ensure that Stewart would be able to participate in the collaborative observational projects that he planned to orchestrate. For the observations of the 1737 eclipse of the Sun, see Colin Maclaurin, 'An Observation of the Eclipse of the Sun, on Feb. 18. 1737. Made at Edinburgh', p. 191, where Maclaurin mentions the observations of Stewart and 'Several Gentlemen who live on the Coast Northwards from *Aberdeen*'. On Maclaurin's network of astronomers, see Roger L. Emerson, 'The Philosophical Society of Edinburgh, 1737–1747', pp. 160–1. The broader context for the interest in solar eclipses is discussed in Alice N. Walters, 'Ephemeral Events: English Broadsides of Early Eighteenth-Century Solar Eclipses'.

[338] AUL, MS 2131/6/III/8, fol. 2r, transcribed in Reid, *Inquiry*, p. 321. On the Great Comet, see David Seargent, *The Greatest Comets in History: Broom Stars and Celestial Scimitars*, pp. 116–21.

[339] A record of Reid's observations is found in AUL, MS 2131/2/I/1, fol. 2r; see the transcription below, p. 63. See also James Douglas, fourteenth earl of Morton, Pierre Charles Le Monnier and James Short, 'An Eclipse of the Sun, July 14. 1748', p. 593. On the network of Scottish astronomers involved in the observation of the eclipse of the Sun in 1748, see Roger L. Emerson, 'The Philosophical Society of Edinburgh, 1748–1768', pp. 137–8.

aberration of light, which served as the basis for an undated calculation by Reid of the time required for light to traverse the diameter of the Earth's orbit (p. 63).[340]

It was the controversy over the shape of the Earth, however, which served as the primary focus for Reid's early work in astronomy. The prominent French astronomer Jacques Cassini sparked the dispute in 1718, when he argued before the Académie royale des sciences in Paris that his reformulation of Descartes' theory of vortices implied that the Earth is slightly distended at the poles.[341] Newton and his disciples, on the other hand, insisted that the Earth's daily rotation meant that it is slightly flattened at the poles and hence is an oblate spheroid. Empirical confirmation of the Newtonian position was announced in Pierre-Louis Moreau de Maupertuis' *La figure de la Terre* (1738), which outlined the findings of the team of academicians dispatched by Louis XV to Lapland to ascertain the length of a degree of latitude near the Arctic Circle.[342] In November 1739 Reid made extensive excerpts from, as well as critical comments on, the English translation of Maupertuis' work that appeared shortly after the publication of the French text.[343] Around the time that he was studying Maupertuis' account of the French expedition to Lapland, he and Stewart were jointly attempting to calculate the shape of the Earth based on the length of its semi-axis and its semi-diameter at the equator. Evidence for their collaboration survives in two documents. The first is a brief letter Reid sent to Stewart in which he provided 'a Demonstration of that beautifull Property of the Elipse upon which your Solution of the Problem is founded'. The second is a related manuscript transcribed below, in which he summarised the mathematical basis for his estimate of the length of the Earth's semi-axis and of its equatorial semi-diameter (p. 62).[344] Moreover, his continuing engagement with the topic is documented in two sets of reading notes. One set consists of undated excerpts from David Gregory's *A Treatise of Practical Geometry* (1745), while the other is a highly detailed and lengthy set of notes taken in January 1751 from the record of the data collected by the French geodetic expedition to Peru published in Pierre Bouguer's *La figure de la Terre* (1749).[345]

[340] Bradley, 'An Account of a New Discovered Motion of the Fix'd Stars'.

[341] On the debate over the figure of the Earth, see especially John L. Greenberg, *The Problem of the Earth's Shape from Newton to Clairaut: The Rise of Mathematical Science in Eighteenth-Century Paris and the Fall of 'Normal' Science.*

[342] On Maupertuis, see especially Mary Terrall, *The Man Who Flattened the Earth: Maupertuis and the Sciences in the Enlightenment*, chs 4 and 5, and Rob Iliffe, '"Aplatisseur du monde et de Cassini": Maupertuis, Precision Measurement and the Shape of the Earth in the 1730s'.

[343] AUL, MS 2131/3/I/2.

[344] Reid to [John Stewart], [1739], in Reid, *Correspondence*, p. 5.

[345] AUL, MS 2131/3/I/12, fol. 1v; compare Gregory, *A Treatise of Practical Geometry*, pp. 42–5. For the notes from Bouguer, see AUL, MS 2131/3/I/7. It appears that after he read Bouguer's book,

The work of Bouguer's team of academicians complemented that carried out earlier by Maupertuis' *équipe*. The group sent to Peru measured a degree of latitude near the equator and, in doing so, provided further corroboration for Newton's claim that the Earth is an oblate spheroid. Reid's detailed notes illustrate the care with which he studied Bouguer's procedures and data, and speak to his recognition of the enormous practical difficulties involved in obtaining accurate measurements through the use of sophisticated surveying techniques and precision scientific instruments. Furthermore, Reid later added to this set of notes information relevant to the determination of the figure of the Earth that was included in the summary of the observations made from 1751 to 1753 at the Cape of Good Hope by the French astronomer Nicolas-Louis de La Caille that appeared in the *Gentleman's Magazine* for November 1755.[346]

James Bradley's mapping of the heavens continued to engage Reid's attention in the 1740s and 1750s. In 1739, Bradley's study of the aberration of the fixed stars was taken a step further by his associate Dr John Bevis, a physician of independent means whose real passion was for observational astronomy. Bevis had built his own observatory in Stoke Newington the previous year, and he was also part of a network of astronomers that included Bradley, the Scot James Short and the Astronomer Royal, Edmond Halley, whose astronomical tables Bevis edited and eventually published in 1749, seven years after Halley's death.[347] According to the mathematician Thomas Simpson, Bevis had 'carefully observed with proper Instruments' the apparent motions of various fixed stars in order to confirm Bradley's findings. And Simpson said of Bevis that he was 'the first of any one that I know of, [to have] experimentally prov'd, that the Phœnomena are universally as conformable to the Hypothesis [of aberration] in Right Ascensions, as the Rev. Mr. Bradley, to whom we owe this great Discovery, had before found them to be in Declinations'.[348] That Reid knew of Bevis' corroboration of Bradley's discovery is clear from Reid's manuscript 'Of the Aberration of

Reid went back and added a paragraph to his reading notes from Maupertuis in which he drew on Bouguer's data to confirm that the Earth has an oblate shape. Reid also added a cryptic note which refers to J. T. Desaguliers on the length of pendulums in London. The reference appears to be to Desaguliers' *A Course of Experimental Philosophy*, vol. II, p. 520.

[346] Reid's additional note appears on the verso of the last folio of AUL, MS 2131/3/I/7; compare La Caille, 'An Account of the Astronomical, Geographical and Physical Observations Made at the Cape of Good Hope, in 1751, 1752 and 1753, by Order of the French King, by the Abbe de La Caille', pp. 512–13.

[347] On Bevis, see Ruth Wallis, 'John Bevis, M.D., F.R.S. (1695–1771): Astronomer Loyal'.

[348] Thomas Simpson, *Essays on Several Curious and Useful Subjects, in Speculative and Mix'd Mathematicks* (1740), p. 10. Right ascension and declination provide coordinates for the location of a star in the heavens. Right ascension is analogous to longitude on Earth, while declination is analogous to latitude. Simpson was incorrect in his assertion that Bevis was the first to confirm the

the fixed Stars', which probably dates from the 1740s or 1750s (pp. 63–5). Reid here drew on Bevis' 'Practical Rules for Finding the Aberrations of the Fixt Stars from the Motion of Light, and of the Earth in its Orbit, in Longitude, Latitude, Declination and Right Ascension' (which were included in Simpson's *Essays on Several Curious and Useful Subjects*) in order to consider the different cases of aberration that Bevis' rules covered.[349] In writing this manuscript, it appears that Reid was attempting to fix in his mind the guiding principles and the practical procedures involved in the interpretation of data on the apparent motions of the fixed stars. There is little doubt that he did so because he was determined to acquire the same level of competence as an observational astronomer as that displayed by Bevis and his circle in England.

At some point in the winter of 1750–51, he learned of another breakthrough made by Bradley. For John Stewart wrote to him at New Machar to alert him to the recent publication of the paper read to the Royal Society in early 1748 in which Bradley announced his discovery of the nutation of the Earth's axis. Bradley had detected a small annual variation in the declination of some of the fixed stars as early as 1727, but it was only in 1747 that he became completely convinced that these variations were produced by the cyclical motion of the Earth's poles. Although we do not know what Reid's immediate response to this paper was, his esteem for Bradley is registered in his graduation oration delivered at King's College in 1753. In his oration, he asserted that the 'recent observations' of 'the renowned Edmond Halley, Bradley, and the French academicians sent to Lapland and Peru' had silenced even 'the most obstinate critics' of Newton's system of the world.[350]

As mentioned in Section 1, Reid's graduation oration provides us with valuable clues regarding the lectures on astronomy he gave as part of the course on natural philosophy he mounted shortly after he began teaching at King's in the session for 1751–52. What is striking about his comments on the rise of the science of astronomy in the modern era is his emphasis on the importance of the invention of the telescope as a catalyst for the discovery of the 'foundation of true physical astronomy'. Reid singled out for praise Galileo, Johannes Hevelius, Gian Domenico Cassini and his son Jacques, Christiaan Huygens and John Flamsteed, whose collective labours in observing the heavens had 'cleared the

aberration of the fixed stars in right ascension. The Bolognese astronomer Eustachio Manfredi had been the first to do so, in 1730; see the entry on Bevis in the *Oxford Dictionary of National Biography*.

[349] Simpson, *Essays on Several Curious and Useful Subjects*, pp. 11–20.

[350] [John Stewart] to Reid, [winter 1750–51], in Reid, *Correspondence*, pp. 8–9; James Bradley, 'A Letter … concerning an Apparent Motion Observed in the Some of the Fixed Stars'; Reid, *Philosophical Orations*, oration I, para. 16.

way' for the construction of a system of celestial dynamics by Kepler and Newton. Moreover, it is significant that he likewise portrayed the advance of the science in the eighteenth century primarily in terms of the notable achievements in the field of observational astronomy made by Halley, Bradley and those associated with the major French geodetic expeditions of the 1730s and 1740s. Reid's brief account of the progress made by astronomers since the turn of the seventeenth century thus highlighted the central role played by observation in driving theory and mathematical analysis. In highlighting this role, he echoed Bradley, who affirmed that 'the Progress of Astronomy indeed has always been found, to have so great a Dependence upon accurate Observations, that, till such were made, it advanced but slowly'.[351] It is also significant that Reid suggested that the gradual acceptance of Newton's system of the world by continental men of science was largely facilitated by the so-called 'Jesuit edition' of Newton's *Principia* published by the Genevan professor Jean Louis Calandrini in collaboration with the Minim friars Thomas Le Seur and François Jacquier in 1739–42. We do not know when Reid first consulted this edition of the *Principia*, but when he did so the editorial apparatus would have introduced him to essays on the tides by Daniel Bernoulli, Leonhard Euler and Colin Maclaurin, to whom the Académie royale des sciences in Paris awarded prizes in 1740.[352] To some extent, then, Reid was aware of work on astronomical topics done by leading men of science in Europe such as Bernoulli and Euler. Reading notes taken in May 1754 from the *Histoire du renouvellement de l'Academie royale des sciences en M.DC.XCIX* (1708) also record his interest in the contributions made to astronomy by the continental academicians eulogised in the volume by Fontenelle.[353]

Various lists of topics that Reid intended to cover in his lectures on astronomy survive among his papers, and these provide some evidence regarding the specific topics that he covered in the classroom.[354] The set of student notes taken from his natural philosophy course during the academic session for 1757–58 provides

[351] Reid, *Philosophical Orations*, oration I, para. 16; Bradley, 'A Letter ... concerning an Apparent Motion Observed in the Some of the Fixed Stars', pp. 1–2. Bradley indicated that it was the observational work done by Tycho Brahe that enabled Kepler to discover his laws of planetary motion, and that Newton was likewise indebted to improvements in 'the *Praxis* of Astronomy' (p. 2).

[352] Reid, *Philosophical Orations*, oration I, para.16; Guicciardini, 'Editing Newton in Geneva and Rome', p. 341.

[353] AUL, MS 2131/3/II/13. Included in this volume were éloges of Galileo's protégé Vincenzo Viviani, the mathematician the marquis de l'Hôpital and Jacob Bernoulli. Reid copied down information about their careers. Reid also read various papers in the *Histoire de l'Academie royale des sciences* for 1700, and noted a proposition on orbital motion in a memoir by l'Hôpital, 'Solution d'un probleme physico-mathématique'; AUL, MS 2131/3/II/13, fol. 3v.

[354] See, for example, AUL, MSS 2131/6/V/31 and 2131/7/VIII/6.

a better sense of his astronomy lectures, although even here there are gaps.[355] Implicitly, Reid defined the scope of the science largely in terms of observational astronomy, partly because he had already dealt with the mathematical and physical analysis of orbital motion in an earlier section of his lectures on centripetal forces. As Reid taught the subject, the focus was primarily on the most conspicuous 'phenomena' observable in the heavens, ranging from the motions of the inferior and superior planets to the phases of Venus and the Moon, eclipses and the parallax of the fixed stars.[356] He did, however, touch on more complex and abstract topics, such as the theory of the Moon and the explanation of the tides, as well as Bradley's discoveries of the aberration of the fixed stars and the nutation of the Earth's axis.[357] His lectures, therefore, reflected his own fascination with the empirical study of the heavens and they provided his students with a sound grounding in the rudiments of observational and physical astronomy.[358]

By far the most significant celestial event to occur during Reid's years in Old Aberdeen was the much anticipated transit of Venus which took place on 6 June 1761. Although astronomers in the early seventeenth century had made careful observations of transits of both Mercury and Venus, it was Reid's ancestor, the mathematician James Gregory, who was the first to recognise that the transits of these planets across the face of the Sun could be used to measure the solar parallax and hence to establish the distance between the Earth and the Sun. In his *Optica promota* (1663), which included an appendix dealing with astronomical subjects, Gregory addressed the problem of calculating the parallax when two planets are in conjunction. In a scholium to his solution of the problem, he noted that 'this problem has a very beautiful application, although perhaps laborious, in observations of Venus or Mercury when they obscure a small portion of the sun; for by means of such observations the parallax of the sun may be investigated'.[359] But Gregory gave no practical hint as to how the requisite observations were to

[355] See especially the section devoted to the tides in Anon., 'Natural Philosophy 1758', pp. 185–92.

[356] Anon., 'Natural Philosophy 1758', pp. 129–39, 143–99; the text of the section on centripetal forces breaks off at p. 139, which might be a sign that Reid's lectures were becoming too technical for the student taking the notes. It may be that Reid's use of the term 'phenomena' in his astronomy lectures was modelled on that found in Book III of the *Principia*; see Newton, *The Principia*, pp. 797–801.

[357] Anon., 'Natural Philosophy 1758', pp. 161–73, 174–84, 193–8.

[358] Reid's emphasis on the observable phenomena of the heavens in his astronomy lectures was also partly a function of the limited mathematical competence of many of his students.

[359] James Gregory, *Optica promota, seu abdita radiorum reflexorum & refractorum mysteria, geometrice enucleata; cui subnectitur appendix, subtilissimorum astronomiæ problematωn resolutionem exhibens* (1663), pp. 128–30. I have used the English translation of the scholium on p. 130 provided in Robert Grant, *History of Physical Astronomy, from the Earliest Ages to the Middle of the*

be made. This practical issue was, however, taken up by Edmond Halley, initially in the context of his voyage to the island of St Helena to observe the transit of Mercury which took place in 1677, and later with reference to the transits of Venus that he predicted would occur in 1761 and 1769.[360] Fundamental to Halley's method for determining 'the distance of the Sun from the Earth' on the basis of solar parallax was the exact measurement of the times of both 'the total ingress of Venus into the Sun's disc' and 'the beginning of her egress from it … that is, when the dark globe of Venus first begins to touch the bright limb of the Sun within', which, he said, could be 'observed within a second of time'.[361] There were thus four measurements required for Halley's method to succeed: the moments of external and internal contact with the 'limb' of the Sun as Venus entered the Sun's disc, and the moments of internal and external contact as Venus exited the disc of the Sun.[362] Doubts were raised about the viability of Halley's method, however, first by the French astronomer Joseph-Nicolas Delisle (who had met Halley in London in 1724) and later by his compatriot Nicolas-Louis de La Caille. In his pamphlet *Avis aux astronomes* (1750), which was distributed to astronomers across Europe, La Caille questioned whether observations of the upcoming transits of Venus could in fact establish the distance between the Earth and the Sun with the degree of precision that Halley had claimed and, *inter alia*, La Caille affirmed that 'the rapid movement of the Sun and Venus in the field of a high-magnification telescope renders it very difficult to tell the exact moment of contact'.[363] Halley's method was thus, in his view, unworkable.

While we cannot be certain that the criticisms of Halley issuing from France prompted Robert Traill to pose his question for discussion in the Wise Club, 'What are the proper Methods of determining the Suns Paralax by the Transit of

Nineteenth Century, p. 428. James Gregory was the brother of Reid's maternal grandfather, David Gregory.

[360] Edmond Halley, 'De visibili conjunctione inferiorum planetarum cum Sole, dissertatio astronomica'; Halley, 'Methodus singularis quâ Solis parallaxis sive distantia à Terra'.

[361] I have used the English translation of Halley's 1716 paper, for which see Edmond Halley, 'Dissertation on the Method of Finding the Sun's Parallax and Distance from the Earth, by the Transit of Venus Over the Sun's Disc, June the 6th, 1761', in James Ferguson, *Astronomy Explained upon Sir Isaac Newton's Principles, and Made Easy to Those Who Have Not Studied Mathematics*, third edition (1764), pp. 316, 319. On Halley's method, see Albert van Helden, *Measuring the Universe: Cosmic Dimensions from Aristarchus to Halley*, pp. 144–6, 152–3, 155.

[362] Woolf, *The Transits of Venus*, pp. 19–21.

[363] Quoted and translated in I. S. Glass, *Nicolas-Louis de La Caille: Astronomer and Geodesist*, p. 32. La Caille's criticisms of Halley's method had some currency in Britain, for they were later summarised in the *Gentleman's Magazine* for 1761 by 'X.Y.Z.', 'Differences in the Observations of the Transit of Venus', p. 318. On Delisle and La Caille, see also Woolf, *The Transits of Venus*, pp. 23–40.

Venus over his Disk in 1761?', the timing of Traill's question strongly suggests that he knew of the doubts raised by the French academicians, especially since there was no public comment in Britain on Halley's method for observing the upcoming transits of Venus until after the Club considered Traill's question on 12 April 1758.[364] What appears to be the earliest treatment of the 1761 transit published in Britain is a letter to the editor of the *Gentleman's Magazine* from one T. Fisher dated 21 August 1758, in which the author used Halley's astronomical tables to compute the positions of the Sun and Venus in the heavens as well as the times of Venus' ingress, egress and conjunction with the Sun.[365] But it was not until the beginning of 1761 that serious attention was given in print to the question of how best to observe the impending transit, with both James Ferguson and Benjamin Martin issuing pamphlets on the question. The *Gentleman's Magazine* also carried a letter in February 1761 from 'Astrophilus' which echoed La Caille's doubts about the accuracy of the results yielded by Halley's method.[366] Hence Reid carried out his preparations for the transit largely on the basis of his own knowledge and expertise, although we do know that he was familiar with some of the relevant work of Halley's French critics and that at some point he read Ferguson's tract, which was also critical of Halley's method.[367]

[364] Ulman, *Minutes of the Aberdeen Philosophical Society*, pp. 83, 189. There was extensive discussion of the usefulness of observing the transits of Mercury and Venus for establishing the distance of the Sun from the Earth in the Académie royale des sciences in 1753.These discussions surfaced in the volume of the *Histoire de l'Académie royale des sciences* that appeared in 1757; see especially Guillame-Joseph-Hyacinthe-Jean-Baptiste Le Gentil de la Galaisière, 'Observation de la conjonction inférieure de Vénus avec le Soleil, arrivée le 31 Octobre 1751, faite à l'Observatoire royal de Paris; avec des remarques sur les deux conjonctions écliptiques de cette planète avec le Soleil, qui doivent arriver en 1761 & 1769'. See also Woolf, *The Transits of Venus*, pp. 50–2.

[365] T. Fisher, 'Transit of Venus over the Sun in 1761'.

[366] James Ferguson, *A Plain Method of Determining the Parallax of Venus, by Her Transit over the Sun: And from Thence, by Analogy, the Parallax and Distance of the Sun, and of All of the Rest of the Planets*, second edition (1761); Benjamin Martin, *Venus in the Sun: Being an Explication of the Rationale of That Great Phænomenon; of the Several Methods Used by Astronomers for Computing the Quantity and Phases Thereof; and the Manner of Applying a Transit of Venus over the Solar Disk, for the Discovery of the Parallax of the Sun; Settling the Theory of That Planet's Motion, and Ascertaining the Dimensions of the Solar System* (1761); Astrophilus, 'Transit of Venus over the Sun'. Two editions of Ferguson's pamphlet were published in 1761, and a lengthy review of the work appeared in the *Monthly Review* in March; Anon., 'Ferguson's Method of Determining the Parallax of Venus, &c.' According to Ferguson's most recent biographers, his pamphlet had a greater public impact than Martin's, although it was also the subject of criticism; see John R. Millburn and Henry C. King, *Wheelwright of the Heavens: The Life and Work of James Ferguson, FRS*, pp. 128, 133–4.

[367] The timing that Reid gives for the transit of Venus in 1769 in his discourse on the 1761 transit (p. 73 below) indicates that he had read Le Gentil de la Galaisière's 'Observation de la conjonction inférieure de Vénus avec le Soleil, arrivée le 31 Octobre 1751'. Le Gentil de la Galaisière's paper was summarised in the *Gentleman's Magazine* for January 1759; see Anon., 'A Summary of the Last

During the lead-up to the transit of Venus on 6 June, Reid scrutinised Halley's astronomical tables[368] and in May he busied himself 'in making an exact meridian'. In doing so, he later confessed that he had 'rashly directed to the sun, by my right eye, the cross hairs of a small telescope'. Immediately afterwards, he suffered from 'a remarkable dimness in that eye', and for 'many weeks' after he experienced in his right eye 'a lucid spot, which trembled much like the image of the sun seen by reflection from water' whenever he was 'in the dark, or shut my eyes'. It seems that his vision was permanently impaired and that his moment of carelessness eventually brought an end to his activities as an observational astronomer.[369] When the day of the transit finally arrived, Reid stationed himself at King's College and had with him a clock, a nine-foot refracting telescope fitted with a micrometer and a board mounted with a sheet of paper upon which the Sun's image was to be projected. He states that he himself made the observations (despite the injury to his eye) and that he was joined by 'several witnesses and particularly three or four Members' of the Wise Club (p. 65).[370] We can only surmise that among the 'witnesses' were John Stewart (unless he was observing the transit at Marischal College), the Marischal Professor of Natural Philosophy George Skene, the Humanist at King's Thomas Gordon, Robert Traill and perhaps the physicians John Gregory and David Skene. Presumably Reid keenly anticipated the test of his skills as an observationalist. But in the event he suffered from one of the occupational hazards of an observational astronomer, namely poor weather. Reid was not alone in being frustrated by cloudy skies, for observers elsewhere in Britain complained of the adverse weather conditions on 6 June. The cloud cover in Old Aberdeen was such that he 'had onely two observations of the Distance of Venus from the nearest part of the Suns Limb' (p. 65). Consequently, when Reid communicated his 'observations on the transit of Venus June 6th at Kings College' to the Wise Club on 14 July, he admitted to his colleagues that 'these observations can signify little for determining the Paralax of the Sun' (p. 66).[371] Because of the inclement weather, his data revealed nothing about the basic question regarding the size of the solar system that Halley's method was intended to answer. And even though he tried to make something of his observations, his

Volume of Memoires of the Royal Academy at Paris', p. 24. His references to 'Ferguson' (p. 70) show that he had read Ferguson's *A Plain Method of Determining the Parallax of Venus*, although he may have done so after the transit had taken place.

368 See above, p. xlvii.

369 AUL, MS 2131/3/I/21; Reid, *Inquiry*, p. 131.

370 It is unclear where at King's College the observations were made. The details of the clock and the telescope and micrometer are also unknown.

371 Ulman, *Minutes of the Aberdeen Philosophical Society*, p. 103.

discourse contributed almost nothing to the literature generated by the transit. Nevertheless, in retrospect his discourse is significant primarily for his defence of Halley's method and for his attempt to explain the discrepancies between his calculation of the conjunction of Venus with the Sun and Halley's tables in terms of the aberration of light (pp. 71, 73).

Like a number of his mathematical manuscripts, the manuscript of Reid's 'An Observation of the Transit of Venus' is stratified chronologically. For there are additions to the original text entered in the period August 1761 to early 1764, and there is material which appears to have been added after he moved to Glasgow in the summer of 1764. His initial additions dating from August 1761 to 1764 document his reading of reports on observations of the transit made elsewhere in Britain and Europe. The entries dating from after his move to Glasgow record his intensive study of the section devoted to the elliptical orbits of comets in Halley's 'Synopsis astronomiæ cometicæ' in the *Tabulæ astronomicæ*, as well as his scrutiny of an essay by the French astronomer Joseph-Jérôme Lefrançois de Lalande on the secular equations for the motions of the Sun, Moon, Saturn, Jupiter and Mars, and his reading of the preface to Nevil Maskelyne's *Astronomical Observations Made at the Royal Observatory at Greenwich, from the Year MDCCLXV to the Year MDCCLXXIV* (1776).[372] The reading notes found in this manuscript are noteworthy for three reasons. First, Reid's perusal of Halley's seminal work on the theory of comets provides evidence for his interest in this area of astronomy. This evidence is especially valuable given that he apparently said little about comets in his astronomy lectures, even though popular and scientific interest in the subject was high due to the return of Halley's Comet in 1758–59.[373] Secondly,

[372] In the period August 1761 to 1764 Reid appears to have read: Anon., 'Observations of the Transit of Venus over the Sun's Disc, June 6. 1761'; Nathaniel Bliss, 'Observations on the Transit of Venus over the Sun, on the 6th of June 1761: In a Letter to the Right Honourable George earl of Macclesfield, President of the Royal Society'; Nathaniel Bliss, 'A Second Letter to the Right Hon. the earl of Macclesfield, President of the Royal Society, concerning the Transit of Venus over the Sun, on the 6th of June 1761'; James Short, 'Second Paper concerning the Parallax of the Sun Determined from the Observations of the Late Transit of Venus, in Which This Subject Is Treated of More at Length, and the Quantity of the Parallax More Fully Ascertained'. He probably read all of the transit reports published in volume 52 of the *Philosophical Transactions*, which appeared in two parts in late 1762 and the first half of 1763. He subsequently read: Halley, *Tabulæ astronomicæ*; Joseph-Jérôme Lefrançois de Lalande, 'Mémoire sur les équations séculaires, et sur les moyens mouvemens du Soleil, de la Lune, de Saturne, de Jupiter & de Mars, avec les observations de Tycho-Brahé, faites sur Mars en 1593, tirées des manuscrits de cet auteur'; Nevil Maskelyne, *Astronomical Observations Made at the Royal Observatory at Greenwich, from the Year MDCCLXV to the Year MDCCLXXIV* (1776), pp. i–vi.

[373] There is only a brief mention of comets in the student notes taken from his astronomy lectures; see Anon., 'Natural Philosophy 1758', p. 146. Halley had earlier outlined his theory of comets in *A Synopsis of the Astronomy of Comets* (1705).

his notes on Lalande show that he periodically consulted the annual volumes of the *Histoire de l'Académie royale des sciences* in order to keep up on the papers produced by the academicians. Furthmore, his notes highlight the fact that Lalande's citations served as a guide to important recent work in astronomy done by leading continental savants such Leonhard Euler, Jean d'Alembert, Paolo Frisi and Tobias Mayer.[374] Thirdly, his notes taken from Maskelyne's *Astronomical Observations* register his fascination with the technical details of observational astronomy. For these notes consist almost entirely of a digest of the details of Maskelyne's array of instruments and methods of observation at the Royal Observatory at Greenwich. But in recording Maskelyne's use of 'elliptic illuminators' on his telescopes and the advantages of the design of this accessory for telescopes fitted with achromatic lenses, Reid made the telling comment, 'A Circumstance to be attended', which speaks to his own preoccupations as an expert observer of the heavens.[375] We see, then, that Reid returned to the study Halley's astronomical tables after he attempted to master their use in 1761 and that in the 1760s and 1770s he continued to familiarise himself with notable contributions to observational astronomy.

Despite the injury to his right eye in 1761, Reid remained active as an observer of the heavens after settling in Glasgow. In 1769 he collaborated with his colleagues at the College in making observations of the transit of Venus which occurred on 3 June. Reid was a member of a team of observers organised by the Professor of Practical Astronomy, Alexander Wilson. Wilson's team included Robert Simson's successor as the Professor of Mathematics, James Williamson (who together with Reid 'managed' a Dollond achromatic telescope), the chemist and botanist William Irvine, and Wilson's son Patrick. In addition, John Anderson observed the transit using a reflecting telescope and an astronomical clock which he had taken up into the College steeple. However, wind caused a 'tremor' in the structure which meant that the times recorded by Anderson varied from those of Wilson and the others.[376] Soon thereafter, in early September 1769, Reid began

374 See Lalande, 'Mémoire sur les équations séculaires', pp. 412, 415, 416, and 427. Lalande refers to: Jean d'Alembert, *Recherches sur la précession des equinoxes, et sur la nutation de l'axe de la Terre, dans le système Newtonien* (1749); Euler, *Opuscula varii argumenti*; Paolo Frisi, *De motu diurno Terrae dissertatio* (1756); Tobias Mayer, 'Novae tabulae motuum Solis et Lunae'. Euler's *Opuscula varii argumenti* contained two important papers on astronomical topics: 'Nova tabulæ astronomicæ motuum Solis ac Lunæ' and 'De perturbatione motus planetarum a resistentia ætheris orta'. See note 203 for Reid's borrowing of Euler's book from Glasgow University Library. He did so in either 1769 or 1770; see 'Professors Receipt Book, 1765–1770'.

375 Maskelyne, *Astronomical Observations*, p. iii; AUL, MS 2131/2/I/7, fol. 12r.

376 Wilson, 'Observations of the Transit of Venus over the Sun', p. 338. Wilson noted that Reid mentioned a 'dark protuberance upon Venus' that made it difficult to ascertain precisely the moment

tracking a comet from the rooms that he and his family occupied in the College. This comet, which is sometimes referred to as 'Napoleon's Comet', had appeared in the sky across Europe the previous month and was also visible as far afield as the South Pacific, where it was seen by Captain Cook on the *Endeavour*. What is not clear from Reid's record of his observations is, first, whether he was working in conjunction with the Wilsons and other colleagues or watching the comet independently and, secondly, whether he was using a telescope or the naked eye (pp. 82–3).[377] After he made his observations of 'Napoleon's Comet' in the early autumn of 1769, however, his career as an observational astronomer seems to have come to an end for there is no evidence that suggests that he continued to observe the heavens during the last quarter century of his life.

While the majority of those who criticised Halley's method for calculating the mean distance between the Earth and the Sun based on the transits of Venus queried the practical feasibility of his proposals, the Edinburgh Professor of Mathematics, Matthew Stewart, rejected the principles upon which Halley's approach was based. According to Stewart, even though 'the celebrated Dr Halley [had] proposed a very ingenious method of solving [the] problem [of ascertaining the Sun's distance from the Earth], by observing the transit of Venus over the disk of the sun in the year 1761', the transit had revealed that this 'method [had] not answered expectation'. Stewart went on to say that he 'was early of opinion, that [Halley's] method would fail' and that he had 'for many years past' endeavoured to develop an alternative method based on the assumption that 'the solar force affecting the gravity of the moon to the earth could be ascertained' on the basis of 'the motion of the moon's apogee, or from the motion of its nodes'. And once the quantity of this force was established, it was a relatively straightforward matter to calculate the distance of the Earth from the Sun on the basis of Newton's theory of gravitation.[378]

Another chronologically layered manuscript headed 'Scholium ad Propositionem 26 Liber 3 Principia Newtoni' shows that at some point in the period

of internal contact between Venus and the Sun and that Reid believed that this phenomenon was not 'mere imagination' (p. 336). Reid had in fact observed what is called the 'black-drop' effect, which was first noticed by some observers of the transit of Venus in 1761; see Woolf, *The Transits of Venus*, pp. 148–9, 193–5. See also Clarke, *Reflections on the Astronomy of Glasgow*, pp. 86–9.

377 The comet is also known as C/1769 P1; see Seargent, *The Greatest Comets in History*, pp. 121–4, and below, p. 193, editorial note 82/13.

378 Matthew Stewart, *The Distance of the Sun from the Earth Determined, by the Theory of Gravity* (1763), pp. v–vi; compare Matthew Stewart, *Tracts, Physical and Mathematical* (1761), pp. v–vi. Stewart's method was based on a suggestion made by John Machin in 1729; see John Machin, 'The Laws of the Moon's Motion According to Gravity', p. 24.

1761 to 1770, Reid scrutinised, and rejected, Stewart's alternative method (pp. 74–80).[379] The initial paragraphs of this manuscript appear to have been taken from an as yet unidentified Latin text related to proposition 26, Book III of the *Principia*, in which Newton dealt with the perturbation of the Moon's motion caused by the gravitational attraction of the Sun.[380] Reid then worked through key propositions in the fourth (untitled) tract included in Stewart's *Tracts, Physical and Mathematical*, and in the process of identifying the points of disagreement between Stewart and Newton he questioned the validity of Stewart's analysis of the effect of the Sun on the motion of the Moon. In particular, he was highly critical of the value for the quantity of the gravitational force of the Sun acting on the Moon designated by Stewart as 'C', noting that 'this value of C is not more but less accurate than that which Sir Isaac Newton gives' (p. 74). Moreover, Stewart's method of determining the distance of the Sun from the Earth presupposed a solution to the 'three-body problem' involving the mutual interactions of the Sun, Earth and Moon. His handling of this problem seemingly prompted Reid to reconsider Newton's mathematical treatment of perturbation theory in proposition 66 and its twenty-two corollaries in Book I of the *Principia*.[381]

Reid's comments on corollaries 14 and 15 of proposition 66 and the section of the manuscript headed 'Of the disturbing force of the Sun upon the Moon' are the only evidence we have for his serious engagement with lunar theory and the three-body problem, which were topics that preoccupied a number of leading mathematical astronomers during the course of the eighteenth century.[382] Unlike

[379] The dating is based on internal evidence. Reid initially refers to Stewart's *Tracts, Physical and Mathematical* and later there is a cryptic reference to the same author's *The Distance of the Sun from the Earth Determined* (below, p. 79). On the same folio Reid also mentions 'Mr Trail'; this is most likely Robert Simson's protégé, William Trail, who matriculated in Glasgow in 1763 and graduated with his MA in 1766. Shortly thereafter he became the Professor of Mathematics at Marischal College. The manuscript ends with material that may have been taken from an as yet unidentified source dated 'Hampstead 21 May 1770'. The *terminus a quo* for the sections of the text dealing with Stewart is thus 1761 and the *terminus ante quem* is 1770, with the most likely date in the period 1763–66. It should also be noted that Reid met Matthew Stewart in Aberdeen while examining William Trail for the Marischal Chair; see above, note 158.

[380] Newton, *The Principia*, pp. 840–3. Propositions 22 and 25–35 outline Newton's theory of the Moon. In this manuscript Reid was returning to an aspect of the *Principia* that he had considered in 1729; see above, p. cxii.

[381] Stewart had earlier solved the 'two-body problem' in his paper 'A Solution of Kepler's Problem', which, as we have seen above (p. xlvi), Reid may have read in March 1757.

[382] Corollaries 14 and 15 deal with the calculation of the disturbing force of the Sun on the Moon and the Earth; on the significance of corollary 14 see I. Bernard Cohen's editorial commentary in Newton, *The Principia*, pp. 355–9. For an overview of the development of perturbation theory in the eighteenth century, see: Anton Pannekoek, *A History of Astronomy*, pp. 299–306; Curtis Wilson, 'Astronomy and Cosmology', pp. 334–7.

Matthew Stewart, Reid did not rely exclusively on the resources of classical geometry to analyse the perturbations of the Moon. Rather, like Newton himself, Reid combined geometrical and analytic techniques in order to calculate the 'accelerative force of the Earth towards the Sun ... and that of the moon towards the Sun' as well as the 'disturbing force of the Sun upon the moon' (p. 77).[383] The most significant feature of his discussion of the corollaries to proposition 66 is the conclusion that he draws from Newton:

> it seems to follow that the Paral[l]ax of the Sun cannot be determined from ... the whole disturbing force of the Sun upon the Moon. And since the Motion of the Moons Nodes, Apogee, &c must be proportional to the causes that produce them, the parallax of the Sun cannot be determined from those Motions. (p. 76)

That is, even though Stewart presented his method of determining the distance of the Sun from the Earth as being founded on Newton's theory of gravitation, Reid inferred from the corollaries to proposition 66 that Newton's theory was incompatible with the basic assumptions of Stewart's method.

The part of the manuscript devoted to establishing a value for the 'disturbing force' exerted by the Sun on the Moon is not directly related to key passages in the *Principia*. It may be that this section is either taken from a published work that has not been identified thus far or was inspired by conversations with members of Reid's circle in Glasgow.[384] Reid here provides his own formulation of the key elements of Newton's analysis of the perturbations in the motions of the Moon caused by the gravitational force of the Sun in proposition 66, Book I, and propositions 25 and 26, Book III, of the *Principia* (pp. 77–9). And, at a later date, he has added two canons for calculating the Sun's parallax, as well as a brief entry at the end of the manuscript signalling his disagreement with estimates made by French astronomers of the parallax of the Moon, the mass of the Earth and the parallax of the Sun (p. 80).

Reid's criticisms of Matthew Stewart are noteworthy historically because they apparently pre-date the debate over Stewart's work which briefly flared up in the late 1760s.[385] In 1767 the theoretical basis for Stewart's estimate of the

[383] Reid's combination of geometry and algebraic techniques is most evident in the section of the manuscript 'Of the disturbing force of the Sun upon the Moon'. See also Guicciardini, 'Thomas Reid's Mathematical Manuscripts', pp. 79–81; Guicciardini, *Reading the Principia*, pp. 92–4.

[384] The reference to William Trail (p. 79) shows that Reid was discussing perturbation theory and the motions of the Moon with Trail, and presumably others, in Glasgow.

[385] Stewart's works were positively reviewed when they first appeared; see Anon., 'Stewart's *Tracts, Physical and Mathematical*', and Anon., 'Steward's [*sic*] Method of Finding the Sun's Distance from the Earth'.

Sun's distance from the Earth was endorsed by the English mathematician and future editor of Newton's writings, the Rev. Samuel Horsley. Although Horsley used a different method to compute the Sun's distance, he nevertheless offered his result 'rather as a verification than an amendment of Dr. Stewart's' estimate, and characterised Stewart as a 'great and able geometrician'.[386] But not every member of the British astronomical community was as receptive to Stewart's ideas as Horsley. In 1769 Stewart was attacked by the Sedbergh medical practitioner and mathematician John Dawson, who declared that Stewart's calculations were 'palpably wrong, and his principles very unsatisfactory'. After cataloguing the problems with Stewart's calculations and identifying six major flaws in the principles upon which Stewart's method rested, Dawson concluded that 'the method of determining the sun's distance, pitched upon by our author, is very ill fitted for the purpose'. He observed that Stewart's 'principles are too complex to make it possible to take every thing in to the account which belongs to it; and unless that be done, since the smallest neglect occasions a very great error, the result will ever be much wide of the truth'. Consequently, he contended that since 'the distance of the sun will never be satisfactorily ascertained by the Theory of Gravity' and that the only way in which the problem of the Sun's distance could be solved was on the basis of the 'careful observation of the approaching Transit of Venus' in June 1769.[387] Dawson's objections were shortly thereafter dismissed by Horsley, who steadfastly defended Stewart's competence as a man of science against Dawson's charges, in a letter written in May 1769 and subsequently published in the *Philosophical Transactions*. Yet Horsley also disclosed that Dawson's pamphlet had prompted him to return to 'Dr. Stewart's Theorems' and, in a remarkable volte-face, he confessed that 'the imperfection' of the methods employed by Stewart and himself was 'much greater than I was at first aware of'.[388] Stewart had thus lost a prominent ally in the Royal Society of London.

The following October, Dawson responded to Horsley in the *Gentleman's Magazine*. In a letter to the editor, 'Mr. Urban', Dawson accused Horsley of failing to offer 'the least shadow of an argument to support' his defence of Stewart. He also reaffirmed his position that Stewart's method of determining the distance of the Sun produced an unacceptably imprecise value and that Stewart's

[386] Samuel Horsley, 'A Computation of the Distance of the Sun from the Earth', p. 179.

[387] John Dawson, *Four Propositions, &c. Shewing, not only, that the Distance of the Sun, as Attempted to be Determined from the Theory of Gravity, by a Late Author, Is upon His Own Principles, Erroneous; but also, that It Is More than Probable this Capital Question Can Never be Satisfactorily Answered by Any Calculus of the Kind* (1769), pp. iii–iv, 12, 33–5, 38–41, 42, 44.

[388] Samuel Horsley, 'On the Computation of the Sun's Distance from the Earth, by the Theory of Gravity', pp. 153–4.

account of the motion of the apses of the Moon was irremediably muddled.[389] A further salvo against Stewart was launched in 1771 by the mathematical practitioner John Landen FRS, who likewise paused to take aim at Horsley. Echoing Dawson's assessment of Stewart's work, Landen stated that he found the Scottish professor's 'principles very exceptionable … [and] his calculation egregiously erroneous'. Landen then proceeded to enumerate the errors and inconsistencies in Stewart's computations, as well as the questionable presuppositions of Stewart's method. Landen identified three flaws in Stewart's calculations: (i) Stewart's value for the mean disturbing force of the Sun on the Moon was incorrect; (ii) he employed two inconsistent values for the force of the Moon on the Earth; and (iii) his method of calculating the Sun's distance yielded a range of values that were all inconsistent with estimates derived from observation.[390] In addition, Landen maintained that the theoretical basis for Stewart's method was compromised by three problems: (i) Stewart failed to see that the angular motion of the Moon's apsides was not the same as their mean motion, even though he had assumed that they were equivalent in his calculation of the mean disturbing force of the Sun on the Moon; (ii) the geometrical construction used to calculate the mean disturbing force of the Sun on the Moon incorrectly represented the real distances involved; and, most important of all, (iii) his theoretical assumptions were contradicted by observational data regarding the distance of the Earth from the Sun.[391] Furthermore, Landen indicated that he was out of sympathy with the style of Stewart's work. He insinuated that his opponent's approach to estimating the Sun's distance from the Earth was driven by theory rather than observation, and he remarked that Stewart betrayed a 'singular fondness' for a purely geometrical method, 'esteeming it more elegant than the algebraic method, without considering, that, however elegant it may be in some instances, it cou'd be of no real use in the computation he proposed to make'.[392]

When we compare Reid's criticisms of Matthew Stewart with the polemics of Dawson and Landen, we see that his comments anticipate those of Stewart's English critics in three key respects. First, like Dawson and Landen, he had reservations about the values involved in Stewart's calculations, although Reid took exception to Stewart's estimate of the magnitude of the disturbing force the Sun exerted on the Moon rather than the Sun's distance from the Earth. Secondly,

389 John Dawson, 'Observations on Dr Stewart's Conclusions, &c. on the Sun's Distance from the Earth', p. 452.

390 John Landen, *Animadversions on Dr Stewart's Computation of the Sun's Distance from the Earth* (1771), pp. 1, 11–12.

391 Landen, *Animadversions*, pp. 12–14.

392 Landen, *Animadversions*, p. 12.

Reid likewise questioned the theoretical grounding of Stewart's method and went even further than Dawson and Landen in maintaining that the premises of Stewart's method were fundamentally flawed. Thirdly, Reid likely shared the view of Dawson and Landen that Stewart's approach to solving the problem of ascertaining the distance between the Earth and the Sun was driven by theory rather than being grounded firmly in observation. Reid was, as we have seen, an observational astronomer in the mould of the Astronomers Royal Edmond Halley, James Bradley and Nevil Maskelyne. His defence of Halley's method in 1761 (p. 73), coupled with his remarks in his 1753 graduation oration, strongly suggest that his understanding of the relations between theory and observation was very different from Stewart's. Moreover, like Landen, he undoubtedly regarded Stewart's strict adherence to the geometry of the ancients as misguided. Hence Reid's critique of Stewart reflected not only their disagreement over theoretical principles and the details of calculating the solar distance but also their divergent interpretations of Newton's mathematical legacy and their methodological differences over the respective roles of theory and observation in the science of astronomy.[393]

Reid's reading of the literature on astronomy after he moved to Glasgow in 1764 can be documented using various sources. The extant evidence reinforces the impression given by the additions to the manuscript of 'An Observation of the Transit of Venus' discussed above that he continued to read works in the field until late in life. Sets of reading notes found in the Birkwood Collection show that in the late 1760s he closely monitored the publications of Nevil Maskelyne. Among his papers, two sets of reading notes survive, one taken from the first volume of *The Nautical Almanac and Astronomical Ephemeris, for the Year 1767* (1766) and the other from four texts generated by the dispute between Maskelyne and John Harrison over the reliability of Harrison's chronometers and their use for establishing the longitude at sea.[394] His excerpts from the *Nautical Almanac* are especially revealing because they illustrate both his knowledge of Tobias Mayer's lunar tables (which were one of the great achievements of eighteenth-century observational astronomy) and his careful consideration of the practicalities of

[393] That is, Stewart treated astronomy as if it were a branch of pure mathematics, a view conditioned by his institutional position as the Professor of Mathematics at Edinburgh. By contrast, Reid conceived of the science of astronomy as a branch of mixed mathematics modelled on the exemplars he mentioned in his 1753 graduation oration at King's College (above, pp. cxvi–cxvii).

[394] On the publication of the *Nautical Almanac* and the 'Harrison affair', see Derek Howse, *Nevil Maskelyne: The Seaman's Astronomer*, chs 8 and 9.

making accurate astronomical observations.[395] As for the dispute over Harrison's timekeepers, Reid's notes attest to his interest in the problem of determining longitude, which was an interest that he shared with the then Glasgow Lecturer on Chemistry, John Robison, who had earlier been involved in the Board of Longitude tests of Harrison's watches.[396]

Two further sets of notes, which are transcribed below, confirm that in Reid's later years he retained his fascination with astronomy, despite the fact that because of failing eyesight and increasing deafness he was probably no longer actively observing the heavens and collaborating with others. His 'Extracts from Pemberton's View of Newtons Philosophy' (pp. 83–4) apparently date from 1784, for he is recorded as borrowing the University Library's copy of Pemberton's 1728 work on 12 March and returning the book three days later.[397] It is unclear why he was reading Pemberton's introduction to Newtonian natural philosophy at this stage in his career; conceivably, he wanted to review Pemberton's *A View of Sir Isaac Newton's Philosophy* because he was in the process of formulating his critique of Joseph Priestley's 'Modern System of Materialism'.[398] Whatever the reason, his extracts from Pemberton cover a limited number of topics, including Pemberton's exposition of Newton's theory of the tides. But they focus primarily on the section of Pemberton's *View* devoted to centripetal forces and, more specifically, on Newton's analysis of orbital motion. In doing so, this set of notes registers the fact that after moving to Glasgow Reid apparently gave careful thought to the fundamentals of Newton's treatment of centripetal forces, as can also be seen in the brief discussion of the subject headed 'Of the Centripetal Forces, Velocities, & Periodical Times of Bodies moving equably in Circles & of the Radij of the

[395] AUL, MS 2131/3/I/18, fols 2v–4v. Folios 1r–2r contain notes taken from an unrelated work that are dated August 1767. Reid's notes from the *Nautical Almanac* thus probably date from the second half of 1767. On Mayer's lunar tables, see Eric Gray Forbes, 'Tobias Mayer (1723–62): A Case of Forgotten Genius', pp. 19–20.

[396] AUL, MS 2131/3/I/6. The texts that Reid read were: Nevil Maskleyne, *An Account of the Going of Mr John Harrison's Watch, at the Royal Observatory, from May 6th, 1766, to March 4th, 1767* (1767); [John Harrison], *The Principles of Mr Harrison's Time-keeper, with Plates of the Same* (1767); John Harrison, *Remarks on a Pamphlet Lately Published by the Rev. Mr Maskelyne, under the Authority of the Board of Longitude* (1767); David Le Roy, *A Succinct Account of the Attempts of Mess. Harrison and Le Roy, for Finding the Longitude at Sea, and of the Proofs Made of Their Works* (1768). On Robison, see especially Jim Bennett, 'The Travels and Trials of Mr Harrison's Timekeeper'.

[397] 'Professors Receipt Book, 1770–[1789]'.

[398] The dating of Reid's response to Priestley is, however, a complex issue; see my editorial introduction to Reid, *Animate Creation*, pp. 38–41.

Circles', which is related to proposition 4, Book I of the *Principia* (pp. 81–2).[399] His 'Facts from Bailly's Histoire de l'Astronomie Moderne', which, given the handwriting, probably dates from the late 1780s or 1790s, suggests that at the end of his life he was drawn to Jean Sylvain Bailly's history, first published in 1779–82, because he wanted to gauge the progress that had been made in the science of astronomy during the eighteenth century and to fill in some gaps in his knowledge of developments on the continent. It may be that, in addition, he hoped to fit his fairly extensive reading over the course of more than fifty years into a coherent historical framework (pp. 84–7).[400]

The surviving professorial borrowing records for the College library shed further light on Reid's reading of works on astronomy after 1764. What immediately stands out in the list of his borrowings is the number of different editions of Newton's works that he consulted. He perused Johann Castillon's collection of Newton's writings entitled *Opuscula mathematica, philosophica et philologica* (1744) and individual volumes (including those containing the *Principia*) from Samuel Horsley's *Opera quæ exstant omnia* (1779–85), while he also made repeated use of the 'Jesuit edition' of Newton's *Principia* published by Calandrini, Le Seur and Jacquier in 1739–42.[401] Complementing his study of Newton, he also borrowed two of the standard introductions to Newtonian natural philosophy. We have just seen that Reid took out Pemberton's *View* in March 1784 and, immediately after returning the Pemberton, he withdrew Colin Maclaurin's *An Account of Sir Isaac Newton's Philosophical Discoveries* (1748).[402] However, the number of other library books on astronomy that he used was limited. As noted in Section 1, in either 1769 or 1770 he signed out Leonhard Euler's *Opuscula varii argumenti*, which contained Euler's solar and lunar tables along with an important paper on the perturbations in the motions of the planets. Shortly thereafter, he took back to his room the library copies of *The Posthumous Works of Robert Hooke* (1705) and Hooke's *An Attempt to Prove the Motion of the Earth from Observations* (1674). While the former collection was something of a miscellany, it included papers on gravity alongside the texts of a discourse on comets and lectures on astronomy and navigation. Hooke's *Attempt* was directly related to Reid's interests as an observational astronomer. Hooke's argument in favour of heliocentrism was

[399] While Reid was in Glasgow, he also reflected on the basics of Newtonian mechanics; see the manuscript 'Of the Laws of Motion', dated 25 December 1770, AUL, MS 2131/7/III/3.

[400] Reid's notes are taken from Jean Sylvain Bailly, *Histoire de l'astronomie moderne* (1779–82).

[401] 'Professors Receipt Book, 1765–1770' and 'Professors Receipt Book, 1770–[1789]'. On 23 August 1784 Reid made a brief set of reading notes from the first volume of Samuel Horsley's edition of Newton works; see AUL, MS 2131/3/I/11, fol. 1r.

[402] No reading notes from Maclaurin's *Account* survive.

based on the observable parallax of the fixed stars. Moreover, in making his argument Hooke contended that the data used to deny the existence of parallax was compromised because the wooden instruments traditionally employed by observers expand and contract, depending on weather conditions. Hence the data were unreliable. Quite apart from the ingenuity of Hooke's case for the truth of Copernicanism, Reid was probably also drawn to the work because it was the starting point for James Bradley's foundational paper on stellar aberration.[403] The last book on astronomy Reid is known to have borrowed from the University library was the translation by the mathematician John Robertson of Nicolas-Louis de La Caille's *The Elements of Astronomy*, although it is far from clear why he wanted to consult this introductory textbook in 1788.

Of greater import for our understanding of Reid's familiarity with the literature on astronomy are the learned journals that he borrowed while he was in Glasgow. Without any additional evidence in the form of reading notes or references in his other writings, we cannot be certain why he read a particular issue or volume of a journal. Nevertheless, the list of his journal borrowings provides invaluable evidence regarding the range of papers on astronomy that he knew of or might have studied. For example, in 1775 Reid took out volume 62 of the *Philosophical Transactions*, which included, *inter alia*, a paper discovered among the manuscripts of James Bradley on the use of micrometers, along with others by Leonhard Euler's son, Johann Albrecht, on the parallax of the Sun, by Peter Dollond and Nevil Maskelyne on recent improvements to Hadley's quadrant and by the late Henry Pemberton on mathematical astronomy.[404] Then, in 1780, he signed out the two parts of the 1772 volume of the *Historie de l'Académie royale des sciences*, which contained a number of *mémoires* on astronomical topics, notably those by Achille-Pierre Dionus du Séjour on calculating eclipses of the Sun, Lalande on the tides, Charles Messier on a comet observed in 1760, Pierre-Simon Laplace on the secular inequalities of the planets and Alexandre-Gui Pingré on establishing the parallax of the Sun on the basis of the observations made of the transit of Venus

[403] Bradley, 'An Account of a New Discovered Motion of the Fix'd Stars', pp. 637–8.

[404] James Bradley, 'Directions for Using the Common Micrometer, Taken from a Paper in the Late Dr Bradley's Hand-writing'; Johann Albrecht Euler, 'A Deduction of the Quantity of the Sun's Parallax from the Comparison of the Several Observations of the Late Transit of Venus, Made in Europe, with Those Made in George Island in the South Seas'; Peter Dollond, 'A Letter from Mr Peter Dollond, to Nevil Maskelyne, F.R.S. and Astronomer Royal; Describing Some Additions and Alterations Made to Hadley's Quadrant, to Render It More Serviceable at Sea'; Nevil Maskelyne, 'Remarks on the Hadley's Quadrant, Tending Principally to Remove the Difficulties Which Have Hitherto Attended the Use of the Back-observation, and to Obviate the Errors That Might Arise from a Want of Parallelism in the Two Surfaces of the Index-glass'; Henry Pemberton, 'Geometrical Solutions of Three Celebrated Astronomical Problems'.

in 1769.[405] His reading of these journals would thus have brought him up to date with recent work on astronomy in Britain and on the continent.[406] Furthermore, Reid seems to have made a conscious effort to monitor the research of continental savants, for he kept up on reviews of the latest European publications by reading the *Bibliothèque des sciences et des beaux-arts*, and runs of the *Nova acta eruditorum* of Leipzig and the *Commercium litterarium* of Nuremburg. He also borrowed multiple volumes of the *Histoire de l'Academie royale des sciences et des belles lettres de Berlin* and the *Nouveaux mémoires de l'Académie royale des sciences et belles-lettres*. In consulting the proceedings of the Berlin Academy, he would have encountered noteworthy essays on astronomical topics by prominent men of science including Leonhard Euler, Jean Bernoulli II, Joseph-Louis Lagrange and Johann Heinrich Lambert.[407] Through his exploration of the learned journals, therefore, Reid was probably better acquainted with the extant literature on astronomy than his modest book borrowings from the College library would otherwise suggest.

To conclude, for much of the eighteenth century, both theory and practice in the science of astronomy were largely driven by questions posed by the *Principia*. Newton's theory of gravitation had profound implications for a range of issues, including perturbation theory and the three-body problem, which came to dominate the work of continental mathematical astronomers. Book III of the *Principia* also raised issues regarding the orbits of comets, the shape of the Earth and the secular inequalities and other puzzling motions of the Moon and the planets. All of these questions served to shape the research agenda of observational astronomers, as did the needs of those searching for a practical solution to the problem of finding longitude at sea. Observers such as Halley, Bradley and Maskelyne set new standards of precision and employed increasingly sophisticated (and expensive)

[405] Achille-Pierre Dionus du Séjour, 'Nouvelles méthodes analytiques pour calculer les éclipses de Soleil, &c.'; Joseph-Jérôme Lefrançois de Lalande, 'Mémoire sur le flux & le reflux de la mer, & spécialement sur les marées des équinoxes'; Charles Messier, 'Mémoire contenant les observations de la première comète qui a paru en 1760, & qui est la cinquante-unième dont l'orbite ait été calculée; observée de l'Observatoire de la Marine à Paris, depuis le 8 Janvier jusqu'au 30 du même mois'; Pierre-Simon de Laplace, 'Mémoire sur les solutions particulières des équations différentielles, & sur les inégalités séculaires des planètes'; Alexandre-Guy Pingré, 'Mémoire sur la parallaxe du Soleil, déduite des meilleures observations du passage de Vénus sur son disque le 3 Juin 1769'.

[406] Reid also borrowed the volumes of the *Philosophical Transactions* covering the years 1777–84 and the *Histoire de l'Académie royale des sciences* for 1757.

[407] See, for example: Leonhard Euler, 'Recherches sur le mouvement de rotation des corps celestes'; Jean Bernoulli II, 'Essai d'une nouvelle méthode de déterminer la diminution séculaire de l'obliquité de l'écliptique, au moyen de l'Étoile polaire'; Joseph-Louis Lagrange, 'Théorie de la libration de la Lune, & des autres phénomenes qui dépendent de la figure non sphérique de cette planète'; Johann Heinrich Lambert, 'Sur les irrégularités du mouvement de Saturne'.

instruments to compile accurate data and to reveal new astronomical phenomena such as the aberration of light and the nutation of the Earth's axis. Moreover, Halley's prediction of the transits of Venus in 1761 and 1769, and his proposals for using the observations of the transits to determine the size of the solar system, inspired not only unprecedented cooperation between men of science from rival nations but also debates which preoccupied astronomers through the middle decades of the century.

In this section we have seen that Reid's work in astronomy registered almost all of these theoretical and practical currents. Beginning with his early study of Newton, Gregory and Whiston, he grappled with the theoretical and mathematical details of Newton's system of the world for the rest of his life. From the late 1730s to the early 1750s he worked through the contributions of the French geodesic expeditions to Lapland and Peru and, along with John Stewart, calculated the shape of the Earth. During this period, he also began to cultivate the craft of celestial observation and to absorb the work of James Bradley, to whom he paid public tribute at King's College. In the 1760s he collaborated with colleagues in Aberdeen and Glasgow in observing the transits of Venus, tracked 'Napoleon's Comet' in 1769 and wrote on perturbation theory and the motions of the Moon in the context of his critical response to the writings of Matthew Stewart (whom he met on at least one occasion). Through his reading, Reid was familiar with the work of the leading British and European astronomers of his day, and knew of the exchanges between the Astronomer Royal, Nevil Maskelyne, and John Harrison over the reliability of Harrison's chronometers. But even though he continued to read books and papers on astronomy until the 1790s, it would seem that his days as a practising observer of the heavens were over by the early 1770s. Moreover, there is no evidence to suggest that he was caught up in the revolution in stellar astronomy precipitated by the research published by William Herschel in the 1780s, even though he was acquainted with Herschel's writings and was a member of the University committee that presented Herschel with an honorary degree at a ceremony held in Glasgow in 1792.[408] Reid's understanding of the science of astronomy was thus firmly rooted in the work of Newton and the astronomers who tried to make sense of Newton's legacy during the course of the eighteenth century. It appears that in retirement he was largely unaware of the fact

[408] 'Minutes of Senate Meetings, 1787–1802', p. 151, minute for 11 June 1792. On Reid's knowledge of Herschel's writings, see below, p. 119. On the degree ceremony, see Constance A. Lubbock, *The Herschel Chronicle: The Life-story of William Herschel and His Sister Caroline Herschel*, pp. 235–6. Herschel was a regular correspondent of both Alexander and Patrick Wilson. We also do not know what Reid made of the cosmological speculations found in Alexander Wilson's pamphlet, *Thoughts on General Gravitation, and Views Thence Arising as to the State of the Universe* (1777).

that Herschel and others were in the process of revolutionising our understanding of the construction of the heavens.

5. Optics

During the course of the Scientific Revolution the science of optics consolidated its position alongside astronomy and mechanics as one of the most theoretically challenging and technically demanding branches of natural philosophy, thanks to the seminal work of such figures as Johannes Kepler, René Descartes, Christiaan Huygens and Robert Hooke. It was Isaac Newton, however, who propelled optics to the forefront of the physical sciences in the early decades of the eighteenth century through his controversial theory of light and colours, announced in 1672, his innovative work on reflecting telescopes and the publication of his *Principia* (1687) and *Opticks* (1704), which both served as exemplars for aspiring opticians throughout the Enlightenment era.[409] Largely because of Newton's achievements, optics was widely cultivated by men of science in eighteenth-century Britain, while Newton's ideas attracted a broad audience outside the scientific community due to the efforts of itinerant lecturers as well as the publication of such popular texts as Francesco Algarotti's *Il Newtonianismo per le dame* (1737), which was quickly translated into English.[410] Reid's graduation oration delivered at King's College in 1762 speaks to the prestige that optics had acquired by the middle decades of the eighteenth century. He informed his auditors that 'several parts of physics … have been constructed on solid foundations, viz. astronomy, mechanics, hydrostatics, optics and chemistry' and he assured them that these sciences 'daily receive additions worthy of the human intellect, without there any longer being dispute among the knowledgeable as regards the principles of these parts of physics'.[411]

In the eighteenth century, optics was defined as a branch of mixed mathematics that encompassed the investigation of a broad range of topics, including vision, the nature of light and colours, atmospheric phenomena such as the rainbow and the laws governing the reflection and refraction of light by mirrors, lenses and

[409] Olivier Darrigol, *A History of Optics: From Greek Antiquity to the Nineteenth Century*, chs 2–4; Cantor, *Optics after Newton*, ch. 2.

[410] Francesco Algarotti, *Il Newtonianismo per le dame ovvero dialoghi sopra la luce e i colori* (1737); Francesco Algarotti, *Sir Isaac Newton's Philosophy Explain'd for the Use of the Ladies* (1739). The English translation was reprinted in Glasgow by Robert Urie in 1765.

[411] Reid, *Philosophical Orations*, oration IV, para.4. As we shall see below, by the end of his life he was no longer as confident about the stability of the foundations of the science of optics as he was while teaching at King's.

prisms. Perhaps the most influential definition of the scope of the science in the period was that given by Ephraim Chambers in 1728:

> *Optics* in its more extensive Acceptation, is a mix'd Mathematical Science, which explains the manner wherein Vision is perform'd in the Eye; treats of Sight in the general; gives the Reasons of the several Modifications or Alterations which the Rays of Light undergo in the Eye; and shews why Objects appear sometimes greater, sometimes smaller, sometimes more distinct, sometimes more confused, sometimes nearer, sometimes more remote.

Furthermore, according to Chambers, optics was 'a considerable Branch of Natural Philosophy' because it 'account[ed] for [an] abundance of Physical Phenomena, otherwise inexplicable' such as 'Light, Colours, Transparency, Opacity, Brightness, Meteors, the Rainbow, Parrhelia, &c', and he affirmed that optics was also 'a considerable Part of Astronomy' insofar as it touched on 'the Nature of the Stars' and other phenomena observable in the heavens.[412]

Reid seemingly accepted Chambers' definition of the science of optics. Although he nowhere in his writings gave an explicit definition of the science, in the *Inquiry* he indicated that optics dealt with the nature of both light and vision. Following Newton, Reid stipulated that the first branch of the science studied 'rays of light'. And, in elaborating on the marvels illustrated by the behaviour of these rays, he provided an implicit definition of the scope of physical optics:

> The rays of light … are the most wonderful and astonishing part of the inanimate creation. We must be satisfied of this, if we consider their extreme minuteness, their inconceivable velocity, the regular variety of colours which they exhibit, the invariable laws according to which they are acted upon by other bodies, in their reflections, inflections, and refractions, without the least change of their original properties, and the facility with which they pervade bodies of great density, and of the closest texture, without resistance, without crowding or disturbing one another, without giving the least sensible impulse to the lightest bodies.[413]

[412] Chambers, *Cyclopædia*, vol. II, p. 667; compare John Harris, *Lexicon technicum: Or, an Universal English Dictionary of Arts and Sciences: Explaining not only the Terms of Art, but the Arts Themselves*, second edition (1708), s.v. 'optics'; Hutton, *Mathematical and Philosophical Dictionary*, vol. II, p. 175.

[413] Reid, *Inquiry*, p. 77; see also the phenomena listed on p. 79. The full title of Newton's *Opticks* can be read as defining the scope of optics – *Opticks: Or, a Treatise of the Reflections, Refractions, Inflections and Colours of Light*. Moreover, the first definition of Book I tacitly asserts that the primary object of the science is the behaviour of 'rays of light' insofar as Newton here defines what he means by that term.

As for the second branch, he implied that the study of vision was grounded on an understanding of 'the structure of the eye, and of all its appurtenances', as well as the laws governing the faculty of sight.[414] Scholars have hitherto focused on Reid's theory of vision and have analysed in some detail his contribution to this branch of optics.[415] However, equally important aspects of Reid's work in the science of optics have been overlooked. This section of the Introduction explores a facet of his research that has largely been ignored in the literature, namely his work in physical optics. In this section we shall examine the lectures on optics he gave at King's College, Aberdeen, the remarkable series of manuscripts on the aberration of light that he composed in the 1770s and 1780s and other related writings from the period. As we shall see, these manuscripts suggest that during the last quarter century of his life the problem of aberration and the related issue of the validity of Newton's law of refraction became principal foci of his researches in mathematics and the sciences of nature.

Reid's interest in physical optics presumably originated in the lectures on natural philosophy given by his regent at Marischal College, George Turnbull. As Christine Shepherd has shown, from the 1680s onwards the salient details of Newton's controversial 1672 paper on light and colours featured in the teaching of the regents at both Aberdeen colleges and, by the time Reid entered Marischal in the autumn of 1722, the science of optics was presented in a thoroughly Newtonian manner. In Turnbull's 1723 graduation thesis, he elaborated on the comments of his predecessors when he proclaimed that Newton had triumphantly employed the method of analysis and synthesis in the investigation of the 'innate differences between rays of light as regards refrangibility, reflexibility and colour, and their alternating phases of readier reflection and transmission, as well as the properties of the bodies, both opaque and transparent, on which the reflections and colours of the rays depend'. Thus it would seem that, at the very least, Turnbull summarised for his students the main experimental and theoretical findings of Newton's *Opticks*, although it is unclear just how much technical detail he included in his lectures.[416] Regrettably, our knowledge of Reid's engagement with physical optics prior to his appointment at King's College in 1751 is limited. Given that he began

[414] Reid, *Inquiry*, pp. 77, 79. The topics covered in chapter 6 of the *Inquiry* further illustrate Reid's understanding of the scope of the study of vision.

[415] See, *inter alia*, Geoffrey Cantor, 'Berkeley, Reid and the Mathematization of Mid-Eighteenth-Century Optics'; Norman Daniels, *Thomas Reid's 'Inquiry': The Geometry of Visibles and the Case for Realism*; Lorne Falkenstein, 'Reid and Smith on Vision'.

[416] Isaac Newton, 'A Letter of Mr Isaac Newton, Professor of the Mathematicks in the University of Cambridge; Containing His New Theory About Light and Colors'; Shepherd, 'Newtonianism in Scottish Universities in the Seventeenth Century', esp. pp. 77–9; Turnbull, 'On the Association of Natural Science with Moral Philosophy', pp. 49–50.

studying Newton's *Principia* while he was a divinity student at Marischal College in the late 1720s, he may well have digested the implications of propositions 94 to 98 in section 14 of Book I that dealt with the reflection, refraction and inflection of light.[417] A few years later, he probably read Robert Smith's highly influential Newtonian textbook, *A Compleat System of Opticks*, not long after its publication in 1738, given that John Stewart was a subscriber to the work.[418] Moreover, it may be that in the 1730s or 1740s Reid was also prompted to consider the problem of the aberration of light given his familiarity with Bradley's work on stellar aberration.[419] However, no direct evidence regarding his reading in the literature on optics or his optical ideas survives from his early years in Aberdeen or his days in New Machar.

As noted in Section 1, Reid taught his initial course of natural philosophy at King's College in the academic session for 1752–53. Although we cannot document the overall structure and specific contents of this course, his 1753 graduation oration provides us with a clue regarding the general orientation of his first sequence of lectures on the science of optics. We have seen in Section 4 that Reid commented at length on the advances made in observational astronomy through the use of the telescope and the ascendancy of Newton's system of the world. As a brief coda to his remarks, he affirmed that 'the very great genius of Newton' had 'uncover[ed] the laws governing the smallest bodies in nature as well as the largest'. In doing so, Newton had, he said, 'provided, in accounts marked in equal measure by their accuracy and acuity, explanations of [the] reflections of light rays, refractions and colours, brought into the open by experiments'.[420] Given these comments, it is likely that in 1752–53 Reid provided his students with an exposition of the fundamentals of Newtonian optics. But it is not clear if he illustrated his lectures with experimental demonstrations patterned on the experiments described in Newton's *Opticks* or in the natural philosophy textbooks of the period.[421]

The surviving set of student notes from the natural philosophy course Reid taught at King's in 1757–58 fills in the details of his subsequent treatment of optics in the classroom. Emulating Newton's *Principia* and *Opticks*, he began his series of lectures by defining the terms he used to describe the basic phenomena

[417] Newton, *The Principia*, pp. 622–9.

[418] See above, pp. xxii.

[419] See above, pp. cxiii–cxiv.

[420] Reid, *Philosophical Orations*, oration I, para. 17.

[421] We do not know what experimental hardware was available to Reid when he began teaching at King's and, as Geoffrey Cantor has reminded me, it may not have been practically feasible for Reid to perform demonstration experiments in his classroom.

of the science. Most of these definitions were unexceptional. He did, however, follow Newton in defining a ray of light as 'the smallest part of [light], which may be separated from its other parts, both successive and co[n]temporary'.[422] While some zealous disciples read Newton's definition as an affirmation that light is material in nature, Reid did not interpret him in this manner. For even though he asserted elsewhere in his course that light was made up of extremely subtle particles, he scrupulously avoided any mention of the physical constitution of light in his lectures devoted to optics and, like Newton, consistently spoke of rays or pencils of rays in a theoretically neutral fashion.[423]

After laying down his basic definitions, he proceeded to state what he called the 'Law of Motion respecting Light':

> The rays of Light when they approach very near the surface of bodies, are attracted to, or repelled from these bodies in a direction perpendicular to the Surface; and this attracting or repelling force is equal at equal distances from the surface of the same body, it is greater in more dense bodies and less in those that are more rare.[424]

Reid attributed the discovery of this 'power of bodies to act upon the rays of light at small distances' to the Jesuit natural philosopher Francesco Grimaldi, but he told his auditors that Newton was the first to have 'fully demonstrated' the properties of this power of 'inflexion' in the *Opticks*.[425] Significantly, Reid eschewed any speculation on the cause of the forces which he invoked to explain optical phenomena, as did earlier Newtonian theorists such as J. T. Desaguliers and Robert Smith. He informed his students that speculations regarding the causal

[422] Anon., 'Natural Philosophy 1758', p. 257; compare Newton, *Opticks*, pp. 1–2.

[423] Anon., 'Natural Philosophy 1758', p. 13; see also AUL, MS 2131/6/V/23, fol. 1r, where Reid cites Edmond Halley, Robert Boyle, John Keill, Bernard Nieuwentijdt and J. T. Desaguliers on the extreme subtlety of matter. Reid elsewhere implied that light is particulate; see, for example, Reid, *Inquiry*, p. 77, and Reid, *Philosophical Orations*, oration I, para. 17. Reid's ambiguity regarding the physical nature of light mirrors that of Newton; see Cantor, *Optics after Newton*, pp. 29–31.

[424] Anon., 'Natural Philosophy 1758', p. 258; compare Newton, *The Principia*, pp. 622–6, and Newton, *Opticks*, p. 79. Reid was not entirely faithful to the letter of Newton's writings on optics, insofar as his 'law' was, for Newton, a supposition. The shift in terminology is noteworthy insofar as Reid's wording reflects the dogmatism among Newton's followers that grew during the course of the first half of the eighteenth century. Geoffrey Cantor has suggested that the increasingly dogmatic presentation of Newton's ideas was rooted in the pedagogical concerns of the Newtonians; see Cantor, *Optics after Newton*, p. 47. Reid's lectures on optics instantiate this development.

[425] Newton was indebted to experiments first described in Francesco Grimaldi, *Physico-mathesis de lumine, coloribus et iride, alijsque sequenti pagina indicatis* (1665); see A. Rupert Hall, 'Beyond the Fringe: Diffraction as Seen by Grimaldi, Fabri, Hooke and Newton'.

mechanisms responsible for the reflection, refraction and inflection of light were more properly a part of 'Pneumaticks' rather than 'Physicks'.[426]

With the requisite preliminaries dispatched, Reid propounded various propositions dealing with the rectilinear propagation of light and the laws governing reflection and refraction.[427] Of the sine law initially stated by Willebrord Snell in 1621, he observed that 'Sir Isaac Newton was the first who demonstrated [this] grand principle in optics … not from an hypothesis but from the general law of Bodies acting on light'.[428] According to Reid, Snell discovered experimentally that there was a given ratio between the angles of incidence and reflection, and from this discovery Descartes had inferred the sine law, without acknowledging his source. After Descartes' publication of the law, he continued, it was recognised as a basic principle in optics and confirmed by further experimentation. He noted, however, that prior to Newton the sine law had been explained by Pierre de Fermat and Leibniz on the basis of 'uncertain and precarious Hypotheses'. Both men, he claimed, had made two unwarranted assumptions: (i) that the velocity of light increases in proportion to the rarity of the medium and conversely decreases in proportion to density; and (ii) that a ray of light either takes the least possible time to traverse its path or follows the shortest possible path. While he allowed that Fermat and Leibniz had succeeded in deriving the sine law on the basis of these assumptions, he nevertheless warned his students that 'there has not perhaps been a more ingenious solution of any natural Phenomenon invented by human wit; and I mention it to show that the most ingenious men when they trust to Hypothesis … have only the chance of going wrong in a more ingenious way'.[429] Although Reid here reiterated a familiar methodological moral, in this instance his use of history served an important didactic function. For in presenting Newton as the discoverer of a genuine law of nature from which the laws of optics could be deduced and portraying Fermat and Leibniz as philosophers deluded by their

[426] Anon., 'Natural Philosophy 1758', p. 10; J. T. Desaguliers, 'Optical Experiments Made in the Beginning of August 1728, before the President and Several Members of the Royal Society, and Other Gentlemen of Several Nations, upon Occasion of Signior Rizzeti's Opticks, with an Account of the Said Book', p. 614; and Smith, *A Compleat System of Opticks*, vol. I, pp. 90–1. On Reid's distinction between 'Pneumaticks' and 'Physicks', see also above, pp. lxi–lxii.

[427] Anon., 'Natural Philosophy 1758', pp. 258–65. Reid here referred to Ole Römer's detection of the velocity of light and James Bradley's discovery of the aberration of the fixed stars, topics which he had earlier discussed in the context of his classes on astronomy; see Anon., 'Natural Philosophy 1758', pp. 193–9.

[428] Anon., 'Natural Philosophy 1758', p. 262; Reid cited the demonstration in proposition 94, Book I of Newton, *The Principia*, pp. 622–3.

[429] Anon., 'Natural Philosophy 1758', pp. 262–3. It may be that Reid derived his historical details regarding Snell's discovery of the sine law from Smith, 'The Author's Remarks upon the Whole Work', in Smith, *A Compleat System of Opticks*, vol. II, p. 1.

own hypotheses, he highlighted the contrast between the truth of the doctrines he was expounding and the falsity of alternative theories. In effect, his potted history served to discredit rival systems and thus satisfied the rhetorical imperatives of effective pedagogy.

After completing his exposition of the simplest cases in which individual rays of light are reflected and refracted, Reid went on to consider topics such as catoptrics and the laws of vision, which need not concern us. What stands out about his optics lectures as a whole is the cursory treatment he gave to physical optics. Apart from attributing to bodies the power to act on light at a distance, he confined himself almost entirely to supplying succinct mathematical descriptions of the phenomena he surveyed. Moreover, he discussed only a limited range of optical phenomena, restricting himself to reflection and refraction while ignoring the inflection or diffraction of light. Nor apparently did he touch on Newton's explanation of colours, let alone outline more esoteric concepts such as Newton's fits of easy reflection and transmission.[430] Pedagogical considerations probably account for these features of his lectures, insofar as he may have felt that a knowledge of the principles of catoptrics, for example, was of more practical value to his students than a familiarity with contested issues of physical theory. Constraints of simplicity and clarity would also have militated against the inclusion of abstruse theoretical concepts and complex mathematical derivations. Admittedly, he examined human vision in considerable detail in the classroom, although he did so because this branch of optics was directly related to his pneumatology lectures and much of his own research had hitherto been devoted to the study of sight. Hence it would seem that his choice of topics was conditioned not only by his own research interests but also by the need to make his lectures intelligible to, and useful for, his adolescent students.[431]

The breadth of Reid's interest in physical optics can be gauged more accurately in the set of reading notes taken in March 1757 from Thomas Melvill's 'Observations on Light and Colours'. Melvill's 'Observations' was one of the

[430] Surprisingly, George Turnbull apparently lectured on Newton's theory of fits of easy reflection and transmission; see Turnbull's comments quoted above on p. cxxxvii. The range of topics covered by Reid was even more circumscribed than that found in some of the standard natural philosophy textbooks. For example, Willem Jacob 's Gravesande discussed reflection, refraction, inflection, the colours of thin plates and of natural bodies, mirrors, lenses, optical instruments, the eye and vision in his *An Explanation of the Newtonian Philosophy*, pp. 215–304; see also 's Gravesande, *Mathematical Elements of Natural Philosophy*, vol. II, pp. 26–148. However, truncated treatments of physical optics similar to Reid's can be found in Rowning, *A Compendious System of Natural Philosophy*, vol. I, part III, pp. 3–17, and Helsham, *A Course of Lectures in Natural Philosophy*, pp. 287–310.

[431] Compare Geoffrey Cantor's analysis of the institutionalisation of projectile optics in Cantor, *Optics after Newton*, pp. 47–9.

most accomplished papers on optics published in the 1750s, and Reid studied it with care.[432] In the 'Observations' Melvill contended that since beams of light can intersect 'without occasioning the least perceivable confusion or deviation from [their] rectilinear course', it followed that light had to consist of exceedingly minute particles projected from luminous bodies and widely spaced along the rays.[433] Apparently impressed, Reid summarised this argument and remarked on Melvill's reference to the Ragusan polymath Ruđer Josip Bošković:

> Here the Author mentions an Ingenious System of Boscovich a Roman Professor in his Dissert[atio] de Lumine & viribus vivis that the Elements of Matter are Indivisible points endowed with an insuperable repulsive power reaching to a finite Distance.[434]

Melvill's explanations of the production of heat by light, the appearance of water droplets on leaves and the colours of bodies also caught Reid's attention, as did Melvill's conjecture that the refractive dispersion of the differently coloured rays of light was linked to the velocity rather than the size of their constituent particles.[435] Reid evidently gave Melvill's conjecture serious thought over the years. An undated fragment written in a late hand survives wherein he suggests that if we assume 'the Rays of Light that give the Different Colours to have Different Velocities', it followed that the blue rays emitted by a fixed star 'shall come to our Eye five minutes later than' the red rays. On the basis of this assumption, he concluded that 'according to the Ptolemaick System', the light coming from

[432] Unfortunately, there is little scholarly commentary on the work of Melvill; see Cantor, *Optics After Newton*, pp. 50–1, 64–6; Jean Eisenstaedt, 'L'optique balistique Newtonienne à l'épreuve des satellites de Jupiter', pp. 138–46; Kurt Møller Pedersen, 'Water-filled Telescopes and the Pre-history of Fresnel's Ether Dragging', pp. 504–12.

[433] Thomas Melvill, 'Observations on Light and Colours', p. 13; Melvill's paper incorporated material from an earlier essay which had appeared in the *Philosophical Transactions*; Thomas Melvill, 'A Letter from Mr T. Melvil [*sic*] to the Rev. James Bradley, D.D. F.R.S. with a Discourse concerning the Cause of the Different Refrangibility of the Rays of Light'.

[434] Melvill, 'Observations on Light and Colours', p. 18; AUL, MS 2131/3/I/10, fol. 1r. In his paragraph on Bošković's 'ingenious system' Melvill refers to 'his Dissert. de lumine et de viribus vivis'. The reference to the work on optics is ambiguous because it could be to either one of the two parts of Bošković's *Dissertatio de lumine* published in 1748 or the uniform edition of the two parts published in 1749; see Edoardo Proverbio (ed.), *Catalogo delle opere a stampa di Ruggiero Giuseppe Boscovich (1711–1787)*, pp. 54–5. The other work Melvill referred to is Ruđer Josip Bošković, *De viribus vivis dissertatio* (1745). Reid also mentions Bošković's matter theory in an undated manuscript related to his teaching; see 'Idea of a Course of Physicks or Natural Philosophy', AUL, MS 2131/2/I/5, p. 4.

[435] Reid took no notes on Melvill's refutation of Leonhard Euler's wave theory of light, his discussion of inflection, or the lengthy set of queries that conclude the paper; see Melvill, 'Observations on Light and Colours', pp. 36–53, 59–90.

a fixed star should 'appear to us as an oblong coloured Spectrum' (p. 123).[436] Apart from illustrating Reid's attempt to determine the consequences of Melvill's hypothesis, this fragment also points to the close connection between optics and astronomy in Reid's thought. We shall see below that after his move to Glasgow in 1764, this connection provided the main stimulus for his optical researches.

Reid's notes from Melvill's 'Observations on Light and Colours', as well as Melvill's widely read work on optics, bring into question aspects of the accounts of Bošković's reception in eighteenth-century Scotland put forward by Richard Olson and Mordechai Feingold.[437] First, Reid's reading notes and Melvill's paper demonstrate that Scottish savants knew of Bošković's theory of point atoms well before the 1770s and 1780s, which Olson and Feingold identify as the pivotal decades in which Scottish men of science began to investigate Bošković's matter theory.[438] Secondly, insofar as Melvill and Reid encountered Boškovićian point atomism in the context of mid-eighteenth-century debates over the nature of light, the distinction Olson draws between Scottish reactions to Bošković's concept of matter, as opposed to his optical theories, is not as straightforward as Olson implies. Thirdly, Melvill's assessment of Bošković's matter theory in terms of its capacity to solve specific problems in optics indicates that empirical considerations were just as important as the epistemological and metaphysical factors that Olson and Feingold both claim encouraged Scottish men of science to study Bošković's works.[439] Thus Melvill's 'Observations on Light and Colours', along with Reid's reading notes, indicate that in order to recount the full story of Bošković's reception in Scotland more attention needs to be paid to the period *circa* 1750 to *circa* 1775, when Bošković's ideas began to circulate among

[436] Compare Melvill, 'Observations on Light and Colours', pp. 40–53, and Melvill, 'A Letter from Mr T. Melvil', p. 262. In speculating on the implications of Melvill's conjecture, Reid ignored James Short's criticisms of Melvill's hypothesis; see Melvill, 'A Letter from Mr T. Melvil', pp. 268–70.

[437] Mordechai Feingold, 'A Jesuit among Protestants: Boscovich in England *c.* 1745–1820'; Richard Olson, 'The Reception of Boscovich's Ideas in Scotland'. Notwithstanding the title of his essay, Feingold discusses the reactions of Edinburgh men of science to the ideas of Bošković in the latter part of the eighteenth century.

[438] Olson, 'The Reception of Boscovich's Ideas in Scotland', pp. 92–6; Olson asserts (p. 95) that Bošković's writings were unknown to the Scots before 1773. Feingold cites Melvill's 'Observations on Light and Colours' but nevertheless focuses on the last three decades of the eighteenth century; Feingold, 'A Jesuit among Protestants', p. 517. His chronology is thus effectively the same as Olson's. Both take their cue from John Playfair's comment that 'Boscovich's theory was hardly known in this country till about the year 1770'; John Playfair, 'Biographical Account of James Hutton, M.D.', p. 88 note.

[439] Olson, 'The Reception of Boscovich's Ideas in Scotland', pp. 92–6; Feingold, 'A Jesuit among Protestants', p. 520.

Scottish savants such as Melvill and Reid, and to the optical context in which Boškovićian point atomism was initially introduced.[440]

While he was drafting the *Inquiry* during the early 1760s Reid apparently had almost no time to devote to physical optics, even though he taught his natural philosophy course at King's in the sessions for 1760–61 and 1763–64. However, his knowledge of the field is registered in the *Inquiry*. As we have seen above, in describing the behaviour of rays of light Reid implicitly defined the scope of physical optics. And although he continued to speak of 'rays of light', the properties he ascribed to them, such as their 'extreme minuteness', implied that light was in fact material and particulate in nature.[441] Moreover, his characterisation of rays of light indicates that he was conversant with some of the standard arguments used in the first half of the eighteenth century to prove that the particles of light are exceedingly small.[442] Reid evidently also shared the theological intent of some of the earlier writers who formulated these arguments, for he emphasised that the behaviour of light manifests the design inherent in the natural order. In addition, Reid's comments on the nature of light in the *Inquiry* speak to his immersion in the details of Newton's *Opticks*. While it would seem that, in his optics lectures, he did not mention the point that Newton ascribed immutable properties to rays of light, it is significant that in the *Inquiry* he echoes Newton when he says that the rays of light have the capacity to interact with 'other bodies' and to undergo 'reflections, inflections, and refractions, without the least change of their original properties'.[443] Elsewhere in the *Inquiry*, he cited Newton's calculation of the size of the particles of opaque bodies on the basis of colour as an example of how we can learn about the constitution of the material world on the basis of the careful investigation of the secondary qualities of matter.[444] Thus, even though Reid's mastery of the literature on the theory of vision was manifest in the *Inquiry*, a careful reading of the text also reveals passages which attest to his firm grasp of the principles of Newtonian optics.

Reid's move to Glasgow in 1764 appears to have rekindled his interest in physical optics. As we have seen in Section 1, in the summer of 1765 Reid informed his friend David Skene that for the first time he was able to study

[440] Feingold's and Olson's discussions of John Robison's engagement with Bošković's ideas have been superseded by the work of David B. Wilson; Wilson, *Seeking Nature's Logic*, pp. 256–69.

[441] Reid, *Inquiry*, p. 77.

[442] Reid echoes writers such as George Cheyne and Bernard Nieuwentijdt; Cantor, *Optics after Newton*, pp. 32–3.

[443] Reid, *Inquiry*, p. 77; compare especially Query 29 in Newton, *Opticks*, pp. 370–4.

[444] Reid, *Inquiry*, p. 47. Reid implicitly refers to proposition 7, Book II, Part iii, in Newton, *Opticks*, pp. 255–62.

the achromatic telescopes manufactured by the London instrument maker John Dollond. Impressed by Dollond's telescopes, he enthused that 'they open a new field in Opticks which may greatly enrich that part of Philosophy'. He also mentioned that he had witnessed experiments with prisms made of various types of glass and crystal and commented that 'the Laws of the Refraction of Light seem to be very different in different kinds both of Glass and of native Chrystal'.[445] In saying this, Reid registered the growing awareness among men of science in Britain that Newton's explanation of the refractive dispersion of colours required either refinement or replacement. For whereas Newton had insisted that refracting telescopes would necessarily produce chromatic aberration, Dollond's achromatic telescopes demonstrated that Newton had been wrong. Moreover, Reid's dissatisfaction with the state of this branch of optics is manifest in his remark that 'the theory' governing the construction of Dollond's telescopes had not been 'prosecuted as it ought'.[446] He subsequently returned to the consideration of Newton's account of refraction in an undated manuscript written in a late form of his handwriting (pp. 111–15) which shows not only that he continued to think about the problem of refractive dispersion but also that he had come to the conclusion that Newton's treatment of the refraction of light was fundamentally flawed.

The body of this manuscript is devoted to the calculation of the angles of incidence and refraction of the 'least Refrangible' (red) and 'most Refrangible' (violet) rays of light passing through different media, as well as glass and water prisms. There is no indication in this part of the manuscript of the weighty theoretical issues at stake. But on an unpaginated leaf following the text in which Reid calculated various sines and tangents he wrote:

> As it is now found that of different Refracting Media, one Scatters the different colloured rays more, another less; May we not hope to find some Medium that refracts without separating them at all[.] Sir Isaac Newton finding in all his Experiments, that the Ratio of the difference between the Sines of Incidence and of Refraction of the Most Refrangible rays, to the difference of the Sine of incidence to that of refraction of the least refrangible was always as 27 to 28

[445] Reid to David Skene, 13 July 1765, in Reid, *Correspondence*, p. 40. Reid indicated that the experiments with prisms replicated 'Sir I Newtons Experiment' and described the different coloured spectra produced by prisms made of 'brazil peeble', 'German Native Chrystal' and a special type of glass. On the optical principles guiding the construction of Dollond's achromatic telescopes, see John Dollond, 'An Account of Some Experiments concerning the Different Refrangibility of Light'; on Dollond's telescopes, see especially Richard Sorrenson, 'Dollond & Son's Pursuit of Achromaticity, 1758–89'.

[446] Reid to Skene, 13 July 1765, in Reid, *Correspondence*, p. 40; Cantor, *Optics after Newton*, pp. 64–9.

> concluded too rashly that this Ratio holds in all Refractions. He conjectured that the disposition of one ray to be more refracted than another depended solely on different inherent qualities of the differently coloured rays.[447]

As this note indicates, Reid questioned the validity of Newton's theory of refraction and even went so far as to suggest that Newton had incorrectly 'conjectured' that refractive dispersion could be explained in terms of the inherent properties of the differently coloured rays of light. Reid had thus come to reject one of the fundamental tenets of Newtonian optics. Moreover, he continued to reflect on the problem of refractive dispersion into the 1790s, for one of his last surviving sets of reading notes was taken in November 1792 from Robert Blair's lengthy paper later published in the *Transactions of the Royal Society of Edinburgh* on the refractive properties of different media, including various types of glass, chemical solutions and the essential oils of lemon and rosemary.[448]

A second topic in physical optics investigated by Reid in Glasgow was the aberration of light. We have seen in Section 4 that Reid greatly admired the work of James Bradley and, in particular, Bradley's discovery of the aberration of the fixed stars. Yet Bradley's account of aberration was the subject of a debate in the newly founded Aberdeen Philosophical Society, for Reid's friend Robert Traill proposed the question 'How far the Motion of the Earth and of Light accounts for the Aberration of the Fixed Stars?', which was discussed on 12 April 1758.[449] We do not know what the members of the Wise Club had to say about stellar aberration, but their discussion presumably made Reid realise that Bradley's theory was not above criticism. Whether Reid and Traill continued their conversations about aberration in Glasgow cannot now be determined, but the issue raised by Traill was taken up by Alexander Wilson's son, Patrick, whose research on the aberration of light inspired Reid to reconsider the question previously posed in the Wise Club. According to John Robison, 'about the year 1767 or 1768, [Patrick] Wilson entertained an opinion that the aberration of the fixed stars indicated the proportion between the orbital v[e]locity of the earth, and the velocity of light in the vitreous humour of the eye'.[450] Robison, who was then

[447] AUL, MS 2131/7/II/11, unpaginated folio, and below, p. 243, textual note 115/9.

[448] AUL, MS 2131/3/I/3; Robert Blair, 'Experiments and Observations on the Unequal Refrangibility of Light'. Blair was the Regius Professor of Astronomy at the University of Edinburgh, although he treated his position as a sinecure; see Emerson, *Academic Patronage in the Scottish Enlightenment*, p. 334.

[449] Ulman, *Minutes of the Aberdeen Philosophical Society*, pp. 83, 189. At the same meeting, Traill and his colleagues also discussed his question regarding the transit of Venus in 1761; see above, pp. xlv–xlvi.

[450] [John Robison], 'Mr Wilson on the Solar System', pp. 28–9.

the Lecturer in Chemistry at Glasgow, must have encouraged Wilson to pursue his idea, because in a memorandum dated 13 October 1770 Reid wrote that 'Patrick Wilson communicated to me an Observation of his own Upon the Aberration of the Rays of Light which tho' very curious Seems to have escaped all who have wrote upon this Subject' (p. 88). According to Reid's summary of their discussion, Wilson contended that the aberration of light depended upon the velocity of rays of light as they strike the retina rather than on their velocity 'in the Air or Ether' (p. 88).[451] Wilson reasoned that if the velocity of light was different in the eye from that in the air, then only the first of these velocities could be deduced from the aberration of the fixed stars. He inferred from this that if the refractive power of the humours of the eye was assumed to be roughly the same as that of water, then according to Newtonian doctrine the ratio of the velocity of light in the eye to that in the air would be 4:3, and hence the velocity of the rays deduced from the aberration of the fixed stars would be in the ratio of 4:3 to their velocity in the air. Wilson therefore concluded (according to Reid) that 'all Equations of the Motions of the Heavenly Bodies arising from the progressive Motion of Light ought to be reduced by a fourth part, if they be grounded upon that Velocity of Light which is inferred from the Aberration of the fixed Stars' (p. 88).

Two days later, on 15 October, a further conversation took place in which Wilson described to Reid a method he had devised to eliminate the errors arising from the aberration of light from all astronomical observations. Unfortunately, Reid did not record the details of Wilson's method. He did, however, mention that it was based on the principle that 'when Light is reflected from any Surface it must have the Motion of that Surface imparted to it'. Reid added:

[451] Reid was probably unaware of the fact that Thomas Melvill had made the same criticism of Bradley's account of the aberration of the light in 1753; see Melvill to James Bradley, 2 June 1753, in Stephen Peter Rigaud (ed.) *Miscellaneous Works and Correspondence of the Rev. James Bradley, D.D. F.R.S. Astronomer Royal, Savilian Professor of Astronomy in the University of Oxford &c.*, p. 485. As we have seen, p. liii, Melvill and Patrick Wilson's father Alexander were close associates in the period *circa* 1748 to *circa* 1750. It is therefore possible that Alexander Wilson knew of Melvill's interpretation of Bradley's observations, although it should be borne in mind that in the last years of Melvill's short life he was apparently a peripatetic tutor and was not necessarily in regular contact with the elder Wilson. Kurt Møller Pedersen contends that Patrick Wilson must have known about Melvill's letter to Bradley, and says that this letter 'circulated in scientific circles, and was later pharaphrazed [*sic*] by Priestley in 1772'; Pedersen, 'Water-filled Telescopes and the Pre-history of Fresnel's Ether Dragging', p. 516. However, Priestley was clearly summarising Melvill's published papers; compare Joseph Priestley, *The History and Present State of Discoveries Relating to Vision, Light and Colours* (1772), pp. 401–4, with Melvill, 'A Letter from Mr T. Melvil' and 'Observations on Light and Colours'.

> As to the first observation [on the theory of aberration], It is the aberration of the Telescope rather than that of the Rays of Light that we observe. There is probably an Aberration of the Axis of the Eye from the fixed star we look at. But we have no means of measuring this Aberration. The Aberration of the Telescope must depend upon the Velocity of the Rays in the Telescope onely. (p. 89)[452]

As we shall see, the questions raised in these exchanges regarding the clarification of the various aspects of the phenomena of aberration and the determination of the behaviour of light when meeting moving surfaces occupied Reid at various times during the following two decades.

Shortly after these conversations took place, Reid listened to Patrick Wilson state his theory of aberration in a formal setting. Although Wilson was not a member of the Glasgow Literary Society, he was given permission to deliver a discourse to the Society on 23 November 1770 'concerning the Velocity of Light shewing that this as deduced from Dr. Bradley's Theory of Aberration of the fixed Stars is erroneous and a new Principle which affects that pointed out and examined'.[453] While the text of his discourse does not appear to have survived and the minutes of the Literary Society do not record the details of the reaction of those at the meeting to Wilson's presentation, a summary of his paper appeared in the *London Chronicle* for 1–4 December 1770.

As summarised in the anonymous report, Wilson's discourse was divided into two parts. In the first section, Wilson challenged Bradley's assumption that we can deduce the velocity of light from the angle of aberration of a fixed star. He countered that on the basis of this angle we can only infer the velocity 'with which the rays [of light] impinge upon the *tunica retina*'. In order to determine the true velocity of light, he argued that we must take into account the refractive power of the humours of the eye, which he said was the same as that of water. Hence the velocity of light in the eye was to its velocity in space as the ratio of 4:3, determined by Newton's theory of refraction. This meant that Bradley's estimate of 8' 12" for the time it took a ray of light emitted by the Sun to reach the Earth had to be corrected to 10' 56", which approximated the figure initially suggested by Ole Römer of roughly 11'. In the second section of the paper, he maintained that if Römer's figure could be verified, this would constitute 'the most cogent of all arguments in favour of the Newtonian doctrine of refraction'. Wilson noted that even though Newton had succeeded in demonstrating 'the grand proposition of optic[s], viz. of the sines of incidence and refraction being in a constant ratio', he had done so on the basis of a 'very simple assumption: "namely that a body

[452] It is unclear whether it was Reid or Wilson who made this point.

[453] 'Laws of the Literary Society in Glasgow College', p. 36.

refracts light by acting upon its rays in lines perpendicular to its surface"'. Leibniz had, however, justifiably questioned this assumption and had asked whether Newton and his followers had in fact proven that light has a 'greater velocity in the denser medium by any direct experiment'. According to Wilson, the comparison between Römer's and Bradley's estimates of the velocity of light provided an 'experimental proof' of the validity of Newton's assumption, although he granted that astronomers had yet to establish the accuracy of Römer's calculation. For Wilson, therefore, the phenomenon of the aberration of light served as an *experimentum crucis* which would determine the truth or falsity of Newton's theory of refraction. And if Römer was shown to be correct, then 'the Newtonian Philosophy will ... have a new triumph' and men of science and the public alike would then realise that 'the more we search into the doctrines of this man [Newton], the more ... the voice of Nature [is] heard on his side; as if he had beheld her mysteries by a sort of intuition bestowed upon Him, her great Interpreter'.[454]

Details of the next phase in Wilson's theorising about the aberration of light are recorded in a draft of a letter of introduction to Richard Price written by Reid on behalf of Wilson in the spring of 1772.[455] Reid's letter suggests that Wilson had continued to refine the argument of the discourse on aberration that he had read to the Glasgow Literary Society, and it also reveals that a copy of the revised text had been sent to the Astronomer Royal, Nevil Maskelyne, who had then lent the paper to Price. Presumably the younger Wilson's essay had been transmitted by his father to Maskelyne, because the two older men were senior colleagues in the British astronomical community who periodically corresponded with one another. Although Price is not known for his work in astronomy, in 1771 he had published a paper on the aberration of light and could thus provide informed criticism of Wilson's ideas.[456] Maskelyne had seemingly returned the essay to Patrick Wilson, along with comments by himself and Price. Both Wilson and Reid

[454] X., 'To the Printer of the London Chronicle', *London Chronicle*, 1–4 December 1770, p. 532; the report was sent from 'Glasgow College' on 27 November 1770. See also Pedersen, 'Water-filled Telescopes and the Pre-history of Fresnel's Ether Dragging', pp. 514–15.

[455] Reid to Richard Price, [1772/73], in Reid, *Correspondence*, p. 63. Reid stated that Patrick Wilson was travelling to London 'about Bussiness', which was most likely related to the type foundry operated by the Wilson family. Wilson was in London in the summer of 1772; see Patrick Wilson to Benjamin Franklin, 3 August 1772, in Benjamin Franklin, *The Papers of Benjamin Franklin*, vol. XIX, p. 228. Reid's letter to Price dates from the spring of 1772, because in the letter (p. 64) he says that 'I have not yet had leisure to examine your Sentiments upon the National Debt, with the Attention I would wish'. Reid refers to Price's *An Appeal to the Public, on the Subject of the National Debt* (1772), which was reviewed in the *Monthly Review* for April 1772. The date of the review suggests that Reid's letter was probably written in the period April to June 1772.

[456] Richard Price, 'A Letter from Richard Price, D.D. F.R.S. to Benjamin Franklin, LL.D. F.R.S. on the Effect of the Aberration of Light on the Time of a Transit of Venus over the Sun'.

disagreed with their critical observations. Reid informed his correspondent that 'Mr Maskelynes Remarks do not satisfy either [Wilson] or me that the Notion he has started is void of foundation'. He therefore encouraged Price to converse with Wilson while the latter was in the metropolis because he was 'very hopefull that you will understand one another more clearly & come to be of one Mind'.[457] We do not know if Wilson actually discussed his ideas with Price during his visit to London or whether he came to accept the validity of Maskelyne's and Price's criticisms of the theory of aberration that he advanced before the Literary Society in November 1770. Nevertheless, we do know that at some point after his trip south, he abandoned his initial theory and began to formulate a revised analysis of the aberration of light.

A few years before his death in December 1811, Patrick Wilson told the editors of the 1809 abridged edition of the collected *Philosophical Transactions* that his new interpretation of aberration had taken shape 'before the end of 1772'. He offered as proof 'original letters in his possession, especially from one gentleman [Maskelyne], of the first eminence as an astronomer and mathematician, who with the greatest liberality and candour honoured, and warmly encouraged him in [his] researches by his correspondence'.[458] Given the comments in Reid's letter to Price, it is unlikely that Wilson completely altered his views immediately after returning to Glasgow in 1772. But the fact that Wilson subsequently read a second discourse to the Glasgow Literary Society, on 31 January 1777, entitled 'Of the Refraction and Refrangibility of Light', indicates that by the time he gave this discourse he had worked out the basics of the revised account of the aberration of light that eventually appeared in the paper that was sent to Maskelyne in late 1781 or very early in 1782 and read at the Royal Society of London on 24 January 1782.[459] In his revamped theory, Wilson jettisoned the claim that aberration was a function of the velocity of the rays of light in the humours of the eye. He did so because he realised that his initial theory implied that 'when the aberration of a star, near the pole of the ecliptic, lay at right angles to a horizontal wire passing through the centre of the field, when the telescope turned in a vertical circle' the image of the star would 'disappear when at some small distance from the wire' and, when the image and the wire coincided, the image would 'appear visible

[457] Reid to Price, [1772/73], in Reid, *Correspondence*, p. 63.

[458] See the editorial note to the 1809 abridged version of Patrick Wilson, 'An Experiment Proposed for Determining, by the Aberration of the Fixed Stars, whether the Rays of Light, in Pervading Different Media, Change Their Velocity According to the Law Which Results from Sir Isaac Newton's Ideas concerning the Cause of Refraction; and for Ascertaining Their Velocity in Every Medium Whose Refractive Density Is Known', p. 193.

[459] 'Laws of the Literary Society in Glasgow College', pp. 45–6; Wilson, 'An Experiment'.

upon it, instead of being hid behind it'.[460] Yet this visual phenomenon had not been observed by Bradley or by any other astronomer. Wilson was thus forced to acknowledge that, contrary to what he had first thought, the optics of the eye plays no role in the aberration of light because 'the aberration of the axis of the eye and that of the telescope must precisely agree, notwithstanding the acceleration of the ray on entering the eye, as resulting from Newton's doctrine of refraction'.[461]

Wilson also realised that if we assume that the eye has the same refractive power as water, then 'a telescope of any length filled with water, or any dense clear fluid, between the object glass and the wires at the focus, would shew the very same aberration with … any other telescope, having air only within it'. He then understood that if this was the case, it meant that 'such an agreement between a water and an air telescope, if actually found by observation, would constitute a proof of the acceleration of light in the dense medium, in the ratio assigned by Newton'. Newton's postulate that the velocity of light increases in relation to the density of the medium through which it travels is what Wilson originally sought to prove in 1770 and what Wilson believed he could demonstrate in his paper published in 1782. Moreover, on the basis of his revised theory, he was led to one further conclusion that subsequently took on considerable significance in the context of Scottish responses to the work on aberration published by Ruđer Josip Bošković in 1785, namely that in the case of terrestrial objects, no aberration would be observed with either a water or a regular telescope. Wilson later said that he was led to this insight because he finally recognised a 'very important circumstance', namely that although 'the real or absolute path of [a] ray [of light], before entering [a telescope], is inclined to the [telescope's] moving axis, and consequently to the surface of the water', the refraction of the ray upon entry alters the absolute path 'so as to make a less absolute angle than it did before with the moving axis and the water telescope'. This change in the absolute direction of the ray of light, which was simultaneously accompanied by an increase in velocity of the ray, meant that 'these two different but concomitant effects must, according to the Newtonian doctrine of refraction, precisely counteract one another, so as to make the aberration of the water and air telescopes to agree, when the star is seen in the axis of both'.[462] Wilson's paper of 1782 thus represented a considerable advance on his earlier views on aberration, for in rethinking his principles he was prompted to consider the phenomena produced by water telescopes and to expand his analysis to encompass the observation of terrestrial as well as celestial objects.

[460] See the editorial note to the abridged version of Wilson, 'An Experiment', p. 192; compare Pedersen, 'Water-filled Telescopes and the Pre-history of Fresnel's Ether Dragging', pp. 516–18.

[461] See the editorial note to the abridged version of Wilson, 'An Experiment', p. 193.

[462] See the editorial note to the abridged version of Wilson, 'An Experiment', p. 194.

Even after his paper was published, Patrick Wilson explored other avenues of enquiry related to the aberration of light. In the spring of 1784, John Robison reported to James Watt that he had recently been in Glasgow, where he learned that the younger Wilson was 'engaged on a most ingenious experiment for discovering whether this planetary System is a set of Satellites to our Sun revolving round another Centre and in what quarter of the Universe that Center is placed. This he hopes to do by means of the aberration of light'.[463] Two years later, as part of the formalities involved in his appointment to succeed his father as the Glasgow Professor of Practical Astronomy, Wilson delivered an oration in March 1786 on the set topic 'De stellarum fixarum parallaxe et distantia', in which, *inter alia*, he suggested that Newton's law of refraction could be tested empirically by using a water telescope to observe the aberration of the fixed stars.[464] As we shall see in what follows, Reid's manuscripts on the aberration of light and related topics in optics register the twists and turns of Wilson's theorising during the 1770s and 1780s.

Reid's memorandum of October 1770 demonstrates that, from the outset, he understood that Patrick Wilson's early theory of the aberration of light differed radically from the commonly accepted account initially formulated by James Bradley, given that Wilson maintained that 'Aberration depends not upon the Velocity which the Rays of light may have in the Air or Ether but upon the Velocity with which they strike upon the Retina of the Eye' (p. 88). The optics of the eye was thus a basic component of the theory of aberration for Wilson, whereas the majority of earlier analysts followed Bradley in simply considering the velocity of the rays of light as they travel from the fixed stars to the Earth.[465] Moreover, as Reid indicated in his memorandum, Wilson's early theory of aberration had significant practical implications for observational astronomers. For if Wilson was correct, at the very least the values for the positions of the fixed stars required serious adjustment. Reid followed up Wilson's suggestion that we need to take account of the optics of the eye in conceptualising the aberration of light, in a lengthy untitled manuscript (pp. 89–101), which may well have served as the basis for a discourse or an occasional paper presented to the Glasgow Literary

[463] John Robison to James Watt, 10 February 1784, in Robinson and McKie, *Partners in Science*, pp. 135–6. The younger Wilson's 'ingenious experiment' was designed to confirm the conjecture put forward in Alexander Wilson's pamphlet, *Thoughts on General Gravitation*, that the solar system is in orbital motion around a centre located elsewhere in the universe.

[464] 'Minutes of Meetings of the Faculty, 1784–1789', GUA MS 26693, minute for 2 March 1786; Clarke, *Reflections on the Astronomy of Glasgow*, p. 100; David Myles Gavine, 'Astronomy in Scotland, 1745–1900', p. 62.

[465] Thomas Melvill was a significant exception; see above, note 451.

Society at some point in the early 1770s.[466] In this manuscript, Reid addressed the problem of the aberration of light as it applied to astronomical observations and surveyed three general cases in which we do not perceive an object in its true location. The first case he examined occurs when 'the Eye is at rest and the Object Moves in a direction making an Angle with the line that Joyns the Object and the Eye' (p. 89). Because light travels at a finite speed, he argued, we see an object moving in the direction it had when the rays of light left it rather than in the direction the object possesses when the rays arrive at the eye. In an apt simile, he compared the rays of light to 'a Post who brings intelligence of the State of things when he set out but can give no intelligence of the changes that have happened since that time' (p. 90). Consequently, since we do not see celestial bodies which are in motion in their true positions, he emphasised that our observations of the motions of the planets and of comets, as well as such phenomena as the eclipses of the moons of Jupiter and Saturn, all required correction for error.

The second case outlined by Reid was also caused by the progressive motion of light, and occurs 'When the Eye moves in a Direction which makes an Angle with the Line joyning the Object & the Eye' (p. 90). Here he tackled one of the basic problems confronting astronomers, namely that of distinguishing between 'real & Relative Motion' and 'between the Real Direction of Motion or of any line & its relative direction' (p. 92).[467] After teasing out the implications of the fact that we perceive relative rather than absolute space (and hence relative rather than absolute direction), he described in elementary geometrical terms the aberration of a fixed star as observed through a telescope. In doing so, he went beyond James Bradley's analysis of aberration in discriminating between instances where the

[466] Reid's comment that 'Some of the cases of this Problem are so complex as to require a Figure & therefore we omit them since they are easily solved by plain Geometry' (p. 92) implies that the manuscript served as the basis for an oral presentation; compare his remark in his discourse on Euclid delivered to the Society (p. 28). Minutes for the Society's proceedings no longer survive for the period 5 November 1779 to 7 November 1794. There are also shorter periods clustered in the 1770s for which minutes are no longer extant. In the 1770s there are gaps in the minutes for February through to May 1771, 29 November 1771 to late January 1773 and 11 February 1773 to May 1776; see 'Laws of the Literary Society in Glasgow College', pp. 38–40. Reid's manuscript may well date to one of the earlier periods not covered by the surviving records of the Literary Society. The handwriting in the manuscript is consistent with this dating.

[467] Patrick Wilson likewise dealt with the issue of distinguishing between the real and relative motions of bodies, as did John Robison in his paper on the aberration of light read to the Royal Society of Edinburgh in April 1788; see John Robison, 'On the Motion of Light, as Affected by Refracting and Reflecting Substances, Which Are Also in Motion', pp. 85, 96. So too did Robison's colleague, Robert Blair, in a paper read to the Royal Society in April 1786 that remained unpublished; see Jean Eisenstaedt, 'Light and Relativity, a Previously Unknown Eighteenth-Century Manuscript by Robert Blair (1748–1828)'.

relative motions of a fixed star and a telescope cancel one another out and those in which the relative motions are such that they produce a shift in the apparent position of the fixed star (pp. 92–3).[468] Again, he drew out the practical consequences of the aberration of light for astronomical observation, and his corollaries are of theoretical interest for he alludes to Thomas Melvill's hypothesis that the rays of light have different velocities according to their colour (p. 94).

Reid's third case addressed the question of whether the aberration of light is a function of the physical structure of the eye. In considering this question, he draws upon his own expertise, as well as the work of such influential contributors to the theory of vision as James Jurin and William Porterfield, to elaborate on Patrick Wilson's suggestion that when we analyse the different scenarios in which the aberration of light occurs we need to take into account the anatomy of the eye.[469] Taking the simplest possible situation, in which the eye and the object are both assumed to be at rest and the path of the rays of light from the object to the eye are in a right line, Reid traced the path taken by the pencils of rays as they pass from the object to the cornea and thence to the retina. On the basis of his reconstruction of the path of the pencils of rays, he inferred three propositions: (i) that, given the above assumptions, we see the object in its true place; (ii) that in these circumstances the object is seen in the direction of a straight line drawn from the optical centre of the eye to the object; and (iii) that 'it is a Law of our Nature that we see every Visible point in the Direction of that Diameter of the Eye which passes from the Image of that point in the Retina' (p. 97).[470] He then demonstrated a fourth proposition, namely that 'If a fixed Star is seen by the Naked Eye in a direction perpendicular to the annual Motion of the Earth, it will not appear in its true place'. To do so, he based his proof on a reformulation of Bradley's method of calculating the aberration of light that utilised Wilson's variable of 'the Velocity of the Rays of Light in the Humors of the Eye'.[471] From a theoretical standpoint, therefore, he had succeeded in showing that what he called the 'Aberration of the Eye' occurs. In a fifth proposition, however, he recognised that unlike the degree of aberration produced by a telescope, the aberration of the eye cannot, in practice, be measured 'as we cannot apply a Micrometer to the Axis

468 Compare Bradley, 'An Account of a New Discovered Motion of the Fix'd Stars', pp. 646–8.

469 Reid refers to Jurin, below, p. 96, and to the *Inquiry*, p. 98. The experiments in the *Inquiry* that he cites were based on those found in Porterfield, 'An Essay concerning the Motions of Our Eyes: Of Their External Motions'; Porterfield, 'An Essay concerning the Motions of Our Eyes: Of Their Internal Motions'; William Porterfield, *A Treatise on the Eye, the Manner and Phænomena of Vision* (1759).

470 Compare Reid, *Inquiry*, pp. 120–31.

471 Compare Bradley, 'An Account of a New Discovered Motion of the Fix'd Stars', pp. 652–3.

of the Eye' (pp. 99–101). It followed from Reid's formulation of Wilson's ideas, therefore, that the observations of astronomers would inevitably contain a small but indeterminate error.

A second undated and untitled manuscript, which appears by the handwriting to have been written in much the same period as the manuscript just discussed, deals in part with the issue of how light interacts with media that are in motion (pp. 101–3).[472] Internal evidence suggests that this second manuscript is a preliminary, and highly tentative, exploration of the problem of the reflection and refraction of light by media in motion that Reid was apparently only beginning to consider.[473] Thus, despite the mathematical trappings of axioms, postulates and corollaries, towards the end of the manuscript he proposed a second problem dealing with the relative path of a ray of light refracted by a prism, only to confess that he was 'discouraged' by the difficulty of solving it and, having drawn a line across the page, he then admitted that his postulate governing the refraction of light 'may be denied' (p. 103). And, given the blank space at the bottom of the verso of the second folio of the manuscript, it would seem that his list of axioms was incomplete. Yet, despite the unfinished state of the manuscript, it is noteworthy, albeit less for what it says about his evolving views on aberration and more for what it reveals about his increasingly critical stance towards Newton's treatment of refraction during the later stages of his career. The fact that Reid recognised that his postulate regarding the refraction of light was flawed is significant because his postulate was framed in terms of the basic principles of Newton's system of optics. In particular, his postulate was based on

> the Supposition that the Rays of Light … in Refraction & Reflection, are acted upon by a Power tending perpendicularly to or from the Refracting or Reflecting Surface, a Power which is equal at equal distances from the Surface, and that the three general Laws of Motion take place in the Rays of Light. (pp. 101–2)

As we have seen, in his optics lectures at King's College Reid had characterised this 'Supposition' as a 'Law of Motion respecting Light'.[474] The change in terminology, as well as his reservations about the postulate, tell us that when he wrote this manuscript he had already begun to harbour the doubts about Newton's

[472] There is some overlap between these two manuscripts insofar as both address relative motion and the aberration of light.

[473] This problem was also taken up by John Robison; see Robison, 'On the Motion of Light'.

[474] Anon., 'Natural Philosophy 1758', p. 258; on this issue see also note 424 above and editorial note 102/4, pp. 204–5, below.

analysis of the refraction of light that were articulated in his late manuscript on the subject discussed above.

Reid continued his study of the problem of the aberration of light in his essay 'Of the Path of a Ray of Light passing through *Media* that are in Motion', which is dated 4 December 1781 (pp. 104–11). In his introductory preamble, he lamented the fact that the topic of his essay had previously been ignored by students of optics, even though Ole Römer's 'noble Discovery of the Progressive Motion and Velocity of the Rays of Light, might have led Writers on Opticks, to trace from Theory the consequences that must follow from that progressive Motion' (p. 104). As a result, 'the Ingenious Dr Bradley' had needlessly expended 'much Labour & Time' on the observation of the heavens, given that the aberration of the fixed stars could have been deduced from the first principles of optics. Hence Reid was convinced that it would be a 'usefull Supplement to the Science of Opticks' to ascertain what the theoretical and observational consequences of the progressive motion of light were in relation to 'the *Phænomena* of Vision', and to systematise the results 'in a proper Number of Propositions adapted to the various Cases & demonstrated from established Principles' (p. 104).[475]

Reid accordingly set himself the problem of investigating the behaviour of a ray of light in the general case wherein an object is at rest and the ray passes through various media (including the eye) which are in motion. To do so, he laid down two postulates: (i) that the 'Power by which the Rays of Light are Reflected or Refracted … acts onely at the Surface which divides two different Media; and Acts onely in a direction perpendicular to that Surface'; and (ii) that a ray of light passes through a uniform transparent medium without any sensible resistance regardless of whether the medium is in motion or at rest (pp. 104–5). Reid's first postulate is a further example of his theoretical quandary over Newton's account of refraction. For if we compare this postulate with his 'Law of Motion respecting Light' quoted earlier, we can see that the wording differs significantly from that of the 'Law' in that the postulate limits the action of the power affecting light to the *surface* dividing the two media, whereas his 'Law of Motion' stipulated that this power acts 'very near' to such surfaces.[476] What motivated this specific change is unclear. But it may be that Reid altered the conditions stated in the postulate

[475] Reid's emphasis on the phenomena of vision reflects his indebtedness to Patrick Wilson.

[476] See above, p. cxxxix. Reid's 'Law of Motion respecting Light' rested on Newton's belief that optical phenomena were to be explained in terms of forces of attraction and repulsion that acted over short distances. As Cantor has shown, in the first half of the eighteenth century, projectile theorists elaborated on Newton's assumption; Cantor, *Optics after Newton*, pp. 28–9, 34–41. Implicitly, Reid's postulate questions this common assumption.

in order to simplify the mathematics involved in analysing the path of the ray of light.[477]

Having stated his postulates, Reid proceeded to trace the path of a ray of light reflected by a body in motion. He showed that the path will be the same as that followed by a ray reflected by a stationary body, except in the case where the reflecting surface either approaches towards or recedes from the ray and the surface is inclined to that ray. In this instance, he said, the motion of the incident ray must be resolved into two motions, one perpendicular to, and the other parallel to, the reflecting surface. Analysed this way, it was clear that only the velocity of the perpendicular motion was affected. Hence he concluded that when the reflecting surface moves towards the ray, the angle of reflection will be less than the angle of incidence, and that the reverse will hold true when the surface moves away from the ray of light. To confirm this theoretical prediction experimentally, he suggested in a scholium that the light of a star in the ecliptic be used to measure the angles of incidence and reflection from an inclined reflecting surface at opposite points of the Earth's orbit when the star is in quadrature with the Sun (p. 106).[478]

In the elaborate commentary on his second proposition, Reid went on to trace the path of a ray of light passing through either a pure vacuum or a refracting medium to the eye, with the eye assumed to be moving at right angles to the ray (pp. 106–8). What is most noteworthy about this section of the essay is that Reid echoes Patrick Wilson in asserting that there is no aberration involved when terrestrial objects are observed through a telescope.[479] In Reid's third proposition, he explained how to derive the relative from the real path of a ray when the ray is supposed to strike the surface of a reflecting or refracting transparent medium travelling in the direction of its surface. And in the fourth proposition of his essay, he analysed the path of a ray of light issuing from a fixed star and passing along the axis of a refracting telescope carried along by the Earth at right angles to the telescope's axis (pp. 109–11). It may be that the original text continued on from here and that only a part of the manuscript survives, given that Reid refers to a no longer extant geometrical figure illustrating the aberration of the telescope discussed in the fourth proposition (p. 110). Moreover, there are small crosses in the margins of the text at various points, which suggests that he intended to

[477] I am grateful to Geoffrey Cantor for this suggestion.

[478] Compare above, p. clv. Reid's prediction that the sines of incidence and reflection would be unequal contradicted the accepted law of reflection, which stated that they would be equal; see, for example, Newton, *Opticks*, p. 5, where the law is stated as an 'Axiom', and Smith, *A Compleat System of Opticks*, vol. I, p. 3.

[479] See the editorial note to the abridged edition of Wilson's 'An Experiment', p. 193.

revise his essay at a future date because he was dissatisfied with his analysis of the different cases that he considered.[480] This manuscript thus most likely represents his first serious attempt to consider the aberration of light in moving media in a systematic fashion.

Although almost all of Reid's extant writings on optics dating from his years in Glasgow deal with the aberration of light and Newton's theory of refraction, one manuscript on vision survives (pp. 115–19).[481] Unfortunately, the first two leaves have been lost, which means that we do not know if the whole of this manuscript was devoted to the subject of indistinct vision, as the surviving folios are. Nor can we be certain as to why it was written. His wording in one passage, however, implies that the manuscript may be a draft of a letter or a set of comments addressed to a colleague who was actively involved in research on the eye and visual phenomena more generally (p. 115). Reid's starting point for the study of indistinct vision, both chronologically and theoretically, was James Jurin's influential essay on the subject appended to Robert Smith's *A Compleat System of Opticks*, and it was probably through Jurin that he learned of the work of William Porterfield, to whom Jurin refers.[482] The writings of Jurin and Porterfield subsequently formed the basis for his detailed treatment of indistinct vision in his optics lectures at King's College, as well his remarks on the topic in the *Inquiry*.[483] This late manuscript, however, suggests that he had struck out in new directions in analysing the phenomena of indistinct vision. For whereas his predecessors had explained indistinct vision primarily in terms of the distance between the object and the eye, he cast his analytical net more widely and suggested that the brightness of the image on the retina could also play a role (p. 116).[484] Moreover, unlike his predecessors, he also clearly distinguished between three different categories of indistinct vision: (i) those cases caused by the faintness of the retinal image; (iii) those related to the distance of the object from the eye; and (iii) those involving double vision (pp. 115–17). The basic theoretical question addressed in this manuscript was posed by the phenomenon of double vision. As he understood it, the question was: 'why the Rays [of light] falling on some parts of the Pupil

480 There are crosses in the margins of pp. 2, 3, 8 and 9 of the manuscript.

481 This manuscript probably dates to the mid-1780s, given Reid's allusions to papers by William Herschel published in the period 1782 to 1785.

482 Jurin, 'An Essay upon Distinct and Indistinct Vision', pp. 133–4. Jurin cites Porterfield's 'An Essay concerning the Motions of Our Eyes: Of Their Internal Motions'.

483 Anon., 'Natural Philosophy 1758', pp. 305–20; Reid, *Inquiry*, pp. 132–66. See also the relevant notes in Reid's hand for his optics lectures, AUL, MSS 2131/5/II/56 and 2131/7/III/2.

484 That is, for Jurin, Porterfield and other writers, if the rays of light transmitted from the object were not focused on the retina, indistinct vision occurred because the retinal image was blurred.

[are] effectual for producing vision while the Rays falling upon other parts at the same time [are] not' (p. 118).

Reid posited three possible causes: 'The Rays which I call ineffectual either are lost in their passage by some opacity in the coats or Humours [of the eye], or they fall upon parts of the Retina that are insensitive, or they are totally reflected' (p. 118). Based on experiments he had performed with pin holes in paper cards, he dismissed the first two explanations because he believed that his experiments demonstrated that 'there is no part of the pupil where the rays are lost in their passage nor any part of the Retina where those Rays fall that is insensible to them' (p. 118). As for the third potential explanation, namely the total reflection of the 'ineffectual' rays of light by the cornea or the humours of the eye, he confessed that he was unable to show that, strictly speaking, this was the 'true cause' and admitted that it was 'beyond [his] ability' to answer the question he had posed (p. 118).[485] Significantly, his suggestion that Newton's controversial theory of fits of easy reflection and transmission might account for the selective reflection of some rays of light by the pupil of the eye signals his indebtedness to Jurin, because Jurin had earlier invoked this theory to explain indistinct vision.[486] Yet even though the manuscript looks back to the researches of his predecessors, in Janus-like fashion it also looks forward in its anticipation of future progress in optics. For Reid maintained that 'there are principles yet to be discovered' in the science and that William Herschel's innovative use of unconventional high-powered telescopes to observe the stars showed that 'by indistinct Vision discoveries may be made in Nature beyond the most sanguine hopes of philosophers' (p. 118).[487]

Reid's last surviving writings related to the aberration of light date from the late 1780s and they register not only his indebtedness to Patrick Wilson's revised account of aberration but also his response to two important papers dealing with the aberration of light and the use of water telescopes published by Ruđer Josip Bošković in 1785.[488] In Book I of the *Opticks*, Newton attributed the 'imperfection' of refracting telescopes to the different refrangibility of the rays of light and,

[485] For Reid on true causes, see above, pp. lxiv–lxvi.

[486] Jurin, 'An Essay upon Distinct and Indistinct Vision', pp. 157–8, 160–8.

[487] Reid's comments on Herschel's telescopes seem to have been inspired by remarks made by Herschel in his 1782 paper 'On the Parallax of the Fixed Stars', pp. 91–2.

[488] Ruđer Josip Bošković, 'De modo determinandi discrimen velocitatis, quam habet lumen, dum percurrit diversa media, per duo telescopia dioptrica, alterum commune, alterum novi cujusdam generis' and 'De calculanda aberratione astrorum orta e propagatione luminis successiva'. For context, see Kurt Møller Pedersen, 'Roger Joseph Boscovich and John Robison on Terrestrial Aberration', and Pedersen, 'Water-filled Telescopes and the Pre-history of Fresnel's Ether Dragging'. Reid probably learned of these papers from Patrick Wilson, who discussed them with John Robison; see Robison, 'On the Motion of Light', p. 85.

in discussing this problem, he suggested ways in which telescopes could be filled with water in order to produce more distinct images.[489] It may be that Newton's suggestion inspired Bošković to contemplate the use of water telescopes in the 1760s, while Wilson, John Robison and later Reid also gave thought to the design and theoretical implications of these instruments.[490] Bošković's description of experiments on the aberration of light involving water telescopes in his *Opera pertinentia ad opticam et astronomiam* of 1785 thus addressed theoretical issues that Reid, Wilson and Robison had been contemplating for the previous fifteen years or more. But what was considered novel in the papers on aberration published in Bošković's collection was his account of experiments based on the observation of terrestrial, as opposed to celestial, objects. It is this scenario envisaged by Bošković that Reid dealt with in an undated paper in which he asserts two propositions: (i) 'There is no aberration of the Axis of a Telescope, whether of Air or water, from terrestrial Objects seen by it' and (ii) that 'The aberration of the Axis of a Telescope from the place of a fixed Star is the same whether the Telescope be of Air or of Water' (pp. 119–20). Moreover, in an incomplete manuscript dated 10 April 1787, Reid demonstrated a related proposition, 'The Axis of a Telescope whether of Air or of Water has no aberration from a terrestrial radiant point' (p. 122). In affirming these propositions Reid drew on his discussions with Patrick Wilson in order to counter Bošković's contention that, in the case of a terrestrial object, a water telescope would produce aberration and, in the case of a fixed star, the amount of aberration was less with a water telescope than with a standard telescope. For as we have seen, Wilson denied that aberration occurs when observing an object on Earth using a water telescope, and he also argued that there was no difference in the degree of aberration of a fixed star viewed through a water telescope or a regular telescope.[491]

[489] Newton, *Opticks*, esp. pp. 101–2, 110; for context, see Zev Bechler, '"A Less Agreeable Matter": The Disagreeable Case of Newton and Achromatic Refraction'.

[490] Bošković wrote to the French astronomer Joseph-Jérôme de Lalande in 1766 proposing an experiment on the aberration of light that involved the use of regular and water telescopes; see Joseph-Jérôme Lefrançois de Lalande, *Astronomie*, second edition (1771–81), vol. IV, pp. 687–8. In 1784 John Robison stated that 'about the year 1775 or 1776' Patrick Wilson thought of using a water telescope to test his suspicion that 'the centre of the solar system was in motion'; Robison, 'Mr Wilson on the Solar System', p. 29. Wilson later mentioned the use of a water telescope in his 1782 paper on aberration; Wilson, 'An Experiment', p. 59. Compare Robison's comment that 'the contrivance of a telescope filled with water, has been long familiar to my thoughts … in consequence of the speculations of my ingenious friend Professor Wilson of Glasgow'; Robison, 'On the Motion of Light', p. 84. See also Pedersen, 'Water-filled Telescopes and the Pre-history of Fresnel's Ether Dragging', pp. 516–33.

[491] See the editorial note to the abridged version of Wilson, 'An Experiment', p. 193–4; Wilson, 'An Experiment', p. 60; compare Robison, 'On the Motion of Light', pp. 92, 94, 110–11.

An undated letter from Reid to Robison written in the period April 1788 to 1790 confirms that even though the two men were in fundamental agreement over the theoretical issues at stake, Reid had formulated his response to Bošković largely on the basis of his conversations with Wilson. While Wilson had shown Reid a copy of Robison's paper, and Reid was apparently sent the proofs or a preprint of the published version along with a covering letter by Robison at a later date, he seems to have learned of the details of Robison's answer to Bošković only *after* he had written the two late manuscripts on aberration just discussed.[492] Moreover, Reid noted that

> Some years ago I had some intercourse with our Friend P Wilson about the Water Telescope, and, after repeated and attentive consideration, we were both convinced, that according to the optical Principles of Newton, it can neither give any aberration to terrestrial Objects, nor any other aberration to celestial than the common telescope gives.[493]

Thus it appears that it was Wilson who had first prompted Reid to consider water telescopes and to think about terrestrial as well as celestial aberration. Furthermore, Reid explained that he had been led to his conclusions by conceiving of the motion of a ray of light along the axis of a telescope to be compounded of two motions, one parallel and one perpendicular to the axis. So conceived, it was clear to him that it was only the parallel motion which was either accelerated or retarded by refraction, since the perpendicular motion was unaffected by the refractive forces of glass, water or air, which acted at right angles to the refracting surface. It followed that if the motion of the telescope was parallel and equal to the perpendicular motion of the ray before it was refracted, then it would continue to be so after refraction. Hence the ray would move along the telescope's axis and there would be no aberration of a terrestrial object.[494] That Reid felt obliged to explain his reasoning in such detail to Robison implies that the two of them had not conversed about the implications of Bošković's papers during the period in which Reid wrote his two last manuscripts on the aberration of light.

In his letter Reid went on to say that even though Bošković had erred in his reasoning, either a water telescope or Robison's ingeniously contrived compound microscope could nevertheless provide 'an *Experimentum Crucis* between

[492] Reid to John Robison, [April 1788–90], in Reid, *Correspondence*, p. 197.

[493] Reid to Robison, [April 1788–90], in Reid, *Correspondence*, p. 198. Reid's comments suggest that his conversations with Patrick Wilson regarding water telescopes began before Wilson published his paper on aberration in 1782.

[494] Reid to Robison, [April 1788–90], in Reid, *Correspondence*, p. 198. For Reid's use of this twofold analysis of the motion of a ray of light, see his late undated paper on aberration, below, p. 120.

Newtons Theory and Eulers'. For if no terrestrial aberration were observed, Newton's account of refraction held and his projectile theory of light would be confirmed; but if, on the other hand, aberration were detected, then Newton's explanation of refraction would be invalidated and his theory of light overthrown.[495] Yet he did not believe that Euler's vibration theory would necessarily have to be adopted 'unless the quantity of [the] Terrestrial Aberration correspond[ed] with [the prediction of] his Theory', and this quantity was, he pointed out, greater than that calculated by Bošković. He therefore encouraged Robison to develop his microscope, because it would 'probably serve to confirm Newtons system & to detect the Error of Eulers', and he urged his friend to examine further cases of the absolute and relative motions of light in reflection and refraction.[496] He also expressed the hope that the two of them might meet again to converse on philosophical subjects, and it may be that despite his increasing deafness they were in fact able to enjoy one another's company in the 1790s, for Robison subsequently recorded that he had performed experiments on the rings of colour in soap bubbles known as 'Newton's rings' with his 'most respectable and intelligent friend, the late Dr Reid'.[497]

While Reid was in Glasgow, he supplemented his writing with a limited amount of reading in the literature on optics. The 'Professors Receipt Books' record that in the late 1760s he took out of the University library two books related to his researches: François de Aguilón's *Opticorum libri sex: Philosophis iuxta ac mathematicis utiles* (1613), which he presumably wanted to reread, given that he had referred to the work in the *Inquiry*; and Leonhard Euler's *Opuscula varii argumenti* (1746), which contained Euler's seminal formulation of the vibration theory of light, 'Nova theoria lucis et colorum'.[498] And it may be that his reading of Euler led him to an earlier version of the vibration theory, namely that outlined in the lectures on the nature of light included in Robert Hooke's *Posthumous Works*, which he consulted in 1770.[499] Reid's continued interest in the subject of indistinct vision discussed above also features in his library withdrawals, for in 1771 he was given permission to borrow from the Simson Collection a pamphlet by James Jurin defending his 'Essay upon Distinct and Indistinct Vision' against

[495] Reid to Robison, [April 1788–90], in Reid, *Correspondence*, p. 198; compare Robison, 'On the Motion of Light', pp. 96–8.

[496] Reid to Robison, [April 1788–90], in Reid, *Correspondence*, pp. 198–9.

[497] Reid to Robison, [April 1788–90], in Reid, *Correspondence*, p. 200; John Robison, 'Impulsion', p. 804.

[498] 'Professors Receipt Book, 1765–1770'; Reid, *Inquiry*, p. 143.

[499] Robert Hooke, 'Lectures of Light, Explicating Its Nature, Properties and Effects, &c.', in Hooke, *Posthumous Works*, pp.71–148; 'Professors Receipt Book, 1770–[1789]', in the listings for 1770.

the ill-tempered criticisms of his fellow Newtonian, Benjamin Robins.[500] A decade later, Reid was again reading in the literature on vision, this time two recently published works, John Elliott's *Philosophical Observations on the Senses of Vision and Hearing* and Peter Degravers' *A Complete Physico-medical and Chirurgical Treatise on the Human Eye*.[501] We have seen above that in November 1792 Reid took notes from a preprint of Robert Blair's paper on refraction that later appeared in the *Transactions of the Royal Society of Edinburgh* and, in June of that year, he penned a hostile set of comments on William Charles Wells' *An Essay upon Single Vision with Two Eyes*.[502] In the *Essay* Wells challenged Reid's account of single vision in the *Inquiry*. It may be that Dugald Stewart wanted his mentor to be aware of Wells' objections, because he lent Reid a copy of the *Essay* shortly after it was published. Reid evidently had no time for his critic. For apart from the negative tone of his reading notes, they indicate that he read at most only the first seventy pages of the text. Writing to Stewart in 1793, Reid dismissed Wells' book with the withering comment that it 'has much learning on the subject, and therefore may be very fit to answer the purpose of one who sets up as a physician in London; but I do not see that it makes any addition to human knowledge'.[503]

To conclude, Reid's writings on physical optics demonstrate that he was an able proponent of what Geoffrey Cantor has called 'the projectile theory of light'.[504] Like other proponents of this theory, Reid accepted that light consists of exceedingly small particles. Yet we have seen as well that in his lectures Reid was closer to Newton in his use of the theory-neutral term 'ray of light' and it is significant that he spoke of rays rather than particles in almost all of his manuscripts.[505] And like his fellow projectile theorists, he maintained that light was subject to the laws of motion and that it was acted on by the attractive and

[500] Jurin, *A Reply to Mr Robins's Remarks*; 'Professors Receipt Book, 1770–[1789]'; 'Minutes of Senate Meetings, 1771–1787', p. 7. Jurin's pamphlet was a rejoinder to Benjamin Robins, *Remarks on Mr Euler's Treatise of Motion, Dr Smith's Compleat System of Opticks and Dr Jurin's Essay upon Distinct and Indistinct Vision* (1739). Robins replied in his pamphlet, *A Full Confutation of Dr Jurin's Reply to the Remarks on his Essay upon Distinct and Indistinct Vision* (1740).

[501] John Elliott, *Philosophical Observations on the Senses of Vision and Hearing; to Which Are Added, a Treatise on Harmonic Sounds and an Essay on Combustion and Animal Heat* (1780); Peter Degravers, *A Complete Physico-medical and Chirurgical Treatise on the Human Eye: And a Demonstration of Natural Vision* (1780); 'Professors Receipt Book, 1770–[1789]', listings for 1781.

[502] The preprint version of Blair's paper is available on ECCO; William Charles Wells, *An Essay upon Single Vision with Two Eyes: Together with Experiments and Observations on Several Other Subjects in Optics* (1792); AUL, MS 2131/3/I/4.

[503] Reid to Dugald Stewart, 21 January 1793, in Reid, *Correspondence*, p. 231.

[504] Cantor, *Optics after Newton*, chs 2–3.

[505] Like Reid, Patrick Wilson claimed to be agnostic about the nature of light; see Wilson, 'An Experiment', p. 68.

repulsive forces associated with ordinary matter. In his explanations of optical phenomena, he merely posited the existence of these forces and did not speculate on their physical nature, believing that this was a question that was outside of the scope of natural philosophy. Moreover, Reid showed comparatively little interest either in Bošković's scheme of alternating spheres of attraction and repulsion or in any of the British variations on the same theme. Moreover, even though, towards the end of the eighteenth century, there was growing disenchantment with Newton's programme of accounting for the behaviour of light in terms of forces analogous to gravitation among supporters of the projectile theory, the *Principia* remained as his exemplar in the field of physical optics. His writings nevertheless reveal that he was aware of some of the problems posed by the supposition that light is particulate, such as collision and lack of momentum. We have seen too that he questioned Newton's account of refractive dispersion, and that even though he was aware of Newton's contentious theory of fits of easy reflection and transmission, he appears not to have thought carefully about it.

More generally, physical optics was closely allied to astronomy for projectile theorists such as Reid, and it was the interplay between these two sciences which gave rise to the fascinating sequence of manuscripts on stellar and terrestrial aberration that he produced in the 1770s and 1780s. This group of manuscripts shows that while he was cognisant of the work of Patrick Wilson and John Robison and that he reached essentially the same conclusions as they did regarding the aberration of light, he nevertheless did so independently of his two friends, on the basis of his analysis of the relative motion of a ray of light as it reaches the body of a telescope. And while he agreed with Wilson and Robison that the phenomena of aberration could provide a crucial experimental test of Newton's law of refraction, we have seen that, unlike his colleagues, he harboured serious doubts about the validity of Newton's law. In addition, as we shall see in Section 7, late in life he also explored the connections between the study of light and chemistry, and it was through his scrutiny of the writings of the 'French Chemists' that he compared light to the matter of heat (caloric) and speculated that 'no phenomenon … shews either of them to gravitate' (p. 140).[506] Furthermore, like his fellow projectile theorists, he employed mathematics extensively in his work on physical optics and, like them, he advocated the use of the inductive method. Considered from an institutional point of view, Reid conforms to the social profile of the projectile theorists Cantor has studied, insofar as he was both university trained and taught optics in an academic setting. In terms of his teaching, his optics lectures were

[506] Reid had earlier suggested in Glasgow lectures on the culture of the mind that light might not be governed by Newton's law of gravitation; Reid, *On Logic, Rhetoric and the Fine Arts*, p. 183.

shaped by pedagogical factors akin to those which Cantor has argued led to the systematisation of the projectile theory prior to 1740.[507] Reid is, therefore, best seen as one of the few projectile theorists who both publicly disseminated the principles of the theory and applied those principles in the context of original research. And while Reid's work in physical optics during his years in Glasgow was dominated by the problems of refractive dispersion and the aberration of light, he remained interested in the theory of vision well into the final years of his life.

6. Electricity

Even though optics enjoyed considerable intellectual prestige within the scientific community and Newton's ideas regarding light and colours had a broad cultural impact, the science of electricity captured the imagination not only of men of science but also the public at large during the eighteenth century. Dramatic demonstration experiments, such as Benjamin Franklin's kites or those involving human chains receiving an electric shock from a Leyden jar, along with electrical phenomena easily reproducible in a domestic setting, ensured that electricity had a wide and socially diverse audience eager to attend lectures or to read the voluminous literature generated by electricians in the Enlightenment.[508] Yet Reid appears not to have shared his contemporaries' widespread enthusiasm for the study of electricity. Despite the ferment of theoretical ideas, his reading of works on electricity appears to have largely been desultory, although we shall see that he took seriously the writings of J. T. Desaguliers and Benjamin Franklin. His manuscripts also suggest that his active engagement with research on electricity coincided largely with his teaching at King's College, Aberdeen. And, as his comments in his 1762 graduation oration quoted above imply,[509] he seems to have thought that, unlike other branches of natural philosophy, the science of electricity had yet to acquire firm theoretical and empirical foundations. His writings on electricity indicate that, at most, he believed that a few basic facts about the power of electricity had been established and that these facts sanctioned the view that there are two forms of electricity. But beyond these facts and this one theoretical principle, there was virtually nothing else in the science of electricity that he regarded as being well founded.

[507] Cantor, *Optics after Newton*, pp. 42–9.

[508] On the science of electricity in the early modern period, see especially the exemplary study by J. L. Heilbron, *Electricity in the Seventeenth and Eighteenth Centuries: A Study of Early Modern Physics*. For a concise overview, see Thomas L. Hankins, *Science and the Enlightenment*, pp. 53–72.

[509] See p. cxxxv.

Because there is insufficient evidence, we cannot determine when and why Reid first studied electricity. He witnessed electrical experiments demonstrated by Desaguliers when he visited the Royal Society in London in 1736, but we do not know if this episode sparked his initial interest.[510] His Aberdeen associate David Fordyce was in contact with the leading English electricians Henry Miles and John Canton, and it may be that until Fordyce's untimely death in 1751 he reported the latest discoveries in the science of electricity to Reid.[511] The real impetus for Reid's study of electricity, however, apparently came from his teaching obligations at King's College following the reform of the curriculum in 1754.

The manuscripts related to Reid's lectures on electricity exhibit an uneasy mixture of theoretical concepts taken from Benjamin Franklin with older ideas derived from the works of Charles François de Cisternay Dufay and Desaguliers.[512] In a set of definitions, propositions and 'Conjectures ... to be confirmed or refuted by future Experiments' which probably dates from 1754 or 1755, Reid defined electricity as a 'Power or Virtue in Bodies, excited by Friction, by which they attract or repel other Bodies' (p. 124).[513] Contained in this definition are three pre-Franklinist conceptions: (i) that electricity is *excited* in a body rather than accumulated in the form of a charge; (ii) that electrical power is aroused by friction; and (iii) that attraction and repulsion are the primary electrical phenomena, rather than charging and discharging.[514] The definitions and propositions which

[510] Royal Society of London, 'Journal Book (copy)', xv (1735–36), pp. 363–4.

[511] See David Fordyce to John Canton, 30 April 1748, Royal Society of London, John Canton Papers, volume 2, fol. 14. It is possible that Fordyce was himself lecturing on electricity when he taught natural philosophy; Wood, *Aberdeen Enlightenment*, pp. 27–8.

[512] Benjamin Franklin's *Experiments and Observations on Electricity, Made at Philadelphia in America* was first published in 1751. An expanded version of the work appeared in 1753 and a second edition was issued in 1754 with the title *New Experiments and Observations on Electricity*. The distinction between vitreous and resinous forms of electricity made in his lecture notes shows that Reid was familiar with the electrical theory of Dufay; see esp. Charles François de Cisternay Dufay, 'A Letter from Mons. Du Fay, F.R.S. and of the Royal Academy of Sciences at Paris, to His Grace Charles Duke of Richmond and Lenox concerning Electricity'. Similarities in phraseology and theory indicate that Reid drew heavily on J. T. Desaguliers' prize-winning text *A Dissertation concerning Electricity* (1742), as well as Desaguliers' other writings on the subject.

[513] Reid taught a course of natural philosophy in the 1754–55 session at King's College. Compare the definition of electricity found in another set of notes that probably date to the same period: 'Electricity is a power in Bodies excited by Friction of Attracting & repelling Alternately light bodies at a considerable distance as several inches or feet'; AUL, MS 2131/8/V/3, fol. 2v. Reid placed more emphasis here on the fact that electricity is a power which acts at a distance, and also indicates that there are alternating spheres of attraction and repulsion involved in electrical phenomena. He may have been indebted to either Desaguliers or W. J. 's Gravesande for this point; Heilbron, *Electricity in the Seventeenth and Eighteenth Centuries*, pp. 291–3.

[514] On the contrast between pre-Franklinist and Franklinist conceptions, see Roderick W. Home, 'Franklin's Electrical Atmospheres', pp. 131–2.

followed were likewise traditional in content. As well as stating some elementary facts about the behaviour of electrified bodies, Reid distinguished electrics from non-electrics and, following Dufay, vitreous and resinous electricities. He also alluded to the connections between electricity, fire and light, and noted that the most dramatic electrical effects were produced in cold, dry air rather than in moist atmospheric conditions. When it came to proposing some conjectures, however, he drew heavily on his close study of Franklin's *New Experiments and Observations on Electricity*.

Following Franklin, Reid postulated that all electrical phenomena were caused by the actions of a subtle, particulate electrical fluid which is diffused equally throughout nature. The particles of this fluid were, he suggested, mutually repellent and attracted by ordinary matter. The fluid existed naturally in a state of equilibrium, but when an electric was rubbed or friction otherwise applied to it, the electric collected a quantity of the fluid around itself drawn from the non-electrics in contact with it. Because the electrical fluid would then attempt to restore its equilibrium, Reid speculated that effects such as shocks, flashes, attractions and repulsions would then occur as the fluid returned to the depleted non-electrics. He indicated, moreover, that bodies have a 'Natural Quantity' of electrical fluid, although it is difficult to see how he reconciled this notion with the idea that the fluid is diffused equally throughout space. When this natural quantity was exceeded, a body was then 'Electrified, positively or Plus', whereas when a body contained less than this quantity, the body was 'Electrified Negatively or Minus' (p. 125). Reid also noted, as did Franklin, that pointed objects 'both receive and communicate the Electrical Fluid at a much Greater Distance & with a less flash & Shock than those that are obtuse', and he too remarked on the special electrical properties of glass (pp. 125–6). Finally, Reid sketched Franklin's famous theory that lightning was caused by the passage of electrical fluid which had accumulated in the clouds to other bodies (p. 126).[515]

Even though Reid had carefully studied Franklin's *New Experiments and Observations on Electricity*, his conjectures imply that he was critical of aspects of the Philadelphia system. A major anomaly recognised by Franklin was that negatively charged bodies repelled one another, whereas according to his theory these bodies lacked electrical fluid and hence did not possess the electrical atmospheres

[515] For discussions of Franklin's theory of electricity see, *inter alia*, Home, 'Franklin's Electrical Atmospheres'; Roderick Home's editorial introduction to F. U.T. Aepinus, *An Essay on the Theory of Electricity and Magnetism* (1759), pp. 77–89; Heilbron, *Electricity in the Seventeenth and Eighteenth Centuries*, ch. 14.

responsible for attractive and repulsive effects.[516] Reid registered this point, and observed that 'When bodies are equally exhausted of the Electrical Fluid they Repell one another which seems not be accounted for' (p. 126). Moreover, Reid reverted to earlier ideas when he hypothesised that there are two electrical fluids, one vitreous and the other resinous, which he thought could equally well be called male and female.[517] He thus redefined the natural state of bodies as one in which 'the two fluids are joyned in such proportion as to saturate each other', and he reasoned that when a body was deprived of some of its vitreous fluid it would then be negatively charged (p. 126). Unfortunately, his musings break off at this point, although he added notes in November 1756 on the electrical properties of sulphur and December 1757 on the snap produced when he removed silk stockings worn over worsted ones in frosty weather (pp. 126–7). Unlike his fellow Scot Robert Symmer, however, he appears not to have explored the theoretical implications of the behaviour of his stockings, which, as it turned out, could be used to support a two-fluid theory of electricity similar to the one that he hinted at.[518]

Reid appears to have incorporated elements of Franklin's system into the substance of his lectures on electricity at King's, although the set of student notes taken from his course of natural philosophy in the session for 1757–58 contain only a few ambiguous hints of his theoretical position.[519] In these notes, Reid is recorded as affirming that electricity 'consists of subtle exhalations, which come to & go from *IdioLectricks* & partly lurks in them, & which repels & attracts all moveable bodies, which they meet'. This characterisation of the physical nature of electricity was firmly rooted in the effluvial theories of the seventeenth and eighteenth centuries, yet there are also references to an 'Electrical fluid' which was 'spread about all bodies' or 'circumfused around them'. These phrases recall Franklin's characterisation of electrical atmospheres.[520] Hence Reid's later lectures on electricity probably incorporated an eclectic blend of theoretical ideas

[516] Franklin, *New Experiments and Observations on Electricity*, pp. 25–6; compare Heilbron, *Electricity in the Seventeenth and Eighteenth Centuries*, p. 337.

[517] In suggesting that the two electrical fluids could be called male and female, Reid hints that he knew something of the theorising of German electricians such as G. M. Bose, who had speculated that there were male and female forms of electrical 'fire'. Reid may have learned about German work on electricity through the detailed account published in the *Gentleman's Magazine* in April 1745; Anon., 'An Historical Account of the Wonderful Discoveries Made in Germany, &c. concerning Electricity'. See also Heilbron, *Electricity in the Seventeenth and Eighteenth Centuries*, ch. 10.

[518] Robert Symmer, 'New Experiments and Observations concerning Electricity'. On Symmer, see esp. Heilbron, *Electricity in the Seventeenth and Eighteenth Centuries*, pp. 431–6.

[519] Anon., 'Natural Philosophy 1758', pp. 217–20. The notes from his lectures on electricity are incomplete.

[520] Anon., 'Natural Philosophy 1758', p. 218. Franklin's positive and negative electricities are not mentioned in this set of student of notes.

similar to the mixture of concepts found the manuscripts discussed above. What is distinctive about the notes from 1757–58 is that they indicate that Reid went into great detail in classifying natural substances in terms of the distinction between electrics and non-electrics, and in relation to their capacity to 'collect', 'receive' or 'imbibe' quantities of electricity.[521] The emphasis on classification found in these notes implies that he may have felt that a natural historical approach was of more practical value to his students than a detailed exposition of electrical theory, especially since there was no theoretical consensus among electricians. Moreover, we have seen that he regarded all of the rival systems of the period as highly conjectural. Pedagogical concerns, therefore, may well have prompted Reid to concentrate on classification and to avoid complex theorising.

One of the most notable electrical discoveries in the 1750s concerned the peculiar electrical properties of the tourmaline crystal. F. U. T. Aepinus was the first to show that the tourmaline could be electrified by heating alone, and news of his discovery spread rapidly to France and England.[522] Because Aepinus explained the behaviour of the tourmaline in terms of positive and negative electricities, Franklin and his allies in the Royal Society of London rushed to confirm Aepinus' results, whereas French savants tried to reconcile the new phenomena with the effluvial theory of Franklin's opponent, Jean Antoine Nollet.[523] Probably in the period 1760 to 1764, Reid acquainted himself with Aepinus' discovery, which he thought might 'convince Naturalists that after all the curious & surprizing Phænomena produced by the Electrical Shock they are very far from having discovered the nature & extent of this Power or the universal Laws to which it is subject' (p. 127). He believed that the most important fact about the tourmaline was that friction was not required for its electrification, which indicated that it was 'susceptible of two Species of Electricity entirely different from each other, one produced by friction the other by heat without friction' (p. 128). Given that he accepted the received wisdom among electricians that friction was necessary for the production of electricity, it is not surprising that he responded to Aepinus' discovery in this manner. But he also noted that the tourmaline could be electrified in boiling water, which contradicted the widely held view that water 'obstruct[ed] the Electrical Virtue' (p. 127). Further, he observed that the 'Electricity of the Tourmaline is constantly both positive and negative', its sides having opposing

[521] Anon., 'Natural Philosophy 1758', pp. 219–20.

[522] On Aepinus' discovery and its dissemination, see Home's editorial introduction to Aepinus, *An Essay on the Theory of Electricity and Magnetism*, pp. 92–5. I draw on Home's account in what follows.

[523] One of the earliest reports in English of Aepinus' work on the tourmaline crystal was written by Franklin's associate John Canton; see [John Canton], 'Electrical Properties of the Tour-malin'.

charges which could be reversed by heating (p. 127). Significantly, he described the electrical properties of the crystal using Franklinist terminology, but there is no evidence that he regarded the tourmaline as providing striking empirical confirmation of the Philadelphia system. And at roughly the same time as he familiarised himself with Aepinus' research, he read Giovanni Carafa's *Lettre du Duc de Noya Carafa sur la tourmaline, a Monsieur de Buffon*, in which Nollet's theory was deployed to explicate the properties of the tourmaline crystal. Here, too, he merely abstracted what he thought were the salient facts about the crystal and did not comment on whether Nollet's theory could account for them (p. 128).[524]

Once Reid was settled in Glasgow his interest in electricity seems to have waned, even though his colleagues John Anderson and John Robison were both keen electricians. In 1754–55 Anderson had attended the lectures of the doyen of French electricians, Jean Antoine Nollet, in Paris, but he seems to have been an early convert to the Philadelphia system.[525] Unfortunately, no record survives of discussions between Reid and Anderson on Franklin's work or on electrical topics more generally. Moreover, Robison may have established experimentally that the force of electricity obeyed the same inverse square law as gravity in the late 1760s, although his claim that he had presented an account of his discovery 'in a public society in 1769' cannot be corroborated.[526] There is no indication, however, that Reid knew about Robison's electrical investigations. But even if Reid did not converse with his colleagues about their electrical researches, for a few years after his move to Glasgow he continued to read new works on electricity. In August 1768, he excerpted a passage from the preface of Joseph Priestley's *The History and Present State of Electricity* on the benefits of specialisation in natural philosophy, and compiled a list of the prominent electricians who featured

[524] Giovanni Carafa, Duca di Noja, *Lettre du Duc de Noya Carafa sur la tourmaline, a Monsieur de Buffon* (1759). Roderick Home argues that the bulk of this pamphlet was written by the French natural historian Michel Adanson; see Home's editorial introduction to Aepinus, *An Essay on the Theory of Electricity and Magnetism*, p. 94.

[525] David Steuart Erskine, eleventh earl of Buchan, 'Concerning John Anderson Professor of Natural & experimental Philosophy in the University of Glasgow', GUL, MS Murray 502/76, fol. 1r; [John Anderson], *A Compend of Experimental Philosophy; Containing Propositions Proved by a Course of Experiments in Natural Philosophy, and the General Heads of Lectures Which Accompany Them* (1760), pp. 34–48; Wilson, *Seeking Nature's Logic*, pp. 186–8. Anderson had travelled with Benjamin Franklin in the Highlands in 1759, and it is said that he was responsible for the erection of a lightning rod on the College building in 1772, the year after Franklin's second visit to Glasgow; Nolan, *Benjamin Franklin in Scotland and Ireland 1759 and 1771*, p. 74, and Murray, *Memories of the Old College of Glasgow*, pp. 55–6.

[526] John Robison, *A System of Mechanical Philosophy*, vol. IV, p. 68; Heilbron, *Electricity in the Seventeenth and Eighteenth Centuries*, pp. 465–8. There is no mention of Robison reading a paper on electricity in the minutes of the Glasgow Literary Society during the period in question.

in Priestley's narrative of the progress of the science from antiquity through to the mid-eighteenth century. He also jotted down various details regarding the electrical properties of bodies but the brevity of these entries suggests that he was primarily interested in the historical material in Part I of the book, rather than Priestley's survey of the factual, theoretical and experimental foundations of the science of electricity contained in Parts II–VIII.[527] Then in June 1769, he took notes from the fourth, and greatly expanded, edition of Benjamin Franklin's *Experiments and Observations on Electricity*, which had been published earlier that year. It appears that in reading this edition of Franklin's work Reid ignored the initial sections of the text outlining the fundamentals of the Philadelphia system because his notes summarise only the contents of letters 33 to 61, which cover a miscellany of topics ranging from harmony in music to improving the design of canal barges. Of the six new letters on electricity included in this part of the text, he merely noted '37 &c to 42 on Electricity'.[528] This set of notes, therefore, implies that even though he wanted to keep up with Franklin's publications, he did not feel the need to reacquaint himself with the details of Franklin's theory of electricity. But there are no further sets of reading notes from works on electricity dating from after 1769, and there are no books or journals listed in the 'Professors Receipt Books' that indicate that he was familiarising himself with the current literature.[529]

Reid's study of electricity thus seems to have been related primarily to his teaching at King's College. His papers on the subject are, nevertheless, significant historically for two reasons. First, they provide valuable information about the reception of Benjamin Franklin's theory of electricity in Scotland. Historians have hitherto concentrated on the reactions of English, French and Italian electricians

[527] AUL, MS 2131/3/I/9; Joseph Priestley, *The History and Present State of Electricity, with Original Experiments* (1767), esp. pp. xiv–xv.

[528] AUL, MS 2131/3/I/14, fol. 2r; Benjamin Franklin, *Experiments and Observations on Electricity ... to Which Are Added, Letters and Papers on Philosophical Subjects* (1769). The numerical sequence of letters in Reid's notes is reversed, which implies that he began reading letter 61 and worked his way backwards through the book.

[529] There is a hint in Reid's late discourse on muscular motion that he was interested in animal electricity insofar as he alludes to the work done in the mid-1770s by Sir John Pringle, John Hunter, Henry Cavendish and others on the electric eel or 'torpedo'; see Reid's 'Of Muscular Motion in the human Body' in Reid, *Animate Creation*, p. 119; Sir John Pringle, *A Discourse on the Torpedo, Delivered at the Anniversary Meeting of the Royal Society, November 30 1774* (1775); John Hunter, 'An Account of the Gymnotus electricus'; Henry Cavendish, 'An Account of Some Attempts to Imitate the Effects of the Torpedo by Electricity'. For context, see W. Cameron Walker, 'Animal Electricity before Galvani'.

to the Philadelphia system.[530] The geographical scope of their accounts can therefore be usefully expanded by examining the responses of Reid and his fellow natural philosophers in Scotland. In Reid's case, we have seen that he was willing to accept Franklin's theory as a working hypothesis and that he may have incorporated elements of the Philadelphia system into his lectures at King's. Moreover, the character of Reid's response differed from those of his English and European colleagues insofar as he neither dismissed Franklin's ideas out of hand nor used them to further his own academic interests.[531] Reid's reaction to the Philadelphia system was seemingly conditioned more by pedagogical considerations than it was by professional rivalries.

A second reason why Reid's manuscripts on electricity are of historical significance is that they show that he was not the dogmatic opponent of fluid theories and conjectures depicted in influential discussions of his methodological views.[532] Reid's papers on electricity demonstrate that he did not rule out theories which invoked effluvia or subtle fluids as a matter of principle and that he was willing to accept conjectures tentatively, with the proviso that they be subjected to further experimental scrutiny.[533] Reid's attack on animal spirits and physiological ethers in the *Inquiry* should not be allowed to obscure the fact that he was less worried by the use of ethers and subtle fluids in the *physical* sciences, provided that the material constituents of the ethers or fluids were not credited with the power to act as efficient causes.[534] Because Franklin's electrical fluid was made up of inert

[530] See, for example, Home's editorial introduction to Aepinus, *Essay on the Theory of Electricity and Magnetism*, pp. 101–6, and Heilbron, *Electricity in the Seventeenth and Eighteenth Centuries*, ch. 15.

[531] Compare Heilbron's analysis of the European reception of Franklin's theory; Heilbron, *Electricity in the Seventeenth and Eighteenth Centuries*, esp. p. 344.

[532] See especially Geoffrey Cantor, 'Henry Brougham and the Scottish Methodological Tradition', pp. 74, 79–80, 81–2; Larry Laudan, 'The Medium and Its Message: A Study of Some Philosophical Controversies about Ether', pp. 169–73; Laudan, 'Thomas Reid and the Newtonian Turn of British Methodological Thought'. For an interpretation of Reid's attitude towards ethers and fluids similar to the one advanced here, see Robert Callergård, *An Essay on Thomas Reid's Philosophy of Science*.

[533] We have seen in Section 1 that Reid sanctioned the use of hypotheses, as long as they were (i) sharply distinguished from empirically established facts and inductive generalisations and (ii) treated as queries requiring empirical investigation; see above, pp. lxiv–lxvi.

[534] On this point, see above, p. lxvi; Reid, *Inquiry*, pp. 161–2; compare Reid, *Intellectual Powers*, pp. 76–87. His attack on physiological ethers and animal spirits notwithstanding, in the *Inquiry* Reid explained how our sense of smell operates in terms of effluvia given off by the bodies around us; Reid, *Inquiry*, pp. 25–6; compare Reid, *Intellectual Powers*, p. 87. Reid's characterisation of these effluvia echoes Stephen Hales' description of 'elastick air', which he maintained was made up of particles possessing a repulsive force. According to Hales, it was the action of this 'elastick air' that sustained the 'beautiful frame of things ... in a continual round of the production and dissolution of animal and vegetable bodies'; Stephen Hales, *Vegetable Staticks: Or, an Account of Some Statical*

particles of matter endowed with immaterial forces acting at a distance, Reid was willing to entertain the Philadelphia system as a conjectural account of electrical phenomena that was subject to further empirical test. And, as we shall see in the next section, Reid's manuscripts on chemistry likewise reveal that his attitude towards subtle fluids was more complex than many commentators recognise.

7. Chemistry

Reid's declaration in his graduation oration of 1762 that chemistry ranked among the branches of natural philosophy that had been 'constructed on solid foundations ... without there any longer being dispute among the knowledgeable as regards [their] principles' and that those principles 'receive[d] additions worthy of the human intellect' on a 'daily' basis speaks to his perception of the science as having a clear disciplinary identity based on a set of widely accepted theoretical concepts and a repertoire of standard experimental practices.[535] How he came to adopt such a view of chemistry, however, remains unclear. Prior to the 1760s, there is little mention of the details of chemical theory or practice in the papers that make up the Birkwood Collection, and there is only one surviving Reid manuscript, headed 'The Chemical History of Salts', which documents his engagement with the science of chemistry before his move to Glasgow in 1764.[536] We can, nevertheless, reconstruct something of Reid's understanding of the disciplinary configuration of chemistry from clues scattered in his extant writings. First, as his comments in his final graduation oration at King's indicate, he regarded chemistry as a branch of natural philosophy. In conceiving of chemistry in this manner, he registered the transformation in the status of chemistry from a practical art to a partly academic science that took place during the course of the seventeenth and early eighteenth centuries.[537] Yet he also believed that chemistry was an independent science because he denied that the phenomena studied by chemists could be explained purely in terms of the attractive and repulsive forces invoked by John Freind and others who drew their theoretical inspiration from Newton's analysis of various chemical processes in the Queries to the *Opticks*.[538] In an undated manuscript

Experiments on the Sap in Vegetables: Being an Essay towards a Natural History of Vegetation (1727), p. 178.

535 Reid, *Philosophical Orations*, oration IV, para. 4.

536 Thomas Reid, 'The Chemical History of Salts', AUL, MS 2343.

537 On this transformation, see, for example, Antonio Clericuzio, '"Sooty Empiricks" and Natural Philosophers: The Status of Chemistry in the Seventeenth Century'.

538 For the application of the Newtonian concept of attractive forces to the explanation of chemical phenomena, see especially John Freind, *Chymical Lectures: In Which Almost All the Operations of Chymistry Are Reduced to Their True Principles, and the Laws of Nature* (1712). For the use of

headed 'Divisions' (which was probably written in the 1750s) he distinguished between the powers 'belonging [to] all Bodies' such as inertia and gravity and the 'Particular powers of Some unorganized Bodies', which for him included 'Chymical Affinities' and 'Transmutations'. He thus maintained that chemistry investigated powers that were different from those analysed in other branches of natural philosophy. He later gave voice to this view when he wrote to Lord Kames in the autumn of 1775 that: 'I suspect there are many things in Agriculture, & many things in Chemistry that cannot be reduced to [the] Principles [of attraction and repulsion], though Sir I. Newton seems to have thought otherwise'.[539] For Reid, therefore, chemistry was an independent discipline because the powers of matter manifest in chemical phenomena could not be reduced to the powers considered in neighbouring sciences, such as mechanics, hydrostatics, pneumatics, optics, magnetism and electricity.

Secondly, Reid's manuscripts suggest that while he recognised that chemistry possessed a distinct disciplinary identity, he was also attuned to a key feature of the science in the eighteenth century, namely that it was 'a discipline without entirely rigid boundaries'.[540] His papers related to his lectures on natural history at King's College show that he used chemical analysis as a basis for the classification of the three kingdoms of nature; in turn, the classification of chemical substances was for him an important component of chemical practice.[541] He was cognisant too of the connections between chemistry and mineralogy, as a set of reading notes from Cronstedt's *An Essay Towards a System of Mineralogy* attests.[542] In addition, he

forces of attraction and repulsion in the study of 'air', see Hales, *Vegetable Staticks*, esp. ch. 6. On Freind and his generation of Newtonian chemists, see Arnold Thackray, *Atoms and Powers: An Essay on Newtonian Matter Theory and the Development of Chemistry*, ch. 3.

539 AUL, MS 2131/6/V/34, fol. 1r; Reid to Lord Kames, 1 October 1775, in Reid, *Correspondence*, p. 93; compare Reid, *Inquiry*, pp. 117, 211. Reid did, however, suggest that chemical affinities resembled the other forces associated with matter, in that they all acted at a distance; see George Baird, 'Notes from the Lectures of Dr Thomas Reid, 1779–80', Mitchell Library, Glasgow, MS 104933, p. 121.

540 Jan Golinski, 'Chemistry', p. 376.

541 AUL, MSS 2131/6/V/2, 6/V/4, 6/V/5, 6/V/6, 6/V/10a, 6/V/21, 6/V/22, 6/V/22a, 7/II/4 and 7/II/18. For Reid's use of taxonomy in chemistry, see 'The Chemical History of Salts'. Because the theory of affinity could be used as a tool for classification, the theory served to strengthen the links between chemistry and natural history; on this point, see Alistair M. Duncan, 'The Functions of Affinity Tables and Lavoisier's List of Elements', pp. 36–41.

542 AUL, MS 2131/3/I/14, fol. 2r; see above, p. lxxi. As Theodore M. Porter has argued, the connections between mineralogy and chemistry were of considerable significance in the eighteenth century, given that the analytical techniques developed by chemists who sought to improve mining helped to shape the concept of chemical elements found in the work of Antoine-Louis Lavoisier, Louis Bernard Guyton de Morveau and Antoine François de Fourcroy; Theodore M. Porter, 'The Promotion of Mining and the Advancement of Science: The Chemical Revolution of Mineralogy'. On the connections between mineralogy and chemistry, see also Evan M. Melhado, 'Mineralogy and

was aware of the medical contexts for chemical pedagogy, not least because his close associates John Gregory and David Skene intended to lecture on chemistry and materia medica as part of their plan in 1758 to establish a medical school at King's.[543] Gregory proposed to teach the chemistry course, and he seems to have cultivated a knowledge of the science because of its relation to physiology, as well as its relevance to his schemes for agricultural improvement.[544] Skene's interest in chemistry was partly pharmaceutical, but his competence as a chemist was also in evidence in his work as a natural historian and in his various improving activities.[545] Like his friends Gregory and Skene, Reid applied his chemical ideas to the improvement of agriculture.[546] The available evidence thus indicates that Reid conceived of chemistry as being closely connected with natural history, medicine and mineralogy, and, furthermore, that he was convinced that chemical knowledge was of utilitarian benefit when applied to medicine, mining and to agricultural improvement.[547]

Reid's writings also tell us something of his knowledge of chemical theory and practice, as well as his familiarity with the instrumental hardware employed by chemists in their laboratories. His allusion to Stephen Hales in the *Inquiry*, along with his assertion in his undated 'Idea of a Course of Physicks or Natural Philosophy' that 'Dr Black has given a very fine Example of Induction in his

the Autonomy of Chemistry Around 1800'. David Skene was alert to the applications of chemistry to the improvement of mining; see B. P. Lenman and J. B. Kenworthy, 'Dr David Skene, Linnaeus and the Applied Geology of the Scottish Enlightenment'; B. P. Lenman and J. B. Kenworthy, 'Dr David Skene and the Applied Geology of the Scottish Enlightenment II: Skene's Study of Contemporary Coalmining Practice'.

543 Wood, *Aberdeen Enlightenment*, p. 71. King's College announced that it would build a chemical laboratory in 1754 but the scheme was abandoned because of a lack of funds. Gregory pressed the College to revive the plan and it agreed to do so in September 1758. The failure of the medical school meant that the plan was dropped.

544 Gregory affirmed that physicians needed 'to be acquainted with the chemical history of our [bodily] fluids, and with the chemical analysis of whatever is taken into the human body as food or medicine, and, in general, of all the substances which can, in any degree, influence it'. According to Gregory, 'this shews the necessity of a knowledge of chemistry, previous to the study of the practice of physic'; John Gregory, *Lectures on the Duties and Qualifications of a Physician*, new edition (1772), p. 75. Chemistry was particularly relevant to Gregory's speculations on the nourishment of plants; see his 'Reflexions on the Principles of Agriculture', AUL, MS 2206/7/18, pp. 5–9.

545 Skene did not attend a course on chemistry when he studied medicine, but he did hear Charles Alston lecture on botany and materia medica when he was in Edinburgh in 1753; see David Skene to Andrew Skene, 19 May 1753 and July 1753, in David Skene, 'Correspondence, 1751–1770', AUL, MS 38, fols 29 and 33.

546 See above, p. xliii.

547 Reid also believed that chemical knowledge should be mobilised for the public good; see his comments in his Glasgow manuscript on the warehousing of grain in Reid, *On Society and Politics*, p. 109.

Paper upon Magnesia', demonstrate that he was familiar with the two works that did the most to stimulate the growth of pneumatic chemistry during the second half of the eighteenth century.[548] Other manuscripts most likely dating from the 1750s reveal that he was familiar with Herman Boerhaave's theories of heat and fire, although it is not clear that he subscribed to the great Dutch professor's explanation of heat. In a paper headed 'Fire or Heat' Reid stated that:

> It is a Question not yet certainly determined whether Heat be a Quality produced by some Motion or Vibration of the parts of the Heated bodie & ceasing to exist when its cause is removed or whether [heat] is some Subtile Element which alwise exists in the same Quantity throughout the Universe & in heated bodies is onely collected from the Circumambient parts & afterwards diffused when they cease to be hot. Arguments on both sides.[549]

In addition, he was aware of the ideas of one of the other major chemists of the early eighteenth century, Georg Ernst Stahl, whose explanation of combustion in terms of the inflammable principle called phlogiston was subsequently taken up in Scotland by William Cullen and Joseph Black, among others.[550] We have seen in Section 1 that when he reported to David Skene on John Robison's chemistry lectures at Glasgow in early 1767, he observed that Robison 'chiefly follows Dr Black and Stahl'.[551] Presumably, his comment was based on his knowledge of Stahl's writings as well as his attendance at Black's lectures on chemistry in

548 Reid, *Inquiry*, p. 25, and AUL, MS 2131/2/I/5, p. 3. Reid refers to Black, 'Experiments upon Magnesia alba, Quick-lime and Some Other Alkaline Substances'. Reid repeated his praise for Black in his Glasgow lectures on the culture of the mind, where he also mentions Stephen Hales' *Vegetable Staticks* as a model of inductive reasoning; Reid, *On Logic, Rhetoric and the Fine Arts*, p. 183.

549 AUL, MS 2131/7/II/4, fol. 2v; compare especially AUL, MS 2131/2/III/3, fol. 1r, as well as MS 2131/4/I/22, fol. 1v (a reference which dates from 1758), MS 2131/7/II/18, fol. 1r, and Reid, *On Logic, Rhetoric and the Fine Arts*, p. 92. Boerhaave's theory of fire is outlined in Herman Boerhaave, *Elements of Chemistry: Being the Annual Lectures of Herman Boerhaave, M.D. Formerly Professor of Chemistry and Botany, and at Present Professor of Physick in the University of Leyden* (1735), vol. I, pp. 78–247. Reid would also have known the restatement of Boerhaave's theory in van Musschenbroek, *Elements of Natural Philosophy*, vol. II. pp. 1–49, along with the account of fire in 's Gravesande, *Mathematical Elements of Natural Philosophy*, vol. II, pp. 1–22. Boerhaave, van Musschenbroek and 's Gravesande all claim that fire is a substance rather than the motion of material particles.

550 See, for example, Arthur L. Donovan, *Philosophical Chemistry in the Scottish Enlightenment: The Doctrines and Discoveries of William Cullen and Joseph Black*, pp. 150–2; Carleton E. Perrin, 'Joseph Black and the Absolute Levity of Phlogiston'; Wilson, *Seeking Nature's Logic*, pp. 145–9, 155–7, 161–2. Reid's conception of the disciplinary identity of chemistry might also have owed something to Stahl; on this point, see Schofield, *Mechanism and Materialism*, pp. 212–14.

551 Reid to David Skene, 25 February 1767, in Reid, *Correspondence*, p. 58, and above, p. lv. Reid may have known about Stahl's system of chemistry through Peter Shaw, *Philosophical Principles of Universal Chemistry: Or, the Foundation of a Scientifical Manner of Inquiring into and Preparing*

the winter of 1765–66. And, as mentioned above, Reid developed an elaborate classification of the powers of matter in which chemical affinities featured as one of the powers possessed by unorganised bodies whose operations could not be explained in terms of the forces of attraction and repulsion. The concept of chemical affinities was first articulated and given diagrammatic form by the French chemist Etienne-François Geoffroy. But, as Georgette Taylor has argued, the impact in Britain of Geoffroy's work was limited until William Cullen organised his system of philosophical chemistry on the basis of the concept of affinity (what Cullen preferred to call 'elective attractions') and turned Geoffroy's diagram into an effective pedagogical tool.[552] As discussed below, Reid knew of Cullen's formulation of the theory of chemical affinities, although his interpretation of the theory differed from Cullen's. For unlike Cullen, he conceived of affinity as a power which occupied a place in a hierarchy that began with the mechanical powers of matter, such as inertia, and gradually ascended to powers also associated with inanimate bodies, such as affinity, through to those powers found in vegetables and ultimately to the 'Powers Relative to Animal Life'.[553] As for Reid's familiarity with the instruments and practices of chemistry, his letters to Andrew and David Skene written after his move to Glasgow display his fascination with the thermometers manufactured by Alexander Wilson and his sons and document his conversations with Joseph Black regarding chemical furnaces.[554] Moreover, Reid also practised the art of chemical analysis. We can see a rudimentary example of his experimental techniques in his notes on the

the Natural and Artificial Bodies for the Uses of Life: Both in the Smaller Way of Experiment, and the Larger Way of Business ... Drawn from the Collegium Jenense of Dr George Ernest Stahl (1730).

552 Etienne-François Geoffroy, 'Des differents rapports observés en chimie entre differentes substances'; Georgette Taylor, 'Marking Out a Disciplinary Common Ground: The Role of Chemical Pedagogy in Establishing the Doctrine of Affinity at the Heart of British Chemistry'; see also Maurice Crosland, 'The Use of Diagrams as Chemical "Equations" in the Lecture Notes of William Cullen and Joseph Black', and Jan Golinski, *Science as Public Culture: Chemistry and Enlightenment in Britain, 1760–1820*, pp. 21–2.

553 Reid's hierarchy of powers is outlined in AUL, MS 2131/6/V/34, fol. 1r; see also Reid, *Inquiry*, p. 211.

554 See above, p. lii. For his conversations with Black regarding chemical furnaces, see Reid to David Skene, 20 December [1765] and 23 March 1766, in Reid, *Correspondence*, pp. 43, 47. In his correspondence with Skene, Reid refers to Johann Andreas Cramer, *Elementa artis docimasticae* (1739), which was shortly thereafter translated into English as *Elements of the Art of Assaying Metals* (1741); on Cramer, see Porter, 'The Promotion of Mining and the Advancement of Science', pp. 551–3, and Andréa Bortolotto, 'Johann Andreas Cramer and Chemical Mineral Assay in the Eighteenth Century'. Black was an expert on chemical furnaces, for he designed an improved portable furnace in the early 1750s and he later devised a laboratory furnace; see Robert G. W. Anderson, *The Playfair Collection and the Teaching of Chemistry at the University of Edinburgh, 1713–1858*, p. 24.

qualitative analysis of potatoes transcribed below (p. 127), and we know from a letter he sent to Joseph Black in 1773 that he engaged in the analysis of spa waters and had once collaborated with the Marischal Professor of Natural Philosophy, George Skene, in examining the water at Pannanich Wells near Ballater.[555] By 1764, therefore, Reid was reasonably fluent in chemical theory and had acquired the basic practical skills of an analytical chemist.

Further evidence for the extent of Reid's grasp of the theoretical and empirical details of the science of chemistry before leaving Old Aberdeen is found in his undated manuscript, 'The Chemical History of Salts'. This manuscript attests to the fact that Reid had access to a set of notes taken from William Cullen's chemistry lectures and perhaps also to a copy of Cullen's paper 'Some Reflections on the Study of Chemistry, and an Essay towards ascertaining the different Species of Salts', although this seem less likely.[556] Internal evidence suggests that the manuscript was written after 1756, for there are a number of references to Joseph Black's seminal paper published in that year, 'Experiments upon Magnesia alba, Quick-lime, and Some Other Alkaline Substances'.[557] It is unclear whose set of lecture notes Reid used as the basis for 'The Chemical History of Salts'. One possibility is that the original notes were taken by his ex-pupil William Ogilvie, who is said to have taken Cullen's course on chemistry while he was in Edinburgh in 1761–62. Given that Ogilvie attended the Glasgow lectures of Joseph Black in 1762–63, he is a plausible candidate for the authorship of the notes initially used

[555] Reid to Joseph Black, 17 January 1773, in Reid, *Correspondence*, pp. 74–5. Black contacted Reid because he had a query about the mineral waters at Peterhead; for other letters on the topic see George Moir to Black and George Skene to Black, 19 March 1773 and 25 March 1773, in Joseph Black, *The Correspondence of Joseph Black*, edited by Robert G. W. Anderson and Jean Jones, vol. I, pp. 278–9. On the chemical analysis of spa waters in Scotland, see especially Matthew D. Eddy, 'The "Doctrine of Salts" and Rev. John Walker's Analysis of a Scottish Spa (1749–1761)'. Reid knew about Walker's activities as a botanist and man of science; see Reid to Andrew Skene, 14 November 1764, in Reid, *Correspondence*, p. 38.

[556] Cullen's paper was read to the Edinburgh Philosophical Society by Dr David Clerk in 1753. A transcription of Cullen's paper appears in Leonard Dobbin, 'A Cullen Chemical Manuscript of 1753'; see also Emerson, 'The Philosophical Society of Edinburgh, 1748–1768', pp. 148–9. The link to Cullen is clear in comparing Dobbin, 'A Cullen Chemical Manuscript of 1753', pp. 144–5, with Reid, 'The Chemical History of Salts', fols 1r–2r. Reid's patron, Lord Deskford, became a member of the Edinburgh Philosophical Society *circa* 1755 and could conceivably have lent Reid a copy of Cullen's paper. Deskford knew of Cullen's work by 1752 at the latest, for he was sent Cullen's paper 'Remarks on Bleaching' by Lord Kames in that year; Donovan, *Philosophical Chemistry in the Scottish Enlightenment*, pp. 80–3. Deskford, and later Kames, were members of the Board of Trustees for Fisheries, Manufactures and Improvements in Scotland.

[557] For references to Black's paper, see Reid, 'The Chemical History of Salts', fols 15v, 17, 37v, 41v, 42v, 43v, 45. Another clue regarding the dating of the manuscript is the reference on fol. 23 to Francis Home, *Experiments on Bleaching* (1756).

by Reid.[558] Reid also collated the lecture notes he utilised with another set taken by his friend David Skene's younger brother George, who studied medicine in Edinburgh in the late 1750s. We know that the younger Skene went to Cullen's classes in the winter of 1758–59 and again in 1759–60.[559] Reid left the verso sides of the folios that make up the manuscript blank so that he could annotate his text, and he clearly compared different sets of lecture notes because on one of the blank sides he wrote: 'In G. Skene's Copy it is said that all of the Neutrals formed by the Nitrous & Muriatic Acids decompound Vitriolated Tartar by a double Elective attraction, but it is doubtfull whether they will compose it'.[560] The evidence we have thus suggests that 'The Chemical History of Salts' dates to the period roughly 1756 to 1763, and that the manuscript is based on notes taken from William Cullen's chemistry lectures by William Ogilvie.[561] As for what the manuscript tells us about Reid's knowledge of chemistry in the period, it demonstrates that he was thoroughly acquainted with Cullen's brand of philosophical chemistry. Moreover, in copying out the notes from Cullen's lectures, he would have been exposed to searching criticisms of standard authors such as Boerhaave and Stahl and he would have been introduced to the work of analytical chemists he had probably not previously encountered, such as Andreas Sigismund Marggraf, Johann Heinrich Pott and Axel Fredrik Cronstedt.[562] Consequently, by

[558] D. C. Macdonald, *Birthright in Land*, p. 158; Macdonald's dating is occasionally unreliable. For Ogilvie in Glasgow, see above, note 276.

[559] See George Skene to Andrew Skene, 13 November 1759, in AUL, MS 38, fol. 66; George Skene said of Cullen that 'he seems indeed to handle his Subject in the most masterly way'.

[560] Reid, 'The Chemical History of Salts', fol. 25v; for other annotations taken from George Skene's set of notes see fols 27v, 59v, 62v, 65.

[561] Comparison with a set of notes from Cullen's lectures taken in 1757–58 by David Carmichael shows that Reid's manuscript is more of a digest than a straightforward transcription of the lecture notes he had access to. Reid seems to have reordered some of his material and the sequence of topics covered in 'The Chemical History of Salts' does not correspond to the one Cullen followed in his course. Whereas Cullen adopted Macquer's classification of bodies as saline, inflammable, metallic, earthy and watery in his lectures and discussed the different categories in that order, in Reid's manuscript the history of earths forms the second rather than the fourth section. Nevertheless, 'The Chemical History of Salts' contains turns of phrase found in Carmichael's notes, and the contents of Reid's manuscript are clearly derived from Cullen's lectures. See [David Carmichael], 'Notes from William Cullen's Chemistry Lectures', Royal College of Physicians of Edinburgh, MS CUL/2/2/3–4, and David Carmichael, 'Dr William Cullen's Chemical Lectures 1757. & 1758. &c', Royal College of Physicians of Edinburgh, MS CUL/2/2/7. Carmichael's 'Dr William Cullen's Chemical Lectures' consists of two volumes (but bound as one) of rough notes taken at Cullen's lectures, whereas the two-volume set of 'Notes' contains a polished version of Carmichael's rough notes. For Cullen's use of Macquer, see Carmichael, 'Notes', fol. 64v.

[562] Reid, 'The Chemical History of Salts', fols 8, 10, 18, 19v, 20, 23r, 24v, 25v, 26, 34, 41, 43v, 45, 48, 51, 53, 56v, 57, 63, 68. On the work of Marggraf, Pott and Cronstedt, see especially Porter, 'The Promotion of Mining and the Advancement of Science'. The lecture notes would also have alerted

the time Reid left Old Aberdeen he had both schooled himself in the fundamentals of chemical theory and practice and immersed himself in the details of Cullen's innovative system of chemistry.[563]

We have seen in Section 1 that in 1763 Reid received intelligence from William Ogilvie about the chemical lectures being given in Glasgow by William Cullen's ex-pupil, Joseph Black. On the basis of Ogilvie's account, Reid declared himself 'very much pleased with Dr Black's Theory of Latent Heat', and it may be that he was subsequently able to acquaint himself more fully with the experimental basis for Black's theory on the basis of the lecture notes taken by Ogilvie during the 1762–63 academic session.[564] His introduction to Black's theory marked the beginning of a new phase in his development as a chemist because from the mid-1760s onwards the science of heat dominated his thinking about chemistry. Moreover, Reid's letters to David Skene document the fact that, from the outset, he was impressed with Black's accomplishments as a chemist. In July 1765 he admitted to Skene that since he had been unable to attend Black's lectures the previous winter he had only 'very imperfect hints of Dr Blacks Theory of Fire'. Nevertheless, he informed his friend that, based on experiments on the calcination of metals and the decomposition of sulphur, Black 'ha[d] a Strong Apprehension that the Phlogistick principle is so far from adding to the weight of Bodies by being Joyned to them that it diminishes, it'.[565] And even though he had not sat in on Black's classes, he observed that 'Chemistry seems to be the onely branch of Philosophy that can be said to be in a progressive State here although other branches are neither ill taught nor ill studied'. But he also regretfully noted that

Reid to the dispute between Charles Alston and Robert Whytt over lime water. See: Charles Alston, *A Dissertation on Quick-lime and Lime-water* (1752); Robert Whytt, 'Of the Various Strength of Different Lime-waters'; Donovan, *Philosophical Chemistry in the Scottish Enlightenment*, pp. 88–90.

[563] On the circulation of lecture notes in eighteenth-century Scotland, see Matthew D. Eddy, 'The Interactive Notebook: How Students Learned to Keep Notes during the Scottish Enlightenment', pp. 114–19; see also pp. 111–12 for a discussion of formatting features found in Reid's 'The Chemical History of Salts'.

[564] Reid to [William Ogilvie], [1763], in Reid, *Correspondence*, p. 26. Ogilvie's set of notes is referred to in Reid to David Skene, 20 December [1765], in Reid, *Correspondence*, p. 44.

[565] Reid to David Skene, 13 July 1765, in Reid, *Correspondence*, p. 39; compare below, p. 135. On Black's theory of phlogiston, see esp. Perrin, 'Joseph Black and the Absolute Levity of Phlogiston'. William Irvine subsequently addressed this problem and gave a discourse based on his experimental investigations to the Glasgow Literary Society in February 1773 entitled 'On the Measure of the Quantity of Matter in Bodies'; see 'Laws of the Literary Society in Glasgow College', p. 39. The text of Irvine's discourse was later published in Irvine and Irvine, Jr, *Essays Chiefly on Chemical Subjects*, pp. 407–21.

'as Black is got into a good deal of [medical] Practice it is to be feared that his Chymical enquiries must go on slowly and heavily in time to come'.[566]

During the 1765–66 session, Reid made the time to attend Black's course and, by December 1765, was able to summarise for Skene the essentials of Black's theory of latent heat as well as the experiments upon which it was based. Of the theory, Reid judged that while it gave 'a great deal of light to the Phenomena of heat that appear in Mixture, Solution & Evaporation', it failed to explain the phenomena 'which appear in Animal heat, Inflammation, & Friction'. Significantly, he commented that Black's 'Doctrine of Latent Heat is the onely thing I have yet heard that is altogether New', although he did say that he counted his colleague's contribution to the science of heat 'as a very important Discovery'. He also suggested that Skene consult Ogilvie's set of lecture notes in order to understand more fully Black's interpretation of the experimental evidence.[567] Skene, however, seems either not to have completely understood Black's theory or had reservations about it, because Reid subsequently supplied Skene with a recapitulation of Black's conception of latent heat, including a description of the key experiments upon which the theory rested.[568] In doing so, he stressed that 'I cannot find that Cullen or the Edinburgh people know any thing of this matter' and asked Skene 'not to make any use' of his summary 'that may endanger the discoverer being defrauded of his Property'.[569] In the end, Reid was fortunate to have attended Black's course in the winter of 1765–66 for the lectures proved to be his colleague's last in Glasgow. As the session was drawing to a close, it emerged that Black was likely to transfer to a chair in Edinburgh. He duly became

[566] Reid to David Skene, 13 July 1765, in Reid, *Correspondence*, pp. 39–40.

[567] Reid to David Skene, 20 December [1765], in Reid, *Correspondence*, p. 44. Reid's comment about Black's lectures suggests that he had more than a superficial knowledge of chemistry. Compare Reid's mildly critical view of Black's course with John Robison's disillusioned assessment of Black's Edinburgh lectures in Robison to James Watt, 23 July 1800, in Robinson and McKie, *Partners in Science*, pp. 343–4.

[568] Reid to David Skene, 23 March 1766, in Reid, *Correspondence*, pp. 48–9. The experiments described in this letter were first performed by William Irvine; see also below, pp. 132–4.

[569] Reid to David Skene, 23 March 1766, in Reid, *Correspondence*, pp. 48, 50. Reid had earlier expressed his concern that Black might be defrauded of his discovery; Reid to David Skene, 20 December [1765], in Reid, *Correspondence*, p. 44. Reid's worry about the dissemination of Black's ideas were well founded because an anonymous unauthorised account of Black's theory of latent heat was published in 1770; see Anon., *An Enquiry into the General Effects of Heat with Observations on the Theories of Mixture* (1770). Black was notoriously reluctant to publish his work; see Black, *The Correspondence of Joseph Black*, vol. I, p. 30, note 26. Reid attributed this trait to his colleague's modesty and caution; see Reid to David Skene, 13 July 1765, in Reid, *Correspondence*, p. 39.

the Professor of Medicine and Chemistry at the Town's College, and resigned his Glasgow post in May 1766.[570]

After Black left Glasgow, Reid collaborated with William Irvine in investigating aspects of the science of heat.[571] As a student at the College and an auditor at Black's lectures, Irvine was part of a circle of aspiring chemists centred on Black that also included William Trail.[572] According to John Robison, Irvine and Trail stood out from Black's other students because of their desire to quantify the study of chemical phenomena. Writing to James Watt in 1800, Robison recalled that Irvine

> had a true mathematical taste, and delighted in reducing every thing to measure by means of Equations. He hunted for opportunities of doing this, and was very quick sighted in discovering the means of procedure. Willy Traill [*sic*] and he were almost constantly occupied in this way.[573]

Moreover, Irvine learned his craft as a chemist while acting as his mentor's assistant. In his edition of Black's lectures, Robison asserted that 'it was with Mr. Irvine's assistance that Dr. Black made his first experiments for measuring the latent heat of steam' and that Irvine had also 'supplied Dr. Black with a vast number of experiments on the equilibrium of heat, on the specific heats of different substances, and on the continued absorption and fixation of heat by glass, sealing-wax, resin, and other substances, which gradually become more fluid'.[574] In addition, Robison recorded that Irvine was especially taken with quantitative problems in thermometry and that, prior to 1770, he had carried out experiments

570 Reid to David Skene, 18 April [1766], in Reid, *Correspondence*, pp. 50–1; Joseph Black to John Black, 30 June 1766, in Black, *Correspondence of Joseph Black*, vol. I, p. 187.

571 James Watt also collaborated with Irvine in the study of heat; see Watt to J. J. Magellan, 29 March 1780, and Watt to Sir Joseph Banks, 1 March 1815, in Robinson and McKie, *Partners in Science*, pp. 87, 420.

572 Irvine attended Black's chemistry lectures in 1761–62; see Joseph Black to James Watt, 15 March 1780, and James Watt to J. J. Magellan, 20 March 1780, in Robinson and McKie, *Partners in Science*, pp. 84, 85.

573 Robison to James Watt, [October 1800], in Robinson and McKie, *Partners in Science*, pp. 359–60; compare Robison's comments in Joseph Black, *Lectures on the Elements of Chemistry, Delivered in the University of Edinburgh, by the Late Joseph Black, M.D.* (1803), edited by John Robison, vol. I, p. xliv. Black himself made a similar assessment and described Irvine as 'a young gentleman of an inquisitive and philosophical mind, of great ingenuity, and peculiarly qualified for [the task of measuring the latent heat of steam], by the habits of mathematical study, and scrupulous attention to all kinds of measurement'; Black, *Lectures on the Elements of Chemistry*, vol. I, p. 171. On Irvine's mathematical bent, see also Anon., 'Biographical Account of Dr William Irvine', pp. 455, 457.

574 Black, *Lectures on the Elements of Chemistry*, vol. I, p. xliv; Robison also stated here that 'the register of these experiments are in my possession'.

on the heat generated by the mixing of fluids such as water and vitriolic acid, which formed the basis for his theory of absolute heat.[575] Thus although Irvine helped to create the experimental foundation for Black's theory of latent heat, he nevertheless disagreed with Black over the interpretation of some of these experiments and developed his own ideas regarding the heat capacities of different substances, absolute heat and the cause of changes of state.[576] Reid's manuscripts on chemistry dating from the period beginning with Irvine's appointment as the Glasgow Lecturer in Chemistry in 1769 to Irvine's death in 1787 register the preoccupations of his younger colleague.

Like his mentor, William Irvine tackled the theoretical and practical problems associated with the use of thermometers. According to Boerhaave, the thermometer provided a reliable measure of the quantity of heat or 'elemental fire' in a body because the quantity of heat was manifest in the expansion or contraction of a body as it heated or cooled. But chemists remained uncertain as to whether the degrees of a thermometer were accurate measures of different degrees of heat. This issue, along with the question of standardisation, remained contentious throughout the eighteenth century.[577] In his lectures, Black discussed at length both the question of standardisation and the issue of 'whether the degrees of [the] scales [of thermometers] express, or point out, equal differences of heat?' While he was confident that instrument makers could produce thermometers that were accurately calibrated to one another, he pointed out that the scale on a thermometer does not 'measure heat itself, but the expansion produced by heat'. He argued that various experiments had shown that there is a 'little disproportion between the degrees of expansion and the degrees of heat', and he thus advised his students that they had to be alert to the 'imperfections that still attend' thermometers.[578] More importantly, Black's theory of latent heat implied that thermometers do *not* measure the quantity of heat in a body, as Boerhaave and other chemists had maintained. Rather, for Black the thermometer measured 'intensity' or temperature rather than the quantity of heat, and he accused his predecessors of confusing

[575] Black, *Lectures on the Elements of Chemistry*, vol. I, pp. xliv, 504–5.

[576] Irvine's concept of heat capacities was related to Black's notion of specific heats. James Watt began to investigate this phenomenon in 1763; see Watt to J. J. Magellan, 1 March 1780, in Robinson and McKie, *Partners in Science*, p. 77; Black, *Lectures on the Elements of Chemistry*, vol. I, pp. 141–2, 504–8; Donovan, *Philosophical Chemistry in the Scottish Enlightenment*, pp. 246–9, 265–71.

[577] On these issues see, for example, Jan Golinski, '"Fit Instruments": Thermometers in Eighteenth-Century Chemistry', and John C. Powers, 'Measuring Fire: Herman Boerhaave and the Introduction of Thermometry into Chemistry'.

[578] Black, *Lectures on the Elements of Chemistry*, vol. I, pp. 56, 59, 60. As Robison noted, Black had also read a paper to the Glasgow Literary Society on thermometry on 28 March 1760. See also Donovan, *Philosophical Chemistry in the Scottish Enlightenment*, pp. 231–5.

the two concepts. He said that they had 'confound[ed] the quantity of heat in different bodies with its general strength or intensity, though it is plain that these are two different things, and should always be distinguished'.[579] Under Black's tutelage, Irvine performed 'thermometrical experiments on the scale of heat' and investigated 'the connection between expansion and variation of temperature'.[580] Later, in reformulating some of Black's basic ideas, he distinguished between the 'relative' and 'absolute' heat of a body, and held that whereas the thermometer served to measure temperature or 'relative' heat, 'absolute' heat could only be determined inferentially, on the basis of calculations involving the temperature of a body in relation to the body's heat capacity.[581]

In 1770 Reid took up these problems in thermometry in a brief entry headed 'Of Heat', which is part of a chronologically layered manuscript that is related to the work on the science of heat carried out by Black, Irvine and Adair Crawford (pp. 129–32).[582] In 'Of Heat', Reid showed himself to be cognisant of the interplay between instrumentation and conceptualisation. He argued that because philosophers had initially relied on their sense of touch to measure the warmth of a body, they believed that heat and cold were two contrary qualities of matter which corresponded to their sensations of hot and cold. When it was discovered that these sensations depended upon the state of our body and were not directly correlated with the qualities of the bodies around us, philosophers were then obliged to choose another measure of heat. They opted to rely on the expansion of bodies as a measure of temperature, but this new method of measurement entailed a change in the meaning of 'heat' because it destroyed the distinction that had been drawn since antiquity between hot and cold as properties of material bodies. Using the thermometer, chemists now spoke of the degrees of heat in a body, whereas they had formerly described a body at 32°F, for example, as having the property of cold (p. 130).[583]

Moreover, Reid maintained that it was impossible to demonstrate experimentally that the thermometer provided a reliable measure of degrees of heat. He

[579] Black, *Lectures on the Elements of Chemistry*, vol. I, pp. 77–8; Golinski, '"Fit Instruments"', pp. 198–9.

[580] Black, *Lectures on the Elements of Chemistry*, vol. I, p. xliv.

[581] See Irvine and Irvine, Jr, *Essays, Chiefly on Chemical Subjects*, pp. 153–9, and the exposition of Irvine's ideas in Adair Crawford, *Experiments and Observations on Animal Heat, and the Inflammation of Combustible Bodies; Being an Attempt to Resolve These Phenomena into a General Law of Nature*, second edition (1788), pp. 2–5. See also Golinski, '"Fit Instruments"', pp. 199–200.

[582] Compare Reid's discussion of heat in AUL, MS 2131/2/III/3, which is entitled 'Of Secondary Qualities'.

[583] On the shift from a reliance on direct sensory evidence to the use of instruments such as the thermometer to measure heat, see Golinski, '"Fit Instruments"', pp. 187–8.

regarded this as being something which had to be taken for granted, since the use of the thermometer for him *defined* what was meant by greater or lesser degrees of heat.[584] Consequently, he thought that the experiments performed by a succession of eighteenth-century chemists in order to establish the connection between the expansion of bodies and degrees of heat were beside the point, just as he had earlier insisted that both Newtonians and Leibnizians were misguided in thinking that they could offer empirical proof for the validity of their respective measures of the force of moving bodies.[585] According to Reid, chemists had,

> taken [it] for granted that two equal quantities of the same kind of body of different temperatures when mixed will have the temperature which is an arithmetical mean between the two. Here one Measure of heat is assumed in order to ascertain another, which may be assumed with as good reason[.] I conceive therefore there is nothing unphilosophical in laying it down as a definition of the degrees of heat that they are greater or less according as they expand the same body more or less. (p. 131)

Hence Reid held that what their experiments had shown was that 'when equal quantities of the same fluid of different temperatures are mixed they produce a temperature which is an arithmetical mean between the two extreams' (p. 131). Following the example set by his colleague Irvine and others, Reid then proceeded to consider the quantitative implications of this fact, although he reached no clear conclusions.

Reid also endeavoured to master the details of Irvine's theory of heat. Irvine probably discussed his concept of heat capacities with Reid in the 1770s, given that there are some undated remarks of Reid's on an unnamed 'discovery now proposed' which appear to be related to Irvine's calculation of the quantity of heat gained when ice is turned into water (pp. 136–8). Reflecting on these remarks in November 1780, Reid identified two presuppositions of the argument that he had earlier summarised: (i) that the quantity of heat is independent of temperature and (ii) that the quantity of heat is proportional to the quantity of matter. Why he did so

[584] Compare Golinski, '"Fit Instruments"', pp. 204–5.

[585] See Black, *Lectures on the Elements of Chemistry*, vol. I, pp. xxxix–xl; Boerhaave, *Elements of Chemistry*, vol. I, p. 85; Edmond Halley, 'An Account of Several Experiments Made to Examine the Nature of the Expansion and Contraction of Fluids by Heat and Cold, in Order to Ascertain the Divisions of the Thermometer, and to Make That Instrument, in All Places, without Adjusting by a Standard'; Martine, *Essays Medical and Philosophical*, esp. pp. 288–9; van Musschenbroek, *Elements of Natural Philosophy*, vol. II, pp. 2, 9–15; Brook Taylor, 'An Account of an Experiment, Made to Ascertain the Proportion of the Expansion of the Liquor in the Thermometer, with Regard to the Degrees of Heat'; and Donovan, *Philosophical Chemistry in the Scottish Enlightenment*, pp. 132–5, on William Cullen.

is unclear for he does not indicate whether he queried either of these propositions. While the first assumption was common to both Black's and Irvine's theories of heat, the second contradicted Black's theory and it may be that he was aware of the contradiction.[586] In addition, he tentatively suggested a method for determining quantities of heat and endeavoured to reconcile his own estimate of the quantity of heat in boiling water with that given by Irvine (p. 138). Later, in December 1787, he employed the distinction first drawn by Irvine between temperature, the relative quantity of heat contained by a body and the absolute quantity of heat in a body. He stated that while temperature was measured by the degrees of the thermometer, the relative quantity of heat was measured by 'the Effect produced by the Mixture of two Bodies of different Temperature and of different Nature, which by their mixture produce no chemical Composition or Resolution nor a change of Form in any of the Bodies mixed' (p. 136). Unfortunately, the text breaks off before he dealt with absolute heat so that we do not know how he understood this key element in Irvine's theory of heat.[587] Due to the fragmentary nature of the evidence in these manuscripts, therefore, the question of whether Reid adopted Irvine's as opposed to Black's theory of heat cannot be answered with any certainty, although the manuscripts clearly indicate that he had a detailed knowledge of the experimental and theoretical components of Irvine's work.

One of the first philosophical chemists to apply the new ideas of Black and Irvine to the problem of animal heat was Adair Crawford.[588] We have seen that Reid judged that one of the weaknesses of Black's theory of latent heat was that it failed to explain the phenomena of animal heat. Irvine may have regarded the work of his preceptor in a similar light, for he addressed the issue of animal heat on 24 February 1769 in a discourse read to the Glasgow Literary Society 'on the Degrees of Heat necessary for supporting the Lives of Animals and Vegetables'.[589] Crawford may thus have been encouraged to study animal heat by Irvine and Reid. Having graduated with an MA from Glasgow in 1770 Crawford returned there in 1776 to continue his studies in order to qualify for an MD degree, which

[586] Black, *Lectures on the Elements of Chemistry*, vol. I, p. 79.

[587] In his notes on Lavoisier's *Elements of Chemistry* Reid stated that 'The whole quantity of caloric is measured by its Temperature, its relative caloric, and its magnitude taken together, or by the product of the three multiplied into one another' (see below, p. 142). By substituting 'heat' for 'caloric' we might approximate Reid's definition of the measure of absolute heat.

[588] On Crawford, see Everett Mendelsohn, *Heat and Life: The Development of the Theory of Animal Heat*, pp. 123–33, 154–9, and Victor Boantza, 'The Phlogistic Role of Heat in the Chemical Revolution and the Origins of Kirwan's "Ingenious Modifications ... into the Theory of Phlogiston"', esp. pp. 322–8.

[589] 'Laws of the Literary Society in Glasgow College', p. 30; the text of Irvine's discourse was later included in Irvine and Irvine, Jr, *Essays, Chiefly on Chemical Subjects*, pp. 191–205.

he duly obtained in 1780.[590] Inspired by Irvine's lectures in the winter of 1776–77, Crawford performed a series of experiments on combustion and animal heat in the summer of 1777 and, 'in the autumn of that year', he presented his results to 'Dr. Reid, Mr. Wilson, and Dr. Irvine'.[591] What Reid thought of Crawford's theory cannot now be determined, but he apparently commented on Crawford's formulation of Irvine's distinction between temperature and the relative quantity and absolute quantity of heat. For in the second edition of his *Experiments and Observations on Animal Heat*, Crawford quoted a cautionary note from Reid on the quantitative formulation of the concept of temperature:

> It has been observed, by my learned friend Dr. Reid, in a very accurate communication, with which he favoured me on this subject, that 'until the ratio between one temperature and another be ascertained by experiment and induction, we ought to consider temperature as a measure which admits of degrees, but not of ratios: and consequently ought not to conclude, that the temperature of one body is double or triple to that of another, unless the ratio of different temperatures were determined. Nor ought we to use the expressions of a double or triple temperature, these being expressions which can convey no distinct meaning, until the ratio of different temperatures be determined'.[592]

And in claiming that the word 'heat' has 'a double signification' in 'common language' insofar as it refers to both 'a *sensation* of the mind' and 'an unknown principle, whether we call it a *quality* or a *substance*, which is the exciting cause of that sensation', Crawford drew on Reid's discussion in the *Inquiry* of the ambiguous meaning in ordinary usage of the terms we use to refer to the secondary qualities of bodies.[593] For his part, Reid would have found the theistic and providentialist elements in the second edition of Crawford's text congenial,

[590] W. Innes Addison, *A Roll of the Graduates of the University of Glasgow*, p. 130. Crawford had earlier been Reid's student, for he matriculated in the moral philosophy class in 1764; see W. Innes Addison, *The Matriculation Albums of the University of Glasgow, from 1728 to 1858*, p. 72 (no. 2299).

[591] Crawford, *Experiments and Observations on Animal Heat*, sig. A3r. Patrick Wilson was Irvine's classmate in Black's chemistry lectures; see Joseph Black to James Watt, 15 March 1780, in Robinson and McKie, *Partners in Science*, p. 84. For evidence of Patrick Wilson's researches related to the science of heat in the 1780s, see his letters to Joseph Black published in Black, *The Correspondence of Joseph Black*.

[592] Crawford, *Experiments and Observations on Animal Heat*, p. 6; compare Irvine and Irvine, Jr, *Essays, Chiefly on Chemical Subjects*, p. 154, and Black, *Lectures on the Elements of Chemistry*, vol. I, p. 64.

[593] Crawford, *Experiments and Observations on Animal Heat*, pp. 1–2; compare Reid, *Inquiry*, pp. 38–43. Crawford used Reid's point to elucidate William Irvine's distinction between absolute and sensible heat.

and these features of Crawford's work serve to remind us of the fact that Scottish chemistry in the late eighteenth century was not uniformly secular in intent.[594]

Irvine's successor as the Glasgow Lecturer in Chemistry, Thomas Charles Hope, was one of the first chemists in Scotland to adopt the theories of Lavoisier. Consequently, it may have been on Hope's recommendation that in May 1789 Reid read and made extensive notes from William Nicholson's English translation, published in 1788, of Antoine François de Fourcroy's *Élémens d'histoire naturelle et de chimie*.[595] These reading notes show that Reid was interested in those sections of Fourcroy's text dealing with the topics that historians have generally recognised as being constitutive of the Chemical Revolution, namely the study of airs, Lavoisier's theory of acids and the rejection of phlogiston.[596] Yet in a second manuscript headed 'Of the Chemical Elements of Bodies', which dates from 1790 (pp. 139–53), Reid discussed a further feature of the Chemical Revolution that has come into focus only in the revisionist literature on the transformation of chemistry in the latter decades of the eighteenth century, namely the changing concept of a chemical element.[597] Portions of this manuscript are summaries of, or random notes from, Parts I and II of Robert Kerr's English translation of Lavoisier's *Elements of Chemistry*, published in Edinburgh in

[594] Crawford, *Experiments and Observations on Animal Heat*, pp. 430, 440–5. For secular readings of Cullen's chemistry and Scottish chemistry more generally, see John R. R. Christie, 'Ether and the Science of Chemistry: 1740–1790', esp. pp. 93–4; Arthur L. Donovan, 'William Cullen and the Research Tradition of Eighteenth-Century Scottish Chemistry', esp. pp. 104–5; Golinski, *Science as Public Culture*, p. 23. Robert Anderson has recently stated that Black 'may well have been an unbeliever'; Robert G. W. Anderson, 'Boerhaave to Black: The Evolution of Chemistry Teaching', p. 238.

[595] Hope's conversion to Lavoisier's chemistry is discussed in: V. A. Eyles, 'The Evolution of a Chemist: Sir James Hall, Bt., F.R.S., P.R.S.E., of Dunglass, Haddingtonshire, (1761–1832), and His Relations with Joseph Black, Antoine Lavoisier and Other Scientists of the Period', pp. 175–6; Carlton E. Perrin, 'A Reluctant Catalyst: Joseph Black and the Edinburgh Reception of Lavoisier's Chemistry', pp. 156–7; and Robert G. W. Anderson, 'Thomas Charles Hope and the Limiting Legacy of Joseph Black', p. 149. For bibliographical details of Nicholson's translation of Fourcroy, see W. A. Smeaton, *Fourcroy: Chemist and Revolutionary, 1755–1809*, p. 214. Fourcroy first endorsed Lavoisier's rejection of phlogiston in the second edition of the *Elemens* translated by Nicholson.

[596] AUL, MS 2131/3/I/16, esp. pp. 1, 3, 4, 13–18, 21–4. Dated 22 May 1789, Reid's reading notes are extremely detailed and amount to some thirty pages.

[597] This feature of the Chemical Revolution was first emphasised in Robert Siegfried and Betty Jo Dobbs, 'Composition, a Neglected Aspect of the Chemical Revolution'; see also Robert Siegfried, *From Elements to Atoms: A History of Chemical Composition*. Joseph Black likewise recognised the significance of this aspect of Lavoisier's system of chemistry, although for him the issue of identifying chemical elements was bound up with question of chemical nomenclature; see Black to Thomas Beddoes, 24 November 1787, in Black, *The Correspondence of Joseph Black*, vol. II, pp. 924–5, and Black, *Lectures on the Elements of Chemistry*, vol. I, p. 489.

1790.[598] In the opening pages, however, Reid attempted to state the essentials of what he took to be Lavoisier's concept of a chemical element and to outline the defining properties of the five substances the French chemist had judged to be 'simple', namely light, caloric, oxygen, azote and hydrogen.

According to Reid, the discovery of the composition of water had shown that the true 'first elements of Bodies' were still unknown.[599] Neverthless, he insisted that it was necessary to retain some working notion of an element, for two reasons: (i) it was important to know how bodies were physically constituted; and (ii) the 'great branch of the Art of Chemistry' devoted to the resolution or analysis of bodies was founded on the belief in the existence of elementary substances. In his view, chemical elements could be defined as

> those principles in the composition of Bodies, which have not hitherto been discovered to be compounded of principles still more simple. Remembring always, that what is now a Chemical Element, may some time hence in a more advanced State of the Chemical Art be found to be a compound. (p. 139)[600]

Defined in these terms, two such elements were light and caloric, although he noted that it was uncertain whether they were two distinct principles or whether they were different modifications of one principle that could be converted into one another. Light he described as 'the instrument of vision' and as a power which 'enlivens the whole Face of Nature'.[601] While acknowledging that Newton had determined the basic properties of the rays of light at the turn of the eighteenth century, Reid maintained that subsequent research in natural philosophy had revealed that light fulfilled a variety of functions in the economy of nature,

[598] On Kerr, see Perrin, 'A Reluctant Catalyst', pp. 155–6.

[599] The composite nature of water was discovered in the early 1780s by Henry Cavendish and James Watt; see Henry Cavendish, 'Experiments on Air'; James Watt, 'Thoughts on the Constituent Parts of Water and of Dephlogisticated Air; with an Account of Some Experiments on That Subject'; James Watt, 'Sequel to the Thoughts on the Constituent Parts of Water and Dephlogisticated Air'.

[600] Compare Antoine-Laurent Lavoisier, *Elements of Chemistry, in a New Systematic Order, Containing All the Modern Discoveries* (1790), pp. xxii–xxiv, 176–7.

[601] Compare Reid, *Inquiry*, p. 77. Reid's language here recalls Stephen Hales' characterisation of 'elastick air'; Hales, *Vegetable Staticks*, p. 178. Significantly, Hales drew a parallel between light and elastic air and cited passages in the Queries to Newton's *Opticks* to justify the comparison. Hales was later quoted in David Macbride, *Experimental Essays on the Following Subjects: I. On the Fermentation of Alimentary Mixtures. II. On the Nature and Properties of Fixed Air. III. On the Respective Powers, and Manner of Acting, of the Different Kinds of Antiseptics. IV. On the Scurvy; with a Proposal for Trying New Methods to Prevent or Cure the Same, at Sea. V. On the Dissolvent Power of Quick-lime* (1764), p. 30. George Berkeley's description of the function of 'pure æther, or light, or fire' in the economy of nature echoes Hales' characterisation of elastic air; see George Berkeley, *Siris: A Chain of Philosophical Reflexions and Inquiries* (1747), p. 82.

ranging from heating opaque bodies to producing the green colour of vegetables and making them release oxygen into the atmosphere.[602] But he thought that only a few of the 'purposes of Light' had been discovered, and he acknowledged that it was also unclear 'How far the matter of Light enters into the composition of bodies as an Element' (p. 140). One fact about light that he believed had been established was that it was imponderable, like the matter of heat. He stated that he knew of 'no phenomenon that shows either of them to gravitate, so as to add to the Weight of Bodies by being conjoyned with them or to diminish the weight of Bodies by being disjoyned from them' (pp. 140–1).[603]

Following the 'French Chemists', Reid defined caloric as 'that Principle, whatever it be, whether material or not, which heats Bodies, & expands them' (p. 141). He said that caloric pervades all bodies, thereby enlarging their dimensions, and that due to its power of repulsion, caloric was capable of changing solids into fluids and fluids into gases. He deemed the repulsive power of caloric to be antagonistic to the attractive power of ordinary matter, and he explained the solid, fluid and gaseous states of matter in terms of the relative strengths of these opposing forces. Regarding the spatial distribution of caloric, he observed that although it might be expected that caloric was diffused equally throughout space because it was an elastic fluid having a constant quantity, this proved not to be the case because different substances could contain caloric in varying capacities. Consequently, he distinguished temperature from the quantity of caloric, and the relative from the absolute quantity of caloric in a body. Moreover, he also speculated that caloric might be fixed in bodies and that there was an equilibrium of caloric in material systems, much as Black and Irvine had done with regard to heat (pp. 142–3).[604]

[602] On the interaction of light and vegetables, see especially Jan Ingenhousz, *Experiments upon Vegetables, Discovering Their Great Power of Purifying the Common Air in the Sun-shine, and of Injuring It in the Shade and at Night* (1779).

[603] See above, p. clxiv, and compare Lavoisier, *Elements of Chemistry*, p. 6. Phlogiston theorists had routinely made similar claims about light and heat being different modifications of one primordial etherial substance; see P. M. Heimann, 'Ether and Imponderables', pp. 75–8. Reid may have known the work of Joseph Black's student, Patrick Dugud Leslie, mentioned by Heimann. Leslie argued that 'phlogiston is fire and light, or a certain subtile elastick fluid, upon the modifications of which the phænomena of heat and light immediately depend' and affirmed that phlogiston is imponderable; see Patrick Dugud Leslie, *A Philosophical Inquiry into the Cause of Animal Heat: With Incidental Observations on Several Phisiological and Chymical Questions, Connected with the Subject* (1778), pp. 104–5, 119–25. Reid was also familiar with the links between fire and light drawn by Hermann Boerhaave and Pieter van Musschenbroek in their highly influential theories of fire; see above, note 549.

[604] Reid was here translating concepts derived from the theories of heat advanced by Black and Irvine into the language of caloric.

The details of Reid's description of the remaining elements – oxygen, azote and hydrogen – need not concern us, for the historical significance of this manuscript lies in the fact that, taken as a whole, it demonstrates his familiarity with, and apparent adoption of, Lavoisier's system of chemistry. Although it is difficult to gauge Reid's reaction to Lavoisier on the basis of a single manuscript, the fact that the passages in which he discussed Lavoisier's concept of chemical elements and his theory of caloric were not merely summaries of sections of the *Elements of Chemistry* indicates that he was synthesising the relevant evidence and endeavouring to master the basic concepts of Lavoisier's system without relying exclusively on the text. Reid's manuscript on Lavoisier also differs in character from the largely derivative notes he took from Fourcroy, insofar as the degree of independent thought contained in this manuscript suggests that he was not simply making an epitome of Lavoisier's *Elements of Chemistry*. Moreover, some of Reid's contemporaries remarked on the seriousness with which he studied French chemistry in the last years of his life, which implies that he was not just keeping up with the latest discoveries in the science but rather was actively engaged in assessing the merits of the new ideas emanating from France.[605] 'Of the Chemical Elements of Bodies' thus strongly suggests that Reid accepted certain features of Lavoisier's chemical system, and he was undoubtedly encouraged to do so by his colleague Thomas Charles Hope, who was himself an early convert to the theoretical doctrines of the French chemists.

As for the contentious issue of the existence of imponderable fluids such as caloric, Reid apparently believed that their existence was confirmed by the available experimental evidence, and it would seem that his scruples about hypotheses and speculations regarding etherial media did not prevent him from tentatively accepting these fluids as legitimate explanatory mechanisms. Even though Reid's published methodological dicta may have predisposed some of his contemporaries to reject the use of ethers and subtle fluids, his own practice was more complex. In the case of caloric, it may be that he was swayed by Lavoisier's methodological rhetoric and quantitative methods. But we should also recognise the fact that philosophical chemistry in the eighteenth century was premised on the existence of subtle fluids, be they Boerhaave's element of fire, Stahl's phlogiston or Black's matter of heat.[606] Furthermore, the similarities between the theoretical ideas of Lavoisier and those advanced by Black, Irvine and Crawford were such

[605] [Cleghorn], *Sketch of the Character of the Late Thomas Reid, D.D. Professor of Moral Philosophy in the University of Glasgow*, p. 5.

[606] John Christie's point 'no ether, no new chemistry' is well taken; Christie, 'Ether and the Science of Chemistry', p. 106. But, as I have indicated, the point applies equally well to almost all of eighteenth-century philosophical chemistry and not just to Cullen and Lavoisier.

that Lavoisier's conception of caloric would have struck Reid as a variation on a theme already stated by his fellow Scots.[607] Hence it is not surprising that Reid evidently embraced the notion of caloric.

In addition to Reid's manuscripts, the 'Professors Receipt Books' shed some light on his reading in the science of chemistry while he was in Glasgow. The lists of his library borrowings show that the number of titles on chemical topics that he consulted was significantly smaller than it was for mathematics or other branches of natural philosophy. Moreover, the books that he borrowed were a curious mixture of standard texts and innovative works that contributed to the significant changes that were taking place in the investigation of chemical phenomena during the course of the second half of the eighteenth century. Among the titles he withdrew from the library, a book on alchemy, Boerhaave's *Elementa chemiae* (1732), Georg Ernst Stahl's *Experimenta, observationes, animadversiones ... chymicae et physicae* and individual volumes from Boyle's *Works* (1744) all reflected the state of chemistry before 1750. However, his interest in the latest developments in the field is registered in his selection of Joseph Priestley's *Experiments and Observations on Different Kinds of Air* (1775), the first two volumes of Torbern Olof Bergman's collected writings, as well as Fourcroy's *Elements of Natural History, and of Chemistry* (1788).[608] And, as we have seen, Reid's copious reading notes taken from Fourcroy's *Elements* attest to the fact that he studied carefully this key text of the Chemical Revolution. Furthermore, some books that straddled different branches of natural philosophy were relevant to his chemical interests. These include Jean-André Deluc's *Récherches sur les modifications de l'atmosphere* and John Elliott's *Philosophical Observations on the Senses of Vision and Hearing*, which contained a lengthy 'Inquiry concerning Combustion' as well as extensive comments on the theories of animal heat advanced by Adair Crawford and Patrick Dugud Leslie.[609] But even though his

[607] The similarities between the ideas of Black, Irvine and Crawford and those of Lavoisier is emphasised in Evan M. Melhado, 'Oxygen, Phlogiston and Caloric: The Case of Guyton', pp. 312, 313–16; see also Robert J. Morris, 'Lavoisier and the Caloric Theory', on the response of Lavoisier to the work of the three Scottish chemists. Carleton Perrin makes a similar point about the response of Joseph Black to Lavoisier; see Perrin, 'A Reluctant Catalyst', p. 164.

[608] 'Professors Receipt Book, 1765–1770' and 'Professors Receipt Book, 1770–[1789]'. Given the publication date recorded for Priestley's *Experiments and Observations on Different Kinds of Air*, Reid evidently borrowed the second volume of the work, which was published in 1775. A set of reading notes taken in the autumn of 1781 from the first volume published in 1774 is found in AUL, MS 2131/3/I/24, fol. 1r.

[609] See above, pp. lxix, clxiii; Elliott, *Philosophical Observations on the Senses of Vision and Hearing*, pp. 85–203, 205–22. For Leslie's theory of animal heat, see Leslie, *A Philosophical Inquiry into the Cause of Animal Heat.* Reid may also have read Richard Kirwan's *Elements of Mineralogy*; see above, p. lxxii.

reading of monographs was limited numerically, he was not necessarily out of touch with the most recent research in chemistry because the learned journals he took out from the library also contained important contributions to the literature. For example, research in eudiometry and the study of airs figured prominently in the volumes of the *Philosophical Transactions* that he consulted. He likely read papers by Joseph Priestley, Henry Cavendish, Felice Fontana and Jan Ingenhousz on these topics, as well as other notable discussions of thermometry, chemical affinities, the composition of water and Lavoisier's critique of the theory of phlogiston.[610] Moreover, his reading of journals from abroad introduced him to the work of prominent foreign chemists. Although there was little on chemistry in the two issues of the *Histoire de l'Académie royale des sciences* that he consulted, the second part of the 1772 volume carried three early papers by Lavoisier, while his extensive reading of the transactions of the academy in Berlin would have acquainted him with the steady stream of contributions by the German analytical chemist Andreas Sigismund Marggraf.[611] Reid was thus better informed about the science of chemistry than the records of his library borrowings initially suggest.

To conclude, even though Reid did not pursue any notable independent research, his engagement with chemistry was neither superficial nor short lived, and the development of his interests in chemical matters tells us a good deal about the activities of his colleagues in Aberdeen and especially in Glasgow. While he was at King's College he was evidently familiar with the theoretical ideas of leading chemists such as Boerhaave, Hales, Stahl, Cullen and Black. Moreover, during this period he appears to have conceived of chemistry as being closely allied with natural history and medicine, and as contributing to the improvement of medical treatment and agricultural practice. Once in Glasgow, his chemical interests

[610] Tiberius Cavallo, 'An Account of Some Thermometrical Experiments; Containing, I. Experiments Relating to the Cold Produced by the Evaporation of Various Fluids, with a Method of Purifying Ether. II. Experiments Relating to the Expansion of Mercury. III. Description of Thermometrical Barometer'; Henry Cavendish, 'An Account of a New Eudiometer'; Felice Fontana, 'Account of the Airs Extracted from Different Kinds of Waters; with Thoughts on the Salubrity of Air at Different Places'; Jan Ingenhousz, 'On the Degree of Salubrity of the Common Air at Sea, Compared with That of the Sea-shore, and That of Places Far Removed from the Sea'; Richard Kirwan, 'Conclusion of the Experiments and Observations concerning the Attractive Powers of the Mineral Acids'; Richard Kirwan, 'Continuation of the Experiments and Observations on the Specific Gravities and Attractive Powers of Various Saline Substances'; Richard Kirwan, 'Experiments and Observations on the Specific Gravities and Attractive Powers of Various Saline Substances'; Joseph Priestley, 'Experiments Relating to Phlogiston, and the Seeming Conversion of Water into Air'; Joseph Priestley, 'Observations on Different Kinds of Air'.

[611] Antoine-Laurent Lavoisier, 'Mémoire sur l'usage de l'esprit-de-vin dans l'analyse des eaux minérales'; Antoine-Laurent Lavoisier, 'Premier mémoire sur la destruction du diamant par le feu'; Antoine-Laurent Lavoisier, 'Second mémoire sur la destruction du diamant, &c.'

shifted primarily to the science of heat, due to the influence of Black, Irvine and Crawford, and his manuscripts attest to the theoretical differences between these three prominent chemists. We have also seen that the arrival of Thomas Charles Hope in 1787 marked a new phase in his development as a philosophical chemist, insofar as he was now exposed to the revolutionary theories issuing from France, which he studied 'with the keen interest of one just entering on life'.[612] Reid's manuscripts dating from the last years of his life reveal that, for him, the Chemical Revolution was as much a matter of a shift in the concept of a chemical element as it was a renunciation of phlogiston or the acceptance of Lavoisier's theory of acids. Finally, it may be that after moving to Glasgow he came to see chemistry more as an autonomous branch of natural philosophy than as an auxiliary to natural history and medicine.[613] Historians of chemistry have hitherto paid scant attention to Reid's chemical pursuits, even though he was an active member of the community of chemists in the Scottish Enlightenment. This section has shown that Reid's correspondence and his papers on chemical topics allow us to chronicle not only the evolution of his own theoretical and practical interests but also the radical and rapid changes which transformed the science of chemistry in late-eighteenth-century Scotland.[614] It is to Reid's manuscripts on chemistry, along with his writings dealing with Euclidean geometry, *vis viva* and the nature of quantity, astronomy, optics and electricity, that we now turn.

[612] [Cleghorn], *Sketch of the Character of the Late Thomas Reid, D.D. Professor of Moral Philosophy in the University of Glasgow*, p. 5.

[613] As Jan Golinski has pointed out, the disciplinary identity of chemistry in eighteenth-century Scotland was unstable; Golinski, *Science as Public Culture*, p. 48. Reid's changing conception of the identity of the science registers this instability.

[614] Historians of chemistry have on occasion used Reid's correspondence with David Skene to shed light on the development of Joseph Black's theory of heat; see Donovan, *Philosophical Chemistry in the Scottish Enlightenment*, pp. 241, 308, n68 and n77; Perrin, 'Joseph Black and the Absolute Levity of Phlogiston', p. 113; Henry Guerlac, 'Joseph Black's Work on Heat', p. 14. They have not, however, used Reid's manuscripts to illuminate the contours of Scottish chemistry in the second half of the eighteenth century, even though Reid's writings provide valuable evidence regarding the cultivation of chemistry in Aberdeen and Glasgow.

THE MANUSCRIPTS

Part One: Euclidean Geometry

I

5/II/47, 1r June 1756*

Observations on the Elements of Euclid

1‹.› Definition 17.* The last clause of this Definition viz that the Diameter divides a Circle into two equal Parts does not properly belong to it being a Proposition and demonstrable by the Application of the two Semicircles to each other. Or if it is thought too evident to need Demonstration it is then to be looked upon as an Axiom but cannot be a Definition or any part of one

2‹.› Definition 27.* It is here & in some of the following Definitions taken for Granted that a Rectilineal figure has as many Sides as angles. Which might easily be demonstrated

3‹.› Axiom 10.* That all Right Angles are equal Among themselves ought to be demonstrated by application

4‹.› Axiom 11.* There are certainly many things Demonstrated that are as evident or perhaps more evident than this Axiom, there are therefore just Exceptions to its being laid down as an Axiom, and yet if it is taken away the Demonstration of Proposition 29* is lame. I apprehend that we might as well lay down the 29 Proposition as an Axiom & Demonstrate this Axiom from it. Since therefore there seems to be a Necessity of making some Alteration in order to Reconcile the Doctrine of Parallel Lines to Geometrical Accuracy I conceive it may be most easily done thus by altering the Definition of Parallel lines

Definition 35.* Parallel lines are th‹o›se which lying in the Same Plain & being cut by a third line make the Alternate Angles equal — this is proposition 27. or the Definition might be the first part of Proposition 28*

Proposition 27 Parallel Lines being produced on both hands never meet. Demonstration ad absurdum. from Proposition 16*

Proposition 28* as it is.

Proposition 29* as it is leaving out onely the first Clause which is the converse of the Definition.

5‹.› The first Definition is indistinct & Superfluous a point being distinctly defined in the 3d, therefore the first is to be left out.* |

5/II/47, 1v Liber 1

Proposition 1* There are two points wherein the Circles drawn do cut each other and no more & accordingly there are two Triangles that Satisfy

Proposition 7* Two cases of this Proposition are ommitted which are rightly supplyed by Dr Barrow.*

NB. Mathematical Demonstrations are of Various kinds*

1 Direct. When the Terms of the Proposition are shewn to agree by the Intervention of one or More Middle terms.
as in the 1 2 3 5 &c of 1 Elements.

2 By Application. As in the 4.1.*

3 Ab Absurdo or ab Impossibili*

4. by Dilemma as 19.1 & most Converses*

5. A Fortiori. as in the 16 & 18 Elements 1*

6. Per partes*

Proposition 44* In the Demonstration of this Proposition the 11 Axiom does again occur as also in Proposition 10 Book 2.* Therefore if we do not receive it as an Axiom we must either Demonstrate it here or refer this and the Subsequent Proposition till it can be demonstrated. I think the easiest way is to Demonstrate the Axiom from Proposition 5 Elements 6* and to Subjoyn the 44 of this book to it.

Proposition 12* The Construction here seems not to be so fully expressed as usual & it is taken for granted that the Circle described will cut the Infinite given right line in two points

The Definition of a Ratio book 5 is vague and good for nothing.* As appears by this that we need another definition to Know when two Ratios are equal or when one is greater or less than the other. |

5/II/47, 2r Sept 1770 Read Euclides ab omni Nævo vindicatus: sive conatus Geometricus quo Stabiliuntur prima ipsa Universæ Geometriæ Principia. Auctore Hieronymo Saccherio Societatis Jesu, in Ticinensi Universitate Matheseos Professore. Opusculum Exmo Senatui Mediolanensi ab Auctore Dicatum. Mediolani 1733 Ex Typographia Pauli Antonij Montani Superiorum Permissi*

The Author in his Dedication mentions his having formerly dedicated to the Same Senate his *Neostatica*. He likewise Mentions his having before published a Book intitled Logica Demonstrativa, forty years ago*

This Work consisting of 142 4to pages & several Plates is divided into two Books. In the first he endeavours to demonstrate the 11 Axiom Liber 1 Elements. In the second Book He Justifies the Definition of Magnitudes that are said to have the Same Ratio, & the Definition of a Ratio compounded of two Ratios. The first Book has 30 Propositions besides Corollaries and Scholia.

The chief of them follow*

1 If two equal Right lines stand upon a rectilinear Base at right Angles, & in the same plain, & their Extremities be joyned by a right line; the Angles at the top will be equal. 2 If in this Quadrilateral Figure the Base and Top be bissected by a right line, both will be cut by that right line at Right angles. 3. Supposing the Same things as in the first proposition; The Top is equal to, or less, or greater than the Base, according as the Angles at the Top are right Angles, or greater, or less than right Angles. Corollary. In every Quadrilaterial Figure which has three right Angles; the Sides adjacent to the fourth Angle will be equal to, less, or greater than the sides opposed to them respectively according as the fourth Angle is equal to, greater or less than a Right Angle. 4 The converse of the 3d

Definition. From the Propositions above Mentioned there result three Hypotheses, for in the Quadrilateral Figure of Proposition 1. Either the Angles at the top are right Angles which is the first Hypothesis & is called the Hypothesis of Right Angles; Or those Angles are Obtuse; which he calls the second Hypothesis or the Hypothesis of Obtuse Angles; Or those angles are Acute which is the third Hypothesis or the Hypothesis of Acute Angles.

Proposition 5 The Hypothesis of Right Angles if it be true in any one instance is true in every Instance. 6 The Hypothesis of obtuse Angles if it be true in one Instance, is true in every instance. In the Demonstration of this Proposition an Axiom is assumed which is not commonly used, although I think it may. It is this. In the quadrilateral figure of Proposition 1. Supposing innumerable Right Lines cutting off equal Length from the two Sides; If one of these is greater than the base and another less there must be some intermediate one that is equal to the base. This however is a New Axiom. 7 The Hypothesis of Acute Angles if it be true in one instance, is true in every instance. The Demonstration of this is drawn by induction from the two former. The Demonstration of the 6th might be made out by means of this Axiom. Supposing the Same things as in the Axiom assumed, If the line cutting off certain equal parts of the Sides of the figure, make with those Sides angles that are Obtuse on the Side of the Base, & cutting off other equal Segments make Acute Angles, it must in some intermediate position make Right Angles on the side of the Base. |

5/II/47, 2v Hitherto I think the Author has gone on very happily. In the Subsequent Propositions I think he may be Mended. I would therefore make this the next Proposition. 8* If two Right lines are cut by a third, So as to make the alternate angles equal, or the external equal to the Internal and opposite on the same side, or the two internal Angles on the same side together equal to two right Angles; I say that a right line may be drawn cutting the two Righ‹t› lines before mentioned at Right angles. This may be done by bissecting the first mentioned cutting line and from the point of bissection drawing a line perpendicular to one of the cut lines & producing it untill it cut the other. it will be demonstrated by 26 Elements 1* that it cuts the other at right angles also. Or rather from the point of Bissection draw perpendiculars to each of the cut lines and by the same 26 & 14 Elements 1* it may be proved that these Perpendiculars make one right line. Hence I conceive we may demonstrate 9 That if the Hypothesis of Right Angles be true it follows that of every triangle the three Angles are equal to two Right Angles; if the Hypothesis of Obtuse Angles be true, it follows that of every triangle the three angles are greater; & if the Hypothesis of Acute angles be true the three angles of every triangle are less than two right angles. 10 Upon the Hypothesis of Right angles the 11 Axiom of Euclid may be demonstrated and all the other properties of Right lines.

What remains to be done is to refute the two Hypotheses of obtuse and of Acute Angles. That of Obtuse Angles I have demonstrated to be absurd in Propositions 4th & 5th of an Essay upon this Subject.* This Author employs many preliminary propositions to this purpose. The Result of these preliminary Propositions is that Upon the Hypothesis of obtuse Angles the 11 Axiom holds true; whence he infers that since the 11 Axiom being granted the Demonstrations of Euclid with regard to right lines and parallels all stand firm, & contradict the Hypothes‹is› of obtuse Angles Hence that Hypothesis confutes it self. He demonstrates the falsehood of this Hypothesis in another way thus Let PA be a perpendicular raised from the point P of the Right line XL; Joyn AX, then the Right Angled triangle PAX will have its two acute angles together greater than one right angle by a preceeding proposition, if the Hypothesis of obtuse angles be true. Let the Angle PAD be as much less than a Right Angle as the two acute angles PAX & PXA are greater; then the three Angles PXA, PAX & PAD will be equal to two right Angles; consequently the Lines AD & XL being cut by a third line so as to make the internal Angles on the same side equal to two right Angles can never Meet. Yet the same lines are cut by the Line AD So as to make the internal Angles on the same side less than two Rights, for the Angle APL is by supposition Right & the Angle PAD less than a Right Angle therefore the Lines AD, PL being produced must meet; but it was before proved that they cannot meet. Therefore the Hypothesis of Obtuse angles from which these contrary conclusions follow contradicts itself.* |

5/II/47, 3r So far the Author proceeds upon sure Grounds But in Refuting the Hypothesis of acute Angles, he is lead into a large field and has recourse to reasonings about infinitesimals. Some of them particularly what he uses in his 37 Proposition I think not Just.*

A Just and Geometrical refutation of this Hypothesis of Acute Angles seems therefore to be still wanting. Perhaps this may at last be obtained. Perhaps by building a System of Geometry upon this Hypothesis we may at last find it contradictory to itself or to some selfevident Principle*

I come now to speak of the 11 Axiom by some called the 12 Which is If a right line falls upon two right lines in the same plane so as to make the two internal Angles on one side together less than two right Angles the right lines being produced will meet on that side.* This Axiom I conceive not to relate to right lines in general, but to parallel right lines. It points out the circumstances in which right lines produced do meet in order that we may infer from it the circumstances in which they never meet that is (according to Euclids Definition of Parallel lines)* in which they are parallel

This Axiom hath given more ado to the Criticks on Euclid than all besides that is contained in his 15 Books. Not that its truth was ever doubted, but whether it has that degree of selfevidence which intitles it to be assumed without proof as an Axiom. It has very generally been Thought that it has not.* The Consequence of this is that it ought to be demonstrated, and that by the aid of the first 28 propositions of the first Book; for in the Demonstration of the 29 it is assumed. Nor can it be imputed to Theon or any other Scoliast.* It enters into many Demonstrations, and is a Proposition that cannot be wanted, & therefore is undoubtedly Euclid's. The great Labour therefore has been to demonstrate it. Ptolemy wrote a Book for this purpose which is lost.* Proclus not satisfied with Ptolemys Demonstration gives one of his own.* The Arabians afterwards attempted to demonstrate it some of their Demonstrations have Reached us.* After all the labours of Greeks & Arabians, the Moderns of different Nations thinking the Field still open have tried their skill to demonstrate this Axiom. Clavius laboured this with all his Might.* Thomas Oliver an English Man wrote a Book upon it.* After these the Learned Sir H Savil who himself wrote Lectures upon Euclid & founded A Profession of Geometry & one of Astronomy at Oxford seems dissatisfied with all that was done upon this Subject for though he does not attempt it himself he earnestly recommends it to the Professors who should afterwards enjoy his foundation to clear those admirable Elements from this blot or supposed Blot.* Accordingly Dr Wallis has attempted it in a Dissertation on the Subject. He thinks indeed contrary to the General Opinion that the Axiom may be admitted without proof upon its own Evidence, yet that it may be demonstrated. And finding that all the Demonstrations of it he had seen erred in assuming something not more evident than the thing they are brought to demonstrate he gives his own Demonstration, which in my humble Opinion is chargeable with that very fault which he finds in the others* Dr D. Gregory another Savilian Prof who with the help of Saviles Mss & Notes Collected for that purpose has given the best Edition of Euclids Works in his own Language, takes another Method to account for this Axiom, which we shall Mention by and by.*

After him Hieronymus Saccherius an Italian Professor of Mathematicks at Pavia thinking the thing not yet effected Wrote a Book which
5/II/47, 3v he calls Euclides ab omni Nævo Vindicatus | published at Milan in 1733. Saccherius was no Novice in Demonstration for he forty years before this published a Logica Demonstrativa and had also wrote on Staticks*

He has laboured to demonstrate this Axiom, in no less than thirty nine Propositions accompanied with Corollaries Scholia & Lemmata, & with some new Axioms. This work of Saccherius Dr Simson must have been acquainted with, as it is in the Collection he left to the University library with large notes in his own hand upon some part‹s› of it, though not upon that part which treats of the 11 Axiom.*

I shall now give a brief Account of what is said upon this Axiom by two of the latest and ablest of those who have treated of it. These Are Doctor D Gregory & our Dr Simson.

I am much disposed to think with Dr Simson that the Axiom does not admit of Demonstration in the way which has been attempted by all that have taken it in hand.* All that have attempted the Demonstration of it as far as I know have taken for granted that it must be demonstrated before the 29 Proposition, which Euclid builds upon it. The 29 Proposition contains the properties of parallel right lines. And these are deduced by Euclid in a very simple & elegant Manner from ‹the› Definition of parallel lines and the disputed Axiom.

Now, if we might presume to alter Euclids method so far, and at the same time are able to demonstrate the properties of parallel lines without the aid of the disputed Axiom, or by means of an Axiom more evident ‹to› beginners in these Studies, Is it not possible that from the properties of parallel lines, The 11 Axiom may be deduced with demonstrative Evidence, though it do not admit of Demonstration in the way that has hither to been taken

May we not therefore try ‹to› deduce the properties of parallel right lines without the aid of the 11 Axiom. As this ha‹s› not been attempted hitherto I hope there is no blameable presumption in the Attempt. I cannot indeed do it without assuming some Axiom different from those which we have in the Elements. That which appears most proper for the purpose is This. That a curve or crooked line in the same plane with a right line, cannot have all its parts at the same distance from the right line. I say A crooked line is distinguished from a curve line by this that it may be composed of right lines put together at angles. whereas a curve line cannot be so composed. The curve and crooked both, are opposed to the right line & between them comprehend every line that is not one right line. This being understood the Axiom is That a curve or a crooked line, in the same plain with a right line cannot have all its points at the same distance from the right line.

Whether this Axiom though not more true, have an Evidence more easily apprehended by young Mathematicians than the 11 Axiom of Euclid, must be left to the Judgment of those who attend to it.

Taking it then for granted, my first inference from it is This. That a line which is in every point at the same distance from a right line & in the same plane must be a right line For if it was either Curve or Crooked it could not by the Axiom be in every point at the same Distance from the right line, therefore being neither Curve nor crooked it is a Right line.

This proposition indeed appears to be evident to all Men without reasoning from the common way in which we draw a right line in practice. Being possessed of a streight ruler &c |

5/II/47, 4r But to proceed it appears evident from what has been said that a right line may be in the same plane with another right line & in every point at an equal Distance from it. I would therefore give this definition of parallel lines That a right line is said to be parallel to a right ‹line› when being in the same plane it is in every point at the same distance from it.*

From this Definition it is easy to demonstrate that if a right line be parallel to a right line every right line which cuts the one at right angles will also cut the other at right angles For if one is cut at right angles & the other at an acute angle the lines must approach to each other if at an obtuse angle they must recede from each other, the line therefore is cut neither at an Acute nor an obtuse angle & consequently it is cut at a right angle

From this proposition it is easy to demonstrate the 29. To wit that if two parallel lines are cut by a right line the alternate angles will be equal, the external angle will be equal to the internal & opposite on the same side; and the two internal on the same side will be together equal to two right angles. These are the properties of parallel right lines which we have deduced from the Axiom assumed without any aid from the 11 Axiom of the Elements

Thus I would demonstrate the properties of parallel lines, & from them I think I am able to demonstrate the 11 Axio‹m›, but as the Demonstration requires a Figure & therefore is not proper for this Discourse, I must take Credit for it.

If after all this attempt should prove unsuccessfull, it is but one added ‹to› a very great number of the kind made by Mathematicians at whose feet the Author of this would be proud to sit and learn

II

3/I/13, 1r

Simsons Euclid Glasgow 1756 4to*

Corollary to Proposition 11.* It seems to be ungeometrical to demonstrate this Corollary from this Proposition because it is supposed in the 8 Proposition upon which this depends as well as in the 4th upon which the 5, 6, 7, & 8. depend therefore if this Corollary is at all to be demonstrated it ought to be demonstrated from Axiom 10 or according to Gregorys Edition Axiom 12 that two right lines do not comprehend a Space*

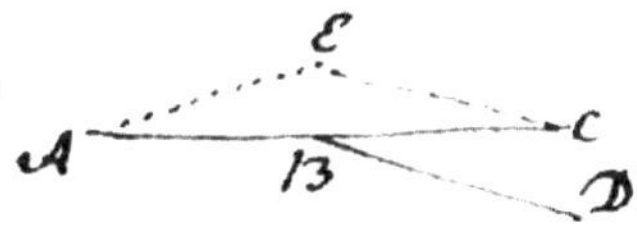

Let the Right lines ABC and ABD have if possible a common Segment AB

Let the right line ABD be turned round upon the point A till BD passes through C then let its position in that case be AEC then it is evident the right lines AEC & ABC would comprehend a Space which is impossible by Axiom 12

But I would rather consider this Corollary as an Axiom since it seems to have the same tittle to be accounted so as the 12 Axiom has

It may be observed in General that there are several things assumed in Euclids Demonstrations which are neither laid down as Axioms by his Editors nor demonstrated Thus in the 1st it is assumed that the circles drawn with the Centers A & B & the interval AB will cut each other in a Point.* In the 2d & 3d that all right lines drawn from the Center of a Circle & produced will cut the Circle.* In the fourth that two right lines cannot have a common Segment* |

3/I/13, 1v That two right lines cannot have a common Segment might also be demonstrated from the 11 Axiom That all right angles are equal one to another*

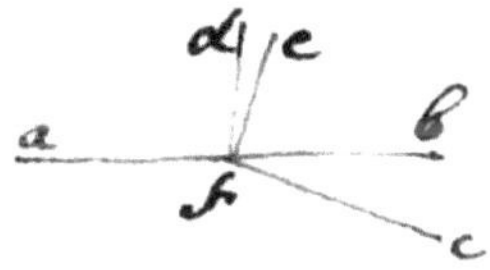

For let afb & afc have a common segment af if possible. let df fall upon afb so as to make the Angles afd & dfb equal & consequently both right and let ef fall upon the right line afb so as to make the angles afe & efc equal and consequently both right. Then since the Sum of the Angles afe & efc is greater than the Sum of the angles afd & dfb, half the first sum will be greater than half the last therefore the Angle afe will be greater than afd, but they are both right contrary to axiom 11.

It may be observed that the Simplest properties of Right lines cannot be demonstrated from Euclids Definition of a Right line which indeed is too vague to found a Demonstration upon.* Therefore these properties are either laid down as Axioms or supposed in the Elements. These properties may be reduced to the following 1 Two Right lines cannot have a Common Section 2 Two right lines cannot include a Space 3 All Right angles are equal. Or which amounts to the Same If a right line falls upon a Right line making Angles the Sum of the angles is always the Same. 4 If two points of a Right line are equally distant from another right line all the points of it will be equally distant from the other. 5 If two Right lines in the same plain converge towards one another they will at last Meet. 6 If two right lines are Drawn from the Same point making an Angle a point may be found in the one whose distance from the other is greater than any given right line. NB The Distance of a Line from a Point is defined Liber 3*

7 A Right line is the Shortest of all lines that go between its extremities this may be demonstrated by Sir Isaac Newtons Method of ultimate Ratios

If a right line cannot cut another in more than one point

If a right line be cut by another in a point the parts of the Cut line will be on the opposite Sides of the Cutting line

A line that is parallel to a right line that is every where equally distant from it must be a right line

III

5/I/1, 1r Proposition 6.* Upon the right Line AK let the unequal right lines AC, BD stand at right angles whereof BD is the greater & let BH, HK be taken equal to AB, and the Perpendiculars HE, KI, be raised meeting CD produced in E, and I. I say what ever is the Excess of BD above CA, the Excess of HE above BD shall not be less.

Take HM equal to AC & AG, HF equal to DB & joyn DM, DG, D‹F›. Then it is evident that the triangles DFM, DGC are every way equal. And because the Angles DFM, DGC are equal to one another and (by proposition 5) are each of them either right angles or less than Right angles Let us first suppose them equal to right Angles, then the Angle DFE will also be a right angle, and the Angles GDB, ‹F›DB being also right angles in consequence of the Same Supposition, GDF will be a right line & the alternate Angles GDC & EDF will be equal therefore the triangles GDC, ‹F›DE will be every way equal and GC the excess of DB above AC will be equal to FE the excess of HE above BD. QED 2 Supposition. Suppose the Angles DFH, DGA, to be less than right angles & consequently BDG, BDF to be also less than right angles. Then GDF will be an obtus‹e› angle being less than two right angles; consequently the Angle EDF will be greater than the Angle CDG. likewise the Angle DFE being greater than a right angle will be greater than the angle DGC which is less than a right Angle. Let Dm be drawn

making the angle ‹F›Dm equal to the angle GDC, & meeting the line FE in m. Then it ‹is› evident that Fm is less than FE because the Angle FDm is less than the angle FDE. Draw likewise Fr meeting Dm in r making the Angle DFr equal to the Angle DCG which has been shewn to be less than the Angle DFE. It is evident therefore that the Angle DFr is less than DFm. & consequently Dr is less than Dm. Now from this Construction it is evident that the triangles DCG, DFr are every way equal & consequently that the angle DrF=∠DGC is less than a right angle consequently the Angle Frm is greater than a right angle; therefore Fm is greater than Fr, & a fortiori FE is greater than Fr; but Fr is equal
5/I/1, 1v to CG therefore FE is greater | than CG. QED.

Corollary Hence it follows that of two indefinite right lines in the same plane if one begins to recede from the other it will recede more and more so that at last its distance from the first shall be greater than any given right line.

May 28 1770 I resolve for the future to give up the Consideration of this Subject; having spent more time & thought in attempting to prove the simple properties of Streight lines from some one definition or Axiom than I can own without shame.* My last thoughts upon this subject are these.

1 I conceive that a Streight line is an Object too Simple to admit of a proper mathematical Definition. Euclid does not attempt to define Magnitude, equality, greater, or lesser. These are common Notions which have Names in all Languages, and these Names have a distinct and determinate Meaning.* The same thing may be said of the streightness or curvature of a Line. The Notion of a Streight Line therefore must be supposed to be already in the Mind of the Learner in Mathematicks and we see that Euclid supposes not onely that he has the Notion of a Streight line but that he is able to draw a Streight line from one point to another. But that the Theory of Streight lines may be delivered Mathematically; since there is no Definition of them to be had, there must be one or more Axioms concerning them upon which all the Reasoning about Streight lines must be built.

For all Reasoning must be drawn either from a Definition an Axiom or a Supposition.

The first Axiom I would lay down about Streight lines is that which is assumed by Euclid Book 11. proposition 1. Εὐδεῖα εὐδεία οὐονμβάλλει χατα ϖλείονα οημεια η καδ᾽ ἕν εἰ δέ μὴ ἐφάρμόσ{ου}οιν αλλήλαις αι ευδειαι.* This is evidently assumed as an Axiom by Euclid. It is fully expressed in this Proposition; but it is tacitly assumed in the 4th of the first Book and in many other places*

From this Axiom it can be easily demonstrated 1st that two streight lines cannot include a Space 2 That two Right lines cannot have a common section. 3 That all right Angles are equal. Therefore there is no occasion for laying down any of these three as Axioms since they can be demonstrated from this one.

From this Axiom we have likewise demonstrated the fourth fifth and sixth Propositions above expressed.* But here we stick. It seems therefore that another Axiom concerning right lines should be assumed

We shall mention several, any one of which being assumed as an |
5/I/1, 2r Axiom the rest will follow from it.

1 If two streight lines be in the same plain and if two points of the first be at the same distance from the second, & on the same side; all the points of the first shall be at the same distance from the Second

2 If two right lines stand upon a third right line & upon the same side of it, & in the same plain, & if they cut the third line at equal angles upon the Same side, every point of one of those right lines shall have the same distance from the other

3 If two right lines in the Same plain are cut by a third right line so as to make the internal angles on the same Side less than two right angles, those streight lines shall at last meet.

4 If two right lines are in the same plain and one of them is nearer to the other at one extremity than it is at the other the right lines being produced will at last meet

Perhaps others of this kind might be mentioned but I conceive there is no other that has a better title to be taken as an axiom, and from which these may be deduced.

Of the four above mentioned the first seems to me to have the best claim to be considered as an Axiom.

Upon second Thoughts a 5th Axiom may be this

5 If a Right line turns, whether upon the same or upon different points untill it comes in to the Same position which it had at first, the Sum of all the Angles it makes in this Revolution will be equal to four right angles.

The third of these axioms is that which Euclid has assumed; and if we were not to assume it but to demonstrate it from some of the others that have been Mentioned we must make a great alteration in the Elements. after the 28 proposition of book first.

Septr 13 1770 I find a Tract upon this Subject Intitled Euclides ab omni Nævo vindicatus: sive Conatus Geometricus, quo stabiliuntur prima ipsæ Universæ Geometræ Principia Auctore Hieronymo Saccheris. S. J. in Ticinensi Universitate Matheseos Professore Mediolani 1733 4o*

Before proceeding farther with regard to right Lines. I would observe that the Definition which we find in Euclid of parallel lines is not mathematical. Parallel lines it is said are those which being in the same plane however produced never meet.* The definition of Par‹a›llel lines ought to be such as that all their other properties may be deduced from it but we find nothing deduced from this. On the contrary this property is deduced from another in the 27 Proposition. So that it would seem
5/I/1, 2v more mathematical to have given what is supposed in that | proposition as the Definition

Definition Parallel Lines are those which being in the same Plane stand at right angles with another right line*

IV

3061/11, 1* The Elements of Euclid have justly been held in admiration by Mathematicians of the best Taste and Judgment in every Age since they were wrote. Their Order Perspicuity and Elegance make them a Pattern of Mathematical Writing, which has hardly ever been exceeded, & very rarely been equalled.*

Our Admiration of these Books ought not however to hinder (as it seems too much to have done) a critical Examination of them. It is laudable to attempt to add to the Perspicuity & Evidence of the first Elements of a Science, which of all others has the best claim to Certainty. Euclid & his Editors were Men as we are, No work of human Genius is above Criticism. And we do more honour to him as well as to ourselves & to human Nature, by examining his Work with Judgment, than by implicit Admiration & blind Submission.

The ingenious Dr Simson has therefore deserved well of the Mathematical World, as he has observed more of the defects of our Copies, and made more Judicious Amendments than all the Criticks that were before him; by which one of the most perfect Works in its kind is made more perfect.

He attributes all the imperfections he has observed to the Editors, particularly to Theon.* And indeed there is good ground to believe that several Additions and Alterations have been made by Persons of less Judgment than Euclid was. However as it is the Work we are chiefly
3061/11, 2 concerned with, | rather than the Author or Editors, I shall beg leave after Dr Simsons Example, to make some Observations which he hath not taken notice of, without concerning my self whether the things I censure are owing to Euclid or to his Editors.

The first Definition, that of a *point*, seems to be exceptionable It is defined to be, *that which hath no parts*.* A Definition which tresspasses against the Rules of Logick in two respects. First it is purely negative. It express not what a point is, but what it is not. And secondly, it is not convertible, as all Definitions ought to be For surely there are many things which have no parts, and yet are not mathematical points

This Definition therefore does not savour of Euclids accuracy, and ought to be ommitted. It is probably an addition of some officious Scholiast, who thought that a Definition of a Point should go before that of a Line. As it is faulty, it is altogether superfluous, points being justly defined by Euclid in the third Definition to be the Extremities, or Terminations of a line;* And it is not probable that Euclid gave two Definitions of the same Term.

We are next to consider the Definition of a right line which is the fourth. A right line is said to be *that which lies evenly between its points*.* This is as much as to say that a right line is a streight line. It is onely giving one synonimous word for another which is not at all what is properly called defining.

A just Mathematical Definition ought to give such a Conception of the thing defined, as that all its properties may be deduced by mathematical reasoning from the Definition. The Definitions which are undoubtedly Euclids yield to none, in accuracy & perspicuity; and by attending to
3061/11, 3 them we may perceive, both | what a Mathematical Definition ought to be, and wherein that under our Consideration is deficient.

When Euclid defines a Square, a Parallelogram or a Circle the Definition comprehends the whole Essence of the thing defined.* Every property of the Figure is drawn from the Definition A just Definition supplies the place of all Axioms with regard to the thing defined. Accordingly we find no Axioms in Euclid with regard to Circles Spheres or Cones, which are properly defined.* To prove their properties, the Definition onely is assumed or what has been proved from the Definition.

It is quite otherways with this Definition of a right line. For first it is never quoted or assumed in the whole Elements. Nothing is nor indeed can be inferred from it. So that it is quite useless. Secondly we may observe, That all the Conclusions in the Elements concerning right lines are grounded upon certain Axioms about right lines which are assumed as selfevident; whereas if we had a just Definition of a right line we should have no occasion for axioms about them to supply its place.

To illustrate farther the Difference between a just Mathematical Definition and a vague popular one, Let us suppose that a Circle had been defined to be a plain Figure that is perfectly round. This Definition could not be applied to demonstrate any of the properties of a Circle. We must in this case have assumed one or more Axioms about circles, such as that all lines joyning the extreme points of circles are equal, or the like. But a proper Definition of a Circle such as Euclid has given make‹s› all Axioms about Circles unnecessary.

It seems evident from what hath been said, That the Definition in the Elements of a right line is unmathematical and useless. Secondly
3061/11, 4 that if we can find a just & mathematical | Definition of a right line, the Axioms about right lines need not be assumed without proof; They will be necessary consequences of the Definition.

These Axioms are in some Copies two in others three, & ought to be three To wit That all right angles are equal: That two right lines cannot include a Space; And that they cannot have a common Segment.*

Though the Truth of these Axioms has not as far as I know been called in Question by Mathematicians; Yet their Evidence would be more Mathematical if they were shewn to follow necessarily from the Definition of a Right line. It must be acknowledged that many things are demonstrated by Euclid which have as much or more self-evidence than they have. Is it not as evident that a right line joyning two points in the circumference of a circle must fall within the Circle, as that two right lines cannot comprehend a Space? Is it not as evident that a right line can touch a Circle onely in one point, as that two right lines cannot have a common Segment? Is it not as evident, that in equal Cir‹cl›es equal Angles at the Center stand upon equal Circumferences, as that all right angles are equal? Yet the former of these are demonstrated by Euclid, the la‹t›ter assumed without proof.

As in Fortification a place ought not to be left weaker on one side than in another, because an Enemy who has Judgment will be sure to make his Attack where it is weakest, & the Expence laid out in making the other sides stronger will be thrown away, So, in a System of Principles which is intended to be impregnable on all sides, Demonstration seems to be lavished on some parts, while other parts less evident or as little evident are left naked and defenceless.

We may observe that the Definition of a plane Surface is equally censurable with that of a Right line and therefore Dr Simson hath left it out and hath given a just & mathematical one in its place* |

3061/11, 5 I should not have taken notice of this, as it has been corrected, but in order to remark that the definitions in Euclid which are exceptionable are those of the simplest things, which inde‹e›d are the most difficult to be defined. As there are first Principles which cannot be demonstrated so there ‹are› first and simple Notions which cannot be defined.* And it requires Judgment to know where to stop in Definition as well as in Demonstration. Some eminent Mathematicians by giving a wrong Definition of Quantity the general Object of Mathematicks, have led others to apply Mathematical reasoning to Subjects that do not admit of it.*

No Man pretends to define Sum or Difference or what it is to be greater or less.

There are therefore some Terms that frequently enter into mathematical reasoning so simple as not to admit of mathematical Definition. The Mathematical Axioms ought to be employed about these and onely about these. Thus the first nine Axioms I apprehend must be assumed without proof. They are indeed all so evident, & so familiar from their dayly application in the affairs of Life, that they seem incapable of greater Evidence even from Demonstration, if it could be had, which perhaps is impossible. It is not so with the other three Axioms about right lines before mentioned. They are neither so familiar in common Life nor have such a degree of Evidence in themselves but that they seem to desiderate more in order to be upon an equal footing with the other first principles of the Science.

Thus I have endeavoured to shew that our Principles of Geometry are not so Geometrical as were to be wished in what concerns the simplest properties of right lines; I have also endeavoured, which I think has not
3061/11, 6 been done before, to shew, to what | cause the defect is to be imputed, namely, to the want of a just mathematical definition of a right line.* And thus knowing the cause of the disease, our Inquiries after a Remedy are reduced within some limits. For the Matter is brought to this short issue: If a right line be too simple an Object to admit of a Mathematical

Definition, then we must even satisfy ourselves with the Axioms that have been assumed about right lines. But if we can find as just and Mathematical a Definition of a right line as Euclid has given of a Square and of a Circle, we shall then find that Geometry shall throw aside the Crutches of those assumed Axioms, & stand as firm & walk as streight upon right lines as upon Circles and Squares.

I shall therefore presume to offer a Definition of a Right line for this purpose. Let it be, *That a Right line is that which cannot meet another Right line in more points than one, otherwise they perfectly coincide, & are one & the same.*

Curve lines that are perfectly similar and equal by being made to meet in more points than one, in a certain position, will coincide; but the property of right lines from which I take their Definition is, that they must coincide in whatever position or circumstances they meet in more points than one. This property therefore belonging to all right lines, and to none but right lines, is so far fit to be made their Definition.

But indeed the chief reason why I propose this Definition of a right line is, that it is assumed by Euclid himself. In the end of the Demonstration of 1. Book 11. which is a Demonstration *ad Absurdum* his words literally translated are these "Therefore of two given right lines there is a common segment, which is impossible‹.› For, says he, a right line does not meet a right line in more points than one, otherwise they coincide."* |

3061/11, 7 We may observe that as this property of a right line is assumed by Euclid without proof & is brought by him in proof of one of the properties which we call Axioms, he must have adduced it either as an Axiom or as his Definition of a Right line; The last seems to me more probable, because all the properties of right lines are deducible from it, without assuming any of them as Axioms.

I must not omit to observe, that Dr Simson to whom we are indebted for so many Emendations of Euclids Elements, thinks fit to omit this assumption of Euclid, which I take to be his Definition of a right line. But we may be permitted to consider the Reason he gives for this omission.

He does not plead the Authority of any Greek Manuscript, or of any ancient Copy whatsoever, & therefore we may presume that he found no such Authority. His Note upon it is in these words, "The words at the end of this "*For a streight line cannot meet a streight line in more than one point*" are left out, as an addition by some unskillfull hand; for this is to be demonstrated, not assumed".*

But I beg leave to observe that if Euclid assumed it, as the Definition of a right line, it is not to be Demonstrated.

His reason for omitting the words in Question appears the more strange, because another assumption which Euclid makes in the very same Demonstration, Doctor Simson takes to be Euclids Definition of a plane Surface, and substitutes it in place of the old Definition in the Greek Copies. His Note upon that Definition is "Instead of this Definition as it is in the Greek Copies a more distinct one is given from a property of a plane Superficies which is manifestly supposed in the Elements"*

The property from which he takes the Definition of a plain Surface is indeed not onely supposed but expressly assumed in this same Proposition 1 Book 11 & I believe no where else in the Elements. But he
3061/11, 8 does | not conclude it to be the addition of an unskillfull hand because it is assumed and not demonstrated, on the contrary he takes it to be Euclids Definition of a plane Surface, and substitutes it as such in place of a vague and indistinct Definition in the Greek Copies. Now when Euclid in the next Sentence assumes, That a right line cannot meet a right line in more points than one, otherways they coincide, have we not the same reason to take this for Euclids definition of a right line and to substitute it in place of a vague Definition given in the Greek Copies. If a distinct Definition of a plane Surface assumed in the Elements ought to be adopted in place of an indistinct one found in the Greek Copies, I see no reason why a distinct Definition of a right line assumed in the Elements may not be adopted in place of an indistinct one found in the Greek Copies, rather than from mere Conjecture conclude it to be the addition of an unskillful hand.

It perhaps may be accidental that Euclid should give us his Definition of a plane Surface and his Definition of a right line in one and the same Demonstration; but indeed there is such an Analogy between them, that the expression of the one might very naturally lead him to express the other, though in many other parts of the Elements it be supposed without being expressed. the Analogy is striking for as a right line cannot meet a plane Surface in more points than one otherwise they coincide, so a right line cannot meet a right line in more points than one oth‹er›wise they coincide the first is the definition of a plane Surface the second of a right line.

If it should be objected to the Definition I have offered that it includes in it the thing defined, a right line being defined by a certain Relation it bears to a right line. This objection might be removed by putting the Definition in the plural Number. Right lines are those that can onely meet in one point or else they coincide. Or it might be expressed thus A right line is that which is determined by two points. Of these different expressions which amount to the very same meaning I have chosen that which has the Authority of Euclid, whose very words I revere too much to change them without a good Reason. |

3061/11, 9 It is true that this Definition expresses onely a certain Relation which right lines bear to one another; But if this be a Relation common to all right lines, & to no other lines, and if all the properties of right lines commonly assumed as Axioms, be deducible from it by mathematical reasoning, it has all the qualities required in a just mathematical Definition

I proceed now to deduce from it as Corollaries the Axioms which are commonly assumed with regard to right lines

Corollary 1 Two right lines cannot have a common Segment.*

For if they have a common Segment they must have many points common and therefore by the definition they must coincide and not be two right lines but one and the same. Euclid as we observed before assumes the Definition as a proof of this Corollary

Dr Simson deduces it as a Corollary from the 11 proposition of Book 1 but in this I humbly apprehend that accurate Writer has made a wrong Step, by not attending to this, that the eleventh proposition is grounded upon the 8th, the 8th upon the 5th & the 5th upon the 4th. And in the 4th Proposition, what he would prove by the 11th is supposed without proof, so that the reasoning turns round in a Circle, a Fallacy which Men of great Genius have often fallen into in attempting to prove things next to selfevident.*

Corollary 2d A second Corollary from the Definition is That two right lines cannot comprehend a Space*

If they meet in one point onely they comprehend no Space; & if they meet in more points than one they must by the Definition perfectly coincide & so can comprehend no Space

Corollary 3 All right Angles are equal.*

A right Angle is formed by a right line falling upon a right line so as to make the angles on both sides equal. Now Since by the Definition a right line meeting a right line in more points than one must coincide with
3061/11, 10 it, it is | evident that in all cases where a right line stands upon a right

line, the sum of the Angles on both sides must be the same & therefore the half of that Sum, which is a right angle must be the same.

These are the three axioms which are commonly assumed without proof. But we see there is no need of assuming them without proof since they evidently follow from the Definition of a Right line given by Euclid himself. |

3061/11, 20 Corollary 4 from the Definition of a right line

If a line be at the same Distance from a right line in every point and in the same plane it is a right line; & lines at right angles to the one are so to both.

Call the first line A & the right line from which it is in every point equally distant B.

The Distance of the several points of A from B is measured by perpendiculars from those points to B

These two lines A & B together with the perpendiculars which joyn them make a figure which may be extended to any length without limit. Of this Figure one part may be laid over another part; or it may be cut into parts at pleasure, by the perpendiculars to B; and any one part applied to any other.

In all these ways in which one part of the figure can be applied to another part, it is evident that as any part of the line B must coincide with any other part because it ‹is› by supposition a right line, so any part of
3061/11, 21 the line A will perfectly coincide with | any other part, and therefore it must be a right line by the Definition.

It may be observed that if B be supposed to be an Arch of a Circle convex towards A. the line A would in that case be an Arch of a larger circle having the same center. And if A was an Arch of a circle concave towards A, the line A would be an arch of a less circle having the same center. And if the arch B were gradually unbent, that is brought nearer to a right line, the arch A will also be brought nearer to a right line. If B be bent the contrary way, so will A. Therefore when B is neither bent to one hand nor the other but is a right line, A will also be a right line.

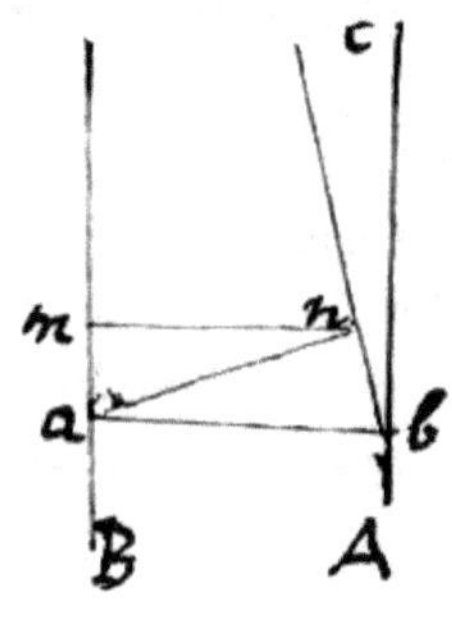

The second part of the Corollary is that lines at right angles to A will also be at right angles to B which is thus proved

Let the right line bn be in every point at the same distance from am I say that any line ab which is at right angles to am will likewise be at right angles to bn. For if bn does not make a right angle with ab it will make an acute angle on one hand and an obtuse one on the other. Let the angle abn be the acute angle, & from a draw an, at right angles to bn; and nm at right angles to am. It is evident that the angle abn being less than a right angle & anb a right angle an must be less than ab; and because nma is a right angle & man less than a right angle nm must be less than an, and therefore less than ab. contrary to the supposition. Therefore the angle abn, is not an acute angle but a right angle QED

This Corollary shews that two right lines in the same plane may be in every point at the same Distance, which may be made the Definition of parallel lines, and from the Definition and the second part of this Corollary the properties of parallel lines are easily Demonstrated. If this Corollary be adopted, the Axiom on page 16 will be superceded |

3061/11, 11 I come now to speak of the eleventh Axiom, by some called the twel‹f›th, which is, “If a right line fall upon two right lines in the same plane so as to make the two internal angles on the same side together less than two right angles; the two right lines being produced will meet on that side.‹”›* This Axiom I conceive to relate not ‹to› right lines in general, as the thre‹e› Axioms before mentioned do, but to parallel right lines. It points out the circumstances in which two right lines in the same plane must meet at last, in order to infer the circumstances in which however far produced they never meet, that is, in which they are parallel, acording to Euclids Definition of parallel lines.*

This Axiom hath given more ado to Criticks upon the Elements than all the fifteen Books besides. Not that its truth was ever doubted, but whether it have that degree of selfevidence which intitles it to be assumed without proof as an Axiom. The general Opinion has been that it has not.* The consequence of this is that it ought to be demonstrated; and that must be done by the aid of the first twenty eight Propositions of the first Book, for in the demonstration of the twenty ninth, it is assumed. Nor can this Axiom be imputed to Theon or any officious Scoliast: It is undoubtedly Euclids, being assumed in many propositions, & so necessary in the whole Fabrick of Geometry that it cannot be wanted. The great Labour therefore has been to demonstrate it, by the aid of the first 28 Propositions

Ptolemy wrote a Book for this purpose which is lost.* Proclus not satisfied with what was done by Ptolemy & others before him gives a Demonstration of it of his own.* After the Greeks had laboured upon it to little purpose the Arabians attempted it & some of their Demonstrations

3061/11, 12 of this Axiom have come down to us.* After all the labours | of Greeks and Arabians, Modern Mathematicians of the different Nations of Europe have tried to make out this Demonstration

Clavius laboured this with all his might.* Thomas Oliver an English man wrote a book upon it, published in 1604.* After these the great Sir Henry Savil who himself wrote Lectures upon Euclid, & founded two mathematical Professions in Oxford one of Geometry and one of Astronomy, appears dissatisfied with all that was done upon this subject. For though he does not attempt it himself, he earnestly recommends it to the future Professors upon his Foundation that they should labour to clear those admirable Elements from this blot or supposed blot.* Accordingly Dr Wallis an early Savilian Professor has wrote a Dissertation on the Subject. He differs from the general Opinion; thinking that the Axiom may be admitted without proof, as having sufficient Evidence in itself. Yet he thinks it may be demonstrated. And observing that all the Demonstrations he had seen, erred in assuming something not more evident than that which they would prove, he gives his own demonstration at great length, which in my humble opinion is chargeable with that very fault with which he charges all that were before him*

Dr David Gregory another Savilian Professor, who with the help of Manuscripts and Books which had been collected by Sir Henry Savil for the purpose, has given the best Edition of all Euclids works in his own Language that is extant, takes another Method to justify Euclid with regard to this Axiom, which I shall mention by and by.*

After him Hieronymus Saccherius an Italian Professor of Mathematicks at Pavia, thinking the Field still open wrote a Book published at Milan in the year 1733 which he calls, *Euclides ab omni Nævo vindicatus*. Saccherius was no Novice in Demonstration, he had wrote a book forty years before which he called Logica Demonstrativa, & had likewise wrote on Staticks.* |

3061/11, 13 He has in order to demonstrate this 11th Axiom given us no less than thirty nine Propositions, accompanied with Corollaries Scholia and Lemmata, and with some new Axioms. This work of Saecherius our Dr Simson must have been acquainted with as it is in his Collection bequeathed to the University Library, with large Notes in his own hand upon some part‹s› of it, though not upon the first Part which treats of the eleventh Axiom.*

I shall now give a brief Account of what has been said upon this Axiom by two of the latest and ablest of those who have treated of it. These are Dr David Gregory and Dr Robert Simson

Dr Gregory thinks that the eleventh Axiom was not prefixed to the Elements by Euclid himself but by Theon or some Scholiast, and that Euclid assumes it in the demonstration of the twenty ninth and other Propositions, not as an Axiom or selfevident Truth, but as the converse of the seventeenth Proposition, from which he says it manifestly follows.*

Gregory seems here to acknowledge that Euclid has assumed a Proposition in his Demonstrations which is neither self evident nor proved by him, a Practice which I think Euclid is not chargeable with on any other occasion.

But says Gregory he assumed it as the converse of the 17 Proposition. We may observe that every converse of a true proposition is not true far less evidently true, & when Euclid has occasion to use the converse of a proposition, he first demonstrates that converse. Gregory says this converse manifestly follows from the 17 Proposition. The 17 Proposition is that any two angles of a triangle are less than two right angles. I confess the eleventh Axiom does not appear to me to follow manifestle from this or indeed to follow at all from it without assuming something
3061/11, 14 not more evident than the Axiom itself | and I wish the Dr may not have been rash in this assertion. Had Euclid added this to his 17th Proposition "That two angles of a Triangle may together be nearer to two right angles than by any given Difference" the eleventh Axiom might very easily have been drawn from that Proposition but without this addition the consequence does not appear to me ‹to follow›

Let us next hear Dr Simson; whose accurate Judgment in Geometry as well as his long Practice and study in the Elements has enabled him to observe more Defects and inaccuracies in our Copies of them than all that were before him, and to correct the Errors and supply the Defects of them; Whose Zeal for the honour of Euclid was such that he attributes every thing that needs to be amended, not to Euclid himself but to Theon or some unskilfull Editor.* So that We may justly say

Si Pergama dextra
*Defendi potuissent dextra hac defensa fuissent.**

I shall therefore transcribe his Note upon the 29 Proposition as far as is necessary for shewing how far he acknowledges a Defect in this part of the Elements, and in what manner he endeavours to supply it.

“The Proposition, says he, which is usually called the fifth postulate, or eleventh Axiom, by some the twel‹f›th upon which this 29 Proposition depends, has given a great deal to do, both to ancient and modern Geometers. It seems not to be properly placed among the Axioms, as, indeed, it is not selfevident; nor does it admit of a Demonstration in a strict sense. But stands in need of some explication to make it clear, which, for the sake of beginners, is given as plainly as we can, as follows.”

“First it may easily be conceived that two straight lines AB, CD in the same plane, which are at right angles to the same straight line AC, are likewise equidistant, that is do neither approach to nor recede from one another however far produced, or rather none can conceive the contrary of such lines. — Next it seems to be exceeding manifest that
3061/11, 15 two straight lines proceding from the same point do separate & | diverge more and more from one another, so that the nearest distance from the end of one of them to the other may become greater than any given length — This property flows from the nature or definition of a straight line, which keeps always the same direction, and therefore cannot be strictly demonstrated by preceeding Propositions.”

“These two things being granted, the Proposition may be thus demonstrated” &c So far Dr Simson 4to Edition of his Euclid 1756*

Here we are to observe, 1st that in Dr Simsons Opinion the 11 Axiom upon which a great part of Geometry hangs, is not selfevident 2ly That it does not admit of Demonstration in a strict sense 3ly In order to make it clear, he assumes two propositions, which he thinks exceeding manifest, but does not pretend to demonstrate them. These two propositions are undoubtedly true, but whether both or either of them be more evident than the eleventh Axiom may be doubted

4ly He gives a reason why the last of them cannot be strictly demonstrated by any of Euclids Propositions preceeding the 29th “This property, he says, flows from the Nature or Definition of a straight line, and therefore cannot be strictly demonstrated by any of the preceeding propositions.‹”›* This reasoning indeed I do not understand. Such a thing flows from the Nature or Definition of a streight Line & therefore cannot be strictly demonstrated by any of the preceding propositions. If he meant that it might be demonstrated by any of the succeeding Propositions it were to be wished that he had pointed out those propositions. They must be such as have no dependance on the 29 otherwise to demonstrate it by their aid would be to reason in a Circle.

We see upon the whole that Dr Simson neither thinks the 11th Axiom selfevident, nor pretends strictly to demonstrate it, nor does he think that any of the attempts made before him to demonstrate it, had been successfull, otherwise he would not have said that it does not admit of strict demonstration so that in the Opinion of this great Geometrician the
3061/11, 16 Efforts | to demonstrate it, carried on for so many Ages past by Greeks Arabians and Moderns, have been so much Labour lost.

I am much disposed to think with Dr Simson, that the Axiom does not admit of Demonstration in the way which has been taken by all who have attempted it. It has been the intention of all who have made the Attempt, to demonstrate it before the 29th Proposition. The 29th Proposition contains the properties of parallel right lines, which are deduced by Euclid in a very simple and elegant manner from the Definition of parallel lines, & the disputed Axiom. Now if we might presume to alter Euclids method so far, and at the same time should be able to demonstrate the properties of parallel right lines, without the aid of the 11th axiom, by means of some Axiom more evident to beginners in these studies, may it not be hoped that having these properties of parallel lines as auxiliaries never before employed in this cause, we might be able to demonstrate the 11th Axiom; though it may not admit of Demonstration without that aid.

I hope there is no blameable presumption in trying to prove the properties of parallel right lines in another manner than Euclid has done. |
3061/11, 17 from what has been said Corollary 4 from the Definition of right Lines* it appears evidently that a right line may be in the same plane with another right line, and in every point at the same distance from it. I would therefore give this definition of parallel lines, in place of that which Euclid gives. A right line is said to be parallel to a right line when being in the same plane it is in every point equally distant from it. From this definition it follows, that if a right line cuts one of two parallel lines at right angles it will cut both at right angles. It is evident that if a right line cut one of two parallel right lines at a right angle and the other at an acute angle the lines must approach to each other, if at an obtuse angle they must recede from each other, but by the Definition they neither |
3061/11, 18 approach nor reced‹e›, therefore the angle at which the line is cut is neither acute nor obtuse, it is therefore a right angle.

From this proposition the 29 of Euclid is very easily demonstrated to wit That if two parallel right lines are cut by a right line the alternate angles will be equal; the external angle will be equal to the internal and opposite on the same side; and the two internal angles on the same side will together be equal to two right angles.

Thus I think the properties of parallel right lines may be demonstrated without borrowing Aid from the 11th Axiom. But as that Axiom is necessary in other parts of Geometry and is thought not to be selfevident it is proper that it should be demonstrated, & this I think I am able to do from the properties of parallel right lines. As the Demonstration however requires a Figure, which is not proper in a Discourse of this kind, I must take Credit for it. In the same way may be demonstrated another Proposition assumed by Dr Simson in his Illustration of the 11th Axiom, which he says depends upon the Nature or Definition of a straight line & therefore cannot be strictly demonstrated by any of the preceeding Propositions The Proposition I mean is, That if two right lines diverge from the same point, their distance will at last become greater th‹an› any given length. This proposition is so nearly allied to the 11th axiom that the same Demonstration serves for both see page 19

If after all this attempt to demonstrate the 11th Axiom should prove unsuccessfull, it is but one, added to a very great Number of the kind made by Mathematicians, at whose feet the Author of this would be proud to sit and learn.

I conclude with two Observations, which I think due to the Honour of the great Father of Mathematicians. The first is That the greatest fault laid to the charge of Euclid, I may say the onely fault which may not be laid to the door of his Editors amounts to this, that he has assumed without proof a Proposition which is undoubtedly true, but is thought not to have all the self evidence required in a mathematical Axiom. Now, if we consider that the Sphere of selfevidence, must enlarge in proportion to the penetration of the Mind that judges; this if it be a fault I think is venial. The second Observation is. That able Mathematicians, during a Succession of two thousand years have been labouring to mend this fault in Euclid, with little if any Success. so that we may apply to him that of Mr Pope

Great Wits sometimes may gloriously offend
And rise to faults true Criticks dare not Mend* |

3061/11, 22 Mr Preses*

There is one Remark which I beg leave to make on this Discourse my self, because it needs an Apology.

The Account given in the Discourse of Dr Simsons Opinion with Respect to the 11th Axiom, I took from the 4to Edition of his Euclid printed at Glasgow in the year 1756. I did not then know, or did not remember, that his Note on this Subject was altered in a Subsequent Edition. His last Opinion onely ought to be imputed to him, & Therefore I retract what I read as his Opinion in the Discourse with my Remarks upon it, and in place of it give this short Acount of his latest Opinion, as it is in the latest Edition of his Euclid I have seen, printed at Edinburgh in the year 1775.*

He continued to think that the Axiom is not selfevident & ought not to be placed among the Axioms, but he does not say that it does not admit of strict Demonstration, but on the contrary says it may be demonstrated thus, then follows his Demonstration of it; which is comprehended in two Definitions, one Axiom, five Propositions, & one Corollary.

His Axiom is this

A straight line cannot first come nearer to another straight line, and then go further from it before it cuts it; and in like manner a straight line cannot go further from another straight line & then come nearer to it; nor can a straight line keep the same distance from another straight line, and then come nearer to it, or go further from it; for a straight line keeps always the same direction.*

If this long Axiom (which consists of three branches & might be called three Axioms.) be taken for granted, I think Dr Simson has from it deduced the Axiom of Euclid justly and geometrically. But whether his Axiom will appear to every Mathematician to have a greater degree of selfevidence than that which Euclid has assumed I will not pretend to judge

It seems more easy in some cases to determine what is true & what is false than to determine what is selfevident and what is not so. 3061/11, 23 Selfevidence | in this respect seems to resemble our Eyesight. One Man by his naked Eye may see what another does not see but by the aid of a Telescope. And that to one Mans Understanding may be self evident which to another needs the aid of reasoning. Even to the same Man when more acquainted with the subject that may be selfevident which at first was not so.

I am not at all confident that what I have now said of Dr Simsons Axiom may not be applied to the fourth Corollary I drew from my Definition of a right line, from which I deduced the properties of parallel lines.

When it is doubted concerning a Corollary or concerning a Proposition, not whether it be true, but whether it be self evident, I believe the common Sense of Men who are accustomed to judge in such Matters is the onely tribunal to which we can appeal for the resolution of this Doubt.* If it can be resolved by finding some thing more evident in proof of it, it is well, but perhaps this cannot always be done

Upon the whole I am very apt to think that Euclid foresaw, that what he assumes as an axiom, would not at first appear selfevident to every one of his Readers, and that he could have said as much for the proof of it as any of his Successors have done to this day; but that in the confidence that by those who had got some practice in geometrical reasoning it could admit of no doubt, he ch‹o›se to assume it without proof rather than go about to prove it in the labourious and doubtfull way which has been attempted by those who came after him. So that I think we may still apply to this father of Geometry the Maxim of Mr Pope before quoted,

Great Wits sometimes may gloriously offend
And rise to faults, true Criticks dare not mend |

3061/11, 19

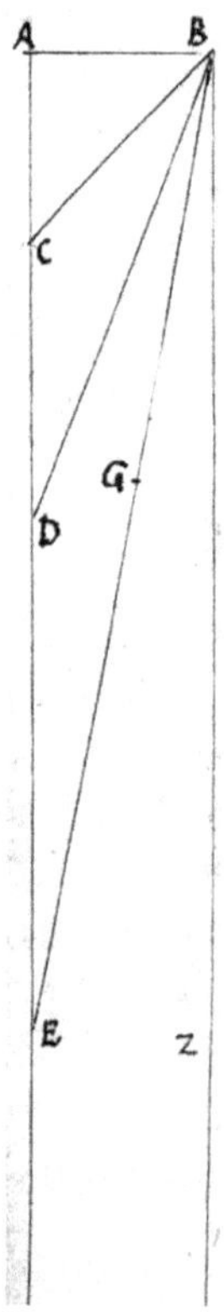

Problem

To construct a plain Triangle upon any given Base, one of whose angles at the Base shall be a right angle, & the other nearer to a right angle than by any given Difference

Let AB be the given Base. From A draw AE at right angles to AB and produce it indefinitely. This will be one side of the Triangle required. From B draw BZ parallel to AE & consequently at right angles to BA. In AE take AC equal to AB, & joyn CB. Then ABC is a Triangle constructed upon the given Base AB & rig‹h›t angled at A. The other Angle at the base differs from a right angle by the angle CBZ. But it is required that this angle shall be less than a given angle, which we shall call M. If then the angle CBZ be greater than M take CD equal to CB and joyn DB, and take DE equal to DB and joyn EB. In like manner EF may be taken equal to EB. and so on as far as is necessary. By carrying on this series of Triangles I say you will at last have a Triangle right angled at A, and having the angle at B nearer to a right angle than by the given Difference M. For in the series of Triangles described, the Differences of whose angles at the Base from a right angle are CBZ, DBZ, EBZ, FBZ, &c respectively, the second is half the first, the third half the second, the fourth half the third and so on. But if from any magnitude you take one half, & from the remainder one half, & so on continually you will at last have a magnitude less than any given magnitude of the same kind. By Proposition 1 Elements 10 Book.* which depends not on any preceeding Proposition. Q.E.F

Corollary 1. A Triangle may be constructed upon any given Base whose angle opposed to the Base shall be less than any given angle

Corollary 2 An angle however small being given, its sides may be produced till the perpendicular from one to the other exceed any given length. |

3061/11, 20 Corollary 3 If a line cut two right lines in the same plane making the internal angles on the same side less than two right angles, those two lines being produced will meet on that side

Let AE & BG be cut by AB so as to make the angles at AB less than two right angles.

Upon AE produced, if necessary, construct a Triangle, by the Problem, upon AB having the angle at B nearer to a right angle than the given angle ABG. The line BG must fall within this Triangle and therefore being produced will meet AE the opposite side of the constructed triangle

Part Two: 'An Essay on Quantity'

I

5/I/20, 1r

Concerning the Object of Mathematicks

The Object of Mathematicks is commonly said to be Quantity & Quantity defined to be what ever is capable of More or less.* But I apprehend there are many things capable of more and less which yet cannot be the Object of Mathematicks. Duration extension and Number Seem to be the primary and direct objects of the Mathematical Sciences and other things onely in as far as they are capable of being measured by these‹.› Velocity is measured by time & space The quantity of Motion, Density, Elasticity, resistance, the Vires, insitæ, impressæ absolutæ motrices acceleratrices must all be measured by space Duration or number or by something which is measured by these in order to make them the object of Mathematicks.* this I think will be evident to any one who considers the definitions given of these several kind of quantitys by those Mathematicians who have treated of them. We can compare two portions of Duration or Space or two Numbers without taking in any other Quantity as a common Measure to both we can directly & immediately perceive a ratio between them as that the one is double treble ½ &c or in some other proportion to the other But other mathematical quantitys cannot be compared so as to perceive any ratio between them without taking some of these as the Measure of them

Corollary 1 That onely is the object of Mathematicks which is capable of Being doubled, trebled, halfed &c that is increased or diminished mult‹i›plied or divided in the ratio of one number to another for Since all Mathematical Quantitys must be measured by time Space or Number‹,› & Space & time may be measured by number therefore all Mathematical Quantitys may be measured by number

There are several things capable of More or less & yet not capable of being measured by number. Tastes Smells, heat cold beauty pleasure pain all the affections | and Appetites of the mind, Probability, Wisdom folly &c &c Altho attempts have been made to apply mathematical reasoning to some of these Subjects and the Merit of Actions has been

5/I/20, 1v

Measured by Simple & compound ratios Yet I do not think that ever any reall knowledge can be struc‹k› out this way.* It may perhaps if discreetly used be a help to discourse on these Subjects by pleasing the imagination and illustrating what is already known. As we use Metaphors & Similes taken from Sensible things to illustrate what is moral or Spiritual but I do not think it can serve any other end

2 In all mathematical reasoning about other Q‹u›antitys besides number time & Space we must first of all assume a certain measure of that quantity & suppose it to be as such a space such a duration or such a number or as some Quantity Measured by these. The Measure of Such an Improper quantity is a part of the Definition of it and unless that is given there can be no mathematical reasoning about it. Thus if one should say that the force of a moving body is as its velocity another that it is as the square of its velocity it seems to me impossible to reason upon this Controversy till some definition of the force of a moving body is aggreed upon and this definition must include some measure of that force.* If the contending parties assume different measures of force they may both think the same way and onely differ about words. Suppose a body moving directly upwards with a force which is continually diminished & at last destroyed by the force of Gravity acting contrary to it it is agreed on both hands that the greater force the body has at the beginning of its Motion the longer it will be before its force is destroyed & the further it will move upward. But one takes one of these for the Measure of its force to wit the time the other the space it moves.

3 Th‹o›se quantitys cannot be compared which have not an common measure thus there can be no ratio between a 1‹s›t & a 2d fluxion yet the ratio of first fluxions may be compared to that of 2d fluxions

4 Mathematical Evidence is an Evidence Sui Generis not competent to any Proposition that does not express a relation of Quantities which may be Measured by lines or Numbers*

Note that wherever I mention Number in this Essay I take it in the largest Sense so as to include not onely whole Numbers but fractions Surd roots the roots of adfected equations and the Sums of infinite Converging Series's*

II

5/I/22, 1r

Essay Concerning the Object of Mathematicks occasioned by reading a piece of Mr Hutchesons wherein Virtue is Measured by simple & Compound Ratios*

The Object of Mathematical Knowledge is justly said to be Quantity. But those who have defined Quantity to be whatever is capable of more and less, have given too wide a Notion of it, which the common acceptation of the word in Language will not bear, and which has led people to apply Mathematical Reasoning to subjects that do not admit of it.

Quantity, as Aristotle long since observed is either Proper or Improper.*

Proper Quantity is that which may be measured by a Quantity of the same Kind. Thus Lines are measured by Lines Surfaces by Surfaces Solids by Solids Number by Number and Time by Time

Improper Quantitys are th‹o›se which are Measured by other Quantities, but not by Quantities of the same kind. Thus Velocity is measured by the Length passed over in a given time.

The Quantity of Motion, Density, Elasticity, Resistance; the Vis Insita, Impressa, Motrix, Acceleratrix* Ratios Fluxions* &c Must all be Measured by Space Duration or Number or by Something which is Measured by these. This I think will be evident to any one who considers the Definitions given of these by those Mathematicians who have treated them with Accuracy particularly by Sir Isaac Newton in the beginning of his principia*

any two portions of Space or Duration, or two Numbers we can immediately compare and perceive a Ratio between them as that one is Double Treble or Half the other and this makes them immediately the objects of Mathematical Reasoning But other Quantitys cannot be compared so as to perceive any such ratio between them without taking some of these as the Measure of them. And ’tis onely when Such a Measure is assigned that they become Mathematical Quantitys.

Thus Velocity being by all Mathematicians taken to be as the Space passed over in a given time; This Measure applyed to it makes it a Mathematical Quantity and capable of being doubled or halfed multiplyed or divided in any Ratio that one Number can bear to another*

Suppose I should take it in my head to affirm that the Velocity of a Body is as the Square of the Space it passes over in a given time. |

5/I/22, 1v I think what is above Said may give som‹e› light to the Controversy about the force of Moving Bodies; Which so long exercised the pens of several Mathematicians and for what I know like other Controversies is rather Sopite* than ended, to the no Smal Scandal of Mathematicks which hath alwise boasted of a degree of Certainty which is not compatible with debates that can be brought to no Issue.* the Philosophers on both sides agree with one another and with the Vulgar thus far that the force of a Moving Body is the same while its velocity is the Same, that when the velocity is increased the force is also increased and diminished if the velocity is diminished. But this Notion of force tho Sufficient for Common Discourse is not Sufficient to make it
5/I/22, 2r an | object of Mathematical Reasoning. In order to that end it must be more accurately defined. And so Defined as that we may know how to measure it and what is a double force or what a treble force. The Ratio of one force to another cannot be perceived immediately but by some common Measure that Measure must be Settled not by Mathematical Reasoning but by a Definition.

One Says the force of A Moving Body is as its Velocity If this is laid down as a Definition as Sir Isaac Newton has done I agree to it entirely because thereby a distinct Measure of force is laid down and the most simple and Natural one that can be assigned for since all men agree that the velocity being the same the force of the body is the same the velocity being increased or diminished this force is so also. Nothing can be more Simple or Natural than to make the Velocity the Measure of the force while the Quantity of Matter is the same.*

Secondly You take it for granted that falling bodies are acted upon by this Uniform power alone without any other conspiring or resisting powers which a Subtile Adversary may always have recourse unto. And these two fortresses he will be able to hold out against all the batteries of Mathematical reasoning or Experiments |

5/I/22, 2v Suppose you attempt to prove your Proposition in this Manner

1 Those Forces are equal which being oposed destroy one another

2 This is the case of Bodies meeting with velocities reciprocally as the quantity of Matter Therefore such bodies have equal force.

3 Since the Quantity of Matter is allowed to be as the force when the Velocity is given. And the force is as the Quantity of Matter and Velocity it follows that the Quantity of Matter being given the force is as the Velocity.

this Argument does also leave the Adversary two Strong holds to which he may have recourse First he may desire you to prove that When bodies meet and stop each other the force of both is destroyed There is an intestine motion raised in each by the Shock this says he employs part of the force the rest onely is employed in Stopping the external Motion. You err therefore in Measuring the whole force by that which is the effect onely of a part of it.*

Secondly he may deny that the Force is as the Quantity of Matter when the Velocity is given.* For it is as easy to deny this and as difficult to prove it as that The force is as the Velocity when the Quantity of Matter is given The two arguments I have Mentioned seemed always to me to have the greatest appearance of Strength and yet we see they take for granted either the thing in Question or Something else that is as far from being evident to reason or experience as the thing in Question. And I believe this will be found to be the case with all other arguments on this Subject.

Sup‹p›ose one takes the other Side of the Question and says that the force of a moving body is as the Square of its Velocity. If he lays down this as a Definition I charge him not with error because a Definition cannot be false and while he sticks to his definition it will lead him into no error in Mathematicks or Mechanicks.

But I charge this Definition with Impropriety. first because the Simple ratio of the velocity will do as well for the Measure of the force and if you leave that you may as well take the Subduplicate or triplicate as the Duplicate

Secondly this Definition does not so well agree with the common Notion of Force as is evident from the Paradoxes it leads to, Mentioned above. Viz that the Power of Gravity does not act uniformly; that the Velocity being given the Force is in the duplicate Ratio of the Quantity of Matter. Or if we will avoid these paradoxes this Notion leads us to Hypotheses. That Falling bodies are acted upon by other powers beside that of Gravity. that unequal Forces may destroy each other without producing any other Sensible Effect. Now tho Reason or Experiment may Support a Paradox or a Hypothesis Yet I thin‹k› either of them have weight enough to overturn a Definition if they have no other Support.

If one advances this that the force of A moving body is in the duplicate Ratio of its Velocity as a Proposition capable of Proof. I ask what he takes for the Measure of the Force. The onely Measure I remember to have been Assigned by the Gentle men of that Side is the height through

Which a body rises by a given force directly before its force is overcome by that of Gravity. this Say they is the whole Effect of the ‹force› & which seems first of all to have led Leibnitz in to his Notion of force* |

5/I/22, 3r 1 A Quantity of any kind may be compared and bears a Ratio to any other Quantity of the same kind but cannot bear a ratio to, or be compared with any Quantity of another kind thus a line can be compared onely with lines a Surface with Surfaces &c

2 Velocity of Motion is always measured by the length which would be passed over in a given time by that velocity, continuing always the same. hence it follows that any Velocity of Motion may be compared with or bear a Ratio to any other velocity of Motion, because all Velocitys of Motion are measured by Quantity of the same kind.

3 Velocity of Increase or Decrease is measured by the Increment or Decrement which would be generated in a given time if that Velocity continued the same. Hence it follows, That Fluxions (which are Velocitys of Increase or Decrease) can be compared when the Increments or decrements generated by them are Quantitys of the same kind, and that they cannot be compared or bear a ratio to one another when the Increments or decrements generated by them are Quantitys of different kinds*

Definition I call that Proper Quantity which may be compared with other Quantity's of the same kind without bringing in any different kind of Quantity as a common Measure of the two Quantitys compared*

4 Lines Surfaces Solids Numbers Portions of Time are Proper Quantitys

5 Velocity's, Quantitys of Motion, Forces Densitys &c can onely be Compared by bringing in some one or more proper Quantitys as a common Measure to compare them by.

6 Improper Quantitys ought to be defined and the measure of them taken into the Definition otherwise there can be no clear reasoning about them

III

3061/7, 1r

An
Essay on Quantity
Read before the Royal Society at London October 1748

Κυρίως δὲ ποσὰ ταῦτα λέγεται μόυα, τὰ {ποσὰ} ἄλλα πάντα κατὰ ονμβεβηκός είς ταῦτα γὰρ ἀλυβλέλουτες, καὶ τα ἄλλα ποσὰ λέγομευ. Aristotle* |

3061/7, 1v

Contents

3061/7, 1

1 What Quantity is?

Since Mathematical Demonstration is thought to carry a peculiar Evidence along with it, which leaves no Room for further Dispute; it may be of some Use, or Entertainment at least, to enquire; to what Subjects this Kind of Proof may be applyed.

Mathematicks is properly the Doctrine of Measure. The Object of this Science is commonly said to be Quantity; and then Quantity ought to be defined, *What may be Measured*.* Those who have defined Quantity to be Whatever is capable of More or Less, have given too wide a Notion of it, which I apprehend hath led some Persons to apply Mathematical Reasoning to Subjects that do not admit of it.*

Pain and Pleasure admit of various Degrees, but who can pretend to measure them: Had this been possible, 'tis not to be doubted, but we should have had as many Names for their various Degrees, as we have for Measures of Length or Capacity; and a Patient should have been able to describe the precise Quantity of his Pain, as well as the Time it begun, or the Part of his Body it affected. To talk intelligibly of the Quantity of Pain, we must have some Standard to measure it by, some known Degree of it so well assertained, that when we talked of it we should all mean the same thing; and comparing other Degrees of Pain with this, we should not onely know whether they exceeded or fell short of it, but by how much, whether by a half a fifth, or in what other proportion

Whatever has Quantity or is Measurable, must be made up of Parts which bear Proportion to one another and to the whole, so that it may not
3061/7, 2 onely be Increased and | Diminished, but may also be Multiplyed and
Divided, and in a word may bear any Proportion to another Quantity of the same Kind, that one Number or Line can bear to another.

That this is essential to all Mathematical Quantity, is apparent from the first Elements of Algebra, which treats of Quantity in General, or of th‹o›se Properties & Relations, which are common to all Quantity: And it is evident that every Algebraical Quantity, is supposed capable not onely of being increased or diminished, but of being exactly doubled, trebled, halfed, and in general, of bearing any assignable Proportion to another Quantity of the same Kind.* This then is the Characteristick of Quantity; whatever has this Property, may be adopted into Mathematicks, and its Quantity and Relations measured with Mathematical Accuracy and Certainty.

2 Of Proper & Improper Quantity.

There are some Quantities which may be called Proper, and others Improper. This Distinction is mentioned by Aristotle, but as a great part of what follows depends upon it, it requires some Explication.*

I call that Proper Quantity, which is measured by its own Kind: Or, which of its own Nature is capable of being doubled or trebled, without taking in any Quantity of a different Kind as a Measure of it. Thus a Line is measured by known Lines, as Inches, Feet or Miles; and the length of a Foot being known, there can be no Question about the length of two Feet, or of any part or Multiple of a Foot. This known Length, by being Multiplyed or Divided, is Sufficient to give us a Distinct Idea of any Length whatsoever.

Improper Quantity, is that which cannot be measured by its own Kind, but to which we assign a Measure in some Proper Quantity that
3061/7, 3 is related to it. Thus Velocity of Motion, | when we consider it by it self,
cannot be measured; We may perceive one body to move faster, another

Slower, but we can perceive no Proportion or Ratio between their Velocities, without taking in some Quantity of another Kind, to measure them by. Having therefore observed, that by a greater Velocity, a greater Space is passed over in the same time, by a less Velocity a less Space, and by an equal Velocity an equal Space; we hence learn to measure Velocity, by the Space passed over in a given Time, and to reckon it to be in exact Proportion to that; and having once assigned this Measure to it, we can then, and not till then, conceive one Velocity exactly double, or triple, or in any other Proportion to another.* We may then introduce it into Mathematical Reasoning, without danger of Error or Confusion; and may use it as a Measure of other Improper Quantities.

All the Proper Quantities we know may I think be reduced to these four. Extension, Duration, Number & Proportion Tho’ Proportion be measurable in its own Nature, & therefore hath Proper Quantity; yet since things cannot have Proportion, which have not Quantity of some other Kind, it follows, that whatever has Proper Quantity, hath it, in one of these three Kinds, Extension, Duration, or Number: These are the Measures of themselves, and of all things else that are measurable. It is to be observed, that Number is applicable to some things, to which it is not commonly applyed by the Vulgar. Thus Lots and Chances of various Kinds, by attentive Consideration, appear to be made up of a determinate number of more Simple Chances, that are allowed to be equal; and by numbring these, the Values and Probabilities of those which are compounded of them, may be Mathematically demonstrated.*

Velocity, the Quantity of Motion, Density, Elasticity, the Vis Insita & Impressa,* the various kinds of centripetal Forces, & different Orders
3061/7, 4 of Fluxions, are all Improper | Quantities; which therefore ought not to be admitted into Mathematical Reasoning, without having a Measure of them assigned.

The Measure of an Improper Quantity, ought alwise to be included in the Definition of it; for it is the giving it a Measure, that Makes it a proper Subject of Mathematical Reasoning. If all Mathematicians had considered this, as carefully as Sir Isaac Newton has done, some Trouble had been saved, both to themselves and their Readers. That great Man, whose clear & comprehensive Understanding appears even in his Definitions, having frequent Occasion to treat of such Improper Quantities, never fails to define them, so as to give a Measure of them, either in proper Quantities, or such as had a known Measure. See the Definitions prefixt to his Principia.*

It is not easy to say, how many Kinds of Improper Quantity, may in time be introduced into Mathematicks, or to what new Subjects measures may be applyed. But this I think we may conclude, that there is no Foundation in Nature, nor can any valuable End be served, by applying Measures to any thing, that has not these two Properties: First, it must admit of Degrees, of greater and less; Then it must be associated with or Related to something that hath proper Quantity so as that when one is increased the other is increased, when one is diminished the other is also diminished; every Degree of the one must have a determinate Magnitude or Quantity of the other, corresponding to it.

It sometimes happens that we have occasion to apply different Measures to the same thing; Centripetal Force as defined by Newton, may be Measured various ways; he himself gives three different Measures of it, and distinguishes them by different Names as may be seen in his Definitions already referred to.*

In Reality I conceive, the applying Measures to things that have not proper Quantity, is onely a Fiction and Artifice of the Mind, for enabling us to conceive more easily, & more distinctly to express
3061/7, 5 and Demonstrate, the Properties and Relations of | these things that have real Quantity. There is not a Proposition in the two first books of Newtons Principia, that might not be expressed, and perhaps Demonstrated without these various Measures of Motion, and of Centripetal and Impressed Forces, which he uses; but this would occasion such Intricate and perplexed Circumlocutions, and such a tedious Length of Demonstrations, as might fright any Sober Person from Attempting to read them.*

3 Corollary First

From the Nature of Quantity we may easily see, what it is that gives Mathematicks such advantage over other Sciences, in Clearness and Certainty; Namely, that Quantity its Object, admits of a much greater variety of Relations, than any other Subject of human Reasoning; and at the same time, every part, and Multiple of a Quantity, and every Relation or Proportion of different Quantities, may, by the help of Lines and Numbers, be so distinctly defined, as to be easily distinguished from all others, without any danger of mistake: Hence it is that we are able to trace its Relations, through a long Process of Reasoning, and with a Perspicuity and Accuracy, which we in vain Expect in Subjects not capable of Mensuration.

Extended Quantities, such as Lines, Surfaces & Solids, besides what they have in common with all other Quantity, have this Peculiar, that their parts have a particular place and Disposition among themselves. A Line may not onely bear any assignable Proportion to another in Length or Magnitude, but Lines of the same Length may vary in the Disposition of their Parts: one may be Streight, another may be part of a Curve of any Kind or Dimension, of which there is an endless Variety; The like may be said of Surfaces and Solids: So that Extended Quantities admit of no less variety with regard to their Form than with regard to their Magnitude. And as their various Forms may be exactly defined & measured, no less than their Magnitudes; hence it is that Geometry which treats of |
3061/7, 6 Extended Quantities, leads us into a much greater Compass & Variety of Reasoning, than any other Branch of Mathematicks‹.› Long Deductions in Algebra are made, for the most part, not so much by carrying on a train of reasoning in the Mind, as by an Artificial kind of Operation, which is built on a few very Simple Principles.* But in Geometry, we may build one Proposition upon another, a third upon that, & so on without ever attaining a Limit which we cannot exceed. The Properties of the more Simple Figures can hardly be exhausted, much less those of the more Complex ones.

4 Corollary Second

Mathematical Evidence is an Evidence *Sui Generis*, not Competent to any Proposition, which does not express a Relation of things Measurable by Lines or Numbers.* All Proper Quantity may be Measured by these; and Improper Quantity must be measured by that which is Proper.

There are many things capable of more and less, which are not capable of Mensuration. Tastes, Smells, the Sensations of Heat & Cold, Beauty & Deformity, Pleasure & Pain, All the Affections and Appetites of the Mind, Wisdom, Folly, & most Kinds of Probability, with many other things tedious to ennumerate, admit of Degrees, but have not yet been reduced to Measure, nor as I apprehend ever can be. I say most Kinds of Probability, because one Kind of it, to wit the Probability of Chances, is properly measurable by Number, as hath been observed above.*

Altho Attempts have been made to apply Mathematical Reasoning to some of these things, and the Quantity of Virtue and Merit in Actions, has been measured by Simple & Compound Ratios;* yet I do not think any real Knowledge has been Struck out this way: It may perhaps if Discreetly used, be a help to Discourse on these Subjects, by pleasing the Imagination & Illustrating what is already known; but untill our Affections and Appetites be themselves reduced to Quantity, and exact Measures of their various degrees assigned; in vain do we essay to measure Virtue & Merit by them; this is onely to ring Changes upon Words, and to make a Shew of Mathematical Reasoning, without advancing one Step in real Knowledge. |

3061/7, 7

5 Corollary Third

I apprehend the Account that hath been given of Proper and Improper Quantity, may also throw some Light upon the Controversy about the Force of moving Bodies, which hath long exercised the pens of many Mathematicians, and, for what I know, like other Controversies, is rather Sopite* than ended, to the no small Scandal of Mathematicks, which hath always boasted of a degree of Evidence, inconsistent with Debates that can be brought to no Issue.*

The Philosophers of both Sides agree with one another, & with the Vulgar, that the Force of a Moving Body is the same while its Velocity is the same, is increased when the Velocity is increased, and Diminished when that is Diminished.

But this Vague Notion of Force in which both Sides agree, tho' Sufficient for common Discourse, yet is not Sufficient to make it Measurable, that is to make it a proper Subject of Mathematical Reasoning: In order to that, it must be more accurately Defined, and so Defined, as to give us a Measure of it, that we may understand what is meant by a Double, or a Triple Force.

The Ratio of one Force to another cannot be perceived but by a Measure; and that Measure must be Settled not by Mathematical Reasoning but by a Definition. Let any one consider Force without Relation to any other Quantity, and see whether he can conceive one Force exactly double to another: I am Sure I cannot till I be endowed with some new Faculty, for I know nothing of Force but by its effects, and therefore can measure it onely by its Effects. Till Force then is defined, and by that Definition a Measure of it assigned, we fight in the dark about a Vague Idea, which is not sufficiently determined to be admitted into any Mathematical Proposition: And when such a Definition is given, the Controversy will presently be at an end.

6 Of the Newtonian Measure of Force

You say the Force of a Body in Motion is as its Velocity: Either you mean, to lay this down as a Definition of Force, as Newton has done, or you mean to affirm it as a Proposition capable of Proof.* If you mean to
3061/7, 8 lay it down as a Definition, | it is no more than if you should say, I call that a Double Force, which gives a double Velocity to the same Body, a triple Force which gives a triple Velocity, & so on in Proportion.

Now if you lay it down as a Definition that the Force is as the Velocity, I entirely agree to it. No Mathematical Definition of Force can be given, that is more clear & simple, none that is more agreable to the common use of the Word in Language. For since all Men agree, that the Force of the Body being the same, the Velocity must also be the same, the Force being increased or diminished, the velocity must be so also; what can be more proper or natural, than to take the Velocity for the Measure of the Force.

Several other things might be advanced, to shew that this Definition of Force, agrees best with the common popular Notion of the Word. If two Bodies meet directly, with a Shock which mutually destroys their Motion, without producing any other Sensible effect, the Vulgar would pronounce without Hesitation that they met with equal Force; and so they do, according to the measure of Force above laid down, for we find by Experience that in this case their Velocities are reciprocally as their Quantities of Matter. In Mechanicks where by a Machine two Powers or Weights are kept in Equilibrio, the Vulgar would reckon that these Powers act with equal Force, and so by this Definition they do.* The Power of Gravity being constant and Uniform, any one should expect, that it should give equal degrees of Force to a body in equal Times, and so according to this Definition it does. So that this Definition is not onely clear and Simple, but it agrees best with the Use of the Word Force in common Language; and this I think is all that can be Desiderated in a Definition.*

But if you are not satisfied with laying it down as a Definition, that the Force of a Body is as its Velocity, but will needs prove it by Demonstration or Experiment; I must beg of you, before you take one

step in the Proof, to let me know what you mean by Force, and what by a double or a triple Force: This you must do, by a Definition which contains a Measure of Force.* Some primary Measure of Force must be taken for granted, or laid down by way of Definition, otherwise we
3061/7, 9 can never reason | about its Quantity. And why then may you not take the Velocity for the Primary Measure as well as any other? You will find none that is more Simple, more Distinct, or more agreable to the common use of the Word Force; & he who rejects one Definition that hath these Properties, hath equal Right to reject any other. I say then that it is impossible by Mathematical Reasoning or Experiment, to prove that the Force of a Body is as its Velocity, without taking for granted the thing you would prove, or something else that is no more evident than the thing to be proved.

7 Of the Leibnitzian Measure of Force

Let us next hear the Leibnitzian, who says that the Force of a Body is as the Square of its Velocity.* If he lays this down as a Definition, I shall agree to it rather than quarrel about a word, and for the future shall understand him, by a Double Force to mean that which gives a Quadruple Velocity, by a Triple Force that which gives nine times the Velocity, and so on in Duplicate Proportion. While he keeps by his Definition, it will lead him into no Error in Mathematicks or Mechanicks; for however Paradoxical his Conclusions may appear, however different in Words from theirs, who measure Force by the simple Ratio of the Velocity, they will in their Meaning be the same. Just as he, who would call a foot twenty four Inches, without changing other Measures of Length, when he says a Yard contains a foot and a half, he means the very same thing as you do when you say a Yard contains three feet.

But tho' I allow this Measure of Force to be distinct, & cannot charge it with Falshood, for no Definition can be false; yet I say in the first place, that it is less Simple than the other; for why should a Duplicate Ratio be used, where the simple Ratio will do as well. In the next place, this Measure of Force is less agreable to the common Use of the Word Force, as hath been Shewn above; and this indeed is all, that the many Laboured Arguments & Experiments, brought to overturn it, do prove: This also is evident, from the Paradoxes into which it has led its Defenders. |

3061/7, 10 We are next to consider the Pretences of the Leibnitzian, who will Undertake to prove by Demonstration or Experiment, that Force is as the Square of the Velocity.* I ask him first, what he lays down for the first Measure of Force. The onely Measure I remember to have been given by the Philosophers of that Side, and which seems first of all to have led Leibnitz into his Notion of Force is this; The Height to which any Body is impelled by any impressed Force, is says he the whole Effect of that Force, and therefore must be proportional to the Cause: But this Height is found to be as the Square of the Velocity, which the Body had at the beginning of its Motion.*

In this Argument I apprehend that great man hath been extreamly unfortunate; For First, whereas all Proof should be taken from Principles that are common to both sides; in order to prove a Proposition which we deny, he assumes a Principle which we think still farther from the Truth; namely, that the Height to which the Body rises, is the whole Effect of the Impulse, and so ought to be the Measure of it.

Secondly, his Reasoning Serves as well against him as for him; For may not I plead with as good Reason thus, The Velocity given by an Impressed Force is the whole Effect of that Impressed Force, and therefore must be proportional to the Cause; the Force therefore must be as the Velocity.

Thirdly Supposing the Height to which the Body is raised, to be the Measure of the Force; This Principle will overturn the Conclusion he would Establish by it, as well as that which he opposes. For Supposing the first Velocity of the Body to be still the same, the Height to which it rises, will be increased if the power of Gravity is diminished, and diminished if the Power of Gravity is increased. Bodies descend Slower towards the Equator, and faster towards the Poles, as is found by Experiments on Pendulums* If then a Body is driven upwards at the Equator with a given Velocity, and the same Body is afterwards with the same Velocity driven upwards at Leipsick, the Height to which it rises in the former case, will be greater than in the latter, but the Velocity in
3061/7, 11 both | was the same: Consequently the Force is not as the Square of the Velocity, any more than as the Velocity.

8 Reflexions on this Controversy

Upon the whole, I cannot but think the Controvertists on both Sides have had a very hard task; whilst mistaking that for a Proposition, which was meant by Newton onely as a Definition; they have lavished much Demonstration & Experiment about the meaning of a Word. For we may I think without the Spirit of Prophecy venture to foretell, that the Controversy will never be brought to an Issue in this way, till both Parties agree in some common Notion or Measure of the Quantity of Force: Till that is done, they have no common Principle to argue from; and we may as well expect, to settle the rules of Grammar by Arithmetick, as to settle the first Measure of Force, by Demonstration or Experiment.

If some Mathematician should take into his head to affirm, that the Velocity of a Body is not as the Space it passes over in a given time, but as the Square of that Space; you might bring Mathematical Arguments and Experiments to Confute him, but you should never by these force him to yield, if he was ingenious in his way; because you have no common Principle left you to argue from in this way; and you differ from one another not in a Mathematical Proposition, but in a Mathematical Definition.

Suppose a Philosopher has considered onely that Measure of Centripetal Force, which is proportional to the Velocity generated by it in a given Time, and from this Measure of it deduces several Propositions. Another Philosopher, in a Distant Country, who has the same general Notion of Centripetal Force, takes the Velocity generated by it, and the Quantity of Matter together, as the Measure of it. From this he deduces several Conclusions, that seem directly contrary to those of the other. There upon a serious Controversy is begun, whether Centripetal Force, be as the Velocity, or as the Velocity and Quantity of Matter taken together: Much Mathematical and Experimental Dust is raised, and yet
3061/7, 12 neither | Party can ever be brought to yield; for they are both in the Right, onely they have been unlucky, in giving the same Name to different Mathematical Conceptions. Had they distinguished these Measures of Centripetal Force, as Newton has done; calling the one *Vis Centripetæ Quantitas Acceleratrix*, the other *Quantitas Motrix*, all Appearance of Contradiction had ceased, & their Propositions which Seem so contrary, had exactly tallied.* |

3061/7, Ar The Intervalls of Musical Notes have in their own Nature certain relations with regard to magnitude. Thus a 4th is to the Ear a less Interval than a 5th & a 5‹th› less than an Octave. Farther we know by the Ear that a 5th & a 4th together make an Octave. These relations are essential & flow from the Nature of those Notes.

The Masters of Musical Composition have thought it necessary to give a Measure to these Intervals in order to represent more distinctly the harmonic Scale. And if we once fix the Measure of one intervall the Measures of all the rest must follow that.

The Ratio of 1:2 has been from very ancient times thought proper to express the Interval called an Octave the Unite representing the Graver Note. & 2 the more acute*

The quicksighted Galileo has put the Question whether the Octave be reasonably expressed by this Ratio: which Question had it been properly improved might I think have given Rise to at least three new Sects in the Theory of Music*

The Old System may indeed be supported by this reason that while the grave Note makes one Vibration the Octave makes two & therefore the ratio of 1 to 2 seems proper to express that interval.

But one who had a mind to make a New System might say that the grave note is to its Octave as 2 to 1 for two Reasons first because the length of a Vibration of the first is double that of the second. & secondly because if the Grave Note is given by a Chord of two feet, the acute will
3061/7, Av be given by a Chord of one foot of equal Tension & thicknes | & he might add 3ly that if Chords of equal Tension & length sound these Notes the Diameters of the Chords will be as 2:1

Another System might Assert that the Octave ought to be exprest by the Ratio of 4:1. Because Chords of the Same length & Matter that sound an Octave that which sounds the grave Note will be four times the weight & bulk of the other.

There is another System behind which may make the Octave as 1:4 for this reason that if two Chords are equal in length & thickness the one streched by a weight of one pound the other by a weight of 4 pounds the second will sound an Octave above the first.

We might dispute to the Worlds end which of these was the true System. For they are all equally true, and a just Scale of Harmonics may be drawn from any one of them. One of them may be more simple & commodious in practice than another, but it cannot properly be said that any one of them is true or false. |

3061/7, 12 To return then to the Controversy about the Quantity of Force, were I worthy to be a Mediator in this Controversy, whereby a Schism hath been made in the Mathematicall World, which, the Experience of more than half a Century may satisfy us, is not to be healed by Mathematical

Debate; I would humbly propose a Treaty, of which the following might be the Preliminaries.

1 In all Books Writings or Discourses, made or held, within great Brittain or any of the Dominions thereto belonging, on any Branch of Mathematicks or Natural Philosophy relating to Motion, the Words *Quantity of Force*, shall after the day of , be understood to mean a Measure thereof, proportional to the Quantity of Matter and Velocity of the body in which that Force is, or on which it is imprest, unless where the contrary is expressly declared.

2 In all Books, Writings, or Discourses, made or held as above, in any other place besides the Dominions above exprest, the Words *Quantity of Force*, shall from the term aforesaid be understood to mean a Measure thereof, proportional to the Quantity of Matter & the Square of the Velocity, unless where the contrary is expressly declared.

3 All Hostilities between Mathematicians on the different Sides shall entirely cease from the term above Specified; and it shall be deemed a breach of this Treaty for either Side to use, or pretend to use, Demonstration or Experiment against the other.

4 The above Articles shall be truly and faithfully keept, ay and untill, the Respective Claims of the two Measures above mentioned, to the Name & Tittle of the Quantity of Force, shall be finally adjusted defined & determined, in ‹the› manner hereafter mentioned. |

3061/7, 13 5 The Determination of the Respective Claims mentioned in the last Article, is and shall be Referred to the Criticks, who are hereby Impowered and Authorised, by common Consent, to Judge and Determine the Right of the Claimants to the above Name and Tittle, and either to give it to both with Proper Distinctions, or to give a new Name to the Claimant that shall be denuded, as they see Cause.

6 If the said Learned Body should not agree in their Judgments, they are hereby authorised and Permitted, to take to different Sides, to muster and levy Critical Forces, to debate & contend, by all Means, and Weapons, competent to be used in a critical War, any thing in this Treaty to the contrary Notwithstanding.

By some such Means as these it Might be hoped that Peace might be restored to the Mathematical World without any Hurt to Truth, and Controversy banished from this Society where it is but a late Usurper to that where it hath undoubted Right and Immemorial Possession.

Finis

P.S. When this Essay was wrote in 1748 I knew so little of the History of the Controversy about the force of moving Bodies as to think that the British Mathematicians onely, opposed the Notion of Leibnitz, and that all the foreign Mathematicians adopted it. The Fact is that the British and French are of one Side, the Germans Dutch & Italians of the other.*

I find likewise that Desaguliers in the second Volume of his Course of Experimental Philosophy published in 1744, is of the Opinion, that the parties in this Dispute put different Meanings upon the Word Force, and that in reality both are in the Right, when well understood.*

IV

505 *An Essay on* Quantity; *occasioned by reading a Treatise, in which* Simple *and* Compound Ratio's *are applied to* Virtue *and* Merit, *by the Rev. Mr.* Reid; *communicated in a Letter from the Rev.* Henry Miles *D.D. & F.R.S. to* Martin Folkes *Esq; Pr.R.S.**

Read Nov. 3. 1748

Sect. 1.
What Quantity is.

Since mathematical Demonstration is thought to carry a peculiar Evidence along with it, which leaves no Room for further Dispute; it
506 may be of some Use, or Enter- | tainment at least, to inquire to what Subjects this kind of Proof may be applied.

Mathematics contain properly the Doctrine of Measure; and the Object of this Science is commonly said to be Quantity; therefore Quantity ought to be defined, What may be measured. Those who have defined Quantity to be whatever is capable of More or Less, have given too wide a Notion of it, which I apprehend has led some Persons to apply mathematical Reasoning to Subjects that do not admit of it.*

Pain and Pleasure admit of various Degrees, but who can pretend to measure them? Had this been possible, it is not to be doubted but we should have had as distinct Names for their various Degrees, as we have for Measures of Length or Capacity; and a Patient should have been able to describe the Quantity of his Pain, as well as the Time it began, or the Part it affected. To talk intelligibly of the Quantity of Pain, we should have some Standard to measure it by; some known Degree of it so well ascertained, that all Men, when they talked of it, should mean the same thing; we should also be able to compare other Degrees of Pain with this, so as to perceive distinctly, not only whether they exceed or fall short of it, but how far, or in what proportion; whether by an half, a fifth, or a tenth.

Whatever has Quantity, or is measurable, must be made up of Parts, which bear Proportion to one another, and to the Whole; so that it may be increased by Addition of like Parts, and diminished by Subtraction, may be multiplied and divided, and in a Word, may bear any Proportion
507 to another | Quantity of the same kind, that one Line or Number can bear to another.* That this is essential to all mathematical Quantity, is evident from the first Elements of Algebra, which treats of Quantity in general, or of those Relations and Properties which are common to all Kinds of Quantity. Every algebraical Quantity is supposed capable not only of being increased and diminished, but of being exactly doubled, tripled, halfed, or of bearing any assignable Proportion to another Quantity of the same kind. This then is the Characteristic of Quantity; whatever has this Property may be adopted into Mathematics; and its Quantity and Relations may be measured with mathematical Accuracy and Certainty.

Sect. 2.

Of Proper and Improper Quantity.

There are some Quantities which may be called *Proper*, and others *Improper*. This Distinction is taken notice of by *Aristotle*; but it deserves some Explication.*

I call that *Proper* Quantity which is measured by its own Kind; or which of its own Nature is capable of being doubled or tripled, without taking in any Quantity of a different Kind as a Measure of it. Thus a Line is measured by known Lines, as Inches, Feet, or Miles; and the Length of a Foot being known, there can be no Question about the Length of two Feet, or of any Part or Multiple of a Foot. And this known Length, by being multiplied or divided, is sufficient to give us a distinct Idea of any Length whatsoever. |

508 *Improper* Quantity is that which cannot be measured by its own Kind; but to which we assign a Measure by the means of some proper Quantity that is related to it. Thus Velocity of Motion, when we consider it by itself, cannot be measured. We may perceive one Body to move

faster, another slower; but we can have no distinct Idea of a Proportion or *Ratio* between their Velocities, without taking in some Quantity of another Kind to measure them by. Having therefore observed, that by a greater Velocity a greater Space is passed over in the same time, by a less Velocity a less Space, and by an equal Velocity an equal Space; we hence learn to measure Velocity by the Space passed over in a given Time, and to reckon it to be in exact Proportion to that Space: And having once assigned this Measure to it, we can then, and not till then, conceive one Velocity to be exactly double, or half, or in any other Proportion to another; we may then introduce it into mathematical Reasoning without Danger of Confusion, or Error, and may also use it as a Measure of other Improper Quantities.*

All the Kinds of Proper Quantity we know, may, I think, be reduced to these four, Extension, Duration, Number, and Proportion. Tho' Proportion be measurable in its own Nature, and therefore hath Proper Quantity, yet as Things cannot have Proportion which have not Quantity of some other Kind, it follows, that whatever has Quantity must have it in one or other of these three Kinds, Extension, Duration, or Number. These are the Measures of themselves, and of all Things else that are measurable. |

509 Number is applicable to some things, to which it is not commonly applied by the Vulgar. Thus, by attentive Consideration, Lots and Chances of various Kinds appear to be made up of a determinate Number of Chances that are allowed to be equal; and by numbering these, the Values and Proportions of those which are compounded of them may be demonstrated.*

Velocity, the Quantity of Motion, Density, Elasticity, the *Vis insita*, and *impressa*,* the various Kinds of centripetal Forces, and different Orders of Fluxions, are all Improper Quantities; which therefore ought not to be admitted into Mathematics, without having a Measure of them assigned. The Measure of an improper Quantity ought always to be included in the Definition of it; for it is the giving it a Measure that makes it a proper Subject of mathematical Reasoning. If all Mathematicians had considered this as carefully as Sir *Isaac Newton* appears to have done, some Labour had been saved both to themselves and to their Readers. That Great Man, whose clear and comprehensive Understanding appears, even in his Definitions, having frequent Occasion to treat of such improper Quantities, never fails to define them, so as to give a Measure of them, either in proper Quantities, or in such as had a known

Measure. This may be seen in the Definitions prefixed to his *Principia Philosophiæ Naturalis Mathematica*.*

It is not easy to say how many Kinds of improper Quantity, may in time, be introduced into Mathematics, or to what new Subjects
510 Measures may be applied: But this I think we may conclude, | that there is no Foundation in Nature for, nor can any valuable End be served by applying Measure to any thing but what has these two Properties. First it must admit of Degrees of greater and less. Secondly, it must be associated with, or related to something that has proper Quantity, so as that when one is increased the other is increased, when one is diminished, the other is diminished also; and every Degree of the one must have a determinate Magnitude or Quantity of the other corresponding to it.

It sometimes happens, that we have Occasion to apply different Measures to the same thing. Centripetal Force, as defined by *Newton*, may be measured various Ways, he himself gives different Measures of it, and distinguishes them by different Names, as may be seen in the above-mentioned Definitions.*

In reality, I conceive that the applying of Measures to things that properly have not Quantity, is only a Fiction or Artifice of the Mind, for enabling us to conceive more easily, and more distinctly to express and demonstrate, the Properties and Relations of those things that have real Quantity. The Propositions contained in the two first Books of *Newton's Principia* might perhaps be expressed and demonstrated, without those various Measures of Motion, and of centripetal and impressed Forces which he uses: But this would occasion such intricate and perplexed Circumlocutions, and such a tedious Length of Demonstrations as would fright any sober Person from attempting to read them.* |

511 SECT. 3. *Corollary first.*

From the Nature of Quantity we may see what it is that gives Mathematics such Advantage over other Sciences, in Clearness and Certainty; namely, that Quantity admits of a much greater Variety of Relations than any other Subject of human Reasoning; and at the same time every Relation or Proportion of Quantities may by the Help of Lines and Numbers be so distinctly defined, as to be easily distinguished from all others, without any Danger of Mistake. Hence it is that we are able to trace its Relations through a long Process of Reasoning, and with a Perspicuity and Accuracy which we in vain expect in Subjects not capable of Mensuration.

Extended Quantities, such as Lines, Surfaces and Solids, besides what they have in common with all other Quantities, have this peculiar, That their Parts have a particular Place and Disposition among themselves: A Line may not only bear any assignable Proportion to another, in Length or Magnitude, but Lines of the same Length may vary in the Disposition of their Parts; one may be streight, another may be Part of a Curve of any Kind or Dimension, of which there is an endless Variety. The like may be said of Surfaces and Solids. So that extended Quantities admit of no less Variety with regard to their Form than with regard to their Magnitude: And as their various Forms may be exactly defined and measured, no less than their Magnitudes, hence it is that Geometry, which treats of extended Quantity, leads us into a much greater Compass and Variety of
512 Reasoning than any other Branch | of Mathematics. Long Deductions in Algebra for the most part are made, not so much by a Train of Reasoning in the Mind, as by an artificial kind of Operation, which is built on a few very simple Principles: But in Geometry we may build one Proposition upon another, a third upon that, and so on, without ever coming to a Limit which we cannot exceed. The Properties of the more simple Figures can hardly be exhausted, much less those of the more complex ones.*

SECT. 4. *Corollary* 2.

It may I think be deduced from what hath been above said, That mathematical Evidence is an Evidence *sui generis*, not competent to any Proposition which does not express a Relation of Things measurable by Lines or Numbers.* All proper Quantity may be measured by these, and improper Quantities must be measured by those that are proper.

There are many Things capable of More and Less, which perhaps are not capable of Mensuration. Tastes, Smells, the Sensations of Heat and Cold, Beauty, Pleasure, all the Affections and Appetites of the Mind, Wisdom, Folly, and most Kinds of Probability, with many other Things too tedious to enumerate, admit of Degrees, but have not yet been reduced to Measure, nor, as I apprehend, ever can be. I say, most Kinds of Probability, because one Kind of it, *viz*. the Probability of Chances is properly measurable by Number, as is above observed.

Altho' Attempts have been made to apply mathematical Reasoning to some of these Things, and the Quantity of Virtue and Merit in Actions
513 has been | measured by simple and compound *Ratio's*; yet I do not think that any real Knowledge has been struck out this Way:* It may perhaps, if discretely used, be a Help to Discourse on these Subjects, by pleasing

the Imagination, and illustrating what is already known; but until our Affections and Appetites shall themselves be reduced to Quantity, and exact Measures of their various Degrees be assigned, in vain shall we essay to measure Virtue and Merit by them. This is only to ring Changes upon Words, and to make a Shew of mathematical Reasoning, without advancing one Step in real Knowledge.

SECT. 5. *Corollary* 3.

I apprehend the Account that hath been given of the Nature of proper and improper Quantity may also throw some Light upon the Controversy about the Force of moving Bodies, which long exercised the Pens of many Mathematicians, and for what I know is rather drop'd than ended; to the no small Scandal of Mathematics, which hath always boasted of a Degree of Evidence, inconsistent with Debates that can be brought to no Issue.*

Tho' Philosophers on both Sides agree with one another, and with the Vulgar in this, That the Force of a moving Body is the same, while its Velocity is the same, is increased when its Velocity is increased, and diminished when that is diminished. But this vague Notion of Force, in which both Sides agree, tho' perhaps sufficient for common Discourse, yet is not sufficient to make it a Subject of mathematical Reasoning: In order to that, it must be more accurately defined, and so defined as to
514 give | us a Measure of it, that we may understand what is meant by a double or a triple Force. The *Ratio* of one Force to another cannot be perceived but by a Measure; and that Measure must be settled not by mathematical Reasoning, but by a Definition. Let any one consider Force without relation to any other Quantity, and see whether he can conceive one Force exactly double to another; I am sure I cannot, nor shall, till I shall be endowed with some new Faculty; for I know nothing of Force but by its Effects, and therefore can measure it only by its Effects. Till Force then is defined, and by that Definition a Measure of it assigned, we fight in the dark about a vague Idea, which is not sufficiently determined to be admitted into any mathematical Proposition. And when such a Definition is given, the Controversy will presently be ended.

SECT. 6.
Of the Newtonian Measure of Force.

You say, the Force of a Body in Motion is as its Velocity: Either you mean to lay this down as a Definition as *Newton* himself has done; or

you mean to affirm it as a Proposition capable of Proof.* If you mean to lay it down as a Definition, it is no more than if you should say, I call that a double Force which gives a double Velocity to the same Body, a triple Force which gives a triple Velocity, and so on in Proportion. This I intirely agree to; no mathematical Definition of Force can be given that is more clear and simple, none that is more agreeable to the common
515 Use of the Word in Language. | For since all Men agree, that the Force of the Body being the same, the Velocity must also be the same; the Force being increased or diminished, the Velocity must be so also, what can be more natural or proper than to take the Velocity for the Measure of the Force?

Several other things might be advanced to shew that this Definition agrees best with the common popular Notion of the Word Force. If two Bodies meet directly with a Shock, which mutually destroys their Motion without producing any other sensible Effect, the Vulgar would pronounce, without Hesitation, that they met with equal Force; and so they do, according to the Measure of Force above laid down: For we find by Experience, that in this Case their Velocities are reciprocally as their Quantities of Matter. In Mechanics, where by a Machine two Powers or Weights are kept in *equilibrio*, the Vulgar would reckon that these Powers act with equal Force, and so by this Definition they do.* The Power of Gravity being constant and uniform, any one would expect that it should give equal Degrees of Force to a Body in equal Times, and so by this Definition it does. So that this Definition is not only clear and simple, but it agrees best with the Use of the Word Force in common Language, and this I think is all that can be desired in a Definition.

But if you are not satisfied with laying it down as a Definition, that the Force of a Body is as its Velocity, but will needs prove it by Demonstration or Experiment; I must beg of you, before you take one
516 Step in the Proof, to let me know what you | mean by Force, and what by a double or a triple Force. This you must do by a Definition which contains a Measure of Force. Some primary Measure of Force must be taken for granted, or laid down by way of Definition; otherwise we can never reason about its Quantity. And why then may you not take the Velocity for the primary Measure as well as any other? You will find none that is more simple, more distinct, or more agreeable to the common Use of the Word Force: And he that rejects one Definition that has these Properties, has equal Right to reject any other.* I say then, that

it is impossible, by mathematical Reasoning or Experiment, to prove that the Force of a Body is as its Velocity, without taking for granted the thing you would prove, or something else that is no more evident than the thing to be proved.

SECT. 7.
Of the Leibnitzian *Measure of Force.*

Let us next hear the *Leibnitzian*, who says, that the Force of a Body is as the Square of its Velocity.* If he lays this down as a Definition, I shall rather agree to it, than quarrel about Words, and for the future shall understand him, by a quadruple Force to mean that which gives a double Velocity, by 9 times the Force that which gives three times the Velocity, and so on in duplicate Proportion. While he keeps by his Definition, it will not necessarily lead him into any Error in Mathematics or Mechanics. For, however paradoxical his Conclusions may appear, however
517 different in Words from theirs | who measure Force by the simple *Ratio* of the Velocity; they will in their Meaning be the same: Just as he who would call a Foot twenty-four Inches, without changing other Measures of Length, when he says a Yard contains a Foot and a half, means the very same as you do, when you say a Yard contains three Feet.

But tho' I allow this Measure of Force to be distinct, and cannot charge it with Falshood, for no Definition can be false, yet I say in the first place, it is less simple than the other; for why should a duplicate *Ratio* be used where the simple *Ratio* will do as well? In the next place, this Measure of Force is less agreeable to the common Use of the Word Force, as hath been shewn above; and this indeed is all that the many laboured Arguments and Experiments, brought to overturn it, do prove. This also is evident, from the Paradoxes into which it has led its Defenders.

We are next to consider the Pretences of the *Leibnitzian*, who will undertake to prove by Demonstration, or Experiment, that Force is as the Square of the Velocity.* I ask him first, what he lays down for the first Measure of Force? The only Measure I remember to have been given by the Philosophers of that Side, and which seems first of all to have led *Leibnitz* into his Notion of Force, is this: The Height to which a Body is impell'd by any impressed Force, is, says he, the whole Effect of that Force, and therefore must be proportional to the Cause: But this Height is found to be as the Square of the Velocity which the Body had at the Beginning of its Motion.* |

518 In this Argument I apprehend that great Man has been extremely unfortunate. For, 1*st*, Whereas all Proof should be taken from Principles that are common to both Sides, in order to prove a thing we deny, he assumes a Principle which we think farther from the Truth; namely, that the Height to which the Body rises is the whole Effect of the Impulse, and ought to be the whole Measure of it. 2*dly*, His Reasoning serves as well against him as for him: For may I not plead with as good Reason at least thus? The Velocity given by an impressed Force is the whole Effect of that impressed Force; and therefore the Force must be as the Velocity. 3*dly*, Supposing the Height to which the Body is raised to be the Measure of the Force, this Principle overturns the Conclusion he would establish by it, as well as that which he opposes. For, supposing the first Velocity of the Body to be still the same; the Height to which it rises will be increased if the Power of Gravity is diminished; and diminished, if the Power of Gravity is increased. Bodies descend slower at the Equator, and faster towards the Poles, as is found by Experiments made on Pendulums.* If then a Body is driven upwards at the Equator with a given Velocity, and the same Body is afterwards driven upwards at *Leipsick* with the same Velocity, the Height to which it rises in the former Case will be greater than in the latter; and therefore, according to his Reasoning, its Force was greater in the former Case; but the Velocity in both was the same; consequently the Force is not as the Square of the Velocity any more than as the Velocity. |

519 Sect. 8.

Reflections on this Controversy.

Upon the whole, I cannot but think the Controvertists on both Sides have had a very hard Task; the one to prove, by mathematical Reasoning and Experiment, what ought to be taken for granted, the other by the same means to prove what might be granted, making some Allowance for Impropriety of Expression, but can never be proved.

If some Mathematician should take it in his head Head to affirm, that the Velocity of a Body is not as the Space it passes over in a given Time, but as the Square of that Space; you might bring mathematical Arguments and Experiments to confute him; but you would never by these force him to yield, if he was ingenious in his Way; because you have no common Principles left you to argue from, and you differ from one another, not in a mathematical Proposition, but in a mathematical Definition.

Suppose a Philosopher has consider'd only that Measure of centripetal Force which is proportional to the Velocity generated by it in a given Time, and from this Measure deduces several Propositions. Another Philosopher in a distant Country, who has the same general Notion of centripetal Force, takes the Velocity generated by it, and the Quantity of Matter together, as the Measure of it. From this he deduces several Conclusions, that seem directly contrary to those of the other. Thereupon a serious Controversy is begun, whether centripetal Force be as the Velocity, or as the Velocity and Quantity of Matter taken
520 together. Much mathematical and | experimental Dust is raised; and yet neither Party can ever be brought to yield; for they are both in the right, only they have been unlucky in giving the same Name to different mathematical Conceptions. Had they distinguished these Measures of centripetal Force as *Newton* has done, calling the one *Vis centripetæ Quantitatis acceleratrix*, the other *Quantitas motrix*;* all Appearance of Contradiction had ceased, and their Propositions, which seem so contrary, had exactly tallied.

Part Three: Astronomy

I

7/VIII/10, 1r

Some observations concerning the Astronomical tables anexed to Whistons Astronomicæ Prælectiones, & his directions how to use them*

1° The third Column of the Table which contains the roots of the mean motions of the Earth & its Apsides has this tittle / Motus Perihelij Terræ ab Equinoctio verno / whereas it should I think be Motus Aphelij &c*

2° All the tables for the Sun agree præcisely with Sir Isaac Newtons Theory Save onely that Sir Isaac makes the Suns Apogeum* move foreward 4' 20" in twenty years; & therefore to make calculations by these tables agree with said Theory we must add to the motion of the Apogeum 4' 20" for each 20 years from the last day of December 1680, & proportionally for lesser spaces*

3° Mr Whiston in his Præcept. 5 for finding the Suns true place for the apparent time page 119 directs us first to find the absolute Æquation of time, then the motion of the Sun correspondent to that Æquation, & this motion he says is to be added to the Suns place for the mean time, if the Absolute Equation of time be positive & substracted from the said place if that Equation be negative; And when he illustrates this precept by an Example wherein the Absolute equation of time is 15' 59" negative he says that this Æquation subtracted from the Apparent time gives the mean time answering thereto. Accordingly in the Same example the Sun being in Scorpio, his Mean anomaly 4 Signs & some odd degrees, he makes his true place for a certain hour & minute of Mean time 40" forward of his true place at the Same hour & minute of Apparent time.*

Now in all this I think Mr Whiston is mistaken for the first Æquation of time is negative when the Sune is in the first Semicircle of his Anomaly & positive when he is in the Second, the Second Æquation of time is negative when the sun moves from an Æquinox to a Solstice
7/VIII/10, 1v & positive when he moves from a Solstice to an | Æquinox & therefore when the Absolute Æquation of time (which is made out of both these)

is negative it must be subducted from the meantime to give the apparent time & e contra

And from this it follows that when the Absolute Æquation of time is negative the Suns place for any given moment of mean time is behind his place for the same moment of apparent time & e contra

And if this be so Mr Whistons direction on this head is the revers of what it ought to be. Whiston is right & I was in the mistake

4º Page 142 Lemma 14* He is determining the middle time of Eclipses or of the nearest approximation of the centers & asserts that it happens neither at the time of full or new moon that is when a line joining the Centers of the Sun & Moon or of the Shaddow of the Earth & Moon is perpendicular to the Ecliptick nor yet precisely when a line joining the Said Centers is perpendicular to the plane of the Moon's orbit, this is all just: But what he adds either I do not understand or it is this that the precise middle time of an Ecclipse happens betwixt the two times & the two perpendiculars mentioned above, that is be twixt the time of new or full moon, & the time when a line joining the centers of the Sun & Moon or the Centers of the Earths Shaddow & Moon is perpendicular to the Moons orbit If this is his meaning I think it is a mistake for this reason that the perpendicular last named falls always betwixt the then nearest node & the place where the true Syzigie happens & therefore any Distance of the Centers betwixt these two must be longer th‹a›n that perpendicular as being at a greater distance from the node I judge there fore that the nearest distance of the Centers happens always in a place which is exceedingly near to the said perpendicular to the moons orbit but lies betwixt that perpendicular & the nearest Node, but that it never falls betwixt that perpendicular & the place of the Syzigie.*

II

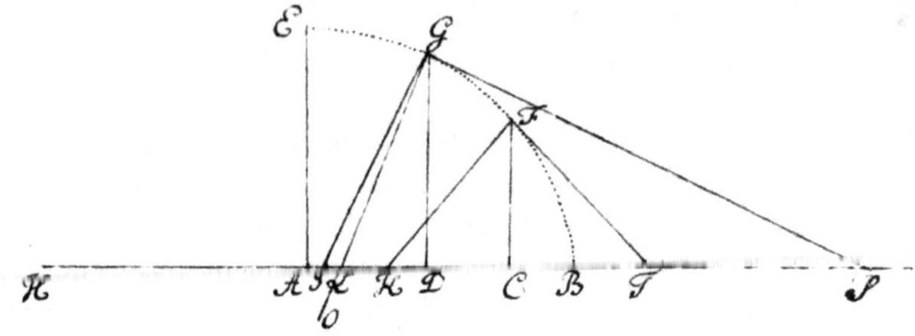

7/II/22, 1r NB. 1° A degree of Latitude by the French Mensuration in the middle of France is 57060 & therefor a degree at the polar Circle is 57437 toises*
2° The cube roots of these numbers are as 384988 to 385833 nearly and making this the ratio of FK to GL the ratio AB to AE comes out to be that of 230 to 228,386 nearly
3° The operation may be Shortned from this Property of the Elipse that AD is to LD as ABq is to AEq Whence taking JD a mean proportional between AD & LD and joining J&G, JG will be equal to AE; from whence these Canons are easily deduced $t/c=\sqrt{p-q/ss-rr}$ and $c=\sqrt{(p+rrtt)/cc}$
4° Having made a Calculation from the first of these Canons and another from the Canon in the letter the result of both was the same namely t:c::230:228,386 and by the other of these Canons I found the Semi-Axis of the Earth to be 3255854 toises and its Equatorial Semidiameter 3278858 toises*
5. I tried the Justice of this last Conclusion upon quite different Principles thus. Let GO be the Radius of a Circle Equi-Curve to the Elipse at G and Let $GL/GD = k$ then $GL = GO - GO/kk \times tt-cc/cc$; which may be easily demonstrated from the nature of the Elipse; taking then the ratio of t to c to be that above mentioned and GO equal to the Radius of a Circle of which one Degree is 57437 toises I found the value of GL calculate from this Equation to differ from that in the other Calculation hardly an unite in the Sixth figure which small Difference might arise from the fractions neglected in the operation

III

2/I/1, 2r

	h	'
Beginning	8	56
Upper Vertical point of the suns Limb covered	9	17
Six Digits Eclipsed	9	40
Beginning of the Annular Appearance	10	19 25
Central to Appearance. but a Glimpse	10	23
Half the Moons limb within the Suns	10	38
Lowest part of the Suns limb emerges	11	1
Six digits eclipsed	11	4
End	11	50

All apparent time at Manse of New Machar being 7' of a degree of Latitude North from aberdeen & 3' of a Degree Longitude west of Said place‹.› Latitude I take to be 57° 19' Latitude west of London 1°49' east of Edinburgh 1°–11'*

IV

7/II/20, 1r The greatest Aberration of the fixed Stars from their true place being according to Dr Bradleys Observations* 20¼" whose Log Tangent is 5.9919689, The Log of the Ratio of that Tangent to the Radius will be 4.0080311 the log of 10185,9 therefore the Mean Velocity of the Earth is to the Velocity of Light as 1:10185,9

If we suppose the Diameter of a Circle to be to its Circumference as 113:355 & that the Earth goes round its Orbit in 365¼ days it will perform the length of the Diameter of its Orbit in 116,263 days. Which Number of Days divided by 10185,9 will give ,0011417 days which turned into Minutes & Seconds makes 16'26" the Time that Light takes to move through a Diameter of the Earths Orbit.*

V

7/VIII/3, 1r

Of the Aberration of the fixed Stars

In the Eccliptick there is no Aberration in Latitude, and the greatest aberration in Longitude, when the Star is in conjunction with or in opposition to the Sun is according to Dr Bevis (see T. Simpson's Essays) is 20", 25.* In other distances of the Suns place from the Star, the aberration in Longitude is as the Sine of the Suns distance from its Quadrature with the Star, to be added in the Semicircle wherein the Sun and Star are opposed, that is when their distance is greater than a quadrant and Subtracted when less.*

The greatest aberration in Longitude increases with the Stars latitude, and is as the Cosine of the Latitude inversely. So that in 60° Latitude the Aberration in Longitude at its Maximum is 40",5

The Maximum of aberration in Longitude is always when the Sun is in Conjunction with the Star, when the Aberration is to be subtracted; and when in Opposition with the Star, in which case it is to be added and in other aspects it is to the Maximum of the Sine of the Suns Distance from Quadrature with the Star to Radius.

The greatest aberration in Latitude is in the Pole of the Eccliptic where it is 20".25 in other Latitudes the Maximum of aberration in Latitude is to 20".25 as the Sine of Latitude of the Star to Radius.

When a Star has its greatest Aberration in Longitude it has none in Latitude and when its aberration in Latitude is greatest it has no aberration in Longitude. And the Aberration in Latitude is to the Maximum of that stars aberration in latitude as the Sine of the Suns Distance from Syzigy, that is from conjunction or opposition with the Star is to Radius.

In order to form a clear conception of this aberration Suppose a small Ellipsis described about every Star as a Center. whose longer Axis is 40",5 and shorter Axis to the longer as the Sine of the Stars latitude to Radius The Star will to appearance be always seen in the Orbit of this Ellipsis, & will go through the whole Orbit of the Elipsis while the Sun goes through the Zodiac. Conceive a Great Circle passing through the Sun & the Center of the Elipse a small part of which makes the Diameter of the Elipse. Conceive also the Diameter conjugate to this to be drawn, and the Star will always appear in one extremity of this Conjugate diameter. Conceive also a Colure passing from the pole of the Eccliptic to the center of the Ellipse, the stars true place; Another to that point in the orbit of the Ellipse which is the Stars apparent place; and conceive a Semidiameter of the Elipse joyning these two Colures. We have then a Spherical Triangle whose vertical Angle at the Pole is the aberration of the Star in longitude & the Difference of the Legs is the aberration in latitude. |

7/VIII/3, 1v In order to conceive the aberration in Right ascension and Declination Conceive a Meridian passing from the Pole of the Earth to the Center of the Elipse the Stars true place, another from the pole to that point in the Orbit of the Elipse which is the Stars apparent place, let these be the legs of a Spherical triangle whose base is that Semidiameter of the Elipse which passes from the true to the apparent place of the Star. Now the angle made by the legs of this triangle at the Pole will be the aberration in Right Ascension and the difference of the legs will be the aberration in declination.

VI

2/I/7, 1r

An Observation of the Transit of Venus June 6th 1761 Made at Kings College Aberdeen*

The Latitude of Kings College from several meridional Altitudes of the Sun taken at different Times I make to be 57° 10' 48". As to Its Longitude I have onely the Authority of Maps which make it about seven minutes in time West of London*

The clock by which the time of the Observations was marked was set to apparent Time by a Meridian and verified by several Altitudes of the Sun taken before and after the transit so that I think the times of Observation may be depended upon within half a Minute.

The Observations were made with a refracting Telescope of nine feet, having a Micrometer fitted to it, each of whose divisions, by many trials was found to subtend an angle of 1'.40". The Sun's Image was received through the Telescope upon a sheet of Paper fixed on a Board

By reason of Clouds we had onely two observations of the Distance of Venus from the nearest part of the Suns Limb. which were marked upon the spot before several witnesses and particularly three or four Members of this Society as follows.*

At 4 hours 2' 12" The Center of Venus was distant from the Suns Limb three Divisions of the Micrometer that is 300"

At 6 hours 44' 36" The Center of Venus was distant from the Suns Limb two divisions and a half of the Micrometer that is 250"

At 8 hours 26' the Transit was quite over, but we did not see the end of it for clouds* |

2/I/7, 1v

Remarks

Although these observations can signify little for determining the Paralax of the Sun. yet we may draw some conclusions from them concerning the Theory of Venus and the Longitude of the Place

In order to this we shall borrow some principles from Halleys Tables*

The Logarithm of the Suns Distance at the time of the Transit I take to be 5.006689. The Logarithm of Venus Distance from the Sun 4.8611914. From these Data taken from Halleys tables we may compute the horary motion both of the Sun, and of Venus as seen from the Sun.* For it is known to Astronomers that the Suns mean horary Motion is to his horary Motion at the time of the transit as the Square of his distance at that time is to the Square of his mean Distance and the Same thing is applicable to Venus. By this Computation I make the Suns horary Motion at the time of the Transit to be 143",35 one hundred and forty three seconds and thirty five hundreds parts of a second. Venus her horary Motion seen from the Sun comes to be 238",23.

Now the Motion of Venus as seen from the Sun will be to her Motion as seen from the Earth as her Distance from the Earth is to her Distance from the Sun

Therefore I make her horary Motion as seen from the Earth to be 598",71 Seconds |

2/I/7, 2r Now let us take one hours Motion of the Sun from the Node of Venus, for one side of a triangle to wit 143",35, one hour's Motion of Venus in her Orbit for another side of our Triangle to wit 238,23. And let the angle included between these two Sides be the Inclination of Venus Orbit to the Eccliptic which by Halleys Tables is 3° 23' 20"*

The Base of this Triangle will be the horary Motion of Venus from the Sun and the angle which this base makes with the Eccliptic is the angle of her apparent Path from the Sun. Hence we find Her Horary Motion from the Sun to be 95",49 Seconds This is the Quantity of it as it appears from the Sun, but as it is seen from the Earth it will be greater in proportion as the distance of Venus from the Sun is greater than her distance from the Earth. And by Computation the Quantity of her horary Motion from the Sun as seen from the Earth is 240". And the Angle which her apparent Path makes with the Eccliptic comes to be 8° 28' 45"

To apply these Principles to the Observations above related. Let us take the Suns apparent Semidiameter at the time of the transit from Halleys Tables to have been 950,6 Seconds,* we shall find that the distance of Venus from the Suns Center at the first observation was 650",6 and at the second 700".6. These distances of Venus from the Suns Center make two sides of a triangle, whose base is the visible path of Venus described in the interval between the Observations that is in two 2/I/7, 2v hours forty two minutes and twenty four Seconds | and since we have shewn that in this path she describes four minutes or 240" every hour it is easy to collect that in 2 hours 42' 24" she must have described 648",16 which is therefore the Base of our Triangle; of which having now the three sides given we find by plain trigonometry the perpendicular to be 591",04 and the segment adjacent to the least Side 271.96. which segment venus will pass over in one hour and eight minutes. So that if we suppose no parallax of the Sun nor of Venus the nearest approach of her Center to that of the Sun was 591",04 and the time one hour & 8 minutes after the first Observation that is at 5 hours 10 Minutes and twelve seconds

We may likewise find that the angle which the lesser side of our triangle makes with the Path of Venus is 65°.17'.30" and the angle at the great Side is 57° 31' 22"

Let us now consider what influence the paralax will have upon this Triangle. And I find by a Computation which those who understand the Doctrine of paralaxes may easily examine that supposing the horizontal Paralax of the Sun to be 10"½. The base of our Triangle has been shortned by the parallax 3",68 seconds the lesser Side lengthned 25",92 and the greater lengthned 19".42. Therefore the two sides of the triangle seen from the Center of the Earth would have been 624,68 & 681,18 but the Base seen from the Center is what we before supposed it. and seen from the Surface was shortned. And we shall now find the segment of the base adjacent to the Shorter side to be 287",16 which will be run over in 1 hour 6' & 47" So that the time of the nearest approximation is 5 hours 8' 59" and the nearest approach of Venus to the Suns Center is 564",68. |

2/I/7, 3r The manner in which I make Allowances for the paralax of the Sun and Venus in this Observation is explained by the following Figure

Let C be the Suns Center A the place of Venus on the Suns Disc at 4 hours 2' 12" as seen from the Center of the Earth. B the place of Venus at 6 hours 44' 36" as seen from the Center of the Earth. a, & b the places of Venus at the same times as seen from the Kings College.

Suppose the Triangle ABC compleated and the Perpendicular to its Base CD, will be the nearest approximation of the Centers as seen from the Center of the Earth. & AD the space that venus has moved in her path from the Sun from the time of the first Observation to the nearest approximation

In order to find out the parts of this triangle by projection (or rather by calculation which I used as more accurate) we have given AB=648",16 We have the Angle which AB makes with the Eccliptic already found to be 8° 28' 45" hence we may find by spherical Trigonometry the angles which AB makes with the two vertical circles Aa, and Bb; I make the angle BAa=103°.33' and the Angle ABb=68°.9'

We are next to find the Quantity of Aa, & Bb and this will be greater or less according as the Suns parallax is greater or less. I suppose the suns horizontal parallax to be 10",5 then the horizontal Parallax of
2/I/7, 3v Venus from the Sun | will be 26",4. Now as the Radius is to the Sine of Venus distance from the Zenith at A so is the Horizontal Paralax of Venus from the Sun to her Paralax from the Sun at A. Which I make to be 26",39, and this is the length of Aa

From the Same principles I find Bb or the parallax of Venus from the Sun at B to be 24".

Having therefore found the Angles BAa, ABb and likewise the lines Aa, & Bb we have the points b, & a, and consequently the line ba

But the lines Ca & Cb were found by observation viz Ca=650",6. & Cb=700",6. Hence we may find the point C and consequently the Sides CA, CB, the Perpendicular CD which is the Nearest approximation of the Centers, the Segment of the Base AD, which being turned into time at the rate of 4' or 240" to an hour gives the Interval betwixt the first observation and the Middle of the transit as seen from the Sun

We may likewise find Dd the parallax of Venus from the Sun at the Middle of the Transit which will give us Dδ the increase of the nearest approximation owing to the parallax.

If the Node of Venus be justly placed by Halleys Tables there must be an error of 6' 2" in the inclination of her Orbit to the Eccliptic so that her inclination should be 3.29' 22" but whether the error be in the place of the Node or in the inclination of the Orbit or partly in the one and partly in the other must be determined by future observations.* |

2/I/7, 4r

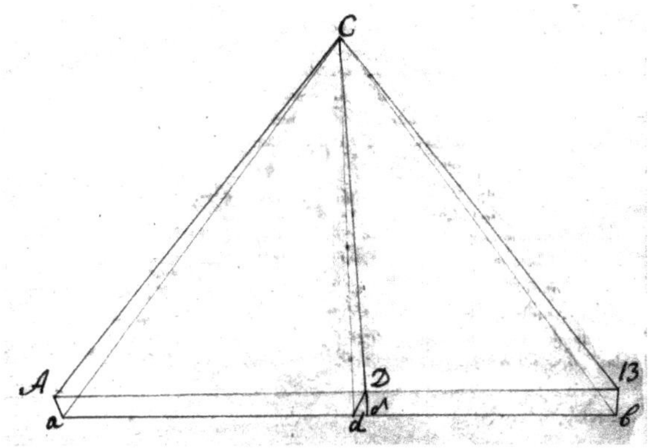

2/I/7, 4v Addition. August

By Observations made in other places it appears that the interval betwixt the external and internal Contact was at a Medium 18' 10" hence the Diameter of Venus will be about 59".* If it is so the Intervall betwixt the nearest approach of the Centers & internal Contact must be 3 hours 2' 1" and the Interval betwixt the middle and external Contact 3 hours 20' 11"

Therefore the Internal Contact at Kings College Must have been 8 hours 11' apparent time & the external Contact 8 hours 29'.10" as seen from the Center

Among the observations made in France that of Mr Ferner Professor of Astronomy at Upsal is nearly a medium and seems to be most distinct and Accurate.* As to the Latitude of Venus it agrees with ours within 2" and within a few seconds as to the time of the nearest Approach allowing for the difference of Longitude. The greatest difference betwixt the Observations made in France Aberdeen Magazine Aug 1761 viz betwixt Maraldi & de la Lande as to the internal Contact is 37".* De la Landes observation differs from his Calculation Above half an hour his observation being so much earlier. He appears to have erred in the Latitude about 22" in his calculation.

The Transit as seen from Kings College seems to have ended 8,27,32. It is not strange that we did not see it at 8 hours 26' since at that time the gap in the Suns Limb would not be 5" |

2/I/7, 5r We shall next consider how the Time and Quantity of the nearest approximation of the Centers of Venus and the Sun comes out by Halleys Tables

By a Computation from those Tables I find that forty one minutes after five in the Morning of the sixth of June new Stile and Mean Time The Sun and Venus are equally distant from the descending Node of Venus. The place of the Sun being at that time II Gemini 15° 35' 42" 56''' The place of Venus in her Orbit Sagittarius 15° 35' 43" and the place of the ascending Node II Gemini 14° 29' 37" So that the distance both of the Sun and of Venus from the descending Node is 1° 6' 6". The distance of Venus from the Center of the Sun is a small Arch which seen from the Sun makes 243",52 Seconds but seen from the Earth it is 598",39

We must conceive this line as seen from the Earth to be the Hypothenuse of a Right angled Triangle, one side of which is that part of the visible path of venus from the Sun, which she hath passed over from the time of her nearest approximation, to the time above mentioned viz 5 hours 41' mean time; The other side of this right angled Triangle is the quantity of her nearest approximation.

This Hypothenuse makes an angle with Venus Orbit of 88° 18' 20" but with her visible or apparent path from the Sun it makes an angle of 83° 12' 55" as may be easily inferred from what has been said above of
2/I/7, 5v the Angle of her Apparent Path. with the Eccliptic | Having therefore the Hypothenuse of this Triangle viz 589",39 & the angle at the Base 83° 12' 55" we find the Perpendicular to be[1] 585",26 which is the nearest approach of the Centers by the Tables. And the Base of our Triangle comes to be 69",63. The time required to pass over this Base at the rate of four minutes to an hour is 17' 14", which being taken from 5 hours 41' gives 5 hours 23' 46" for the mean time of the nearest approximation and if we add to this the Equation of time which is one minute fifty three seconds we have[2] 5 hours 25' 39" for the apparent Time of the nearest approximation at Greenich.

If we subtract from this the apparent time by our observation at Kings College the Difference is 16' 40". By our common Maps Aberdeen is one degree forty five minutes of Longitude west of Greenich.* This

[1] Ferguson makes it 583"*

[2] Ferguson makes it 5 hours 24' probably taking the mean time & not the apparent.*

answers to seven Minutes in time which being taken of‹f› leaves 9' 40" by which the observation differs from the Tables.

This difference of the Tables from the Observation cannot be owing to the paralax of the Sun. For it hath been shewn above that if there was no parallax at all the middle of the transit would have been little more than one minute later and if the parallax be greater than we have supposed this would make the middle of the transit earlier. Therefore we must ascribe this difference of 9' 40" between the observation and the
2/I/7, 6r tables to one of these three causes, either to an error of | the observation or to an error in the maps with regard to the Longitude of the Place or to an Error in the tables

I have already mentioned the care taken to make the Observations exact & shall say no more upon that subject. As to the Maps if this Difference was owing to them they must err two degrees and near 18' in the Longitude of Aberdeen which is not probable; but if the error lies in the Maps we shall be able to rectify it when we have accounts of the observations made of this transit in other places particularly at London.*

As to the Tables I think it is much more to be wondred that they should exhibit this transit so accurately than that there should be an error of so small magnitude. But perhaps the greatest part of this Error is to be attributed ‹to› an Equation arising from the progressive motion of Light, which Experience hath taught us to apply to the Eclipses of Jupiter & Saturns Satellites, but it is equally applicable to all the Heavenly bodies which are some times near and sometimes far off from us, although hitherto neglected*

Let us therefore suppose that the Tables exhibit the true place of Venus as she appears in the heavens at her middle distance from the Earth which is the most favourable supposition for the Tables that can possibly be made. It necessar‹i›ly follows from this that these tables at
2/I/7, 6v the transit must give her place more backward | than it appears in the heavens and the Error of the Tables must be equal to the time that Light takes to move through a Semidiameter of Venus Orbit. which I take to be 5' 48". Taking off this there remains 3' 52" not accounted for but if we suppose that the tables exhibit the place of Venus as she is seen in the heavens when near her greatest distance from the Earth. The whole 9' 40" may be imputed to the progressive motion of Light. However it be this 9' 40" ought to be taken from the time of the next transit by the tables.* if there is no error of our Longitude

Having considered how far our observation differs from the tables with regard to the Longitude of Venus in her Orbit it remains onely that we consider how far they differ as to her Latitude. The tables making the nearest approach to be 585",26. The observation makes it 564",68. So that it is 20",58 nearer by observation than the tables make it if the Suns paralax be 10",5 but if there is no paralax the distance of the Centers will be 5,78 greater by the observation than the tables make it. There is therefore probably a small error of the Tables with regard to the Latitude of Venus at the time of the Transit. This error will be greater than we have made it if the suns parallax be greater and less if the suns Parallax be less than 10",5 and for every second we add to or take from the Suns paralax we must increase or diminish this error 2",5.

Taking it therefore for granted that the suns parallax is 10"½ till 2/I/7, 7r we know more certainly from the observations of this transit | the real Quantity of it we shall consider how far this error affects the Tables and what alteration it requires. 20.58 seen from the Earth make onely 8",19 seen from the Sun.

We may therefore conceive a small Triangle whereof one side reaching from the Eccliptic to Venus Orbit is 8",19. another side is part of venus Orbit, and a third is part of the Eccliptic. the first side makes with Venus Orbit an angle of 95° 5' 25", the Eccliptic makes with her Orbit an angle of 3° 23' 20". Hence we may find that side of the triangle which represents part of the Eccliptic to be 137",6 that ‹is› 2'.18". And so much do the tables make Venus Node more forward than it really was at the time of the transit. If the paralax of the Sun be greater than 10",5 this Error will be greater if less it will be less.

Upon the whole it appears that if our Longitude from London be what the maps represent it 7' west, and if our observation be found to correspond with those made in other parts we must in calculating the next transit of Venus by halleys tables add 38" to her Longitude in her Orbit or take off 9' 40" from the time of the Middle of it which the tables give us. And if mor‹e›over the suns parallax be 10",5 we must take off 2' 18" from the place of her node at the Transit.

I shall onely further observe that the Effect of the Parallax in this Transit behoved to be very remarkable at Aberdeen in increasing the 2/I/7, 7v least Distance of the Centers | of the Sun and Planet, as it appears from what hath been already said that for every second of the Suns parallax, the least Distance of the centers of the Sun & Venus must have been increased two and a half Seconds

But in the next transit of Venus and the onely one that can be seen by any person now alive, which begins June 3d 1769 about Eight at Night;* The effect of the parallax at Aberdeen will be very remarkable in forwarding the time of the beginning of the transit. In so much that for Every second and a half of the Suns parallax the beginning of the transit will be accelerated at aberdeen one minute in time. But I cannot help thinking that the Method proposed by Doctor Halley, is by far the most proper for determining the Suns parallax by these transits Namely to observe the duration of it in two opposite Meridians; in one place about 23° North Latitude & where the middle of the transit will be at midday. and in the opposite meridian where it will happen at Midnight but so far North as that the beginning may be seen before sun setting and the End After Sun rising. If this method be taken at the next transit we may at last know the Distance of the Distance of the Sun without erring above an 800th part of the whole.*

Supposing an error of 5" that is a twentieth part of one Division of the Micrometer in both the observations of the Distance of Venus from the Suns Limb, which I think is the greatest that can reasonably be supposed, & supposing also an error of half a minute of time in our Meridian; There might happen by these suppositions an Error of 3' of time as to the time of the nearest approximation; if the errors all leaned the same way. And the same Causes concurring might produce an error of three or four Seconds as to the Quantity of the nearest approximation |

2/1/7, 8r By the Observations made at the Royal Observatory and at Sherburn house, and the calculations founded upon them taking a medium of the observations which agreed best; the middle of the transit at the Royal Observatory was 5 hours 20'.5" the internal Contact was 8.19' that is 8' later than at Kings College* Whether there is all this difference of Longitude betwixt the two places or whither My Observation may not have erred a Minute as to the time I cannot tell. The nearest Approximation of the centers as seen from the center of the Earth agrees to a part of a second with the Observations at London and in france

1764 As the parallax of the Sun from a medium of the observations of the late transit of Venus is found to be 8" ½ the error which is observed above in halleys tables with regard to the place of the Node will be lessned one fifth and therefore will not be 2' 18 but 1' 52"*

VII

7/III/14, 1r Scholium ad Propositionem 26 Liber 3 Principia Newtoni
vide figuram apud Authorem*

Sit S vis accelerativa Terræ versus Solem, vel Lunæ in mediocri sua distantia; T vis acceleratrix gravitatis Lunæ versus terram. Et L designet vim LT qua lunæ motus circa terram a vi Solis pertubatur. Resolvitur vis L seu LT in vires duas LE. ET quarum prima in directione Orbitæ agens Orbitatis dice potest & designetur per O

Secunda Centrales est quam designamus litera C.

Quoniam PT : TK : : PL(3PK) : LE erit recta LE = 3PK × TK/PT; Recta PE = 3PKq/PT et Recta ET = PT − 3PKq/PT eritaquæ ET positiva vel negativa sicut PT major est vel minor recta‹e› PE.

Sunt vero vires S, O, C rectis ST, (distantiæ scilicet Soles a Terra) LE, TE, proportionales Igitur C : S : : PT − 3PKq / PT : ST. Et C = PTq − 3PKq × S/PT × ST.

Sit Tempus Periodicum Terræ circa Solem P, Lunæ circa Terram π et Quoniam C : S : : PTq − 3PKq : ST × PT Est vero S : T : : ST/P^2 : PT/π^2 erit ex Æquo C : T : : PTq − 3PKq × ST/P^2 : ST × PT × PT/π^2 : PTq − 3PKq × π^2 : PTq × P^2 : ⅓PTq − PKq × π^2 : ⅓PT × P^2

Rectæ ET quantitas in Quadraturis est = +PT in Syzigijs est −2 PT, mediocris igitur quantitas −½PT. Rectæ LE quantitas maxima est in octantibus ‹scilicet?› 3/2 PT mediocris igitur ¾PT* |

7/III/14, 1v By Proposition 6 Tract 4 Stewarts Tracts S : C : : ⅓ST − PK‹?› × PT : PTq/3 − PKq : : ST ± 3PK × PT : PTq − 3PKq whereas according to Proposition 26 Liber 3 Principia S : C : ST × PT : PTq − 3PKq.*

By Proposition 7 Tract 4 T : C : : PTq × P^2 : PTq − 3PKq × π^2 in this he agrees with Newton.* By proposition 8 Tract 4 the right line LN which answers to ET. Newton. will be equal to PTq − 3PKg / TP × STq / STq − gPKq whereas according to Newton it is Simply PTq − 3PKq / TP*

By the 6 Proposition of the tracts C = PTq − 3PKq / PT × S / ST ± 3PK.* Now it may be observed first that this value of C is not more but less accurate than that which Sir Isaac Newton gives. And secondly that if the distance of the Sun be doubled or increased in any other Ratio the Force C may notwithstanding be the same as before. All that is necessary for this purpose is that the fraction S / ST ± 3PK be of the same value as before which will happen if when ST is increased or diminished

S be so increased or diminished at the Same time as that ST ± 3PK shall alway‹s› have the same ratio to S

In proposition 21 Tract 4.* The Author proposes to demonstrate that if the Ratio of the Moons force towards the Earth to the mean disturbing force of the moon upon the Sun be given; the ratio of the moons distance to the suns distance will be given. In the Demonstration it is said that
7/III/14, 2r because the Rectangle KTF | is given therefore KT is given. and the 16 proposition is quoted for this.* Now I see nothing in proposition 16 from which this can be inferred if KT is given because KT × TF is given this must be because TF is given but I do not see how TF is given |

7/III/14, 5r

Observation upon Corollaries 14 & 15 Proposition 66 Liber 1 Principia Newtoni. see the figure in the Book*

If the body S, be very distant and the Orbit PAD nearly circular; then ST and SN may be conceived equal; SL & SM parallel; PK = ½KL; & consequently NM = 3PK. Supposing these things LM will always be equal to PT, and MN when greatest will be equal to 3PT, and in all cases MM will be equal to thrice the Sine of the Arch PC by which the body L is distant from the nearest Quadrature

And hence it appears that if the force LM be increased or diminished the force MN, in any given aspect of the body P, will be increased or diminished in the same Ratio. And if the Force LM, the Radius PT, & the aspect of the body P with regard to S be given the force MN will also be given. The force MN varying from nothing to 3PT; its mi‹d›dle Quantity will be 3PT/2, and it will have this quantity when P is thirty degrees from its quadrature with S; and at 41° 48' from the Quadrature it will be equal to the Force LM

Let the Force SK be the accelerative force by which the Sun at S attracts the Earth at T, or the moon at her middle distance SK, & call this Force S Let the force LM be called L. then upon the suppositions made above S : L : : ST : PT therefore L=S × PT/ST

Corollary Hence it follows that if the product S × PT be increased or diminished, and at the same time ST be always increased or diminished
7/III/14, 5v in the same Ratio; the disturbing force L will remain | the same

Corollary 2d If PT the distance of the moon from the Earth be a fixed quantity and the Force S, and distance of the Sun ST, be both increased or diminished in the Same Ratio, the disturbing force L will remain the Same.

Corollary 3d Since the Paralax of the Sun is reciprocally as ST, let that paralax be called P; and because L = S × PT / ST therefore L will be as S × P × PT. Consequently when PT is fixed the disturbing Force L may continue the same although the Paralax of the Sun should be doubled or tripled or decupled, providing the Suns force S is diminished in the same ratio as the paralax is increased.

Corollary 4 Hence it seems to follow that the Paralax of the Sun cannot be determined from the Quantity of the Force L nor consequently from the mean Quantity of the force MN which is always equal to 3L/2. that is, it cannot be determined from the whole disturbing force of the Sun upon the Moon. And since the Motion of the Moons Nodes, Apogea, &c must be proportional to the causes that produce them, the
7/III/14, 6r parallax of the Sun cannot be determined from | those Motions.

Corollary 5 The onely inequalities of the Moons Motion which seem to depend upon the distance or parallax of the Sun, are those which Sir I Newton takes notice of Corollary 12 Proposition 66 Liber 1. which arise from the difference betwixt the force MN in the conjunction of the moon, and in her opposition, but these are so small that neither in Newtons Theory of the moon nor in halley's tables is any allowance made for them.*

Corollary 6 The Force MN may be resolved into two forces one of which is directed from the center of the Orbit T, the other touches the Orbit. The first will be to the force MN as the Sine of the moons distance from the Quadrature, to the radius. And the second will be to the force MN as the cosine of the Arch above mentioned to the Radius. Therefore let the Sine of that Arch be x, its cosine y the radius unity Then the force MN = 3x × PT which will be resolved in to the two forces $3x^2$ × PT, & 3xy × PT. The first of these diminishes the centripetal force of the moon; the second diminishes her velocity in her Orbit in the 1 & 3 quarter & equally increases it in the 2d & 4th |

7/III/14, 6v Corollary 7 Hence the whole disturbing force of the Sun upon the Moon consists of these two viz L−$3x^2$ × PT which is directed to or from the center of the Orbit according as it is positive or negative; and the force 3xy × PT which is directed in a tangent to the Orbit

Corollary 8 The for‹c›e L−$3x^2$ × PT will be equal to nothing when $3x^2$ is equal to unity or when x^2 = ⅓ that is when the moon is distant from the Quadrature 35° 16'

Corollary 9 Since L = S × PT/ST the whole disturbing force as far as it is directed to or from the Center will be S$3x^2$ST × PT/ST

Corollary 10 Let T be the force of ‹the› Gravitation of the Moon towards the Earth π her periodical time and P the periodical time of the Earth round the Sun then $S:T::ST/P^2:PT/\pi^2$ consequently $S = T \times ST \times \pi^2/PT \times P^2$ substitute this value in place of S in Corollary 9 and the central disturbing force upon the Moon will be $T \times \pi^2/P^2 - 3x^2 \times PT$ and Since PT in this Expression is in reality the force L the central disturbing force will be $T \times \pi^2/P^2 \times 1 - 3x^2$ |

7/III/14, 7r Of the disturbing force of the Sun upon the Moon

1 Let D be the Suns Distance from the Earth d the difference betwixt the Earths distance from the Sun and the moons distance from the Sun then $D - d$ will be the distance of the Moon from the Sun in the first and last Quarter and $D + d$ will be the distance of the Moon from the Sun in the 2d & third Quarter

2 If the accelerative force of the Earth towards the Sun be called F and that of the moon towards the Sun be called Φ then $F:\Phi::D \pm d^2:D^2$ therefore $F = \Phi \pm 2d\Phi/D + d^2\Phi/D^2$ consequently $\Phi = F/1 \pm 2d/D + d^2/D^2 = F \pm 2dF/D + 3d^2F/D^2 \pm 4d^3F/D^3 \pm$ &c Now the force F being taken from this leaves the disturbing force of the Sun upon the moon, the quantity of it being $\pm 2dF/D + 3d^2F/D^2 \pm 4d^3F/D^3 + 5d^4F/D^4 \pm 6d^5F/D^5$ &c The negative Signs are to be taken when the moon is in her second and third quarter, and in the first & fourth the signs are all positive.

3 Let D the suns distance from the Earth represent the accelerative force of the Earth to the Sun‹.› Then $F:\Phi::D:\Phi D/F$ therefore $\Phi D/F$ will represent the accelerative force of the moon to the Sun, and (see figure of Proposition 66 Liber 1 Principia)* $\Phi D/F \pm D = KL$ likewise $\Phi D/F + d \sim D = PL$ consequently $PL = \pm 3d + 3d^2/D \pm 4d^3/D^2 + 5d^4/D^3$ &c |

7/III/14, 7v 4 Let the force SL be resolved into the two forces ST & TL of which ST affecting equally the Moon and the Earth can give no disturbance to the motion of the moon round the Earth. Therefore The force LT is the whole disturbing force of the Sun upon the Moon‹.› Again let this disturbing Force LT be resolved into the two forces LE & ET (see Figure of Proposition 26 Liber 3 Principia)* the first of these being in the direction of the Moons motion in the second and last quarter and in a contrary direction in the first and third quarters, alternately accelerates and retards her motion and makes her path more streight or more curve. The other being directed towards the center at the quadratures and from the center at the Syzigies increases her centripetal force in the former case and diminishes it in the lat‹t›er, but diminishes it in the latter case about three times as much ‹as› it is increased in the former and therefore in a whole revolution of the moon diminishes the centripetal force

5 In the triangle PEL (see the figure quoted in last article) right angled at E. the Angle EPL = ATP + TSP in the first and last Quarter and in the
7/III/14, 8r Second & third Quarter | EPL = BTP − BSP. Let the Sine of this angle be called s & its cosine c the radius r then $r : s :: PL : PL \times s/r = LE$ and $r : c :: PL : PL \times c/r = PE$ and if the distance of the moon from the Earth be called δ, then towards the Sy‹z›igies $PE - \delta$ and towards the Quadratures $\delta - PE$ will be equal to TE the disturbing central force

6 In Article 3 we considered PL as positive in the first and last quarter of the moon and negative in the second and third: but if we consider it as positive in both cases; it will in the first and last quarter be $3d + 3d^2/D + 4d^3/D^2 + 5d^4/D^3$ &c and in the second & third it will be $3d - 3d^2/D + 4d^3/D^2 - 5d^4/D^3 +$ &c therefore by the 5 Article LE in the first and last quarter will be $= s/r \times 3d + 3d^2/D$ &c and in the second and third LE will be equal to $s/r \times 3d - 3d^2/D$ &c consequently half the Sum of the disturbing forces LE in two opposite points of the moons orbit will be $s/r \times 3d + 4d^3/D^2 + 6d^5/D^4 + 8d^7/D^6$ &c and this may be considered as the disturbing force LE by which the moon is accelerated in the second & last quarter and retarded in the first and the third. More over the difference between the disturbing forces LE in opposite parts of the moons Orbit is $= s/r \times 6d^2/D + 10d^4/D^3 + 14d^6/D^5$ &c

7 From the 5 Article it appears also that PE in the first and last quarter $= c/r \times 3d + 3d^2/D + 4d^3/D^2$ &c and in the second & third $PE = c/r \times$
7/III/14, 8v $3d - 3d^2/D + 4d^3/D^2 - 5d^4/D^3$ &c | therefore half the sum of the right lines PE in two opposite points of the moons orbit is $= c/r \times 3d + 4d^3/D^2 + 6d^5/D^4 + 8d^7/D^6$ &c and the difference of the right lines PE in opposite points of the moons orbit is $c/r \times 6d^2/D + 10d^4/D^3 + 14d^6/D^5$ &c

8 Hence it appears that the half Sum of the central disturbing forces in opposite points of the moons orbit is $c/r \times 3d + 4d^3/D^2 + 6d^5/D^4$ &c$-\delta$. So that when δ is equal to the Series then the disturbing central force is nothing: When the series is greater than δ, which happen‹s› towards the syzigies then the central disturbing force is directed from the center of the Earth: And when the Series is less than δ then the central disturbing force changes its sign and is directed towards the center of the Earth. At the quadratures $c = 0$ and likewise $d = 0$ therefore the Series vanishes and $TE = -\delta$. At the Syzigies $c = r$ & $\delta = d$ therefore the force TE is in a contrary direction and is $= 2d + 4d^3/D^2 + 6d^5/D^4$ &c And the

arithmetical mean between its greatest positive and its greatest negative magnitude it will be $\frac{1}{2}\delta + 2\delta^3/D^2 + 3\delta^5/D^4$ + &c This I take to be what Dr Stewart calls the mean disturbing force of the Sun upon the Moon.* and the series last mentioned expresses the Quantity of it supposing the accelerative force of the Earth to the Sun to be expressed by D.

9 If the Accelerative force of the Earth to the Sun be called F as before;
7/III/14, 9r the accelerative force of the Moon towards the Earth f and | let the periodical Time of the Earth round the Sun be P, the periodical Time of the Moon round the Earth be π then $F:f::D/p^2:\delta/\pi^2::D:\delta P^2/\pi^2$ consequently supposing D to represent the force of the Earth to the Sun $\delta P^2/\pi^2$ will represent the force of the Moon to the Earth. Therefore the three forces of the Earth to the Sun, The moon to the Earth, and the mean disturbing force above mentioned will be respectively as D, $\delta P^2/\pi^2$ and $\frac{1}{2}\delta + 2\delta^3/D^2 + 3\delta^5/D^4$ + &c Let this Mean disturbing force be M. and let the ratio of f to M be given so that M/f is equal to a given quantity A, then $Af = M$ that is $\delta P^3 A/\pi^2 = \frac{1}{2}\delta + 2\delta^3/D^2 + 3\delta^5/D^4$ &c

10 If D is supposed to be the onely unknown quantity in this Equation let us consider whether its value may be found by the resolution of the Equation

1 Suppose that all the terms after the Second are so small that they may be neglected the Equation reduced will be $\delta P^2 A - \frac{1}{2}\delta\pi^2 \times D^2 = 2\delta^3\pi^2$ consequently $D^2 = 4\delta^3\pi^2/2\delta P^2 A - \delta\pi^2 = 4\delta^2\pi^2/2P^2A - \pi^2$ Therefore $D = 2\delta\pi/\sqrt{2P^2A - \pi^2}$ |

7/III/14, 10r 11 From the 9 Article it appears that if D be the force of the Earth to the Sun, & if the periodical time of the moon be called Unity & her distance from the Earth be also Unity then the three forces mentioned in that Section will be D, P^2, & $\frac{1}{2} + 2/D^2 + 3/D^4 + 4/D^6$ &c Let the last be called M as before & $M - \frac{1}{2} = N$ then $N = 2/D^2 + 3/D^4 + 4/D^6 + 5/D^8$ &c and by the reversion of this Series we shall have $2/D^2 = N - \frac{3}{4}N^2 + \frac{5}{8}N^3 - 33/64N^4 + 57/64N^5$ whence $\sqrt{2}/D = N^{1/2} - \frac{3}{8}N^{3/2} + 31/128N^{5/2} - 169/1024N^{7/2} + 11603/32768N^{9/2}$ &c |

7/III/14, 9v According to Dr Stewart the Ratio of the centripetal force of the Moon to the Earth is to what he calls the mean Solar force affecting the Gravity of the Moon to the Earth as 357,43365 to 1*

$P^2:\pi^2::178{,}725:1$*

By applying these data to the Theorem of §9 Mr Trail* brought out D to 295,72 Distances of the Moon from the Earth. Supposing the Mean distance of the Moon to be 60½ Semidiameters of the Earth the Suns paralax by this computation would be 11½" |

7/III/14, 11r

Canon I

Let the Ratio of the Sun and Moons Periods; that of the Masses of the Earth & Moon; that of the Forces of the Sun upon the Earth & of the Earth upon the Moon; That of the Distances of the Sun & Moon from the Earth; be $Q:1$, $M:1$, $F:1$, $x:1$ respectively, and the Suns Distance will be to the moon as $Q^3:\sqrt{M}$ or $x = Q^3/\sqrt{M}$. also $F = Q \times \sqrt{M}$.

Another Canon for the Suns Parallax

Let S be the Number of Seconds in the Suns Semidiameter. $D:1$ the Ratio of the Densities of the Earth & Sun, M & Q denoting as in the former Solution, & I say the Parallax sought will be in Seconds $S \times M^{\frac{1}{2}}/D^{1/3} \times Q^{7/3}$

Thus if S be 966" & its Logarithm	2.98497710
M = 39,788 & Log $M^{\frac{1}{2}}$	<u>0.79987605</u>
	<u>3.78485315</u>
And D being 4,0559 add to 1/3 of its Log	<u>0.20269573</u>
7/3 of the Log of Q	<u>2.62754952</u>
	<u>2.83024525</u>
The Difference	.95460790

the Log of 9".0076 the Parallax sought. The Answer by the other Rule is the same to the 10.000 part of a Second

And the same canon may be reduced to a form more convenient still, namely writing p for the Parallax sought; P for the Moons Horizontal Parallax in Seconds (which in this Example is 3412") we have $p'' = P\sqrt{M}/Q^3$ |

7/III/14, 11v

Log P =	3.5330090	
Log M½ =	0.7998760	
	4.3328850	
Subtract Log of Q^3 =	3.3782780	
	0.9546070	Log of 9".0076.

Hampstead 21 May 1770*

Hence the Inconsistency of the French Estimate
They make the Moon's parallax 57'.3"
Earths Mass 50 instead of 65½
Sun's Parallax 10" instead of 11,"58*

VIII

7/III/1, 1r

Of the Centripetal Forces, Velocities, & Periodical Times of Bodies moving equably in Circles & of the Radij of the Circles

A centripetal Force is a Force that acts constantly and is directed to a certain point. Such a Force may be always the same, or it may continually increase, or continually decrease.

If the Force be always the same, the spaces from the beginning of the bodies Motion will be as the force & the square of the times So it is in the descent of bodies by gravity on the Earth‹'s› Surface

If the force be variable & constantly increasing the Space will be greater than in proportion to the square of the time & the force at the beginning of the Motion & less than in proportion to the square of the Time & the Force at the end of the Motion. The contrary will happen when the Force is continually decreasing.

If we take the Motion in its nascent state onely, the Force whether variable or constant may be considered as constant: And the velocity generated will be as the time & force taken together; & the Space as the Force & the square of the Time, or the square of the last velocity taken together.

As Bodies by their *Vis insita** can onely move in a streight line; it follows that a body moving in a curve line is constantly urged by a centripetal force, tending to some point on the concave side of the Curve.

If the centripetal force should cease in any point of the Curve the Body would by its *vis insita* continue to move uniformly in a tangent to the curve at that point. And hence it follows that the centripetal force at any point of the curve is to be measured by the deviation of the curve from its tangent in a given very small time. This Deviation being directed towards the Center of the force, is called the *sagitta*;* and if the curve be a circle whose center is the center of Force, the sagitta is equal to the versed Sine of the Arch described, and therefore equal to the square of the Chord of that Arch divided by the Diameter. And as in a nascent Arch the Ratio of the Chord to the Arch is a Ratio of equality it follows that in bodies moving equably in a circle the centripetal force is directed to the Center, & may be measured by the Square of the arch described in a given very small Time divided by the Diameter of the Circle. The Arch described in a given Time is as the velocity and the

Diameter is as the Radius, therefore calling the centripetal Force, F, the Radius, R, & the Velocity V. we shall have $F : VV/R$.

If we call the periodical Time of a Body moving equably in a circle, P. the Space passed over in that time is the circumference of the Circle, which is as the Radius, therefore $PV : R$ consequently $P : R/V$. And $PP : RR/VV$ And $F : R/PP$

From these Canons the Corollaries of Proposition 4. Newton Principia easily follow.*

IX

7/VIII/8, 1r

Glasgow College September 1769*

From the West window of my Room the South point lies over the East Chimney tops of Provos‹t› Eaton's House*

September 4th Monday. This day at one in the Morning the Comet appeared in the East in Orion.* The Nucleus was so near to the α of Orion that Star might be mistaken for it, the rather as the Nucleus was faint but it was really 10' or 15' southward from the α and a little higher. The tail reached to the α Ceti which I take to be about 35° of the Heavens.* Its direction was nearly parallel to the Equator and some what bent downward. The Air was unfavourable and the Comet was seen onely about an hour & a half for Clowds

September 7 Three in the Morning. I saw the Comet very faintly the Air being dusky the place may be thus defined Draw a Line between the two great Stars in the two Shoulders of Orion. From the east most of the two Stars draw another line South ward making an angle with the line above mentioned of 95° and let the length of this line be 4/5 of the former. The Comet appeared in the South extremity of this second line.

I could trace the tail but a little way for Clouds and the Nucleus was rather like a little bright Cloud than like a Star.

September 10 Three in the Morning The Comet was seen about twice as far from the ‹blank› of the two Stars in Procyon as these Stars are from each other and the two Stars of Procyon Canis Minor with the Comet made a Right Angle 69 18° Latitude 22 Degrees.* About half an hour after two the Comet was in a Right Line with Venus and Procyon but by four it was very sensibly below that line |

7/VIII/8, 1v September 12 About three in the Morning. the Comet was seen very faintly. Its place 69 28½°. Latitude 23° South

August 5 1770 Being Sunday The Day was excessively hot at 5 afternoon the heat in the shady side of my Room window was 82° of Farenheits Scale at 8 at night it was 75°. I apprehend the Heat was greater at Midday than at 5 but I did not examine it*

X

3/I/20, 1r

Extracts from Pembertons View of Newtons Philosophy*

page 90 The time of a whole Vibration of a pendulum moving in a Cycloyd, is to the Time in which a hea‹v›y Body would by its Gravity fall through half the length of the Pendulum as the Circumference of a Circle to its Diameter, that is, nearly as 355 to 113. But if the Pendulum move in the Arch of a Circle, equal to one sixth of the whole Circumference, it will swing onely 115 times, while it ought according to the Proportion before mentioned to have swung 117 times. If it swing onely through one tenth of the Circle it will lose one vibration in 160. If it swing in 1/20 of the Circle, it will lose about one vibration in 69. If its swing be confined to 1/40 part of the Circle it will lose very little more than one vibration in 2600. And if it take no greater Swing than through 1/60 part of the Circle, it will not lose one Swing in 5800 that is in a Second Pendulum, one Minute in four days and forty minutes.*
130 &c A pop‹u›llar Objection is here very properly answered by popular Reasoning. It has been said that if a Planet by the force of Gravity is made to approach the Sun in its Perihelion, the same forces acting upon it can never raise it up again to its Aphelion. The Author shews by very simple reasoning, that if the Planet at its Perihelion were turned back, with the same Velocity with which it moved forward and in a contrary Direction it would trace backward its own path, & therefore rise to the Aphelion, & therefore will do so no less by going forward with that velocity.*

The same Objection & others of that Nature may be obviated by this Consideration, That when a Body is acted upon by a centripetal Force, & by a *vis Insita** in a different Direction, the Centripetal Force will draw it gradually out of that Direction, so that it will in all cases describe a Curve to which the Direction of the *vis Insita* is a Tangent. This Curve is in all Cases concave towards the Center of the Centripetal 3/I/20, 1v Force. But whether it | approach the Center or recede from it, & whether the Curvature be greater or less depends upon Circumstances. And to determine these Events the Quantity of both Forces and the Direction of the *Vis insita* must be taken into consideration. If the Direction of the *Vis insita* make an obtuse Angle with the line drawn to the Center of Force the Body must recede from the Center. If that Angle be acute it must approach the Center; if it be a right Angle the Body may either approach the Center or reced‹e› from it according as the centripetal force is greater or less compared with the *vis insita*

When the Space through which the Body would be carried, in the least Moment of Time, by the centripetal Force alone, is to the space through which it would be carried by the *vis insita* alone, as the last space is to twice the Distance of the Center of Forces, then the Body will neither recede from nor approach the Center. When the Centripetal Force is greater, the body will approach when less it will recede. Supposing always the Angle above mentioned to be a right Angle.*

129 It is very properly observed that Sir I. Newton first introduced the Method of reasoning from a figure made up of a Number of streight lines which may be shortened & their Number increased without end to a figure of a Continued Curvature; and the Method of reasoning from a Succession of impulses acting upon a Body at small intervals to a continued Centripetal Force, to the great Improvement of Geometry & natural Philosophy.* Did not archimedes & even Euclid do the first*

The Account of Newtons Doctrine of the Resistance of Fluids deserves to be read carefully & compared with the Principia.*

The Account of the Tides to be also read with Care.*

XI

3/I/19, 1r

Facts from Bailly's Histoire de l'Astronomie Moderne*

Forreigners now generally allow that Sir I. Newton was the Inventor of the Method of Fluxions both direct & inverse, though they give the name of the Differential Method, or Differential Calculus to it. This name, given by Leibnitz, who for a long time, upon the continent past for the Inventor, does not appear so proper to signify the velocity of increase or decrease in variable quantitys as the name given by Newton. But the Name and algorithm invented by Leibnitz having been adopted when he was conceived to be the Inventer of the Method, has been continued by forreigners, and Newton's p‹h›raseology & Symbols seem to be used onely by the british Mathematicians.

Bailly in his History of Astronomy says that Newton left the inverse Method, or what they call the Integral Calculus imperfect, and that great Additions have been made to it by the French Germans and Italians.*

The Theory of the inverse Method as far as it has been carried by the british Authors is I think contained in Newtons Quadrature of Curves & Cotes Harmonia Mensurarum, as every Problem in the inverse Method may be reduced to Quadrature of Curves.* And whether any thing Material has been added to this Theory by forreigners I know not.* I apprehend however that several of the French and other Nations have applied the integral Calculus to Problems in Physical Astronomy more than the British have done.

Three Geometers who appeared almost at one time. Clairaut & d'Alembert in France and Euler at Berlin under took the Famous Problem of the Actions of three Bodies upon one another, according to the Laws of Gravitation. Their computations from Theory agreed with observation as far as concerned the chief Equations of the Moons Motion, & they determined many small Equations, which together had an amount too important to be omitted, and which brought the calculations of the Moons place to tally better with the observations than before they had done. At the same time all the three without mutual Communication made the Motion of the Moons Apogee to be according to the Laws of Gravity onely 20° in the year, whereas from observation it was known to be 40°. So fully perswaded all the three were that they had drawn a just Conclusion from the Laws of Gravity that several Theories were proposed to account for so remarkable a difference between the calculated motion of the Apogee and that which was observed, the Law of Gravity, which in every other instance had corresponded so exactly with observation, but in this instance, deviated so remarkably from it, was brought into doubt. The Cause of this Rebellion of the Moons Apogee against the Law of Gravitation was disputed for two years from 1747 to 1749. The Dispute ended by Clairaut's finding that in his Approximation he had omitted a small quantity which being taken in brought the Theory to agree with Observation. d'Alembert & Euler acknowledged the very same mistake, and so the Law of Gravity was restored to its Honour & confirmed beyond all future Contradiction*

Bailly says that Clairaut's Tables of the Moon constructed upon his
3/I/19, 1v Solution of the | Problem of three Bodies answer to observation with a precision difficult to be surpassed. Mayer compared these Tables with his Observations and endeavoured to improve them from a great Number of Observations. Bailly thinks Mayers Tables in no respect better than Clairauts. More than 500 places of the Moon observed by M. Lemery have been calculated & found generally to differ from Mayer & Clairaut's Tables generally less than 1'.* It is remarkable that where Clairaut had taken the help of Observation to correct some of his Coefficients, Euler found out the same Corrections from Theory by carrying his approximation a little farther. All depends on finding the best methods of Approximation, which, says Bailly, Euler, d'Alembert, la Grange, Condorcet, & la Place, are now employed in perfecting.*

Bailly thinks that d'Alembert has gone beyond Newton in his determination of the Præcession of the Equinoxes & the Nutation of the Earth‹'s› Axis, and had the honour to rectify some Errors of that great Man*

M d'Alembert discovered a new branch of the *integral Calculus*, that of Equations for partial Differences.* This Invention is a new Instrument in the hands of the Analyst & the source of many discoveries. In consequence of this Improvement M de la Place in 1777 thought fit to examine anew in their full extent the Questions about the tides & the Nutation of the Earths Axis. Some phenomena had not been accounted for particularly the near equality of the two tides which happen on the same day. Hitherto the Theories had given a considerable difference between these, though by observation they had no sensible difference. De la Place found that the depth of the Sea had influence on this Phenomenon, and that to make the two tides of the same day apparently equal the depth of the Sea must be about four Leagues, & a little more towards the poles.*

Euler applied his solution of the Problem of three bodies to investigate the Effects of Jupiter and Saturn upon each other. Mayer compared his Theory with observation and found it agreable to the Phenomena. The inequalities produced by those two planets on each others Motions were however found to be comprehended in a period of twenty or thirty years, and no cause was found of the Secular Equation of these Planets.*

Similar inequalities in the Earths Motion must be produced by the Actions of Jupiter & Venus

The Nodes of the Eccliptic with the Orbits of these planets has a smal retrograde Motion but this is so small that it will require Ages to ascertain it by observation. The quantity cannot be ascertained by Theory because the Mass of Venus is uncertain. The same Causes must produce a small diminution of the Obliquity of the Eccliptic. But this diminution has a period after which it will increase by the same degrees.*

M. de la Caille's Tables of the Sun highly commended*

Clairaut calculated the return of the Comet 1759 within ‹a› Month of its Appearance*

M du Sejour applied the Integral Calculus to the Calculation of Ecclipses*

M l'abbé Frisi on Gravitation commended.* |

3/I/19, 2r Mr Murdoch has from the Theory of Gravity computed the Paralax of the Moon. Philosophical Transactions 1764 & the Paralax of the Sun Philosophical Transactions 1767. Horseley has also computed the Paralax of the Sun Philosophical Transactions 1768*

Bailly has laboured on the inequalities of the Motions of Jupiters Satellites and on a method of discovering the proportion of Light coming to us from the various heavenly Bodies*

Part Four: Optics

I

7/II/5, 1r October 13 1770 Patrick Wilson communicated to me an Observation of his own Upon the Aberation of the Rays of Light which tho' very curious Seems to have escaped all who have wrote upon this Subject.* It is this

If the annual Motion of the Earth be supposed to be in a line perpendicular to the Ray coming from a fixed Star; If a line representing the Velocity of Light be made Radius, and a perpendicular raised at one extremity of this Radius, represent the velocity of the Earth in its orbit, This perpendicular will be the tangent of the Angle of the Aberration of that Star, according to the common theory of Aberration.* So that the velocity of the Ray bears the same Ratio to the Velocity of the Earth as the Radius bears to the Tangent of the angle of the greatest aberration. The Defect which Mr Wilson observed in this Theory is That the Aberration depends not upon the Velocity which the Rays of light may have in the Air or Ether but upon the Velocity with which they strike upon the Retina of the Eye.* If therefore the Velocity of Rays in the Vitreous Humor of the Eye* be different from their Velocity in Air or Ether, it is the first of these Velocities onely that we can infer from the aberration & not the last. Now if we suppose the Refracting power of the Vitreous Humor to be nearly the same with that of Water, the Velocity of the Rays in the vitreous Humor will be to their velocity in the air or Ether as 4 to 3 according to Newtons Doctrine, consequently the velocity of the Rays deduced from the Aberration of the fixed Stars will be to their velocity in the Ether as 4:3.* Hence all Equations of the Motions of the Heavenly Bodies arising from the progressive Motion of Light ought to be reduced a fourth part, if they be grounded upon that Velocity of Light which is inferred from the Aberration of the fixed Stars

October 15 The Same Gentleman suggested to me a Method which he had thought of, of freeing Astronomical Observations from all Errors arising from the Aberration of Light; grounded upon this Principle that when Light is reflected from any Surface it must have the Motion of that Surface imparted to it.*

N.B. As to the first observation, It is the aberration of the Telescope rather than that of the Rays of Light that we observe. There is probably an Aberration of the Axis of the Eye from the fixed star we look at. But we have no means of measuring this Aberration. The Aberration of the Telescope must depend upon the Velocity of the Rays in the Telescope onely.

II

5/II/55, 1 As all Visible objects are seen by Rays of Light which pass from the Object to the Eye; it is onely by the direction of the Rays that we perceive the direction of the Object from the Eye. Hence it follows that when the direction of the Rays is turned out of a Streight line by Reflexion Refraction or inflexion, we see the Object, not in its real Place, but in the last direction of the Rays when after all their deviations from a right line they fall upon the Eye. This is the General Doctrine of the Writers in Opticks, and is in all cases very near the truth.* But late discoveries have taught us, that, now when Optical Instruments have been brought to such a degree of Perfection as to enable us to discern to a small fraction of a minute the direction of Objects, we must admit some small Exceptions to the General Rule above laid down.

The Exceptions arise First from the Progressive Motion of the Rays of Light, & Secondly from the Structure of the Eye and the Laws of Vision

The progressive Motion of Light produces no Exception to the General Rule above laid down in the following Cases. 1 When the Object and Eye are both at rest. 2 When they move parallel to each other & with the Same Velocity 3 When they move either towards each other or from each other in the line that joyns them 4 When the Motion of the Ray is compounded of two Motions, one of which is parallel to the Motion of the Eye and of the same Velocity, the other Motion of the Ray being directly towards the Eye. But other Motions either of the Eye or of the Object may make the Object appear in the place where it is not

1 Case

When the Eye is at Rest and the Object Moves in a direction making an Angle with the line that Joyns the Object and the Eye. Here it is evident that the Rays will come to the Eye in that direction which the

Object had when they set out and not in the Direction which it was when they arrive at the Eye. The Rays in this Case resemble a Post who brings intelligence of the State of things when he set out but can give |
5/II/55, 2 no intelligence of the changes that have happened since that time The consequence of this is that every one of the heavenly bodies which really changes its place appears to us in the place where it was when the Rays by which we see it, set out upon their Journey. The Sun continuing in the Same place, his Rays when they come to us have their Direction towards his place at that moment as well as when they set out. Jupiter & his Satellites give intelligence to us of their Motions sometimes in 33' sometimes not in less than 48'. And this Phænomenon has been observed in the Eclipses of Jupiters Satellites.* The same Equation on account of the Progressive Motion of Light must be applicable to the Ecclipses of Saturns Satellites and indeed to the Motions of all the Superior Planets. Possibly a much larger Equation of this kind may be required to adjust the apparent Motions of Comets to the laws of Gravitation. If they be ever seen at a greater distance than that of Saturn If the fixed Stars had any Real Motion we could not receive intelligence of it in less than nine Months and should onely see them in the place where they were nine Months ago or more. The Superior Planets must take about half an hour more to pass from their opposition to the Sun, to their apparent conjunction, than in passing from their Conjunction to their opposition. The Inferior planets will also take more time to pass from their Inferior to their Superior Conjunction than to pass from the last to the first. In general When the Object changes its place sensibly during the time which the Rays of Light take up in passing from the Object to the Eye; the Object will appear not in the place it occupies at the time of observation but in the Place which it occupied when the Ray issued from it

2 Case

When the Eye moves in a Direction which makes an Angle with the Line joyning the Object & the Eye.

Real & absolute Motion signifies a change of Place with regard to the Parts of Absolute Space which are fixed & immoveable.* But as the Parts of absolute Space are not perceiveable by our Senses, they cannot Serve us as fixed points from which we are to judge of the Motion of
5/II/55, 3 Bodies. So far therefore as we Judge by our Senses of the | motion of Bodies. we must take some Body which we conceive to be at rest as a

fixed Point. & then other bodies which change their distance or position with regard to this fixed point must appear to the senses to move. When two Bodies change their distance or Situation with regard to each other, one of them at least must have really been moved this is self evident. But we do not perceive immediately by our Senses whether the Real Motion was in this Body or that or whether it was partly in one and partly in the other. As in all other Relations, so in Relative Motion, the Relation may be changed by a change in either of the Related things. Thus if two variable Magnitudes A & B be just now Equal & some time hence A becomes double to B. This change of the Relation cannot possibly happen without real change of Magnitude either in A or in B. But it may happen either by A being doubled while B continues the same or by B being halfed while A continues the same, or by A being increased and at the same time B diminished. In like Manner when the Body A seems to me to move ten foot westward; this appearance is grounded upon the Supposition that I am at rest. For in reality all that my Senses inform me is that the Body has changed its situation and distance with regard to me. And this change may be produced either by the Body A being really moved ten foot westward while I remain in the same place; or Secondly by my being moved ten foot eastward while the Body A remains in its place or thirdly by my being moved five foot eastward while the body A is at the same time move five foot westward To which of these causes the apparent Motion is owing my Senses do not inform me. And when I judge the Motion to be onely in the Body A, this Judgement is formed upon the Supposition that I am at Rest. But my Senses cannot perceive whether I be absolutely at Rest or not. We perceive in some cases that we are moved but we may be moved when we have no perception of it. When we have no perception of our own Motion we rashly conclude that we are at Rest, and thence conclude that other Bodies are in motion meerly because they are perceived to change their Situation with regard to us. And as many instances occur in common Life where in we find our Selves deceived in such conclusions, this makes it necessary for us
5/II/55, 4 to distinguish between that Motion which is absolute | and Real from that which is apparent & Relative. The motion which we perceive by our Senses is Relative And when we pretend to Judge of Real Motion from the Testimony of Sense we use our Senses in a Manner that Nature never intended and for which they are nowise fitted. There may be other Beings who have the power of discerning the parts of absolute Space, & by that means of discerning immediately what Motions are real & what

are onely relative but we have no such power. We may in some cases discern what motions are real & what are relative by their properties by their Causes or by their Effects. But we have no power of discerning immediately what is real from what is relative.

When the Relative Motion of two Bodies is known if it ‹is› afterwards discovered that one of them is at rest it is easy to discover what must be the real Motion of the other. If it is known that both move we must know the direction and comparative Velocity of each in order to deduce the real Motion from the Relative. Some of the cases of this Problem are so complex as to require a Figure & therefore we omit them since they are easily solved by plain Geometry.* There are other cases which may easily be understood without a figure & we shall briefly mention some of them. When A is supposed at rest the Body B has an apparent Motion from B to C. If now it is found that B was at rest in reality and that the relative Motion was wholly caused by the Motion of A. In order to produce the relative Motion A must move in a contrary direction and with the same Velocity which B was conceived to have when the Motion was imputed to it. If it is known that A & B both move but in contrary directions & with equal Velocities. The consequence will be that B really moves in the direction which it was supposed to have but with half the velocity; & A moves in a contrary direction with the same velocity

It is necessary to distinguish not onely between real & Relative Motion but between the Real Direction of Motion or of any line & its relative direction. The Direction of Motion which is real and Absolute is determined by the points of absolute Space which it passes through. But as we see not the points of absolute Space, we must determine the direction of a Line by the Angle it makes with some other Line which we take to be fixed. Whence it is evident that the direction of a line may be the same with relation to another line by which its direction is measured, when it is changed with regard to absolute Space. The former may be called the relative direction of the line, the lat‹t›er its Absolute Direction.

A Ray of Light which passes along the Axis of a Telescope while the
5/II/55, 5 Tele | scope is supposed to be at rest. moves relatively in the direction of the Telescope, and the object will be seen by it in that direction. But if the Telescope having its axis in the real direction of the Ray really moves at right angles to the direction of the Ray, so as to perform a Space of one inch while the ray proceeds twenty foot, it is evident that the Ray

in its passage will not go along the axis of the Telescope but will cross that axis at a small Angle, an Angle which having twenty foot for its Radius will have one inch for its Tangent. In order therefore that the Ray may pass along the Axis, the Telescope must be inclined to the direction of the Ray in the Angle above mentioned. & consequently the Star which in this case appears in the Axis of the Telescope is not seen in its true place but in a place which is distant from its true place by the Angle above mentioned & which is on that side of the true place towards which the Telescope is moved. Let x be the place of a Star seen from the point b. Let ca be the Axis of the Telescope when the ray enters into it at c. & ab the space which the Telescope moves parallel to it self while the Ray passes through its axis, so that the velocity of the Ray is to the velocity of the Telescope as cb to ba.

It is evident in this case, that as the Ray enters at the Axis of the Telescope at c when the lower end of the Axis is at a, so it will be in the lower end of the axis when it arrives at b, the axis having in the same time moved from a to b. It is no less evident that if the Motion of the Ray and of the Axis are both equable, the ray in every point of its motion from c to b, will be found in the axis of the telescope; for the Axis ca moving parallel to it self its intersection with cb. will be carried down from c to b with an equable motion, which will exactly correspond with the Motion of the Ray. The Star therefore will appear to the Observer in the axis of the Telescope that is in the direction ac and the aberration of the apparent place from the true will be measured by the Angle acb. which is easily found when the velocities ab, & bc are known & the Angle abc.

Corollary 1 If the Star z in the time in which the Ray comes from z to b performs an angular Motion in the Direction ab, equal to the angle of aberration, it will be seen in its true place. For here the Error arising from the Motion of the Star according to Case 1 & the Error of aberration compensate one another. But if the Star has an angular Motion in the direction ba. the error arising from the Motion of the Star will be added to that of the aberration. & the apparent place will be distant from the |
5/II/55, 6 real Place by the Sum of both errors. But it is to be observed That the angular Motion we speak of in this Corollary is that onely which arises from a real Motion in the Star; and that we must exclude the relative Angular Motion which arises from the real Motion of the Eye. Hence

this corollary is not to be applied to the Motion of the Sun in the Ecliptick which is relative onely.*

Corollary 2 The Moon, Sun, Planets, & Comets have all the same aberration with the fixed Stars when seen in the same part of the heavens, and therefore the apparent Distance of any body of our System from a fixed Star is not sensibly affected by aberration when they are near to it; but if we were to take the distance of the Moon in her Quadratures from a Star in opposition to the Sun their Distance may err about 20" on account of the aberration of the Star.

Corollary 3 There are three Elements from which the Aberration of a Star is deduced to wit the velocity of Light, the velocity of the Eye or Telescope and the Angle which the direction of the Ray makes with the direction of the motion of the Eye.* The two former represent the Sides of a plain Triangle & the last the angle included by them. The Angle opposed to the second of these sides is the aberration Sought. which from the Elements mentioned is determined by Plain Trigonometry

Corollary 4 When the greatest Aberration is known which according to Dr Bradley is 20", the aberration in any given case whether in Longitude or Latitude, Right Ascension or declination, Altitude or Azimuth is easily determined by Spherical Trigonometry*

Corollary 5 As the onely motion of the Eye or Telescope which can occasion a Sensible aberration is the diurnal or Annual Motion of the Earth, the diurnal Motion even at the Equator being not above the twel‹f›th part of the velocity of the Annual will never cause an aberration of a full two seconds. Yet the difference of the greatest Abberation when both Motions concurr from that which happens when they are contrary will be near one sixth of the whole. The change in the Earths velocity in the different parts of its orbit is never one thirtieth part of the whole, & therefore needs not be regarded.

Corollary 6. The third Element upon which the Aberration depends is the Velocity of the Rays. Here we may observe that if the velocity of the Red making Rays was sensibly less than that of the blue a Star when its aberration is greatest should appear oblong the End towards which the Earth moves being Red and the other prismatick colours succeeding in order.* But if the difference of Velocity in the rays do not amount to one twentieth part of the whole it will not be observable. |

5/II/55, 7 Corollary 7. The velocity of the Rays of Light in Air may be collected from the aberration & is found to be such as that Rays pass from the Sun to us in about 8' of time.* The velocity in Vacuo may be somewhat less but by Sir Isaac Newtons Hypothesis as the Sine of Incidence from a Vacuum into common Air is little more than one four thousand part greater than the Sine of Refraction it is probable that the increase of Velocity in the Rays in Air beyond what they have in Vacuo is not Sensible* |

5/II/55, 10

Case 3d

We have hitherto considered a Ray of Light as passing along the Axis of A Telescope, in which case it undergoes no Refraction, and yet the Star from which it comes is not in the Axis of the Telescope. The apparent place of the Star is in the Axis of the Telescope, but its real place is about 20" distant from that Axis, when the aberration is greatest. We are next to consider whether there be any similar aberration in the Eye itself.* In order to this it will be proper to premise some Definitions, in which to prevent confusion as much as possible we shall Suppose the Eye at rest, & several Visible points within its field seen by distinct Vision; so that the Rays coming from each visible point are collected accurately in one point of the Retina we shall suppose these points at rest & that the Rays coming from them undergo no refraction or inflexion but in passing through the cornea and humors of the Eye.

These things being supposed, it is well known that a pencil of Rays diverging from a visible Point, fall upon all parts of the Pupil of the Eye and by the refracting powers of the Eye are made to converge again, so as to meet in one point of the retina; & by this means every pencil of Rays forms a double Cone.* The cone of Rays which has the visible point for its Vertex and the Cornea of the Eye for its Base we shall call the outward Cone. That which has the Cornea for its Base & the point of the Retina where the Image is formed for its Vertex, we shall call the inward Cone of that Pencil. But as the Rays may undergo some small Refraction after they pass the cornea, we would be understood to give the name of the inward cone especially to that part of it which reaches from the Retina to the back of the Chrystaline humor, in which part there is no refraction.

As the Rays of every Pencil first diverge from one point to all parts of the Pupil, & then converge from all parts of the Pupil to one point in the Retina, most of them must have one direction in the outward Cone and another direction in the inward Cone. But there is, in every pencil, one Ray, which must have the same direction in both cones; either having gone in a streight line from the visible point to its image in the

Retina, or by equal and contrary refractions being brought into its first direction. This is evident from the Nature of the thing, and is well |
5/II/55, 11 known to Opticians. Let us therefore call the Path of this Ray which in every Pencil retains its original Direction, the Axis of the Pencil, or the Axis of the outward & Inward Cone formed by that Pencil.* The Axis of the two cones will be either one streight line passing from the visible point to its image on the Retina, or they will be two parallel lines which very nearly coincide. Let it be remembered therefore that every point of the Retina on which an image can be formed, has an inward cone which is proper to it, & an axis of that cone which is also proper.

The Axis or Principal Diameter of the Eye is a Right line in which a Ray falling perpendicularly upon the Cornea and upon all the humors passes to the Retina without any Refraction. If the Eye is rightly centered, which is probably the Case in all perfect Eyes, the Axis will pass through the Middle of the pupil*

That part of the Retina where it is cut by the Axis or principal Diameter of the Eye, we shall call the center of the Retina. It is evident that the Axis or Principal Diameter of the Eye is also the Axis of that inward Cone which is proper to the Center of the Retina*

The Axis of that inward Cone which is proper to any other point of the Retina shall likewise be called a Diameter of the Eye. Whence it follows that every Point of the Retina on which an image can be formed has a proper Diameter of the Eye which terminates in that point. And that a Ray which passes along this Diameter has the same Direction, when it falls upon the cornea, and when it falls upon the Retina.

If the Humors of the Eye were equal in their Refracting Power to each other and to the Cornea, the Rays of Light would all pass in Right lines from the Cornea to the Retina. And in that case every Diameter of the Eye would pass precisely through the center of that Spherical Surface of which the Cornea makes a part. And this Center might with perfect propriety be called the Optical Center of the Eye, being the point in which all the Diameters of the Eye cut the Axis, and cut each other. But ‹it is› probable, from the Experiments quoted by Dr Jurin in his Essay on Distinct and Indistinct Vision, as well as from other Circumstances,
5/II/55, 12 that the Humors of the Eye differ somewhat in their refracting | power, though that difference is very small compared with the difference between the refracting power of the Air & of the Cornea.* Hence it follows that the Rays will receive almost the whole of their Refraction at the Cornea, & that they will pass from thence to the Retina nearly in

a right line. And the consequence of this will be that all the Diameters of the Eye will meet, though not accurately yet nearly at the Center of the Arch of the Cornea, & that that Center may be called the Optical Center of the Eye, physically though not Mathematically.

Having premised these things that we may be able to express ourselves more distinctly & intelligibly upon this Subject I proceed to some Propositions.

Proposition 1 When the Object and the Eye are at rest, & the Rays come in right lines from the Object to the Cornea, we see the Object in its true place, that is, in that Direction from the Eye which it really has; And when several Objects which are within the field of Vision we see them at that Angular distance from each other which they really bear.

This is confirmed by universal Experience, & if it were not accurately true our Eyes would be irremediably fallacious.

Proposition 2 The Direction in which we see an Object, in the circumstances supposed in ‹the› last Proposition, is that of a Right line drawn from the Center of the Eye to the Object. This admits of no doubt.

Proposition 3 From these Propositions we may I think draw this conclusion with all the Evidence that is to be expected in Physical Matters. That it is a Law of our Nature that we see every Visible point in the Direction of that Diameter of the Eye which passes from the Image of that point in the Retina.* For 1. It is not the Objects being in such a Direction that makes us see it in that Direction: For we know that unless Rays of light come from the Object to the Eye we do not see it at all. We know likewise ‹that› the Rays of Light may undergo such changes between the Object and the ‹E›ye as may make us see it in a very different direction from its true one as we‹ll› as of a different visible magnitude and figure from that which really belongs to it. We are therefore led from the Object to the Rays of Light in seeking after the Laws of Vision. 2 We know that
5/II/55, 13 the Rays of Light never cause Vision | in any kind or degree unless they Enter the Eye and Strike upon the Retina When either the Cornea or any of the Humors are opaque so as not to suffer the Rays to strike upon the Retina there is no Vision. This therefore is evidently a law of Nature that without a Picture upon the Retina there is no vision.* And as the Object led us to the Rays of Light, in investigating the Laws of Vision, so these lead us to the Picture formed on the Retina

3 It is not probable that we shall ever be able to give a Reason why a picture upon the Retina should cause Vision any more than a Picture upon the cornea or Chrystalline. As little is it probable that we shall be

able to shew the reason why a picture on this point of the Retina causes us ‹to› see the object in such a Direction while the Picture on another point shews the Object in a different Direction. All we can do here is to find out the General Rule that Nature follows in producing Vision, without pretending to shew the efficient Cause of that Rule, Just as the great Newton investigated the Laws of Gravitation without pretending to shew its Cause.* Now I say the general Rule observed by nature in this Matter is that every visible point is seen by us in the direction of a Diameter of the Eye which passes from the image of that point upon the Retina 1 That this Rule is observed in all Cases where the Object and Eye are both at Rest & the Rays come in Right lines to the Eye is evident from Proposition 2. For as the Object is seen in the Direction of a right line passing from the Center of the Eye to the Object, and this right line is either a continuation of the Diameter of the Eye which passes from the picture upon the Retina or is parallel to it and to all Sense coincides with it, It is evident that these directions are one and the same. 2 The Rule holds no less when objects are seen by reflected inflected or refracted Rays, for it is well known that such Objects are seen in the last direction of the Rays passing through the center of the Eye which agrees with the Rule. 3 As all the ordinary cases of vision whether with the Naked Eye or with optical Glasses agree with this Rule so it holds in the most uncommon cases both of distinct & indistinct vision; as where by a certain artifice one object is seen double or triple with one Eye, where an Object is seen Single by two or three different pencils of rays passing through different holes to the Eye. See the Experiments that prove this in Reids Inquiry ch 6 §12.* 4 I know of no one instance where this Law is not observed in sound Eyes. So that it seems to have all the Evidence that can be had in a matter of this kind. |

5/II/55, 14 Corollary 1 The apparent Angular distance of two visible points seen at the same Time is measured accurately by the Angle made by the Diameters of the Eye which pass from the Images of those visible Points in the Retina

When two Points are seen by the Same Eye, & at the same time, it is known from Experience that they allways appear in that Plain which passes through them & through the middle of the Eye. Hence it follows that the two lines of Direction in which they are seen, cut each other and form an Angle which measures their apparent angular Distance from one another: But by Proposition 3d The lines of Direction in which they are seen are the two Diameters of the Eye which proceed from the

Images of the visible Points. Whence it follows that any two Diameters of the Eye meet in a point & form an angle. and that this angle measures the Apparent angular Distance of the visible Objects whose pictures are formed in the Retina at the extremities of these Diameters.

Corollary 2 The apparent Angular Distance of two Visible Points seen at the same time will be very nearly equal to the Angle made by two lines drawn from the Center of the Arch of the Cornea to the pictures of those visible points upon the Retina. For it has been shewn that this Corollary would be Geometrically true if the Rays underwent no Refraction but in the Cornea. It appears likewise from all the Experiments that have been Made on the Refracting powers of the Cornea & of the different humors of the Eye, that this difference is very Small See Dr Jurins Essay on Distinct & Indistinct Vision* whence it follows that almost the whole Refraction of the Rays is performed at their Entring the Cornea, & that the Rays which fall near the axis of the Eye, which are the onely Rays that give distinct vision, may without sensible Error be conceived to proceed in Right lines from the Cornea to the Retina |

5/II/55, 15 Proposition 4 If a fixed Star is seen by the Naked Eye in a direction perpendicular to the annual Motion of the Earth, it will not appear in its true place but in antecedentia, and the Aberration will be an angle whose tangent is to the Radius as the velocity of the Earths motion to the Velocity of the Rays of Light in the Humors of the Eye*

Let az be the real direction of the Star from the Eye; ab, the axis of the Eye which is carried along by the Earth in the direction bc, while the Ray passes from the Cornea to the Retina So that ab is the Space passed over by the Ray in the time which the Eye takes to move from b, to c. These things being Supposed it is easy to see that the Ray za will move along the Axis of the Eye while that Axis is carried parallel to it self through cb. For it is found in one extremity of the Axis at a in the beginning of that time; & in the other extremity at c at the end of that time, and since both the Motion of the Ray & the Motion of the Eye are supposed to be uniform, it is evident that the Ray will be found in the Axis of the Eye in every intermediate point of the Time. It therefore is carried along the axis of the Eye & cuts the Retina where the Axis cuts it, and therefore the Star will be seen in the direction of that Axis by Proposition 3. That is, if ab is produced to o, the Star will appear not in its true place at z but at o. And it is evident that the angle of aberration oaz, or bac has ab for its Radius and cb for its tangent, therefore its Tangent is to the Radius as the Velocity of the Earth to the Velocity of Light in passing through the Eye. Q.E.D.

The deviation of the Axis of the Eye from the true place of a Star which is necessary to make the Rays coming from that Star pass along the Axis of the Eye may be called the Aberration of the Eye And the deviation of the Axis of a Telescope from the true place of a Star which is necessary to make the Rays coming from that Star pass along the Axis of the Telescope may be called the aberration of the Telescope

Corollary from Proposition 4 If the Velocity of Light in the Eye be to its velocity in Air as 4 to 3* The aberration of the Telescope will be to the Aberration of the Eye as four to three very nearly. This is evident from
5/II/55, 16 comparing the | analogies by which those Aberrations are respectively determined.

Proposition 5 Supposing the Velocity of Light in the Eye to be to the Velocity of Light in the Air as 4 to 3 to describe the Phenomena of Aberration of a Star seen through a Telescope.

Let us Suppose the Star to be really in the Zenith at z, so that its true direction from the Eye is the line az. Let the aberration of the Telescope be the angle zap, & that of the Eye zao, which from what has been shewn will be about three fourth parts of the aberration of the Telescope.

1 Here it is evident that the Axis of the Telescope must be in the position ap when the Rays pass along it, & consequently when the image of the star made by the object Glass falls upon the cross hairs. And if at that moment the aberration of the Axis of the Telescope from the vertical line be observed it will be equal to the angle zap.

2 But it is to be observed that although the Ray in the Telescope has the Relative direction pa, and continues in that relative direction while it is in the Medium of Air, yet as soon as it enters the Eye at a it changes its velocity, & on that account changes its relative direction from pa to oa; so that that very Ray which came along the Axis of the Telescope in the direction pa shews the star not in the direction ap but in the direction ao. This appears to me to be a necessary conclusion from the change of Velocity in the Ray but I shall offer another Reason that leads to the same conclusion

The Ray by passing through the Axis of the telescope suffers no change in its direction, whence I conclude that the star will be seen in the same place by this Ray as if the Telescope were taken away and the star were seen by the naked eye; but it was shewn by Proposition 4 that the Star to the naked eye would appear in the direction ao, therefore it will appear in that direction through the telescope.

That the Ray from the Star may pass along the Axis of the Telescope, that Axis must be placed in the Direction ap, & the angle zap which is the Aberration of the Telescope may be found by the Micrometer But that the Ray may pass along the Axis of the Eye, that Axis must be in the Direction ao, & the angle zao is the Aberration of the Eye which differs from the aberration of the Telescope but five seconds at Most, but we have no mean of measuring this difference as we cannot apply a Micrometer to the Axis of the Eye

III

7/II/15, 1r

Axiom

If two bodies either approach each other, or recede from each other in contrary directions, the relative velocity with which they approach, or with which they recede, will be equal to the Sum of their real Velocities. And if they approach or recede, moving in the same direction, the relative velocity with ‹which› they approach or recede will be equal to the difference of their real Velocities.

Corollary. Hence if a Prism be carried towards a fixed Star, with a velocity equal to that of the Earths annual Motion, & having its Axis perpendicular to the direction of its Motion, the relative velocity of the Rays from the fixed Star falling upon the Prism will be equall to the real Velocity of the Ray and that of the Earth taken together And if the prism be carried from the fixed Star with the same Velocity, the relative Velocity of the Ray falling upon it will be equall to the Difference of the real Velocity of the Ray & of the Earth.

Postulate. In different velocities of the Same Ray the Sine of Refraction will be as the velocity of the Ray reciprocally, and if the refracting Surface move in the same direction with the Ray or in a contrary Direction, it is the relative Velocity with which they approach or ‹r›ecede from each other that determines the Refraction. This is grounded on the Supposition that the Rays of Light, whether we call them bodies or not, in Refraction & Reflection, are acted upon by a Power tending perpendicularly to or from the Refracting or Reflecting Surface, a Power which is equal at equal distances from the Surface, and that the three general Laws of Motion take place in the Rays of Light.*

Problem 1. Suppose a Prism at rest, and placed in a certain position with regard to a fixed Star so as to refract a Ray coming from that Star, and that the Angle of that Refraction is given; it is required to find the Angle of Refraction of the same Ray when the Prism is moved with a great Velocity towards the Star or from it. |

7/II/15, 1v To simplify the Problem as much as possible, let us suppose that in both cases the Ray passes through the same angle of the Prism, that it is perpendicular to the first surface, and that the axis of the Prism is perpendicular to the path of the Ray; the Prism being in the second Case carried to, or from the Star parallel to itself & keeping the position mentioned.

Let the real Velocity of the Ray be M. the real velocity of the Prism l. Then when the Prism moves towards the Star, the relative velocity of the Rays approach will be M + l by Corollary. And by postulate the Sine of the Refraction when the Prism is at rest will be to the Sine of Refraction when it moves towards the Star as M + l is to M.

For the same Reason the Sine of Refraction when the Prism is at rest will be to the Sine of Refraction when it moves from the Star as M – l is to M.

And the Sine of Refraction when the Prism moves towards the Star will be to the Sine of Refraction when the Prism moves from the Star as M – l is to M + l. Or when l is very small compared with M, as M is to M + 2l.

Example‹.› Suppose the refracting angle of the Prism to be 30° then the Ray will have an angle of incidence upon the second Surface of the Prism of 30° whose Sine is half the Radius. And the Sine of Refraction will be ¾ of the Radius which is the Sine of 48°.35'.25,4" This therefore is the Angle of Refraction when the Prism is at rest. When the Prism moves towards the Star, this Sine will be diminished in the Ratio of M to M – l that is in the ratio of the Radius to the Radius diminished by the Sine of 20" or of 10,000 000 to 10,000,000-969. This Sine therefore will be 7,499,273 which is the Sine of 48° 35' 2",6

Example 2d Suppose the refracting Angle of the Prism to be 40° whose Sine is 6,427876. The Sine of Refraction will be to this as 3 to 2 It will therefore be 9,641814 which is the Sine of 74°.37'.7" when the

Prism is at rest. When the Prism moves towards the Star, this Sine must be diminished in the Ratio of 10,000 000 to 10,000 000-969, which makes it, 9,6407800 the Sine of 74°.35'.54",5 |

7/II/15, 2r In this second Example a Telescope to receive the refracted Ray must make an Angle with the path of the incident Ray of 34°.35'.54",5. for so much the Ray is turned out of its way by Refraction.

A Second Problem should be to determine the relative path of the Ray after Refraction. & thereby its aberration from the real path above determined. This I am discouraged to attempt, because the refraction of the Ray would turn its Image into a long coloured Spectrum, unless we could procure an Achromatick Prism.

The postulate on which the preceeding proposition is founded, may be denied. Perhaps it would be more reasonable to assume that while the angle of incidence is the same the angle of refraction will be the same, whether the velocity of the incident ray be greater or less. And that the Sines of Incidence & of Refraction will have the same Ratio whether the velocity of the incident ray be greater or less. If this be so when the velocity of a Ray in any one Medium is increased, its velocity in any other medium in to which it is refracted will be increased in the same proportion, & when diminished, will be diminished in the same proportion |

7/II/15, 2v

Axioms

1 If the Eye be at rest and the motion of Light instantaneous every object will appear in its true visible place & to have its true visible motion

2 Supposing still the motion of Light instantanteous, if the object be at rest, its apparent Motion arising from an unperceived Motion of the Eye, will be the same as if, the eye being at rest, the object had a motion equal to that of the Eye, but in a contrary direction.

3 When the Eye is at rest, objects seen by light moving progressively in right lines from the object will be seen in the place where the object was when the Ray issued from it; and not in the place where it is when seen, if by any motion it has changed its place during the progress of the Ray.

4 When the Eye moves unper‹c›eptibly, the object being at rest, and the Light progressive, the Ray in order to meet the moving Eye must have a motion which may be resolved into two motions; the first parallel and equal to the motion of the Eye; the second directed to the place

of the Eye at the moment the ray set out. The first of these motions is unperceived, the Eye being imagined to be at rest. Therefore the onely motion of the ray that is perceived is that which is directed from the object to the place in which the Eye was when the ray set out.

IV

2/I/3, 1 Of the Path of a Ray of Light passing through *Media* that are in Motion*

December 4 1781

This is a branch of Opticks which has not as far as I know been considered by any Writer on that Science. The noble Discovery of the Progressive Motion and Velocity of the Rays of Light, might have led Writers on Opticks, to trace from Theory the consequences that must follow from that progressive Motion.* But this seems not to have been done. Had it been done, what is called the Aberration of the fixed Stars would have been discovered from Theory, and much Labour & Time would have been saved to the Ingenious Dr Bradley which he employed in tracing that Aberration from Observation.*

I apprehend therefore that it would be an usefull Supplement to the Science of Opticks, if the Consequences that follow from the progressive Motion of Light, in the *Phænomena* of Vision, were laid down in a proper Number of Propositions adapted to the various Cases & demonstrated from established Principles.*

The following Propositions relate to one Case of this kind, which however branches out into several more particular Cases. The more general Case is when the Object is at Rest, and the Ray by which it is seen passes through different Media which together with the Eye itself are in Motion.*

Postulatum 1. The Power by which the Rays of Light are Reflected or Refracted, or by which they have their Velocity increased or diminished, acts onely at the Surface which divides two different Media; and Acts
2/I/3, 2 onely in a direction | perpendicular to that Surface*

Postulatum 2 A Ray goes on in a uniform pellucid Medium without any Sensible Resistance, whether the Medium move or be at rest. This seems to be a necessary consequence of the first Postulate, but is particularly mentioned, to prevent mistake

Proposition 1 Problem to describe the path of a Ray Reflected from a Body in Motion. A Ray of Light will be Reflected from the Surface of a Medium in Motion in the same way as if the Medium was at Rest and in that Position which it has at the moment of the Reflection, excepting in the fourth of the following Cases

Case 1 Suppose the Reflecting Surface to be a plain Surface and that it moves in the same plain & at right Angles to the direction of the Ray.

In this Case the Reflected Ray will return in the path of the Incident. For the Reflecting Power acting onely in a direction perpendicular to the Reflecting Surface by the first Postulate it cannot give any Motion to the Ray but in that Direction

Case 2 If the Ray falls obliquely upon the reflecting Surface it will be reflected so as to make the Angle of Reflection equal to the Angle of Incidence for the same Reason.

Case 3 If the Reflecting Surface have any such Motion as that its inclination to the path of the Incident Ray is continually changing, there will be a certain inclination of the plain to the incident Ray at the Moment of its Incidence, & in that position it will be equally inclined to the reflected Ray. Or if the Reflecting plain should act upon the Ray for some very short interval of time, the Angles of Incidence & Reflection will be equal in the position which the plain has in the Middle of that Interval

Case 4 If the Reflecting Surface move towards the incident Ray or from it, the reflected Ray should, according to the Laws of Motion, be reflected in the first case with a greater velocity than the incident Ray had, & in the second case with a less. If the incident Ray strikes at right Angles, the reflected Ray should return in the same path but with a velocity equal, in the first Case, to the Sum of the Velocities of the incident ray and of the reflecting Surface, & in the second Case, equal to the Difference of those Velocities. |

2/I/3, 3 If the incident Ray be inclined to the Reflecting Surface, we may resolve its Motion into two Motions, one perpendicular and the other parallel to the reflecting Surface. And in the same manner we may resolve the Motion of the Reflected Ray. That Motion which is parallel to the reflecting Surface will not be at all affected by the Motion of the Surface but will continue the Same after Reflection as before. But the Velocity of the Motion perpendicular to the reflecting Surface, should be, after Reflection increased, or diminished by the velocity of the Surface towards the Ray or from it. Hence it should follow that the angle of Reflection should be less than the Angle of Incidence when the Surface moves towards the Ray and greater when it moves from the Ray.*

Scholium. To see whether it be possible to examine this by experiment, we may observe that the Velocity of Light is to the Velocity of the Earth in its course round the Sun as Radius to the Sine of 20".* Hence the difference between the greatest relative velocity wherewith the light of a Star in the Eccliptick strikes the Earth, & the least is to the whole Velocity of light as the Sine of 40" to the Radius. Hence the Velocity of a Ray reflected perpendicularly from the Earth's Surface in the different parts of its Orbit should be different, & the greatest difference is to the velocity of the incident Ray as the Sine of 40" to Radius.

And if the Ray be inclined to the reflecting Surface so that the Sine of the Angle which its path makes with the Surface, be onely the tenth part of the Radius, then the Velocity of that Motion of the Ray which tends perpendicularly to the Surface, will be onely the tenth part of what the perpendicular Ray had; At the same time the Action of the Reflecting Surface upon it will not be diminished. Hence if we suppose a Reflecting Surface exposed to the Light of a Star in the Eccliptick in those opposite points of the Earths Orbit wherein the Star is in Quadrature with the Sun, and that the incident Ray is inclined to the reflecting Surface in the Angle above mentioned, the Reflected ray will not be the same at these opposite points, but at the one its Sine will ‹be› as the tenth part of
2/1/3, 4 Radius added to the Sine of 20" and | at the other it will be as the tenth part of Radius diminished by the Sine of 20"

Proposition 2d Problem To describe the path of a Ray passing through Media that are in Motion.

Case 1 Suppose the Ray passes from the Object through a pure Vacuum to the Eye, & that the Eye is carried in a Direction at Right Angles to the path of the Ray.

What is to be observed in this Case is that the apparent path of the Ray is not the same with its real path, but deviates from it on account of the real Motion of the Eye, which not being known to the Spectator, who judges of the Appearance as if he was at rest, the appearance is the same as if the Eye was at Rest, and the Ray had the same Motion in a Contrary Direction.*

Suppose in Fig 1 That the Ray is at z when the Eye is at a The path of the Ray is zb and while it moves through that Space, the Eye moves through the Space ab meeting the Ray at b. Draw bc parallel to az, and I say that the Eye at b will see the Object in the Direction bc and not in its true direction bz. For supposing a Tube reaching from the Eye at a to the Ray at z; and that this tube is carried along with the Eye parallel to it self, untill it comes to the position bc. This tube has the appearance to the Spectator of being at rest. And the Ray entering the Axis of the tube at z continues to move along that axis untill it comes to b; therefore the Ray will appear to have come in the Direction cb; & not in the Direction cz.

Let the Angle cbz or bza which is equal to it, be called the Aberration of the Tube. It is obvious that when the Velocity of the Ray, and the Velocity of the Eye are known the Aberration of the tube may be found by plain Trigonometry. For let zb, the Space passed over by the Ray, & ab the Space passed over by the Eye in the Same time, represent the Velocities of the Ray and of the Eye respectively. If zb be the Radius of a circle, ba will be the tangent of the Angle of Aberration of the tube. |

2/1/3, 5 Corollary 1. If the Motion of the Eye be not at right Angles to the path of the Ray but at some other known Angle the Aberration of the Tube will be different but may be found by the Resolution of the plain Triangle zab. Of which we have the sides zb & ab given and the Angle included between them zba. whence we may find the angle azb which is the aberration of the tube.

Corollary 2d If the Object moves with the same Velocity as the Eye and in the same Direction, it will appear in its true place. Let z be the Object seen by the Ray passing in the path zb. And let that Object, as soon as the Ray issues from it in the Direction zb, pass to the point c, while the eye passes in a parallel Direction and with equal Velocity from a to b. The Eye being come to b meets the Ray from z, which appears to have come in the Direction cb. And therefore the Object appears at c where it really is; though it is seen by a Ray coming from the place where it was some time before, but now is not.

Hence it may be inferred that when the Ray from a terrestrial Object passes along the Axis of the Tube, there is indeed an Aberration of the Tube from the path of the Ray but no Aberration from the true place of the Object, at the moment it is seen*

Corollary 3. Supposing the Ray to move in a streight line and with an uniform velocity in any medium whatsoever, from z to b while the Eye moves from a to b the Aberration of the Tube from the path of the Ray will depend upon the three Elements before mentioned, to wit, the

Velocity of the Ray, the velocity of the Eye, & the Angle which the path of the Ray makes with the path of the Eye, and from these three Elements the aberration of the Tube from the path of the Ray may be found by resolving the triangle zba, or its equal bzc. And the Object seen by that Ray will appear in the Direction bc. |

2/I/3, 6 Corollary 4 Whatever Refractions the Ray may have suffered before its entrance into the tube at z, these may cause an aberration of the path of the ray zb from the place of the Object but they make no change in the aberration of the tube from the path of the Ray. And the Object will still be seen in the direction bc, found as above.

Case 2d When the Ray after passing through a Vacuum falls at right angles upon a refracting Medium in which the Eye is placed, and which Medium is carried along with the Eye in a direction perpendicular to the path of the Ray

In this Case the Ray enters into the Refracting Medium without any deviation from its rectilineal path because by Postulate 2 its Motion is the same as if the Medium was at rest. Supposing therefore the Tube placed in this refracting Medium at za and that the Ray passes in the direction zb. & that the ray passes along the axis of the Tube while the Eye with the Tube passes from az to bc. The aberration of the Tube from the path of the Ray & from the object will be the angle azb as before, & will therefore depend upon the Velocity of the Ray, the velocity of the Eye and the angle comprehended under the path of the one and of the other.

If the Ray fall at an oblique angle upon the refracting Surface it will undergo the same refraction as if that Surface was at Rest by Postulate 2d, but the aberration of the Tube from the path of that refracted Ray which goes along its Axis will be the same as before, providing the velocity of the Ray be the same. If the velocity of the Ray is increased the Aberration of the Tube will be diminished. If the Velocity was increased to infinity there would be no aberration.

But the Aberration of the Tube from the path of the Ray which goes along its Axis ought to be distinguished from the aberration which the tube may have from the true place of the Object For if the Ray have been bent out of its rectilinear Course by refraction, before it comes into the Axis of the tube, its path will not be in the direction of the Object, and in that Case the deviation of the Tube from the Object will be compounded
2/I/3, 7 of two | parts one of which is the Aberration of the Tube from the path of the Ray that passes along its Axis; the other is the bending of the Ray by refraction before it came into that Axis

Proposition 3 Problem. To determine the real and relative path of a Ray of Light which impinges upon the Surface of a Reflecting or Refracting pellucid Medium, which Moves in the Direction of its Surface

The Real Path of the Ray will be the Same if the Medium was at Rest, whether it be reflected or refracted. For as the Action of the Medium upon the Ray is onely in the Direction of a line perpendicular to its surface & is the same in every part of that Surface, and as the Ray has no sensible Resistance in a pellucid Medium, the Ray will be acted upon by the same force and in the same Direction whether the Medium be at rest or whether it move in the direction of its Surface

If therefore the Ray impinges upon the Medium at right Angles, it will either be reflected back in the same real path it had before, or go forward without any Refraction and with the same velocity as if the Medium was at Rest.

If the Ray impinges at any oblique Angle it will either be reflected so as to make the Angle of Reflexion equal to the Angle of Incidence, or it will be refracted in the same degree with the same Velocity and in the same plane as if the Medium was at rest. All this follows from the first and second Postulate

The Relative Path of the Ray with respect to the Medium may be deduced mathematically from its real Path & Velocity and the Path & Velocity of the Medium. The Simple and general Solution is this

Suppose the Medium at Rest, and that the Ray has a Motion given it (and compounded with its former Motion) equal to the Motion which the Medium had before, but in a Contrary Direction, whatever is the Real Motion of the Ray, upon this Supposition, the Same will its relative Motion with regard to the Medium be upon the first Supposition.

The application of this Solution to particular Cases is so easy that to avoid being tedious it is omitted. |

2/I/3, 8 Proposition 4 Problem. To trace the path of a Ray coming from a fixed Star and passing along the Axis of a Refracting Telescope when the Motion of the Earth is at right Angles to the Axis of the Telescope.

I apprehend the easiest way to solve this Problem is to trace the Ray backwards from the Cross hairs of the telescope to the Star; it being a known Principle in Opticks that the Ray would take the same Course backward as it did before and land at the same Point.

Suppose then the Ray to issue from the cross hairs of the Telescope and to pass along its Axis untill it enters the lower Surface of the Object Glass. In order to this the Telescope must be carried back parallel to itself with the same Velocity it had before but in a contrary direction. For as the Ray is supposed to trace back its first Course; the Earth with the Eye and Telescope must be supposed to do the Same. It is likewise to be observed that although the real Path of the Ray makes an Angle of 20" at most with the Axis of the telescope, yet in its relative path it coincides with that Axis and therefore will impinge upon the lower Surface of the Object Glass perpendicularly and go on in the same streight line untill it meets the upper Surface of the Object Glass.

If the Ray continued to have the same Velocity as it had before, it would go along the Axis of the Object glass and issue from it without altering its direction. But we shall suppose that its Velocity is one third greater within the glass than in the Air. To estimate the Effect of this increase of Velocity in the Ray. See Fig 2d*

Let ncb be the real path of the Ray, which is supposed to be in the point c when the Axis of the Telescope is in the line mca

Let ab be the upper Surface of the Object glass, cb the Lower

It has been already observed that if the Ray, after entering into the Object Glass at right Angles at c, went on with the same velocity, it would reach the point b at the same time that the point a of the Axis reaches to b, but the velocity of the Ray being increased in the Ratio of two to three, it will reach the point b when the point a of the Axis has
2/I/3, 9 performed | onely two thirds of its motion from a, to b, therefore when the Ray is in b, a will be distant from it a third part of ab. Now as the Angle acb is 20", at most the line ab will be equal to the tangent of 20" the thickness of the Object glass being Radius. And when the Ray is at b the point of the Axis a, will be distant from the Ray the third part of that tangent, which distance we shall to avoid fractions call the tangent of 7" to the same Radius.

If the thickness of the Object Glass is half an Inch, the tangent of 7" to the Radius is altogether without the reach of our Senses. And if the upper Surface of the Object Glass is plain & perpendicular to its Axis the Ray will impinge upon it at Right Angles, for the Reason before mentioned, and proceed without any Refraction in that real path which it had at first. So that when the upper Surface of the Object Glass is plain the Ray has not the least Refraction in passing through it. It has onely its celerity increased for half an inch of its way.

But let us suppose in the last place that the upper Surface of the Object Glass is spherical, to a Radius of twenty feet, which I apprehend is the least Radius we can suppose for this Observation

The Radius therefore is 480 times the thickness of the Object Glass, and that which was the tangent of 7" to the last radius will be the tangent of about the 70th part of a second to the first radius The Ray falling upon the Spherical Surface at this distance from the Axis of the Sphere & in a Direction parallel to that Axis will be refracted from the Axis; and its deviation from its former real Path will be less than the 40th part of a Second.

The Ray has now got free of all action of the Telescope upon it & if it is not refracted in passing through the Air or Ether may reach the fixed Star without deviating from its path when it first set out from the cross hairs the fortieth part of a second

And as it would follow the same tract in coming as in returning we may conclude that if there is no refraction of the ray before it reaches the Telescope, the aberration of the Tube from the path of the Ray carried along its Axis is the same with the aberration of the Tube from the real place of the Star.

V

7/II/11, 1 I suppose the Sine of Incidence to the Sine of Refraction from Water to Air, in the least Refrangible light to be as 81:108. In the most Refrangible as 81:109*

From Glass into Air least Refrangible as 50:77: most Refrangible as 50:78*

From Glass into Water least Refrangible as 174,2:201,2: most Refrangible as 174,2:202,2*

Log of the Ratio of 81:108 is 0.1249388
of 108:109 is 0.0040027
of 50:77 is 0.1875207
of 77:78 is 0.0056039
of 174,2:201,2 is 0.0625798
of 201,2:202,2 is 0.0021532

Problem 1‹s›t

Conceive the most refrangible & the least refrangible to be refracted from Glass into Water the common angle of Incidence being 30° Required their angles of Refraction?

Log Sine of 30°	9.6989700
Ratio of 174,2:201,2 add	0.0625798
Sine of 35.16'29"	9.7615498 Refraction of the Violet
Ratio of 201,2:202,2 ad‹d›	0.0021532
Sine of 35° 28'.35"	9.7637030 Refraction of the Red

Angle which the violet makes with the red 12'.6"

Problem 2d

Conceive the most refrangible & the least refrangible ray, at the same Incidence to be refracted from Air into Water, in such manner that those two Rays shall after Refraction diverge at an angle of 12'.6". Required the different angles of Refraction & the common angle of Incidence?

In this Problem the difference of two angles of Refraction together with the Ratio of the Sines of those angles, to wit that of 108:109. is given.

Let 108, & 109 be the two sides of a Triangle, comprehending the given angle 12'.6". One of the Angles at the base of this triangle, & the supplement of the other angle at the Base will be the angles of Refraction sought.

Therefore the Sum of the Sides 217. is to the difference of the sides, 1 as the Tangent of half the Sum of the angles at the base viz 89°53' 57" is to the Tangent of half their difference |

7/II/11, 2

Log Tangent 89° 53' 57"	12.7537749
Log of 217 Subtract	2.3364597
Log Tangent of 69° 3' 57"	10.4173152

The Sum of these two angles being 158°.57' 54" is the greatest of the angles at the base of the Triangle: And its Supplement 21°.2'.6" is one of the Angles of Refraction sought.

The difference of the two angles whose Tangents are above set down being 20.50'.0" is the other angle of Refraction sought, & we see their difference is just 12' 6"

Sine of Refraction of the most Refracted ray viz of 20° 50' is	9.5510237
Add the Ratio of 81:108	0.1249388
The Sum is	9.6759625
Add to this the Ratio of 108:109	40027
	9.6799652

The Sum is the Sine of the Common Incidence Sine of 28° 16' 36"

Problem 3

Let us now suppose, that a Ray of the Suns Light falls from Air upon a Water Prism* at angle of Incidence of 28.16' 36"; that it is there refracted as mentioned in the last Proposition; the angle of Refraction of the most refrangible being 20° 50' 0" of the least refrangible 21° 2' 6". So that these extream rays diverge in the Water from the point of Refraction, in an angle of 12' 6"; Suppose farther that in this State of divergence they fall at a proper distance upon a glass prism each of whose angles is 60° & whose base is parallel to the horizon; That the most refrangible ray falls on the glass prism above in an angle of Incidence of 35° 28' 35", the least refrangible below in an angle of Incidence of 35° 16' 29", the interval between the least refrangible & the most refrangible being Unity upon the Surface of the Glass prism. We suppose all the refractions mentioned, and to be mentioned in this proposition, to be made in the same vertical plain, to which all the refracting surfaces are perpendicular. It appears from Proposition 1. that the red & violet rays will have equal angles of emergence from water into the glass, & that they will pass through the glass parallel to each other, & emerge into the water again at the same distance of unity on the farther side of the glass prism, & having angles of emergence equal to their angles of Incidence from water upon the glass prism, respectively. So that in the Water they will again converge in an angle of 12' 6" & meet
7/II/11, 3 in a point at the | same distance and in a similar situation with regard to the glass prism, with the point from which they first diverged. And if at this point they be again refracted from the water into air having their angles of incidence respectively equal to their angles of emersion at first out of air into Water, it is evident from the second proposition, that at this last Emersion the angles of emersion of the most and of the least refrangible Ray will be the same

These things being supposed it is required to determine the Angles which the incident and refracted ray, & which the refracting Surfaces of the water prism make with the Horizon, & to determine the situation of the glass prism inclosed in the water prism.

1 It is evident that in the glass prism the rays move parallel to the Horizon. In passing through the water to the glass prism the most refrangible ray is elevated above the horizon in an angle of 5° 28' 38" and depressed below it in the same angle on passing through the water on the other side. This Ray makes with the perpendicular to the refracting Surface of the water prism an angle of 20° 50' and consequently with the refracting

Surface of water an Angle 69° 10'. If we take from this the elevation of this Ray above the Horizon the remainder is the inclination of the refracting surface of the water prism to the Horizon viz 63° 41'.25"

2 Subtracting the angle of Refraction of the most refrangible ray viz 20° 50' from its angle of Incidence 28° 16' 36" we have the angle which the incident ray produced makes with the refracted viz 7°.26' 36". Taking from this angle the elevation of the refracted ray 5° 28' 35" there remains 1° 58' 1". by which the incident ray dips below the horizon. And as the emergent ray on the other side will have an equal dip, these rays produced, would not be parallel but make an angle of 3° 56' 2"

3. As to the situation of the glass prism, that is determined by resolving the small triangle, two of whose sides are the most refrangible & the least refrangible ray, lying between the point of their first emersion in water, & their incidence on the glass prism, the angle comprehended by these two sides is 12' 6" & the side upon the glass prism subtending this angle we have called unity. The angle at the base opposed to the least refrangible ray appears by Proposition 1 to be 54° 31' 25". And thence the side opposed to it, to wit the path of the least refrangible ray in the water till it meet the glass prism will be found to be 231,44 |

7/II/11, 4 Suppose now a Ray of the Suns Light to fall horizontally upon a water prism, the angle of Incidence being 30° & the Refraction upwards

Sine of Incidence 30°	9.6989700
Log Ratio of 108:81	0.1249388
Sine of 22° 1' 27½"	9.5740312 Refraction of Red
Ratio of 109:108	40027
Sine of 21° 48' 42½"	9.5700285 Refraction of violet

Difference of Refraction of red & violet 12' 45"

The most refrangible ray has now an elevation of 8° 11' 18" And we are to find an angle of Incidence & one of Refraction from Water into glass which have this Difference

This is to be done by finding the angles at the base of a triangle whose sides are as 202,2:174,2 & which comprehend an angle of 8° 11' 18.

Sum of the sides 376,4: to their difference 28 : : Tangent of half the Sum of the angles at the base viz 85° 54' 21": to Tangent of half their difference

Log Tangent of 85° 54' 21".	11.1452178	85 54 21
Log of 28+	1.4471560	46 8 3
	12 5923738	132 02 24 greater angle at base

Log of 376,4	25751878	47 57 36 Supplement
Tangent 46° 8' 3"	10.0171860	39 46 21 less angle at the base
Log Tangent of 85° 54' 21".	11.1452179	85 54 21
Log 27+	1.4313638	45 8 16
Sum	12.5765817	131 2 37 Sum
Log 375.4	2.5744943	48 57 23 Supplement
Log Tangent of 45°.8'.1"6	10.0020874	40 46 5 Angles of Incidence & Refraction

VI

7/III/13, 5* in their Eye was not circular but oval, the longest diameter being vertical, and this indicated that the surface of the Cornea in those Eyes was not part of a perfectly spherical Surface; but of such a figure that supposing one line horizontal another vertical, to divide the Surface of the Cornea into equal parts; the vertical line had less convexity than the horizontal.

But I lay so little Stress upon this Observation that I would not have mentioned it but that you may take notice if any thing of this kind that falls in your Way. I perceive nothing of it in my own Eyes nor indeed in any Eye I am now acquainted with.*

As to the real Cause of the Appearances above mentioned, I think, it may be concluded from them that the Rays falling upon certain parts of the Aperture of the Eye, produced little or no Effect in Vision, while the Rays falling upon other parts produced the Effect which from the known laws of Opticks we would expect. If it be asked why those Rays did not produce Vision I can onely think of three Causes possible and am unable to say which of them is true either the Rays falling on those parts are totally reflected by the Cornea or some of the Humours, or secondly they are lost in th‹ei›r passage by some opacity in those parts or thirdly the parts of the Retina on which they fall have lost their sensibility.

To show, that the Rays falling upon some parts of the Aperture of the Eye produced no sensible Effect in the Vision. Let us suppose the Aperture of the Eye divided by imaginary lines into many small Spaces, some above and others below, some to the right and others to the left, & others about the middle.

1 By the known Laws of Opticks the Rays passing through each of these little Spaces, form an image which is distinct in the Focus, though faint

in proportion to the Aperture of the little space through which the Rays must pass. When these Images are received by a plain in the Focus they all coincide in all their parts, having the same size the same position and being all perfect images of the Object. And therefore they all together make one bright Image, to every part of which, every little space of the aperture contributes its share of Rays according to its particular aperture. If any one of the little Spaces is shut so as to admit no Rays, no part of the image is lost, but every part has fewer Rays and consequently is less bright. The half of a lens or the tenth part gives an image as large and as perfect as the whole lens, the difference lies onely in this that the Image made by a part of the Lens is faint in proportion to the Number of Rays which form it.

2 Suppose now the plain on which the Images are received to be nearer than the Focus, as the Retina is in the case of Objects seen indistinctly by being to‹o› near the Eye. In this Case the Images from the different little Spaces do not perfectly coincide. If the Object be very small the Images formed by two little spaces most distant from each other, perhaps may not touch, but have some interval between them. The images formed by two other little Spaces that are nearer to each other will interfere more or less according to the distance of the little Spaces from one another. From this it ‹is› easy to see that the Effect of all the Images taken together will be one indistinct and ill defined Image of the Object which will be very faint in the Margin and stronger towards the middle. If the Object be of one uniform Colour, that Colour will be deeper towards the middle of the image and fainter towards its Margin. If it be of different Colours these Colours will be more mixed towards the middle of the Image, & purer but very faint towards the Margin. |

7/III/13, 6 But though we may distinguish one of these Images from another in our Imagination, it seems impossible to do it by Sight. No Eye can trace any one through such a Number that are mixed with it, especially when taken separately they are so very faint. The appearance must be one large indistinct image of the object such as we have described, without any appearance of duplicity or multiplicity.

3 Suppose two little Spaces of the Aperture to be open, one on the right and another on the left part of the pupil, & all the rest to be shut. The Rays passing through these two spaces will form two Images on the corresponding sides of the Retina, which two Images if they have light enough to make them visible and are at such a distance as not to interfere will shew the Object double. The two Objects standing to the right and

left in the contrary order to that of the Spaces through which the rays that form them pass.

This is finely illustrated by the pin holes.* I make in strong white Paper four pin holes in the form of a Square or even Six in the form of a Hexagon, all which fall within the Aperture of the Pupil. I apply them almost close to the Eye. Then placing a small object well illuminated, like the head of a pin, beyond the nearest limit of distinct Vision, I see one Object distinct and Strong. The four or six Images all perfectly coincide upon the Retina, and make one strong Image.

Then I bring the Object within a foot or eight inches of my eye, and I perceive four or six Objects, each compleat in all its parts but more faint, having onely the fourth or sixth part of the Rays that the former had.

I turn the Paper round on its own Plane, the Images go round in the same direction, & make a pretty kind of Dance; for each of them keeps its position while it changes its place; so that what was at first the upper part of the Image continues to be so through its whole Revolution.

You see therefore that in indistinct Vision a very small Object may appear double triple, or even sextuple to one Eye, providing it be seen by Rays passing through as many small Spaces of the Pupil as distant from each other as may be, and no Rays pass through the other parts of the Pupil to disturb & confound the Images, by filling up their intervals.

From these principles which follow from the known Laws of Opticks and Are confirmed by Experiment, I conclude that in the Cases of double Vision in my Eyes, the Rays falling upon certain parts of the Pupil were ineffectual to produce Vision, while the Rays falling upon other parts had all the Effect we would expect.* |

7/III/13, 7 Thus when I see a vertical black line upon paper, double, a strong image to the left and a faint one to the right as in the 4 Observation above,* it seems to me evident that the strong image was formed by the Rays falling on the right part of the Pupil & the faint by those falling on the left. But if the Rays falling on the other parts of the Pupil had been effectual to cause Vision they would have filled up the interval of the two Images and perfectly confounded them.

Again when I place the line Horizontal as in the same Observation, and see two lines one above the Other, it is plain that the upper line is formed by the Rays that fall on the lower part of the Pupil and the lower line by the Rays that fall on the upper part. But the Rays that fall in the middle have not any effect.

So far I think we go upon sure Ground.

It seems evident likewise that the faintness of one of the lines seen by the left Eye in the fourth observation, was owing to some peculiarity in that Eye, because it was not common to both Eyes

The thing that remains to be accounted for is, why the Rays falling on some parts of the Pupil were effectual for producing vision while the Rays falling upon other parts at the same time were not.

I mentioned above, three Causes as all the possible ones that occur to me of this

The Rays which I call ineffectual either are lost in their passage by some opacity in the coats or Humors, or they fall upon parts of the Retina that are insensitive, or they are totally reflected.

The two first of these Causes seem to be excluded by the Experiments of the pin holes. I look at a pin head through a small pin hole. I carry the pin hole successively up and down, to the right and to the left over all the parts of the pupil The Object is seen equally distinct, by what ever part of the pupil the rays are admitted I make three four or six pin holes all within the Aperture of the pupil. I then at eight inches distance see as many pin heads as there are holes, without any remarkable difference as to faintness. I turn round the holes in the plain of the paper; the pin heads turn round and are visible through the whole rotation.

From this I conclude that there is no part of the pupil where the rays are lost in their passage nor any part of the Retina where those Rays fall that is insensible to them, in this Experiment.

So far I think I may go in excluding certain Causes of this Phenomenon that might be assigned But to find the true Cause, *hic Labor hoc Opus*,* and indeed it is beyond my ability.

Sir Isaac Newtons Doctrine of the Fits of easy Reflection & easy Transmission of the Rays of Light, may for what I know give some Light to this Subject, but I have not that doctrine so clear in my head at present as to be able to say whether it can give aid in the Solution of this Phenomenon or not.* If it does not, I suspect the Phenomenon must
7/III/13, 8 depend on some Law of nature | hitherto unknown.

Though no part of Natural Philosophy has furnished so many Noble Discoveries as Opticks, yet I apprehend there are principles yet to be discovered in that Branch, and particularly that Attention to the Phenomena of what is called indistinct Vision may open some new Lights. Herschels fine discoveries in the fixed Stars shew that by indistinct Vision discoveries may be made in Nature beyond the most sanguine hopes of philosophers* The Laws of indistinct Vision therefore deserve to be inquired into on account of its Utility, & perhaps when those Laws are better understood they may throw Light upon other Subjects.

When Mr Herschel by his Telescopes magnifies a fixed Star three of four thousand times, the Vision is extreamely indistinct, in the Sense of Opticians; that is, the rays from one point of the Object, are so far from meeting in one point of the Retina, that they cover a large Circle upon the Retina. Yet this Vision, so indistinct in the sense of Opticians, is so distinct in the common popular Sense, that it shews a fixed Star to be double, triple, or quadruple, which with the best telescope fitted for distinct Vision appears onely to be single

I have made an Experiment with the pin holes, so often mentioned, which would lead me to believe Herschels discoveries in the fixed Stars, even if I was disposed to doubt of them, which I am not.

In a thin plate of Brass I have several small holes made, fit onely to admit a human hair, & so near to each other that the pupil can take in several of them at the same time In a good Light I apply these as close to my Eye as possible, that is, within little more than a quarter of an inch. I see them as so many pretty large circles well defined, even when they interfere with one another for more than half the diameter.

I can perceive their Figure whether perfectly circular or not, their comparative Magnitude, the distance of the Centers whether equal or unequal, & their position whether in a streight ‹line› or otherwise; All this I perceive with the naked Eye, better than with a microscope which magnifies them greatly & gives distinct Vision.

I add that the little holes seen in this way resemble very much the figures which Herschel has given us of the Appearances of double triple & quadruple fixed Stars as seen by him*

VII

7/VIII/2, 1r Proposition 1. There is no aberration of the Axis of a Telescope, whether of Air or water, from terrestrial Objects seen by it.*

1 Let AE be the path of the Earth in its orbit, AB an air Telescope standing at right angles to AE & directed to the lucid point O, which is supposed to move along with the Earth and Telescope, so as to be always in the axis of the Telescope

Let OD be a Ray from O, which passes from O to D in the time that the Telescope moves from AB to CD, and enters into the Axis of the Air

Telescope at D. It is evident that, abstracting from the acceleration of the Ray in passing through the thickness of the object glass, the Ray will go on uniformly, arriving at E when the Telescope is in the position EH & the lucid point at I, & that in the whole of its passage from D to E it will be in the axis of the telescope; & therefore when the Telescope is in EH the lucid point will appear to be at I where it really is. Whence it appears, that the Axis of the Air Telescope has an aberration from the path of the Ray equal to the Angle OEI but the lucid point having moved from O to I while the Ray moved from O to E, the axis continually points to the real place of the lucid point, and it is seen in that axis by the Ray OE; so that there is no aberration. which was first to be proved.

2 Let CD be now supposed to be a water Telescope into whose axis the Ray enters as before at D, and that its velocity in the water is accelerated. If we resolve the motion of the Ray in the water in to two motions, the first in the direction DC perpendicular to the surface of the water, the second in the direction DH parallel to that surface; the acceleration of the Ray is wholly in the first of these directions, & no part of it affects the motion in the second direction. The Ray therefore, |
7/VIII/2, 1v continues to move in the direction DH with the same velocity as before, & consequently will arrive at the line HE produced, in the same time as when it moved in Air.

Let the Velocity of the Ray in Air, in the direction DC or HE be to its velocity in Water in the same direction as HE to HG; it follows that the path of the Ray in the Water will be DG, which it will perform in the same time that in Air it passed through DE. Let the water Telescope be lengthned at the lower end so as to reach from H to G. It is evident that the Ray DG will be still in the axis of the water Telescope while it is carried from C to E and that the lucid point will likewise be still in that axis produced, and that there will be no aberration of the axis of the water Telescope from the lucid point O. But the Ray by which that point is seen will have the direction OD while it moves in air & the direction DG when it moves in the water.

Proposition 2. The aberration of the Axis of a Telescope from the place of a fixed Star is the same whether the Telescope be of Air or of Water.*

The same Figure will serve for this proposition, the same suppositions, & the same reasoning, with this difference onely that the point O be supposed a fixed Star and at an infinite distance.

If the Ray OE coming in a right line from the fixed Star moves from D to E while the Earth, and air Telescope moves from C to E, then the

Angle HED will be the aberration of the axis of the Telescope from the
7/VIII/2, 2r place of the fixed Star. | In the water Telescope the path of the Ray will be ODG as before, which will be always in the Axis of the Telescope while it is carried parallel to itself from C to E. Therefore the aberration of the Axis of the Telescope from the absolute direction of the Ray in its passage through the water will be the angle HGD. But its aberration from the place of the fixed Star will be measured by the angle which the axis GH makes with the Right line drawn from G to the fixed Star, which because of the infinite distance of the fixed Star will be equal to the angle HEO. Q.E.D.

Corollary 1 If the incident Ray at D were accelerated even to infinity it would still move along the Axis of the Telescope & therefore be seen in the direction CD parallel to EH and therefore the aberration would be the same as in the Air Telescope.

Corollary 2 However the Ray is accelerated or retarded by the Action of the Medium in the Telescope, the aberration will be the same; so that the aberration depends solely upon the velocity of the Ray before its Incidence at D.

Corollary 3 When the Ray is not accelerated or retarded by the medium in the Telescope its absolute path is not changed. When it is either accelerated or retarded by the Medium in the Telescope its absolute path is changed.

Corollary 4 In the Air Telescope the Velocity of the Ray to the Velocity of the Telescope was as DE to EC; in the Water Telescope it is as DF to FC or as DG to EC.

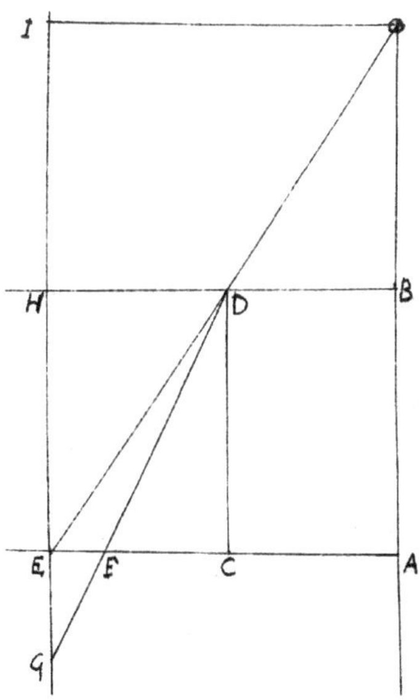

VIII

3/III/11, 1r April 10 1787.

Problem

To find the aberration of the Axis of a Water telescope from a fixed Star when the Earths Motion is at right Angles to the Axis of the Telescope.

Solution. As the velocity of Light in the Telescope, is to the velocity of the Earths Motion, so is Radius, to the Tangent of the Angle of Aberration

Corollary. As the velocity of Light in a water Telescope is supposed to be a fourth greater than in a common Telescope* it follows that the aberration of the water telescope is one fourth less than that of the common Telescope; & if the aberration of the Water Telescope be found to be a fourth less than that of the common Telescope, it follows that the velocity of the rays of Light in water is one fourth greater than in Air. |

3/III/11, 1v Proposition The Axis of a Telescope whether of Air or of Water has no aberration from a terrestrial radiant point.

Let AB be the Axis of an Air Telescope at right angles to the Motion of the Earth in its orbit AC

While the Earth moves uniformly from A to C and the radiant point from B to F, let the Ray move uniformly from B to C; it is evident that Ray will be all along in the Axis of the Telescope; and therefore when the Axis is in CF the Radiant point will appear to be in F, where it really now is. Had the radiant point remained at B, where it was when the Ray BC issued from it, the Axis of the Axis of the Telescope would have had an aberration from it of the Angle BCF. But the Radiant point having moved through that angle during the passage of the Ray from B to C there is no aberration.

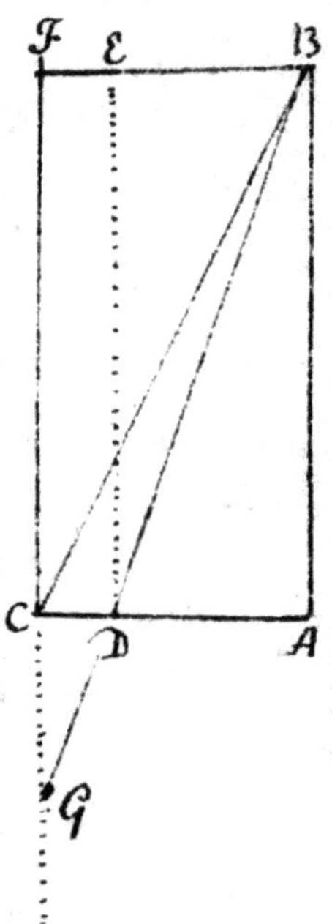

Suppose secondly that the Telescope is filled with water and that thereby the velocity of the Rays is increased so that a Ray BG reaches the point G in the same time that in the Air the Ray would have moved from B to C; the point G being in the right line FC produced.

Since BG is to BC as the velocity of a Ray in Water to its velocity in air, and since the velocity of the Ray in the direction AC is the same as before, it follows that the point G will be in FC produced

Let BG meet the line AC in D; then while the Telescope is carried from AB to DE. The Ray BD will move along its Axis and at DE will shew the radiant point at E where it really is QED

IX

7/VIII/9, 1r Suppose the Rays of Light that give the Different Colours to have Different Velocities, so that Red and Blue Rays coming from a Star at the same time, the last shall come to our Eye five minutes later than the first

Suppose also that according to the Ptolemaick System the Earth is at rest, & the fixed Stars move round it in twenty four hours; Should not that Star appear to us as an oblong coloured Spectrum.*

Part Five: Electricity

I

6/V/11, 1r

Electricity*

Definition 1 Electricity is a Power or Virtue in Bodies, excited by Friction, by which they attract or repel other Bodies*

2 That State in which bodies do thus Attract or repell others is called an Electrified State

1 Proposition Some bodies are electrified of themselves by friction others are Electrified by touching or approaching an electrified Bodie

Definition The first I call Electric Bodies the second Non Electric*

2 Proposition Amber Rosin Sulfur Sealing Wax Silk Hair Cat Gut. Glass Christal when dry and free from all Moisture are Electric Bodies

All Liquids Moist or Wet Substances & Metals are Non electrics.

3 An Electric cannot be Electrified to any Considerable Degree by the touch of an Electrified Body. but onely by friction patting &c

4 A Non Electric cannot be electrified by friction but onely by touching or approaching an Electrified body.

5 An Electrified body Repells Electrics but Attracts Non Electricks till it electrifies them & then immediately it repells them.

6 An Electrified Body tends to comunicate Electricity to the circum-Ambient Non Electricks to their utmost Extremities in a very Short time and thereby ceases to be in an Electrified State.

7 The Electrical Virtue is communicate from one Bodie to another by a Flash of Fire & a Snap. Some times Sufficient to kill birds & other weak Animals

8 A Globe or Hollow Tube of Glass exhausted of Air does not by Friction attract Non Electricks from without but is filled with a purple Light. & exerts its Electric force inwardly

9 Vitreous & Resinous Electricks seem to differ in this that bodies Electrified by the One are Attracted by the other & vice versa.*

10 An Electrified body does not sensibly attract light Non electrics when it is nearer to heavy ones.

11 The Electrical Virtue can hardly be raised or made to produce very sensible Effect‹s› in Moist Air

12 It produces the Greatest Effects in cold & dry Air*

13 The greatest Shock from Electricity seems to be when two Nonelectricks in an Electrified State touch one another.|

6/V/11, 1v

Conjectures concerning Electricity to be confirmed or refuted by future Experiments*

1 Electricity seems to arise from ‹a› very Subtile Fluid which in its Natural State is equally diffused through all Bodies. & so is in a kind of Equilibrium*

2 By the Friction of Electric Bodies this Fluid seems to be collected in a greater Quantity about the Body rubbed from the nonelectric Bodies that touch it. But as soon as the Friction ceases it diffuses it by means of the non Electrics that touch it or come near it

‹3› Bodies that are long & sharp pointed both receive and communicate the Electrical Fluid at a much Greater Distance & with a less flash & Shock than those that are obtuse*

‹4› When a Body has a greater than the Natural Quantity of the Electrical Fluid it is said to be Electrified positively or Plus*

‹5› When a Body has less then the Natural Quantity of the Electrical Fluid it is said to be Electrified Negatively or Minus

What is said in the propositions above relates to positive Electricity. The Effects of Negative Electricity Correspond to them.

‹6› The onely way we know by which the Natural Equilibrium of the Electrical Fluid is destroyed is by Friction & all the Shocks & Flashes Attractions & Repulsions that Occur in Electrical Experiments are owing to the Effort of the Electrical Fluid to recover its Equilibrium, by getting in to Bodies that have less of this Fluid from those that have more of it*

‹7› Glass has this peculiar Quality of all the Bodies we know that in a Glass Vial or a pane of Glass in proportion as the inside of the Vial or the one Side of the Pane is positively Electrified the other is negatively Electrified & the Equilibrium is restored by a Shock when there is a Communication by non electrics between the Inside & the out Side* |

6/V/11, 2r The parts of the Electrical Fluid seem to repell each other but to attract all other Bodies.* Glass Bodies attract this fluid so strongly as to retain always their proportion of it so that one side of the Glass can be deprived of it onely by accumulating it in the other‹.› Glass by friction is disposed to draw a greater Quantity of the fluid from circum-ambient Conductors but as soon as the friction ceases this surcharge of the Electric fluid is ‹again› drawn of‹f› by the nearest Conductors. The Fluid passes Silently & imperceptibly through a continued Conductor & never shews it self but in passing or endeavouring to pass from one conductor to an other at a smaller or greater distance. In this Case it is that it flashes snaps attracts the body it is passing in to & produces other Surprizing Effects. Which are the Greater the More the Fluid is accumulate in the one Body & the more the other Body is destitute of it

Query. Whether the Difference between Resinous & Vitreous ‹Electrics› does not ly in this that A Globe of Sulphur or Rosin ‹drives› the fluid from the Barrel & throws it off by the Action a Globe of Glass draws by the Cushion & throws it out by the barrel.*

The Electrical Fluid in passing from one Conductor to another will sometimes pierce a hole through a Quire of Paper. it ‹will› Melt Metals change the polarity of Magnetic Needles or give a polarity to th‹o›se that had none before*

There appears to be an Accumulation of the Electric fluid in some Clouds especially those that give Thunder, & lightning is probably caused by the passage of this Fluid from the Clouds to other Bodies.*

When bodies are equally exhausted of the Electrical Fluid they Repell one another which seems not to be accounted for*

Is it not probable that there are two Electric fluids the Vitreous & the Resinous. That when a body is thought to be negatively electrifyd it is onely deprived with the vitreus but is by that means the more charged with the resinous fluid. & that the natural State of Bodies is that where the two fluids are joyned in such proportion as to saturate each other. Might not the one be called Male the other Female* |

6/V/11, 2v 1756

November 16 A Piece of common Sulphur which had lain long in a Drawer being taken in my warm hand in a remarkably Rainy day Snapped several Times and left a smart feeling in my Arm like a touch of the Electrified Barrel.* Dr David Skene felt the same who took it in his hand after I had carried it through two Rooms in Mine.* And a third person taking it it snapped Still but left no pain.

1757
December 23 A pair of black silk Stockings when drawn of‹f› my legs snapped very much. NB the Weather was frosty & there was a pair of Worsted Stockings below the silk ones which were not drawn off at the same Time*
January 1758 Pittatoes burnt yield so great a Quantity of Vegetable Salt that one cannot hold his tongue to their Ashes without pain.* And those ashes become quite wet in a few Hours
February 22 1758 Starch made of Pittatoes burnt in a Crucible in a Coal Fire for about an hour Smoaked bubled up and at last Continued Red for half an hour without any Alteration when cooled it was perfectly black and Spungy broke with a very Smooth Shining Surface, like coal. And being twice burnt again in an Open fire it yielded no ashes nor had any Saltish Taste, nor indeed seemed to undergo any alteration in the Fire

II

6/V/20, 1r Mr Epinus a worthy member of the Accademy of Berlin, has made some new Electrical Experiments tending to convince Naturalists that after all the curious & surprizing Phænomena produced by the Electrical Shock they are very far from having discovered the nature & extent of this Power or the universal Laws to which it is subject.* A very rare Gem transparent and of a brownish Tinct, which has three times the specific Gravity of Water, and is found upon the shores of the Island of Ceylon and called Trip or Tourmaline has given this Philosopher an opportunity of making surprising and curious observations.* This stone is strongly electrified by heat onely without any friction & when heated on Coals, it alternately attracts and repells the Cinders, from this remarkable property the Du‹t›ch have given it the name of Aschen trecker or ashes drawer* The Electricity of the Tourmaline is constantly both positive and negative, for when one side is positively Electrical the other Side is negatively so. This is proved by putting it for a few minutes into boyling water, for it is observed that the Electricity of this stone is rendred more
6/V/20, 1v active | by boyling water, whereas in all other Cases water is known to obstruct the Electrical Virtue* A Third Observation of Mr Epinus is That the negative side may be rendred positive & vice versa. by laying the Tourmaline on a plate of Metal Glass or live coals but afterwards the stone of it self returns to its natural State. It is always in this State

when its two sides are heated to a degree nearly equal but if one be considerably more heated than the other the Tourmaline varies from its natural State.* In fine it is also electrified by friction, so that Tourmaline is susceptible of two Species of Electricity entirely different from each other, one produced by friction the other by heat without friction.*

I must not omit the Duke de Noya Caraffa's letter to Mr Buffon.* This Nobleman so highly distinguished for his consummate knowledge in Natural History & Philosophy having bought two Tourmalines from a picture dealer in Amsterdam, made repeated Experiments with them & among other particulars informs us. That the Tourmaline which when electrified by heat alternately repells and attracts light bodies; is simply attractive without the least repulsive power when electrified by friction:
6/V/20, 2r That it differs from other Electrified bodies in these six | particulars 1 In being electrified by heat alone and much more than by friction. 2 When electrified it gives no Light nor sparks. 3 Its virtue acts through water. 4 it looses not its electricity by points nor any of the means practicable in the Electrical machine. 5 Instead of being repelled by an Electrified Tube it is attracted 6 Two Tourmalines suspended on threads & heated are seen in mutual Attraction, whereas other electrical bodies repel each other.*

Part Six: Chemistry

I

7/III/6, 1 1770

Of Heat

Heat is a quality or State of bodies of which we get the first Notice by th‹o›se feelings of heat and cold which follow upon the application of other bodies to our body. And all that we learn of it by feeling, is that the same body at one time gives us the feeling which we call heat at another time that which we call cold, and at another time neither the one nor the Other. As the Sensations of heat and cold have various degrees we in like manner attribute various degrees to the quality or State of the body which produced it and which is known onely by this effect of it. By more attentive observation we find that the State of a body which we call heat is accompanied with an Expansion of that body in all its dimensions and that the degrees of Expansion in the Same body when all other circumstances are the Same correspond to the degrees of heat. Thus while my hand has the same temperature, if twenty different vessels of Water which expand the Quicksilver in the thermometer in as many various degrees in order, are set before me by feeling one after another in order I perceive that every vessel of water has a greater or a less degree of heat according as it expands the Quicksilver more or less.

Therefore as the Expansion corresponds to the feeling of cold and heat we conceive the unknown cause of both to be one and the Same and call it by the name of heat.*

As a Cause which is known onely by its Effects can onely be measured by its Effects; We are under a necessity of pitching upon some one or other of the Effects of heat by which we may measure its various degrees. At first Men measure heat by the feelings of heat and cold which are produced by the heated body when applied to theirs. And this way of Measuring the degrees of heat continued to be used untill thermometers were invented As the feeling of heat and the feeling of cold are contrary feelings it seems most natural while we measure heat by the feeling to conceive contrary states or qualities in bodies corresponding to the contrary feelings, and a state intermediate between them wherein the body is neither hot nor cold. In this way the Peripateticks considered heat and Cold, but as they observed that some bodies by being taken into the Stomach as Spices and Spirituous liquors, others applied to the Skin as Nettles produced the feeling of heat more slowly and more permanently than the feeling of heat is usually produced by others, they distinguished between actual & potential heat. Ascribing to the various kinds of food and Medicaments different degrees of potential heat & potential Cold.*

According to this System therefore the feeling of touch must be the measure of actual heat, & the feeling of warmth in the blood and over the body occasioned by the Use of any kind of food or Medicine is the
7/III/6, 2 Measure of the potential heat | and it would be a contradiction to say that any body was actually hot which did not feel hot or that any meat or drink was potentially hot which did not produce some more permanent feeling of heat in the blood or in some part of the body.

More accurate observations discovered that our feelings of heat and cold do not correspond always to any state or quality of the bodies which occasion those feelings, but rather to a certain relation between other bodies and our body. Thus when one hand is warm and the other very cold, water which is of a middle temperature between the two feels cold to the one and warm to the other. There must therefore be a certain state or temperature of the Water and a certain State or temperature of the hand, and heat in the Sense we have hitherto used the Word is neither the one nor the other but something depending upon both.

As soon as this came to be observed it was natural for men to seek after some Measure of the various degrees of that State or temperature both of our Body and of other bodies upon which this feeling of heat depends. The expansion of bodies seemed to offer it self as the most proper measure of this temperature and has accordingly been very generally received by philosophers.*

But it is to be observed that this new Measure of heat really changes the Notion of it or leads us to give another meaning to the Name heat that body may have a considerable degree of heat according to this new notion of the Word which according to the former Notion had no degree of heat but a great degree of Cold.

It would seem however that this Measure of the degrees of heat by the thermometer is not a thing capable of proof nor does it need it. It is very justly taken for granted. And when I affirm that to be a greater degree of heat which raises the thermometer higher or expands mercury more this is onely to be considered as a definition of what I mean by a greater and what by a less degree of heat. As therefore it would be improper to offer proof of a Definition; It seems to be improper to offer any proof of this Measure of heat. As when Mankind measured heat by their feeling it would have been an impropriety to offer to prove that what feels hotter is really hot‹t›er. So now when we assume the Expansion of Quicksilver as a Measure of the degree of heat it seems to be an impropriety to pretend to prove that heat is justly measured by that Expansion.

But let us consider the Argument brought to prove that the expansion of bodies is a just measure of the degrees of heat‹.› I think it amounts to this. That if you mix equall quantities of water of different temperature the whole Mixt must have a temperature which is an Arithmetical Mean
7/III/6, 3 between the two Extreams | Now it is found by Experiment in such cases that the heighth of the thermometer in the mixt Water is always an arithmetical Mean between its heights or the temperatures of the ingredients.

In this Argument we may observe first that it is taken for granted that two equal quantities of the same kind of body of different temperatures when mixed will have the temperature which is an arithmetical mean between the two. Here one Measure of heat is assumed in order to ascertain another, which may be assumed with as good reason‹.› I conceive therefore there is nothing unphilosophical in laying it down as a definition of the degrees of heat that they are greater or less according as they expand the same body more or less. And that the Experiment before Mentioned enables us to draw a conclusion in the philosophy of heat viz that when equal quantities of the same fluid of different temperatures are mixed they produce a temperature which is an arithmetical mean between the two extreams.*

Query. Whether unequal Quantities of the same Matter and of different temperatures. have a temperature which divides the interval between the two temperatures in the ratio of the Quantities. Thus if one pint of Water of 40 degrees be mixed with two pints of 70 degrees will the mixt be 60 degrees. Or if two pints 40 degrees be mixed with one pint 70 degrees will the mixt be 50 degrees & so in other cases? If this holds in all the variety of cases in which it may be tried it leads to another Query

Query May we not find a Geometrical Ratio in heats, thus? Two pints of 40 degrees to one pint of 70 degrees make the mixt 50° Ex hypothesi. But two pints 60 degrees to two pints 40 degrees make the mixt 50° Here we have two mixts of 50 degrees each; the one consisting of three pints the other of four, therefore the heat of the one to the heat of the other is as 3:4. Therefore the heat of two pints of 60 degrees + 2 pints of 40 to the heat of one pint of 70 + 2 pints of 40 is as 3:4 consequently the heat of one pint of 60 degrees to the heat of one pint of 70 is as 3:8. Another Example 3 pints of 40 to 1 of 80 make the mixt 50. But 3 pints of 40 to 3 pint of 60 make the mixt 50 The heat of the first mixt is to that of the second as 4 to 6 therefore the heat of 3 pints of 60 is to the heat of one pint of 80 as 2:6 consequently the heat of one pint of 60 degrees to the heat of one pint of 80 is as 4 : 18 : : 3 : 13½ Another Example in Spirit of Vitriol 3 Pints 70 degrees to 3 Pints 30 the mixt is 50 degrees. 3 Pints 70 to 1 Pint -10 the mixt is 50 degrees the whole heat of the first mixt is to that of the second as 6:4 and in equal quantities as 4:4 this reasoning to be farther Examined |

7/III/6, 4 1766*

Experiment by Dr Black. Two florence flasks* had 6 oz Water put in each which in the one was made to freeze, in the other brought as near as possible to the freezing point without freezing that is to about 33 degrees. Both were set to warm in a warm large room. the unfrozen Water soon came to the temperature of the room, the frozen Water took 11 or 12 hours to dissolve and for the greatest part of this time was not sensibly heated. Supposing that the frozen water had as much heat communicated to it every half hour as the unfrozen had the first half hour it was found by calculation that about 136 or 140 degrees of heat were necessary to give fluidity to the ice before it received any accession of sensible heat.*

Experiment 2. by the Same 6 oz of Ice had 6 oz of boyling water poured upon it. the ice was melted immediately. the Mixt was about 52 degrees.*

From these Experiments it appears that Water in passing from ice to water absorbs about 140 degrees of heat which it retains while in a fluid form without communicating any part of it to the ambient bodies. Whatever more heat it acquires it communicates to the bodies around it till it comes to an equilibrium with them, but the heat necessary to fluidity is not to be reckoned in this equilibrium.*

Experiment 3 by Mushenbroeck. repeated & improved by Dr. Black. When the air is ten degrees below the freezing point set a vessel of Water as a beer glass which is deep and Narrow in the Air to freeze. let the water remain perfectly stagnant without any Motion. I‹t› will descend regularly below 32 degrees even to 22 degrees without freezing. but as soon as it is disturbed immediately ‹a› number of Icy spiculæ are formed in it and the heat of it at the same moment rises to 32. & continues so till all is frozen.*

Experiments 4, 5 Sperma Ceti loses of its sensible heat about 150 degrees in passing from a Solid to a fluid form. Bees wax about 160°. As these bodies do not pass instantly from a Solid to a fluid state as water does the heat which renders them fluid is not absorbed all at once but by degrees as they pass through greater and less degrees of hardness to perfect fluidity.*

Experiment 6. Several of the Metals have also been found by Experiments to have this quality of absorbing heat in becoming fluid particularly Silver & Cop‹p›er. And the former has something similar to what has been observed in Water Experiment 3 being capable by perfect rest to be brought below the temperature at which it usually becomes hard and then upon being agitated becoming suddenly hard with so great haste as to separate the parts from each other* |

7/III/6, 5 Fluidity the Effect of heat. All bodies have been rendred fluid by heat excepting Spirit of Vitriol & Air. The temperature that freezes Quicksilver of‹f› Farenheits Scale*

Supposing 10 oz of Spirit of Vitriol of O degrees be mixed successively with 1,2,3 &‹c› oz of Spirit of Vitriol of 100 degrees and that the heat of the Mixt divide the difference of heat of the two Extremes in the ratio of the quantities of Spirit of Vitriol reciprocally So that with the mixture of one, two, three, &c pints the mixt is severally 9°,1. 16,6. 23. 28,5. 33,3. 37,5. 41,1. 44. 47,3. 50 In general let a be the quantity of the fluid of the lowest temperature, b the quantity of the highest and let d & δ be their temperatures respectively, the temperature of the mixt will be $d + (\delta\text{-}d)/a \times a/a + b$ or making d the lowest Temperature & δ the difference it will be $d + \delta a/a + b$ and this multiplied into the quantity will give the whole quantity of heat in both bodies viz $da + db + \delta a$

$DA + DB + 2\delta A$

$D\tfrac{1}{2}A + D\times B + \tfrac{1}{2}A + 2\delta\tfrac{1}{2}A$* |

7/III/6, 6 Experiment 7 When Water is mixed with Salt or any other body which hinders its freezing it cools in frosty Air regularly to the temperature of the Air, and gives out none of its latent heat untill it begins to freeze*

Experiment 8 The frigorfick Mixtures compounded of pounded Ice and Salts are such as Suddenly convert the Ice into Water. And as in this conversion a great deal of heat must be absorbed and become latent hence appears the reason why such mixtures seem to produce so much cold.*

Experiment 9 The length of time which a Quantity of Water takes to freeze wholly when the Air is much below 32 shews the quantity of Latent heat contained in the Water. And the length of time which Ice takes to thaw wholly in Ice houses and in cold Air though much above 32 shews the great Quantity of heat necessary to fluidity.*

Of Vapour*

Every liquid has a certain degree of heat at which it is converted into an Elastic fluid. Water in this State is called Steam. But the degree of heat required to convert water into Steam is some what different according to the pressure upon it by the Air. The boyling point in Water is 212 when the weight of the Air is of a certain quantity but according as the weight of the Atmosphere is increased or diminished the boyling ‹point› is higher or lower So as to vary 4 or 5 degrees of farenheits Scale by the common variations of the Atmosphere, & in Vacuo water boyls with a heat of about 100 degrees*

As Ice is converted into Water by receiving about 140 degrees of heat into its composition which it retains while it retains the form of water without communicating any part of it to the Ambient bodies, so water is converted into Steam by receiving about 800 degrees more of Latent heat into its Composition, which it retains while it is in the form of Steam, and communicates to the surrounding bodies as soon as it is again converted into Water*

Experiment 10 8 oz of Water of 52 degrees temperature. Set upon a plate of cast Iron heated by a stove under it to a red heat. came to the boyling heat in 5', it took 20' to evaporate without taking any more sensible heat. Now as it received 160 degrees in five minutes, it is presumable that it received four times that quantity in 20' that is* |

7/III/6, 7 Experiment 11.*

Experiment 12 A Glass vessel with a Small quantity of Water closse stopt was heated to 222 degrees upon opening the Stopper a small quantity flew out at the Mouth partly in the form of Steam partly in the form of froth and immediately the temperature of the water that remained fell down to 212.*

Experiment 13 in Papins Digester Water was heated to 140 degrees above the boyling point. upon opening a Small Orifice the Steam flew out with vast force and struck the Ceiling, & soon formed a large Cloud but after rushing out for a small time ceased and the water that remained was found at the boyling point.*

Experiment 14 cooling of bottled liquor by dipping the bottle in water and Whirling it round, especially if the bottle is inclosed in a woolen bag

Experiment 15 Cooling Rooms by hanging wet cloaths in them

Experiment 16. Ether boyls violently in vacuo with the heat of summer and will freeze water in which it Stands.

Experiment 17 In several liquors that have been tried the difference between the heat that boyls them under the pressure of the Atmosphere and that which boyls them in Vacuo is found the same namely about 120 degrees thus Water boyls in the Air at 212 in Vacuo at 100 or under Ether boyls in the Air at 100 and therefore probably in Vacuo 10 degrees or 20 degrees below 0. And when it boyls the heat absorbed by that part which is converted into vapour reduces the rest down to the boyling point which Explains Experiment 16*

The Spontaneous Evaporation of Liquors particularly of Water cannot be accounted for by any Chymical ‹reaction› between the Water and Air because this spontaneous Evaporation is as great or rather greater in Vacuo than in Common Air So that the pressure of the Air seems to impede Evaporation rather than forward it* |

7/III/6, 8

Inflammability*

That there is a principle in the composition of many bodies upon which their being capable of inflammation depends. It is the same in all bodies Is not inflammable by itself. Does not increase the weight of Bodies to which it is joyned but on the contrary diminishes their weight.*

December 1787*

To speak of heat distinctly it is necessary to distinguish the Temperature of Heat in Bodies from the Relative Quantity of Heat they contain; & both these must be distinguished from the Absolute Quantity of their Heat.*

The Temperature of a Body is indicated by the Thermometer So that all Bodies that raise the Thermometer to the same degree, whatever be their Form, whatever their quantity of Matter, have the same Temperature. And the difference of Temperature is measured by the Number of degrees of the Thermometer which lie between one Temperature and another.*

The Relative Heat of Bodies is measured by the Effect produced by the Mixture of two Bodies of different Temperature and of different Nature, which by their mixture produce no chemical Composition or Resolution nor a change of Form in any of the Bodies mixed. In such a Mixture, one body is cooled and the other heated till they come to a common Temperature; and it is supposed that the heat which passes from the one goes into the other. In this Case if, when equal Measures of the two bodies are mixed, the Temperature of the mixture be an arithmetical Mean be‹tween› the Temperatures of the bodies before mixture; they are said to have equal Relative heat. But if in the same Case the temperature of the Mixture be nearer to the temperature of one body before mixture than to that of the other in any proportion or Ratio; the relative heat of the first is said to have the same Ratio or Proportion to that of the ‹second›. Thus if a pint of Water Temperature 70 be mixed with a pint of Quicksilver temperature 40; & the Mixture be found to have the temperature 60; The water is said to have double the Relative heat of Quicksilver. From what has been said it may easily be concluded how the Relative heat of two Bodies is to be measured, when they differ in Magnitude or Measure, other things being the same as before. Thus if two pints of Quicksilver Temperature 40, mixed with one pint of Water Temperature 50 make the Mixture of Temperature 45; we conclude as before that the Relative heat of Water is double that of Quicksilver*

II

7/III/11, 1r The discovery now proposed seems to be drawn by just reasoning from principles formerly known and acknowledged.*

The principles are onely two The first that Ice at 32 requires a heat of 140° of Farenheits Thermometer to convert it into water of the same temperature. The Experiment may be made thus Take a pound of water at 32 & a pound of water at 172. Mix them. You will find the temperature of the mixt to be the middle point between the two to wit 102. Here there is no heat lost for 2 pound of the Temperature 102 Contain the same Quantity as one pound at 32 & another at 172. Again take a pound of pounded Ice at thirty two and a pound of Water at 172. Mix them. You will find the temperature when the Ice is quite dissolved to be 32. Where it is evident the Ice by being turned into water has gained no sensible heat, the water however has lost 140° in turning the Ice into Water

The second Principle is that the Quantity of heat in Ice at 32 is to the Quantity of heat in an equal Bulk of water at 32 As 7:8. Which may be tried thus. Take 8 measures of pounded Ice & 14 of Quicksilver the first of 32 the second of 22 Mix them the Mixt will be found at 27. By which it appears that 8 measures of pounded Ice contain as much heat as 14 of Quicksilver. By a like Experiment we find that 14 Measures of Quicksilver contain the same Quantity of heat as 7 Of Water whence it follows that 7 Measures of Water contain as much heat as eight of Ice of the same temperature. & of Equal Bulks the heat in Water is to that of Ice of the Same temperature as 8 to 7. QED.

From these two Principles it follows that Ice of 32 by being converted into Water gets one seventh part of heat more than it had but the quantity it gets is 140° therefore 140° is one seventh part of the heat of ice at 32. consequently the whole heat of ice at 32 is 7 × 140 = 980° of farenheits Thermometer. or 948 below 0 |

7/III/11, 1v November 1780 Observation on the Experiments & Reasoning in ‹the› last page

Suppose that ⅞ of a pound of Water Temperature 212 dissolves a pound of Ice at 32 and brings the mixture to 32. The quantity of heat lost by the Water is 180. Will it not follow that a pound of water of 140 degrees contains the same quantity of heat as ⅞ of a pound of 180 degrees Consequently that the quantity of heat in a pound of Water of 140 degrees is to the quantity of heat of a pound of Water of 180 as 7 is to 8.

This reasoning supposes that the same quantity of heat is lost in turning a pound of ice into water, whether it be done by hot‹t›er water in a less quantity or by water less hot in greater quantity.

It supposes also that the quantity of heat in a body when the temperature is the same is proportional to the quantity of Matter.*

The Numbers in the above observation are assumed arbitrarily, and therefore the conclusion is onely hypothetical, but the Numbers may be determined by Experiment. And the conclusion drawn by the same reasoning from such Numbers will be categorical

The quantity of heat in different bodies of the same temperature, might I think be determined by finding the change of temperature produced in a certain quantity of each by turning a pound of Ice at 32 into a pound of water at 32.

To make the result agree with Dr Irvines conclusion nearly.* I suppose that of the water at 212 there is 24/25 parts of a pound. The heat lost of this quantity is 180. The heat lost of a whole pound at 172 is 140. The Quantitys of heat lost in these two cases are equal

The Quantity of heat lost by bringing 24/25 parts of a pound of water from 212 to 32 is to the quantity of heat lost in bringing a whole pound of water from 212 to 32 as 24 is to 25. Therefore the quantity of heat lost in bringing a pound of water from 172 to 32 is to the quantity of heat lost in bringing a pound of water from 212 to 32 is as 24 to 25. That is the heats of equal quantities of water at 172 & 212 are as 24 to 25. The 40 degrees of difference of temperature between 172 & 212 make a 25th part of the whole heat at 212 consequently the heat of Boyling Water is 1000 degrees of farenheits Scale. by Dr Irvines Method it is 1500 so that to make the Methods tally 36/37 parts of a pound of boyling water should onely & hardly be sufficient to turn a pound of ice at 32 into a pound of water at 32. |

7/III/11, 2r As there are numberless causes which destroy the æquilibrium of heat in contiguous bodies so there is from some Law of Nature a constant endeavour to restore it.* It passes continually from those bodies whose temperature is above the æquilibrium into those whose temperature is below it; and rests onely when that æquilibrium is attained‹.› |

7/III/11, 2v But it is to be observed that the Equilibrium of different contiguous bodies is not when they have equal quantities of heat in the Same bulk. For it is evident from what has been already observed that a pint of Quicksilver has but half the Quantity of heat which is in a pint of Water of the same temperature. and a pint of Ice of 32 has but ⅞ of the heat that is in a pint of Water of 32. When these three bodies Water Ice & Quicksilver raise the Thermometer to the same height, they have unequal Quantities of heat in the same bulk, which the thermometer does not discover‹.› whence this heat which is not discovered by the Thermometer is called by the ingenious Discoverer Latent Heat.* This

III

2/I/4, 1 1790

§1
Of the Chemical Elements of Bodies

What may really be the first elements of Bodies, we know not. Nothing seemed to have a better claim to the Name of an Element, or was more universally in all Ages accounted such, than pure Water. Yet it is found to be a compound which can be resolved into its principles and again compounded of them.*

We ought never therefore to affirm of any body whatsoever that it is truly elementary, that is, void of all composition. As little can we upon any rational ground determine the Number of Elements, or perfectly simple ingredients, of which the various concrete Bodies that fall within our Notice are composed.

Yet as it is a very important part of natural knowledge to know what human Industry has been able to discover of the composition of Bodies; and as it is one great branch of the Art of Chemistry to resolve Bodies, seemingly simple, into those that are more simple; we may give the Name of Chemical Elements to those principles in the composition of Bodies, which have not hitherto been discovered to be compounded of principles still more simple. Remembring always, that what is now a Chemical Element, may some time hence in a more advanced State of the Chemical Art be found to be a compound.*

Among the Chemical Elements there is a couple which are so nearly allied that it is still uncertain whether they be different principles or different modifications of one and the same Principle, & may according to circumstances be converted one into the other. These are Light & Caloric.*

1 Light, which is the instrument of Vision & enlivens the whole Face of Nature*

That the Light of the Sun the great fountain of Light to our System is composed of Rays which differ in refrangibility & Reflexibility, & in the colour which they give to bodies that reflect them to our Eye was shewn by the great Newton.* The properties of the different kinds of Rays seem to be inherent and invariable by any means we can use.* Nor has it ever been found that by any other ignited Body, any kind of Rays are produced which have different properties from those of the Sun.

This Element, so wonderfull in the minuteness of its parts, in the velocity of its Motion, in its penetration of the densest diaphanous bodies without any sensible obstruction or retardation, and in the beautifull variety of colour it gives to bodies, is now found to answer many other purposes in the operations of Nature.* |

2/I/4, 2 Its Rays concentrated by Refraction or Reflexion produce in opaque Bodies on which they strike greater degrees of Heat than can be produced by any other means yet known; and to what degree the Rays of Light may be still more concentrated by improvements in Mechanism, & thereby still greater degrees of heat produced than have yet been by human Art, we may conjecture, but do not know. Far less do we know what farther discoveries may be made in the composition and decomposition of bodies by this most powerfull instrument.* Query. Has it ever been tried what Effect the Concentrated Rays of the Sun have upon the most diaphanous bodies; such as a thin plate of the purest Glass. If the Rays of Light produce heat onely by being absorbed, will it not follow that the heat produced was before in the Rays and is given out on their being decomposed or fixed in the heated Body. May not the Name *Heat* be given to Caloric uncombined*

Other purposes of Light are now partly known and partly conjectured, though perhaps but a few of those to which it is subservient.

If Light be a material Substance, as it seems to be, we may distinguish its *Matter* from its *Form*. How far the matter of Light enters into the composition of bodies as an Element, we do not know; but all Bodies, when heated to a certain degree give out light, and some do so by being exposed for a short time to the suns Rays; others even in the common heat of the atmosphere, shine & flame, & others in various degrees of heat

Whether in these cases the Matter of Light was before in the body that shines or in any of the surrounding Elements we do not know. Query Has this been examined, by trying them in Vacuo

It is known however that light produces the green colour in Vegetables, & probably the oyls we find in their Composition

In bodies that are coloured by Art, the Colour generally fades by being much exposed to the light, or even to pure Air or Oxygen gas The Rays of the sun make growing vegetables emit oxigen gas*

Though many of the phenomena of Light & Heat lead us to conceive and to speak of them as material Substances, yet I know of no phenomenon that shews either of them to gravitate, so as to add to the

Weight of Bodies by being conjoyned with them or to diminish the weight of Bodies by being disjoyned from them.*

2 Caloric

2 Heat. We give the name of *Heat* to that sensation which we feel when any part of our Body is heated. The same name is given to that Principle which heats bodies, & is the Cause of this Sensation. But, though in common Life it may not be found necessary that the cause & the Effect
2/1/4, 3 should have different Names, yet philosophical precision requires | that they should; and the rather because other sensible effects besides this feeling are produced by the same principle.*

It is sufficiently ascertained by Experience, that this Principle in proportion as it is accumulated in any body whatsoever, expands the body in all its dimensions, and lessens the cohesion of its parts.*

The French Chemists have therefore given the Name of *Caloric* to that Principle, whatever it be, whether material or not, which heats Bodies, & expands them.*

All the bodies we are acquainted with are penetrated by Caloric and have their dimensions enlarged by it. What their dimensions would be if totally deprived of caloric we know not. But it appears that the hardest bodies, by imbibing a certain Quantity of Caloric become Fluid, & by a still greater quantity are turned into an elastic Fluid or Gas. And as hardness in Bodies is produced by a certain attraction of their parts, so the effect of Calorick is a repulsion of the parts of Bodies. And these two Power‹s› of Nature Attraction & Caloric act as antagonists. According as the degree in which one or the other is prevalent, every body in Nature assumes the different forms of Hard, or Soft, or Liquid, or Gas.*

The Cold we feel upon touching or approaching an external body indicates the passage of calorick from our body into the external body; & the warmth or heat we feel upon touching or approaching an external body, indicates the passage of caloric from the external body in to our body. We feel neither Heat nor cold from the touch or approach of a body which neither gives out caloric to our body nor receives caloric from it.*

In all inanimate bodies, the imbibing or emitting caloric, though it does not produce the sensations of heat or cold, produces another effect, to wit that of expanding or contracting them in their bulk.

Thus the Mercury or Spirit of wine in the Thermometer, when it rises imbibes heat, & gives out of its heat when it falls; and this expansion by imbibing caloric, & contraction by emiting it, appears to be common to all the bodies we know. |

2/I/4, 4 Whether Light and Caloric be different Principles or the same Principle under different Modifications we know not. Neither of them is known to gravitate, or to give impulse to bodies. The Rays of Light heat all bodies which do not give them a free passage or reflect them totally, & all bodies heated to a certain degree emit Light. All inflammation is produced by the sudden disengagement of Caloric from Vital Air*

Caloric seems to be essentially fluid, and to pervade all bodies. No vessel can confine it. We know of no Operation of Nature or Art by which the quantity of Caloric upon the whole may be increased or diminished or its Nature changed. One would be apt to think that a fluid so subtile elastic and unchangeable should be equally disseminated through all Space so as that equall spaces should contain equal quantities of it, or that if at any time it was condensed in one part, it would never be at rest till its density was eq‹u›al in every place. And this indeed would be the case if it were not differently affected by different bodies. But such are its relations to different bodies, that even when at rest one body contains more caloric in the same space than another; and every body has its peculiar relation to caloric in this respect. Thus when a pint of Water and a pint of Mercury remain in contact, or mixed, till neither receives any caloric from the other; whatever be the sum of caloric in both, two thirds of it remain in the Water, & onely one third in the Mercury. The Water & Mercury are said in this case to have the *same Temperature* or the same *sensible Heat* while each has a different quantity of Caloric.

All bodies have the same Temperature when being applied to each other no caloric passes from one to the other. It is therefore the temperature of bodies, not their quantity of caloric that is indicated by the Thermometer.

The proportion of Caloric in bodies of equal bulk and of the same Temperature may be called their Relative or Specific Caloric. And this Relative Caloric is the Same whether the Temperature be high or low.* Crawford & other Authors have given Tables of the Relative Calorick of most Bodies; but not that of Space void of Body The whole quantity of caloric is measured by its Temperature, its relative caloric, and its magnitude taken together, or by the product of the three multiplied into one another.*

From what has been said it appears that even when Caloric is at rest
2/I/4, 5 it is in very unequal Quantities in different Bodies, & | that if its quantity

in a System of contiguous bodies be augmented or diminished, it will not be at rest till each have the same proportion of the whole caloric of the System as it had before.

It must still be understood notwithstanding what is said above that when the form of a Body is changed, from being Solid to be liquid or from being liquid to be an elastic fluid or Gas; its Relative Caloric in these different forms is greatly changed Thus Water has much more relative Caloric than Ice and Steam more than Water.*

Query. May not other gases, by their bases being suddenly combined produce heat and Flame

Answer. If there be any instance in which Caloric is a fixed principle in Bodies, which is not increased or diminished by the change of temperature, while the body is not decomposed, which Lavoisier seems to apprehend, this ought to be called fixed caloric.*

Query. Whether is there any fixed caloric in the nitric, or in the oxygenated muriatic acid.

If any Body by its decomposition yields more caloric, than in proportion to its temperature, relative caloric, and magnitude taken in compound proportion, must not this be caloric which before decomposition was fixed in the body?

3 Oxygen

Oxygen raised by caloric into a Gas makes *oxygen gas* which is a principal part of our Atmosphere, being about 0,27 of the whole, & the onely part that is fit for respiration. It was formerly called *vital Air & Dephlogisticated Air*. All inflammation is a sudden decomposition of oxygen gas by the combustible or inflammable body; which suddenly attracting the oxygen, the caloric is disengaged with heat and flame. Water is a composition of oxygen, & hydrogen or inflammable air, 85 parts by weight of oxygen & 15 of hydrogen. or 17 of the first to 3 of the last. Hence it appears that oxygen makes more than a fourth part of the atmosphere, & more than five sixth parts of all the water on our Earth, besides the share it has in other bodies.*

Oxygen seems to combine with all the simple Substances we know in Nature, excepting the five Simple Earths, Lime, Magnesia, Barytes, Argil, and Silex. It is thought that its combination with various Radicals constitutes all the various Acids we know. Some of these Radicals are hitherto unknown, as in the Muriatic, the Fluoric & Boracic Acids; some

Acids have Radicals compounded of several simple Substances. Of one or other of these kinds of Acids names have been given to no less than 48 Acids there being so many bases either found to be or supposed to be acidifiable. Of which some belong to the mineral, some to the vegetable, and some to the Animal kingdom, according as their Bases or Radicals are taken from these different kingdoms.* |

2/I/4, 6 Substances combined with Oxygen are said to be oxygenated; and many of them are found capable of different degrees of oxygenation whereby their appearance and properties are much changed. Lavoisier distinguishes four degrees of oxygenation. Radicals which have onely the first or second degree he calls *Oxyds*. In the third degree they are *Acids* and in the fourth *oxygenated Acids**

Onely one degree of oxygenation of Hydrogen is known, to wit, that by which Water is produced.*

Oxigen combined with Azote, has three known degrees of oxygenation in the first degree it makes the base of nitrous gas, or nitrous Air. In the second it is Nitrous Acid, or smoaking nitrous Acid, in the third it is Nitric Acid, or pale not smoaking nitrous Acid.*

Oxygen combined with charcoal makes *Carbonic Acid*, or fixed Air*

The Calces or Metals are Oxyds of the Metals, in the first or second degree of Oxygenation*

Onely three of the metallic bodies are known in the third degree of oxygenation, Arsenic, Molybdena, & Tungstein.*

The marine, or muriatic Acid, onely, is known in the fourth degree of oxyg‹e›nation, and is then named *Oxygenated muriatic Acid* formerly the dephlogistigated marine Acid*

Sugar, Starch, Mucus &c are Vegetable Oxyds*

Lymph, red part of the blood, animal Secretions are animal Oxyds*

4th Azote

This is one of the most aboundant Elements. With Caloric it forms azotic Gas, which makes near three fourths of our Atmosphere. It is one of the essential ingredients of animal Bodies; in which it is combined with charcoal & hydrogen, and sometimes with phosphorus. These are united together by a certain portion of oxygen, by which they are formed into oxyds or acids according to the degree of oxygenation.

When combined with Oxigen onely it forms the nitrous and nitric oxyds & acids, when with hydrogen, ammoniac is produced.

Azote may be procured 1 from atmospheric Air. 2 By dissolving animal substances in diluted nitric acid, very little heated. 3 by deflagrating Nitre 4 By combining ammoniac with metallic oxids. The hydrogen of the ammoniac combines with the oxygen of the oxyd & forms Water, while the Azote e‹s›capes in ‹the› form of gas.* |

2/I/4, 7

5th Hydrogen

With Caloric, forms Hydrogen gas: With Azote, Ammoniac, or volatile alkali; With Oxygen, Water. With Carbone it forms the fixed and Volatile Oyls, and the Radicals of a considerable part of the vegetable & animal oxyds & acids.*

The Elements hitherto mentioned are so disposed to combination with other Elements even in the Temperature of our Atmospher‹e›, that they never are found single, but onely in a state of combination*

The muriatic, fluoric, & boracic acides, from Analogy are thought to consist of a Radical or Base combined with Oxygen, but as these Acids have never been resolved into different ingredients their bases, whether they be simple or compound are unknown and therefore are not reckoned among the chemical Elements*

It may also be questioned whether Soda & Potash are to be accounted Elements as it is uncertain whether they existed originally in the Plants from which they are got, or whether they *be produced* from other Elements in the Operation of burning*

6 Sulphur 7 Phosphorus 8 Charcoal Inflammable*

The Combinations of these with other Substances by the Termination *et* as *Sulphuret*, *Phosphoret*, *Carbonet*. In like manner the Combinations of Azote with other Simple Substances are called *Azarets & those with* Hydrogen *Hydrurets*.*

To these are to be added 17 metallic Bodies, to wit, eight metals and nine semimetals, And five simple Earths in all 30*

The five Earths viz. Lime, Magnesia, Barytes, Argil, Silex, are conjectured by Lavoisier to be compounds. They are the onely simple Substances known which do not unite with oxygen, perhaps because they are already saturated with it, & are really oxyds of metals hither to unknown. This in particular is suspected to be the case with barytes.* |

2/I/4, 8

Facts from Lavoisier's Elements

Phosphorus 100 parts requires 154 of oxygen by weight for its saturation, & this combination produces 254 parts of concrete phosphoric acid, in ‹the› form of white fleecy flakes; which attract Water very strongly*

The Combustion of one pound of Phosphorus, melts a little more than 100 pounds of ice. One pound of Hydrogen melted 295 pounds 9 oz 3 ½ gros Ice*

Of Charcoal 28 parts by weight require for saturation 72 parts of oxygen. One pound of Charcoal in Combustion melted 96 pounds eight ounces of ice.*

Nitrous gas consists of nearly two parts by weight of oxy‹g›en combined with one part of azote is not miscible with Water. Less than three parts of oxygen combined with one part of azote, make *nitrous acid*, formerly *fuming nitrous acid* which is of a red colour and emits copious fumes of nitrous gas. Four parts of oxygen combined with one of azote, make the *nitric acid*. This is clear and colourless, more fixed in heat, has less odour, than the nitrous acid, its constituent parts being more firmly united.*

From the experiment page 87. we may infer that 100 parts of distilled Water are formed of 84 of oxygen & 16 of hydrogen*

From the Experiment page 89, 100 parts of water contain 85 parts of oxygen and 15 of hydrogen.*

In the combination of oxygen with azote in the nitrous & nitric acids The oxygen seems to lose but little of its caloric, which explains the great disengagement of caloric during the deflagrations of nitre, or upon all occasions of the decomposition of the nitric acid.*

Fixed Oyls are composed of 21 parts hydrogen by weight, combined with 79 parts of carbon.*

The constituent parts of all vegetables are hydrogen, oxygen and carbone These ingredients in the temperature of our Air remain in æquilibrio. The hydrogen is not combined with the oxygen so as to form Water, nor is the carbone combined with the hydrogen so as to form oyl, nor the carbon with the oxygen so as to form carbonic acid. But an increase of temperature destroys this æquilibrium. When the temperature does not exceed the heat of boyling water, one part of the hydrogen combines with the oxygen, and forms water, the rest of the hydrogen combines with a part of the carbone, & forms volatile oyl,

whilst the remainder of the carbone being set free remains in the bottom
2/I/4, 9 of the distilling vessel. On the contrary when we employ a red | heat, no water is formed, or any that was formed in the first application of the heat is decomposed; the oxygen having a greater affinity with the charcoal at this degree of temperature, combines with it to form carbonic acid, and the hydrogen being left free unites with the caloric, & escapes in the state of hydrogen gas.*

Sugar consists of hydrogen 8 parts, oxygen 64 parts & carbone 28 parts*

In distilling Sugar, so long as we employ a heat but a little below that of boyling water, it onely loses its water of chrystilization, but retains all its properties. Upon raising the heat but a little above that degree it becomes blackned, a part of the charcoal separates from the combination; water slightly acidulated passes over accompanied by a little oyl, and the charcoal which remains in the retort, is nearly a third part of the weight of the sugar*

In vegetable Substances which contain azote such as the cruciferous plants, a small part of the hydrogen combining with the azote forms ammoniac, or volatile alkali. And in those that contain phosphorous, the phosphorus seems to combine with the charcoal, and acquiring fixity from that Union, remains behind in the retort.*

Animal Substances, by distillation, give the same products as the cruciferous plants, with this difference, that as they contain a greater proportion of hydrogen & of azote, they produce more oyl & more ammoniac*

Dippels oyl, when procured by a first distillation in a naked fire, is brown, from containing a little charcoal almost in a free state; but by rectification it becomes quite colourless. Even in this state, the charcoal in their composition has so slight a connection with the other elements as to separate by mere exposure to the Air. For if a quantity of this oyl well rectified & consequently transparent & clear be put in a bell glass filled with oxygen gas over Mercury; in a short time, the gas is much diminished, the oxygen, combining with the hydrogen of the oyl, forms water, which sinks to the bottom, at the same time the charcoal, which was combined with the hydrogen, being set free, renders the oyl black. If we use large vessels, & a considerable degree of heat, by successive distillations the oyl is at last totally decomposed, & converted into water and charcoal.* |

2/I/4, 10 In 100 parts by weight of dry yeast, there are of

oxygen	60
hydrogen	10,
carbone	28,5+
azote	1,5-

dry acetous Acid 100 parts by weight

Oxygen	68.75
Hydrogen	6.25
Carbone	25.00

In 100 parts of carbonic acid, oxygen 74, carbon 26.
In 100 parts of dry alcohol, oxygen 55, hydrogen 16, carbone 29*

When alcohol is made to pass through a red hot tube of glass or porcelain, it is resolved into water and carbonic acid.* |

2/I/4, 11 Of the Combination of Combustible bodies with each other

Almost all the Metals are capable of combining with each other and forming *alloys*. The combinations of Mercury with other metals have been, and are still called *amalgams*.*

Sulphur Phosphorus & Carbone readily unite with metals. Combinations of Sulphur with Metals are usually called pyrites. Lavoisier calls the combinations of Sulphur with metals *Sulphurets*. Those of phosphorus, *phosphurets*, & those with carbone *carburets*. These Denominations are extended to all the combinations into which the above three substances enter, without being previously oxygenated. Thus the combination of Sulphur with potash is called *sulphuret of potash* that which sulphur forms with ammoniac, *sulphuret of ammoniac*.*

Hydrogen gas dissolves sulphur, carbone, & phosphorus, & several Metals These combinations are called *sulphurated hydrogen gas*, *carbonated hydrogen gas*, *phosphorated hydrogen gas*. The sulphurated hydrogen gas was called hepatic air, fœtid air from sulphur. The virtues of several mineral Waters, & the fœtid smell of animal excrements, chiefly arise from the presence of this gas*

The p‹h›osphorated hydrogen gas is remarkable for the property of taking fire spontaneously on getting into contact with atmospheric air, much more with oxygen gas. This gas has also a strong flavour resembling that of putrid fish.*

Hydrogen & carbone combined, without the intervention of caloric to bring the hydrogen into the form of gas, form oyl, which is either fixed or volatile according as the carbone prevails more or less.*

Fixed or fat oyls drawn from vegetables by expression, contain an excess of carbone, which is separated when the oyls are heated above the degree of boyling water. Fixed oyl is composed of 21 parts by weight of hydrogen combined with 79 parts of carbone*

It has not been proved that hydrogen ever combines with metals in a solid state. In all the experiments on metals which are attended with the production of hydrogen gas, it seems to be owing to the decomposition of water.*

In the solution of metals by the sulphuric acid, the metal does not attract all the oxygen the acid contains, they therefore reduc‹e› it, not into sulphur but into sulphurous acid, which is a gas in the usual temperature* |

2/I/4, 12

Of the Formation of Neutral Salts

Earth & Alkalis unite with Acids to form neutral Salts with out the intervention of any medium; whereas metallic substances are incapable of forming this combination without being previously less or more oxydated. Strictly speaking therefore, metals are not soluble in acids, but metallic oxyds. When we put a metal into an acid it must first be oxydated, by the decomposition either of the acid or of the water*

The effervescence which often attends metallic Solutions is owing to the disengagement of gas. When hydrogen gas is disengaged, it is owing to the decomposition of the water, the other ingredient of the water to wit the oxygen combining with the metal & oxydating it. In solutions with the sulphuric acid, it is either hydrogen gas or sulphurous acid that is disengaged according as the oxydation of the metal Happens to be made at the expence of the water, or of the sulphuric acid.*

A second phenomenon in the solution of metals is, That when they have been previously oxydated, they all dissolve in acids without effervescence; having now no occasion for combining with oxygen they neither decompose the water nor the acid*

A third phenomenon is, That none of the metals effervesce by solution in oxygenated muriat‹i›c acid. The metal is oxydated by carrying off the excess of oxygen in the acid, & reduces it to the state of ordinary muriatic acid, which here finds water sufficient to keep it in the liquid form*

The fourth phenomenon is that metals are absolutely insoluble in such acids, whose bases have a stronger affinity to oxygen than the metal has.*

From these phenomena it appears that oxygen is the bond of union between the metals and acids; and from this we are led to conjecture, that oxygen is contained in all substances, which have a strong affinity with acids. Supposing the Earths to be metallic oxyds which have a stronger affinity to oxygen than carbone has, they would not be reducible by any known means*

The neutral Salts are combinations of any of the Acids with an alkali, 2/I/4, 13 an Earth or a Metal. And as 48 acids have names, and | may possibly combine with 24 salifiable bases, this would produce 1152 neutral Salts. To give distinct names to all these without burthening the Memory; The neutral salts of one & the same Acid have one appellative name taken from the name of the acid, and another taken from the Name of the base of the neutral Salt. Thus *Sulphat of Soda*, *Sulphate of potash*, *sulp‹h›at‹e› of ammoniac*, *Sulphate of Iron* &c are neutral Salts, of which the acid is the sulphuric. In like manner *nitrates Muriats*, *Carbonates*, *Acetates*, are neutral Salts of which the acid is the nitric, the muriatic, the carbonic, the Acetic*

In a lower degree of oxigenation they are called *Sulphites*, *Phosphorites* &c & in the highest degree *oxygenated muriates* &c*

Of compound oxydable and acidifiable Bases.

We are onely acquainted with one compound radical from the mineral kingdom, the *nitro-muriatic*, or base of aqua regia, which is formed by the combination of azote with the muriatic radical.*

The compound Radicals from the vegetable kingdom are formed by the combination of hydrogen and carbone combined in such a way as to form single bases, & differ from each other by the proportions in which these two ingredients enter into the composition of their bases, & by the degree of oxygenation which th‹eir› bases have received. But as these proportions are not yet perfectly known, their Names must at present be taken from the substances which yield them, such as, the *Tartaric radical*, the *oxalic*, the *acetous*, &c*

For the same reason the same rule is followed in giving names to the compound oxydable, & acidifiable Radicals from the animal kingdom, such as the *Lactic*, the *Sebacic*, the *Prussic Radical* &c*

We know onely that the Radicals from this kindgom, and even some of those from vegetables, besides hydrogen & carbone, contain azote, & sometimes phosphorous.*

Lavoisier says Elements page 183. "We know that caloric in certain cases becomes fixed in bodies, so as to make a part of their Solid Substance"*
Quere, What instances can be given of this?*
Quere, Whether all the caloric retained in the nitric, & oxigenated muriatic acids, be relative caloric, or whether part of it be not fixed caloric? |

2/I/4, 14 Of the oxygenation of simple Substances

Some Substances have such a strong and constant affinity with oxygen that we cannot procure them in their unoxygenated state. Such are the radicals of the muriatic, the fluoric, & the boracic acids. Perhaps there may be other substances in the mineral kingdom, which by a like strong and constant affinity with oxygen seem to be simple elements, while they are in reality oxyds which we are not able to decompose. It is suspected that the five Earths, Lime Magnesia, Barytes, Argil, & silex may be of this kind, and that their being already saturated with oxygen, prevents their farther action upon that Element.*

There are two means of oxygenating substances 1. Exposing them to Air especially to oxygen gas in a certain temperature. 2 Placing them in contact with oxigenated Substances, which have a weaker affinity with oxygen*

The heat required for oxygenating different Substances in Air is different Lead, Mercury & Tin require a heat but a little higher than that of the atmosphere.* Rust in metals seems to be a slow and gradual oxygenation And when we find this to be so much aided by moisture, is it not probable that the water contained in moist air is decomposed, its oxygen entering into the rusted metal while its hydrogen is set free. When moist linnen is smoothed by hot irons is there not a decomposition of water discernable by the smell of hydrogen gas in the Room.

When oxygenation is very rapid it is accompanied with heat and flame*

The oxygenated bodies fittest to oxigenat‹e› others with which they are mixed are those that have a weaker affinity with oxygen, and at the same time do not combine with the body which they are intended to oxygenate. The red oxyd of mercury, the black oxyd of manganese. All the metallic oxyds yield their oxygen to carbone. All the metallic reductions are operations of this kind*

All combustible substances may likewise be oxygenated by mixing them with nitrat of potash or of Soda, or with oxygenated muriat of potash, and subjecting the mixture to a certain degree of heat. This species of oxygenation requires to be performed with extream caution, & in very small quantities, because as the oxygen enters into the Nitrates and more especially into oxygenated muriates, combined with almost as much caloric as is necessary for converting it into oxygen gas; this immense quantity of caloric becomes free the instant of the combination of the oxygen with the combustible body & produces an explosion perfectly irresistible.*

In the humid way most of the oxyds in the three kingdoms of Nature may be converted into acids. For this purpose the nitric acid is chiefly
2/I/4, 15 employed, because | it holds its oxygen by a weak affinity.*

Azote, its binary combinations

Azote combined with caloric forms azotic gas, of which near three fourths of our atmosphere consists.* As so great a proportion of the Air we & all animals breathe is azote, is it not probable that it is usefull to animal life even by being breathed? It does not indeed contribute to respiration, but it may be fixed in the blood & by that means in the whole body by respiration. The air we exspire ought to be examined. Perhaps it may be found that the air inspired has lost not onely a part of its oxigen, but that it has also lost a part of its azote. If this be so we may see why azote enters into the composition of all animal substances, though it is found onely in very few vegetables & even in these in very small quantity. How should those animals who live onely upon vegetables, have so considerable a quantity of azote in their composition, if they have it not from the Air they breathe.

May it not be dubious, or rather improbable that pure oxigenous gas is the most salubrious air to breathe? If azote be a necessary element of the animal Substance, & be supplied chiefly by breathing, it is possible that the want of azotic gas in the Air we breath, may kill as effectually though not so soon as the want of oxigen gas. Perhaps, after all, a greater proportion of oxygen gas may be salutary to persons in pulmonic disorders; but to the bulk of Men and of other animals I am disposed to think that the proportion of the ingredients of the air has been best fitted by nature, and that a man in health needs not mind much the report of his eudiometer.*

It is now known that azote is the base of the nitrous & nitric acids; and that combined with hydrogen it makes the volatile alkali or Ammoniac. It is from this suspected to be ‹an› ingredient in the fixed alkalis of Soda and Potash. Other binary combinations into which it enters are hardly known*

Carbone

Whether the fixed Salts of Potash, and Soda, found in charcoal, exist in it before combustion, or are formed by some unknown combination during that process we are uncertain.*

If these alkalis exist in the charcoal before combustion they must be ingredients in all vegetable substances |

2/I/4, ‹16› If the alkalis are produced by combustion, it should be enquired whether the combustion can be performed without any contact of azotic gas.*

Composition of Acids

Azote 20½ parts, oxygen 43½ parts combined make 64 parts of nitrous gas. If 36 parts of oxygen be added, we have 100 parts of nitric acid.*

Mr Bertholet by one Experiment found that 69 parts of sulphur united with 31 of oxygen make 100 of sulphuric acid. By an experiment made in a different manner, he makes the proportion to be 72 to 28.*

Sulphurous acid in a very low temperature condenses & becomes fluid*

Water absorbs more of this gas than of carbonic acid but much less than it does of muriatic acid gas.*

Phosphoric acid consists of 28½ parts of Phosphorus combined with 71½ of oxygen.*

NOTES

Editorial Notes

Part One: Euclidean Geometry

5/II/47: Observations on the Elements of Euclid

3/1 The first section of this manuscript predates the publication of Robert Simson's Latin and English editions of Euclid's *Elements*, which both appeared in late 1756; see Philip Gaskell, *A Bibliography of the Foulis Press*, pp. 206–7. It is unclear which of the many different editions of the *Elements* circulating in the eighteenth century served as the basis for Reid's comments. However, his remark about proposition 7 in Book I (4/7–8) indicates that even though he was familiar with Isaac Barrow's 1705 version of the *Elements* it was not the one that he was using. Given Reid's high regard for the Latin and Greek edition of Euclid's writings published by his kinsman David Gregory (see above, p. xxxvi–xxxvii), in what follows I will use the annotated English translation of Gregory's text of the *Elements* prepared by the Scottish mathematician Edmund Stone.

3/3 'A diameter of a circle is a right line drawn through the centre, and terminated on each side by the circumference, and divides the circle into two equal parts'; Edmund Stone, *Euclid's Elements of Geometry, the First Six, the Eleventh and Twelfth Books; Translated into English, from Dr Gregory's Edition, with Notes and Additions* (1752), p. 3. Reid's comments on the flaws of definition 17 echo those in Stone, *Euclid's Elements*, p. 3 note l.

3/9 '… amongst three-sided figures, that is a right-angled triangle, which has a right angle'; Stone, *Euclid's Elements*, p. 4.

3/12 'All right angles are equal to one another'; Stone, *Euclid's Elements*, p. 5.

3/14 'If a right line falling upon two right lines does make the internal angles on the same side less than two right angles, those right lines, being infinitely produced, do meet on that side where the angles are less than two right angles'; Stone, *Euclid's Elements*, pp. 5–6.

3/17 'If a right line falls upon two parallel right lines, it makes the alternate angles equal to one another; the outward angle equal to the inward and opposite angle on the same side; and the inward angles on the same side together equal to two right angles'; Stone, *Euclid's Elements*, p. 34. The demonstration of proposition 29 in Book I presupposes the truth of the disputed eleventh axiom.

3/23 'Parallels are right lines, which being in the same plane, and produced infinitely either way, will not meet one another either way'; Stone, *Euclid's Elements*, p. 4.

3/26 Proposition 27 in Book I reads: 'If a right line falling upon two right lines makes the alternate angles equal to one another, these right lines will be parallel to one another'. Proposition 28 reads: 'If a right line falling upon two right lines makes the outward angle of the one equal to the inward opposite angle of the other, on the same side; or the inward angles on the same side together equal to two right angles: these right lines shall be parallel to one another'; Stone, *Euclid's Elements*, pp. 32–3.

3/28 For proposition 27 see the previous editorial note. Stone translates proposition 16 thus: 'Any one side of every triangle being produced, the outward angle is greater than either of the inward opposite angles'; Stone, *Euclid's Elements*, p. 22.

3/29 For proposition 28, see editorial note 3/26.

3/30 For proposition 29, see editorial note 3/17.

4/2 Definition 1 states, 'A point is that which has no parts', while definition 3 states, 'The extremes of a line are points'; Stone, *Euclid's Elements*, p. 1.

4/4 See the demonstration of proposition 1in Stone, *Euclid's Elements*, p. 7.

4/7 'Two right lines cannot be constituted on the same right line equal to two other right lines, each to each, at different points on the same side, and having the same ends with the first right lines'; Stone, *Euclid's Elements*, p. 13.

4/8 Isaac Barrow, *Euclides Elements; the Whole Fifteen Books Compendiously Demonstrated* (1705), pp. 12–13. Barrow's version of Euclid's *Elements* first appeared in English in 1660.

4/9 In Reid's 1753 graduation oration delivered at King's College, Aberdeen, he insisted that mathematicians do not employ syllogistic reasoning and in his logic lectures he apparently catalogued the different forms of demonstration to be found in mathematics; Reid, *Philosophical Orations*, oration I, para. 18, and Thomas Reid, 'A System of Logic, Taught at Aberdeen 1763', Edinburgh University Library, MS Dk.3.2, pp. 79–82. See also his later comments in Reid,

On Logic, Rhetoric and the Fine Arts, pp. 128–9, 159, 178, and Reid, *Intellectual Powers*, pp. 545–7, 555–6.

4/13 See the demonstration of proposition 4 in Stone, *Euclid's Elements*, pp. 9–10.

4/14 Demonstrations 'from absurdity' or 'from impossibility'.

4/15 For the demonstration of proposition 19, see Stone, *Euclid's Elements*, p. 24.

4/16 'with stronger reason'; for propositions 16 and 18, see Stone, *Euclid's Elements*, pp. 22–4.

4/17 'by parts'.

4/18 See the demonstration of proposition 44 in Stone, *Euclid's Elements*, pp. 47–8.

4/19 For the demonstration of proposition 10 in Book II, see Stone, *Euclid's Elements*, pp. 87–9.

4/23 'If two triangles have their sides proportional; those triangles will be equiangular; these angles being equal which are opposite to the homologous or co-rational sides'; Stone, *Euclid's Elements*, pp. 247–8.

4/24 Proposition 12 states the problem, 'From a given point, without a given infinite right line, to draw a right line perpendicular to it'; Stone, *Euclid's Elements*, pp. 18–19.

4/27 Definition 3 in Book V states: 'Ratio is a certain mutual relation of two magnitudes to one another of the same kind, according to quantity'; Stone, *Euclid's Elements*, p. 205. Like Reid, Stone was highly critical of Euclid's wording in definition 3; see Stone, *Euclid's Elements*, pp. 205–6.

4/35 Girolamo Saccheri, *Euclides ab omni nævo vindicatus: Sive conatus geometricus quo stabiliuntur prima ipsa universæ geometriæ principia* (1733); see also below, p. 15, ll. 1–4.

4/39 Girolamo Saccheri, *Neo-statica* (1708); Girolamo Saccheri, *Logica demonstrativa* (1697). The 1697 edition of Saccheri's *Logica demonstrativa* was previously thought to be the first, but see Paolo Pagli, 'Two Unnoticed Editions of Girolamo Saccheri's *Logica demonstrativa*'. Pagli has discovered two hitherto unknown editions of the work dating from 1696 and 1699.

5/7 In the next three paragraphs Reid summarises propositions 1–7 of Book I of Saccheri's *Euclides ab omni nævo vindicatus*; Gerolamo Saccheri, *Euclid Vindicated from Every Blemish*, ed. Vincenzo De Risi and trans. George Bruce Halstead and Linda Allegri (2014), pp. 71–81.

6/7 Compare Saccheri, *Euclid Vindicated*, pp. 81–3.

6/15 'If two triangles have two angles of the one equal to two angles of the other, each to each; and one side of the one equal to one side of the other, either that side which is between the equal angles, or that which is opposite to one of them; then will the remaining sides of the one triangle be equal to the remaining sides of the other, each to each; and the remaining angle of the one will be equal to the remaining angle of the other'; Stone, *Euclid's Elements*, p. 30.

6/17 'If at a point in any right line two right lines drawn contrary ways do make, with the first line, the adjacent angles equal to two right angles; those right lines will be directly placed to one another, or both fall into one right line'; Stone, *Euclid's Elements*, p. 20.

6/28 Reid's 'Essay' is no longer extant. What is probably a draft of part of the 'Essay' survives in AUL, MS 2131/5/I/1; see below, p. 12, l. 15, to p. 13, l. 15, and editorial note 12/15.

7/11 Reid summarises proposition 14, in which Saccheri refutes the hypothesis of obtuse angles; Saccheri, *Euclid Vindicated*, p. 93.

7/15 Saccheri devoted propositions 17–33 in the first part part and the whole of the second part (propositions 34–39) of Book I of *Euclides ab omni nævo vindicatus* to the refutation of the hypothesis of acute angles. For proposition 37, see Saccheri, *Euclid Vindicated*, pp. 178–83.

7/19 The remainder of this manuscript is written in Reid's late hand and presumably dates from the early 1790s, when he was drafting the discourse on Euclid he read to the Glasgow Literary Society in the period *circa* 1790 to 1793. Compare the paragraphs running from l. 20 to p. 8, l. 38, with those below in AUL, MS 3061/11, p. 23, l. 14, to p. 25, l. 3, and the paragraphs running from p. 9, l. 1, to p. 10, l. 29, in the present manuscript with those on p. 27, l. 8, to p. 28, l. 20.

7/23 In David Gregory's edition of Euclid's *Elements* the axiom is numbered 11, whereas in editions derived from the Latin translation published by Federico Commandino the axiom appears as number 12. Compare Euclid, *ΕΥΚΛΕΙΔΟΥ ΤΑ ΣΩΖΟΜΕΝΑ. Euclidis quae supersunt omnia*, ed. David Gregory, p. 3, with John Keill, *Euclid's Elements of Geometry*, trans. and rev. Samuel Cunn (1723), p. 4; and Robert Simson, *The Elements of Euclid, viz. the First Six Books, Together with the Eleventh* (1756), p. 7.

7/27 Definition 35 reads: 'Parallels are right lines, which being in the same plane, and produced infinitely either way, will not meet one another either way'; Stone, *Euclid's Elements*, p. 4.

7/33 On this point see, for example, Simson, *Elements of Euclid*, p. 360; Stone, *Euclid's Elements*, p. 6; Claude François Milliet Dechales, *The Elements of Euclid, Explained and Demonstrated in a New and Most Easy Method*, trans. Reeve Williams (1703), pp. 24–5; Edmund Scarburgh, *The English Euclide, Being the First Six Elements of Geometry, Translated Out of the Greek, with Annotations and Useful Supplements* (1705), p. 43.

7/36 Robert Simson attributed most of the serious flaws in Euclid's *Elements* to the editorial intervention of 'Theon, or whoever was the Editor of the present Greek Text'; see the unpaginated 'Preface' to Simson, *Elements of Euclid.* See also Edmund Stone, *Euclid's Elements. Volume II* (1731), sig. A3r–v.

8/1 According to Proclus, Ptolemy devoted a book to the problem of parallel lines entitled 'That lines produced from angles less than two right angles meet one another'; Proclus, *A Commentary on the First Book of Euclid's Elements*, trans. Glenn R. Morrow (1970), p. 285.

8/3 Proclus, *A Commentary on the First Book of Euclid's Elements*, pp. 150–1, 284–92.

8/4 One of the 'Arabians' Reid knew of was the Persian polymath Nasīr al-Dīn al-Tūsī, whose work on parallel lines was discussed in John Wallis, 'De postulato quinto; et definitione quinta Lib. 6. Euclidis; disceptatio geometrica', in Wallis, *Opera mathematica* (1693–99), vol. II, pp. 669–73. Wallis also mentions (p. 669) the tenth-century commentator on the *Elements* 'Anaritius' or Abū'l-Abbās al-Fadl ibn Hātim al-Nairizi.

8/7 For Clavius' approach to the problematic axiom, see Christopher Clavius, *Euclidis elementorum libri XV*, second edition (1589). Clavius' commentary on Euclid's *Elements* is discussed in Vincenzo De Risi's editorial introduction to Saccheri, *Euclid Vindicated*, pp. 12–14, and Frederick A. Homann, S.J., 'Christopher Clavius and the Renaissance of Euclidean Geometry'.

8/8 See the brief tract 'De rectarum linearum parallelismo & concursu doctrina geometrica', in Thomas Oliver, *De sophismatum præstigiis cavendis admonitio ...* (1604). Each of the four tracts in this collection is separately paginated. Oliver is mentioned in Wallis, 'De postulato quinto', in Wallis, *Opera mathematica*, vol. II, p. 669.

8/13 Sir Henry Savile, *Praelectiones tresdecim in principium Elementorum Euclidis* (1621), especially pp. 140–4.

8/20 Wallis, 'De postulato quinto', in Wallis, *Opera mathematica.* Wallis'

'Dissertation' incorporated the text of a lecture delivered at Oxford on 11 July 1663.

8/24 Euclid, *ΕΥΚΛΕΙΔΟΥ ΤΑ ΣΩΖΟΜΕΝΑ*, sig. a 2r–v.

8/29 See editorial notes 4/35 and 4/39 for the publication details of these works.

8/35 Robert Simson's copy of Sacherri's *Euclides ab omni nævo vindicatus* has not survived. In the editorial notes to his edition of Euclid's *Elements*, Simson criticises Saccheri's demonstration of proposition 18 in Book V (which deals with the theory of proportion) but makes no reference to Saccheri's extended treatment of the problem of parallel lines; see Simson, *Elements of Euclid*, pp. 383–4.

9/3 Simson, *Elements of Euclid*, pp. 360–2.

10/8 Compare Reid's earlier discussion of the definition of parallel lines on p. 3, ll. 23–6.

3/I/13: Simson's Euclid

10/30 Simson, *Elements of Euclid*; for the bibliographical details of Simson's English translation of the *Elements*, see Gaskell, *Bibliography of the Foulis Press*, pp. 206–7.

10/31 Proposition 11 in Book I poses the problem, 'To draw a straight line at right angles to a given straight line, from a given point in the same'. The corollary reads: 'By help of this Problem it may be demonstrated that two straight lines cannot have a common segment'; Simson, *Elements of Euclid*, pp. 17–18.

11/3 As translated by Simson, axiom 10 states: 'Two straight lines cannot inclose a space'; Simson, *Elements of Euclid*, p. 7. For the axiom in David Gregory's edition of the *Elements*, see Euclid, *ΕΥΚΛΕΙΔΟΥ ΤΑ ΣΩΖΟΜΕΝΑ*, p. 3.

11/16 Proposition 1 in Book I is the problem, 'To describe an equilateral triangle upon a given finite straight line'; Simson, *Elements of Euclid*, p. 8.

11/18 Proposition 2 in Book I is the problem, 'To draw a straight line from a given point that shall be equal to a given straight line'. Proposition 3 is also a problem, 'To cut off from the greater of two given straight lines a part equal to the lesser'; Simson, *Elements of Euclid*, pp. 8–9.

11/19 Proposition 4 is the theorem: 'If two triangles have two sides of the one equal to two sides of the other, each to each; and have likewise the angles contained by these sides equal to one another: they shall likewise have their bases, *or third sides*, equal; and the two triangles shall be equal; and their other angles shall be equal, each to each, viz. those to which the equal sides are opposite'; Simson, *Elements of Euclid*, p. 10.

11/21 Reid follows the wording in Simson, *Elements of Euclid*, p. 7.

11/33 In Simson's translation, Euclid's definition reads: 'A straight line is that which lies evenly between its extreme points'; Simson, *Elements of Euclid*, p. 1.

12/6 An apparent allusion primarily to definition 4 in Book III: 'Straight lines are said to be equally distant from the centre of a circle [i.e. a point], when the perpendiculars drawn to them from the centre are equal'. Definition 5 states: 'But the straight line on which the greater perpendicular falls, is said to be further from the centre'; Simson, *Elements of Euclid*, p. 74.

5/I/1: Proposition 6

12/15 Line 15 to p. 13, l. 15, is most likely a draft for a section of the 'Essay' referred to in AUL, MS 2131/5/II/47, above, p. 6, l. 28. See also his reference to the two preceding propositions below, p. 14, ll. 8–9.

13/19 See also Reid's comments in Reid to [William Ogilvie], [1763], and Reid to [Robert Simson], [1764], in Reid, *Correspondence*, pp. 23, 25, 33.

13/25 In affirming that magnitude and the relations of equality, more and less are 'common Notions which have Names in all Languages' and that 'these Names have a distinct and determinate Meaning', Reid is claiming that these basic mathematical concepts, along with those of straight and curved lines, are similar to the 'common notions' involved in the 'natural' judgements of the human mind, grounded on the principles of common sense. Reid, however, maintained that even though basic mathematical notions such as magnitude have their origins in sensory experience, the concepts employed in geometry and the other branches of mathematics are the product of a process of reflection, abstraction and the exercise of judgement (unlike the principles of common sense). Reid's 'Learner' begins with an imprecise notion of straight lines gained through experience. By working through the propositions and demonstrations in the *Elements* the 'Learner' will acquire an 'accurate and scientific' notion of straight lines; Reid, *Intellectual Powers*, pp. 406–23 (especially p. 420), and Reid to James Gregory, [spring 1786], in Reid, *Correspondence*, pp. 182–3.

13/39 'one right line cannot meet another in more points than one; otherwise the right lines will coincide'; Stone, *Euclid's Elements*, p. 315. In his edition of Euclid, Robert Simson stated that this passage was 'an addition by some unskilful hand; for this [property of straight lines] is to be demonstrated, not assumed'; Simson, *Elements of Euclid*, p. 415. Compare below, p. 19, l. 34, to p. 20, l. 2.

14/2 In Simson's edition of the *Elements*, Book I, proposition 4, reads: 'If two triangles have two sides of the one equal to two sides of the other, each to each; and have likewise the angles contained by these sides equal to one another: they shall likewise have their bases, *or third sides*, equal; and the two triangles shall be equal; and their other angles shall be equal, each to each, viz. those to which the equal sides are opposite'. The demonstration of this proposition rests on axiom 10, 'Two straight lines cannot inclose a space'; Simson, *Elements of Euclid*, pp. 7, 10–11.

14/9 Propositions 4 and 5 of Reid's 'Essay' refuted the so-called 'Hypothesis of Obtuse Angles'; see above, p. 6, l. 28.

15/4 For Reid's reading notes from Saccheri, which were presumably taken after 13 September 1770, see above, p. 4, l. 30, to p. 7, l. 19.

15/9 Definition 35 in Simson's edition of Euclid's *Elements* reads: 'Parallel, or equidistant, straight lines, are such as are in the same plane, and which being produced never so far both ways, do never meet'; Simson, *Elements of Euclid*, p. 5.

15/16 Compare Reid's definitions of parallel lines above, p. 10, ll. 7–8, and p. 27, ll. 28–9.

3061/11: The Elements of Euclid

15/17 Internal evidence indicates that this manuscript is the text of a discourse Reid read to the Glasgow Literary Society. Because there is a gap in the Society's minutes from 5 November 1779 until 7 November 1794 we do not know the exact date of the meeting at which Reid delivered his discourse. The manuscript is written in Reid's characteristic late hand, which is also found in the manuscripts of his Glasgow Literary Society discourses dating from the 1790s, 'Some Thoughts on the Utopian System' (1794) and 'Of Muscular Motion in the human Body' (1795), as well as a number of his papers on mathematics that date from the same period (see above, pp. lxxv–lxxvi). It is unlikely that the manuscript was written in the

1780s because Reid appears to have read a series of discourses to the Society on materialism and on the freedom of the will; for a list of these discourses see Paul Wood, 'Thomas Reid, Natural Philosopher: A Study of Science and Philosophy in the Scottish Enlightenment', p. 386. Evidence regarding dating also comes in the note Reid added at the end of the manuscript (see textual note 31/22). Reid's note implies that he sent the manuscript to the Edinburgh Professor of Mathematics, John Playfair, via Thomas Charles Hope. Hope began teaching at the University of Glasgow in 1787 and remained there until 1795. Moreover, Playfair refers to the manuscript in the notes to his edition of Euclid published in 1795; John Playfair, *Elements of Geometry; Containing the First Six Books of Euclid, with Two Books on the Geometry of Solids* (1795), pp. 351–2. 1787 is thus a plausible *terminus post quem* for the composition of the manuscript and 1793–94 the *terminus ante quem*. Given the sequence of Reid's discourses, however, it more likely that Reid composed the manuscript in the period *circa* 1790 to 1793.

15/21 Writing to William Ogilvie in 1763, Reid referred to Euclid as 'the Patriarch of Mathematicians'; Reid to [William Ogilvie], [1763], in Reid, *Correspondence*, p. 23. For Reid Euclid's *Elements* served as the exemplar of the axiomatic method; see Reid, *Philosophical Orations*, oration II, para. 20. Reid also claimed that Euclid's *Elements* provided 'the firm foundation of all future superstructures' in geometry, although he suggested that the very perfection of the *Elements* obscured the fact that Euclid had systematised and improved 'the inventions in geometry, which had been made in a tract of preceding ages'; Reid, *Intellectual Powers*, pp. 62, 512–13.

16/2 Simson, *Elements of Euclid*, 'Preface' (unpaginated).

16/10 Although Reid comments at length on Simson's edition of Euclid's *Elements* the wording of the passages he refers to does not necessarily follow Simson's. The differences in wording suggest that either he relied primarily on his memory and his own translation of the text, or that he continued to use another edition of the *Elements*, even though he was commenting on Simson. For example, Simson's translation of definition 1 reads: 'A point is that which hath no parts, or which hath no magnitude'; Simson, *Elements of Euclid*, p. 1. Compare Edmund Stone's translation, which reads: 'A point is that which has no parts'; Stone, *Euclid's Elements*, p. 1.

16/20 Definition 3 reads: 'The extremities of a line are points'; Simson,

Elements of Euclid, p. 1. Stone has 'extremes' rather than 'extremities'; Stone, *Euclid's Elements*, p. 1.

16/24 Definition 4 reads: 'A straight line is that which lies evenly between its extreme points'; Simson, *Elements of Euclid*, p. 1. Compare Stone's translation: 'A right line is that which lies evenly between its points'; Stone, *Euclid's Elements*, p. 1.

16/34 Definition 15 states that 'A circle is a plain figure contained by one line, which is called the circumference, and is such that all straight lines drawn from a certain point within the figure unto the circumference, are equal to one another'. Definition 30 states that 'Of four sided figures, a square is that which has all its sides equal, and all its angles right angles'. A parallelogram is defined in proposition 34 as 'a four sided figure whose opposite sides are parallel. [A]nd the diameter is the straight line joining two of its opposite angles'; Simson, *Elements of Euclid*, pp. 3, 5, 39.

16/38 Spheres and cones are defined in Book XI of the *Elements*. Definition 14 states that 'A Sphere is a solid figure described by the revolution of a semicircle about the diameter, which remains unmoved'. Definition 18 states that 'A Cone is a solid figure described by the revolution of a right angled triangle about one of the sides containing the right angle, which side remains fixed'; Simson, *Elements of Euclid*, p. 232.

17/23 Reid's statement of these axioms differs those found in Stone's and Simson's editions of the *Elements*; compare Stone, *Euclid's Elements*, pp. 5–6, and Simson, *Elements of Euclid*, p. 7.

18/7 Reid refers to definition 7: 'A plain superficies is that in which any two points being taken, the straight line between them lies wholly in that superficies'. Simson in fact includes the definition but states in his editorial notes: 'Instead of this Definition as it is in the Greek copies, a more distinct one is given from a property of a plane superficies, which is manifestly supposed in the Elements, viz. that a straight line drawn from any point in a plane to any other in it, is wholly in that plane'; Simson, *Elements of Euclid*, pp. 1, 355.

18/12 Compare above, p. 13, ll. 21–33, and see editorial note 13/25.

18/17 Reid here echoes the argument of his 'An Essay on Quantity', for which see pp. 50–9, and, for analysis, pp. xcviii–cviii. Colin Maclaurin was one of the mathematicians Reid had in mind in this sentence; see the Introduction, p. xxxii. Reid did not know that Maclaurin had himself had applied mathematical reasoning to moral theory in an essay written in 1714, 'De viribus mentium bonipetis'; see Judith V. Grabiner, 'Maclaurin and Newton: The Newtonian Style and the Authority of Mathematics', pp. 145–6.

18/36 Reid's preoccupation with the definition of right lines may have originated in his early reading of David Hume's *A Treatise of Human Nature*, wherein Hume observed that 'mathematicians pretend to give an exact definition of a right line' and then proceeded to point out the inadequacies of their definition; see the Introduction, pp. lxxxvi–lxxxvii, and David Hume, *A Treatise of Human Nature*, ed. David Fate Norton and Mary J. Norton, (2000), pp. 37–8 (1.2.4, 26–7).

19/23 Reid's wording of the last sentence of the demonstration of proposition 1, Book XI, is similar to that found in Stone's edition of Euclid: 'Wherefore two right lines … have both one common segment … which is impossible; for one right line cannot meet another in more points than one; otherwise the right lines will coincide'; Stone, *Euclid's Elements*, p. 315.

19/39 'The words at the end of this, "for a straight line cannot meet a straight line in more than one point," are left out, as an addition by some unskilful hand; for this is to be demonstrated, not assumed'; Simson, *Elements of Euclid*, p. 415.

20/10 'Instead of this Definition as it is in the Greek copies, a more distinct one is given from a property of a plane superficies, which is manifestly supposed in the Elements, viz. that a straight line drawn from any point in a plane to any other in it, is wholly in that plane'; Simson, *Elements of Euclid*, p. 355.

21/16 This corollary is simply affirmed in the demonstration of proposition 1, Book XI, in Stone's edition of Euclid, whereas it is proved as a corollary to proposition 11, Book I, in Simson's edition; compare Stone, *Euclid's Elements*, p. 315, and Simson, *Elements of Euclid*, pp. 17–18.

21/28 Simson, *Elements of Euclid*, pp. 10–12, 15–16, 17–18.

21/30 Axiom 12 in Stone's edition and axiom 10 in Simson's; Stone, *Euclid's Elements*, p. 6, and Simson, *Elements of Euclid*, p. 7.

21/34 Axiom 10 in Stone's edition and axiom 11 in Simson's; Stone, *Euclid's Elements*, p. 5, and Simson, *Elements of Euclid*, p. 7.

23/18 Axiom 11 in Stone's edition and axiom 12 in Simson's; Stone, *Euclid's Elements*, pp. 5–6, and Simson, *Elements of Euclid*, p. 7. See also editorial note 7/23.

23/23 In Stone's translation, definition 35 reads: 'Parallels are right lines, which being in the same plane, and produced infinitely either way, will not meet one another either way'; Stone, *Euclid's Elements*, p. 4.

23/28 See editorial note 7/33.

23/36 See editorial note 8/1.

23/38 Proclus, *A Commentary on the First Book of Euclid's Elements*, pp. 150–1, 284–92.

24/1 See editorial note 8/4.

24/4 Clavius, *Euclidis elementorum libri XV*.

24/5 Oliver, 'De rectarum linearum parallelismo & concursu doctrina geometrica', in Oliver, *De sophismatum præstigiis cavendis admonitio*.

24/11 Savile, *Praelectiones tresdecim in principium Elementorum Euclidis*.

24/19 Wallis, 'De postulato quinto', in Wallis, *Opera mathematica*, vol. II.

24/24 Euclid, *ΕΥΚΛΕΙΔΟΥ ΤΑ ΣΩΖΟΜΕΝΑ*.

24/30 Saccheri, *Euclides ab omni nævo vindicatus*; Saccheri, *Logica demonstrativa*; Saccheri, *Neo-statica*.

24/37 As mentioned in editorial note 8/35, Simson's annotated copy of Saccheri's *Euclides ab omni nævo vindicatus* has not survived.

25/8 Euclid, *ΕΥΚΛΕΙΔΟΥ ΤΑ ΣΩΖΟΜΕΝΑ*, sig. a verso. Proposition 17 states: 'Any two angles of every triangle taken together, are less than two right angles'; Stone, *Euclid's Elements*, p. 23.

25/33 Simson, *Elements of Euclid*, unpaginated 'Preface'.

25/35 'if Troy's towers could be saved by strength of hand, by mine, too, had they been saved'; Virgil, *Aeneid*, Book II, lines 291–2.

26/21 Reid paraphrases Simson, *Elements of Euclid*, pp. 360–1.

26/33 'This property flows from the nature or Definition of a straight line which keeps always the same direction, and therefor[e] cannot be strictly demonstrated by the preceding Propositions'; Simson, *Elements of Euclid*, p. 361.

27/25 Reid's fourth corollary is stated above, p. 22, ll. 8–10.

28/34 Alexander Pope, *An Essay on Criticism* (1711), Part I, lines 152–3, in Alexander Pope, *Poetical Works*, ed. Herbert Davis (1985).

28/35 Meetings of the Glasgow Literary Society were chaired by a 'President'. The position rotated among the members. After delivering a discourse or leading the discussion of a question, a member would then preside over the next meeting. The laws of the Society stipulated a number of rules governing the President's responsibilities; see 'Laws of the Literary Society in Glasgow College', Glasgow University Library (hereafter GUL), MS Murray 505, pp. 2–7.

29/9 Robert Simson, *The Elements of Euclid, viz. the First Six Books,*

Together with the Eleventh and Twelfth, fifth edition (1775), pp. 302–7. Simson revised his note on proposition 29, Book I, in the second edition of his translation, published in 1762. Reid had, in fact, written to Simson regarding this note, stating that 'I think The note on Prop 29 Lib. 1. greatly improved in this edition'; Reid to [Robert Simson], [1764], in Reid, *Correspondence*, p. 33.

29/21 Simson, *Elements of Euclid*, fifth edition, p. 303.

30/5 On self-evident propositions and the 'tribunal' of common sense, see Reid, *Intellectual Powers*, pp. 141–2, 407, 426, 432–3.

31/6 Proposition 1, Book X of the *Elements* states: 'Two unequal Magnitudes being given, if from the greater be taken away a greater than the half, and again from that remaining one greater than its half; and so on continually; then at length will be left a Magnitude less than the lesser of the given Magnitudes'; Stone, *Euclid's Elements. Volume II*, p. 83.

Part Two: 'An Essay on Quantity'

5/I/20: Object of Mathematicks

32/3 Compare the definitions given in Ephraim Chambers, *Cyclopædia: Or, an Universal Dictionary of Arts and Sciences* (1728), s.v. 'Mathematics' and 'Quantity', vol. II, pp. 508, 934, where mathematics is said to be 'the Science of Quantity' and quantity is defined as 'any thing capable of Estimation, or Mensuration; or, which being compared with another thing of the same kind, may be said to be greater, or less, than; equal, or unequal to, it'. According to Helena Pycior, the conception of mathematics as the science of quantity was the 'prevailing' one in seventeenth- and eighteenth-century Britain; Helena M. Pycior, *Symbols, Impossible Numbers and Geometric Entanglements: British Algebra through the Commentaries on Newton's* Universal Arithmetick, p. 4. See also editorial note 38/24.

32/12 Inertia (*vis insita*), impressed, absolute and accelerative forces. The measures of these forces, as well as the other quantities mentioned in this sentence, are given in the definitions stated at the beginning of Book I of Newton's *Principia*; Isaac Newton, *The Principia: Mathematical Principles of Natural Philosophy* (1999), pp. 403–15.

33/2 Reid alludes to Francis Hutcheson's *An Inquiry into the Original of Our Ideas of Beauty and Virtue in Two Treatises*. In two versions of the first edition of the *Inquiry*, published in 1725, the title page states that the work includes 'an Attempt to introduce a *Mathematical*

Calculation in Subjects of *Morality*'. This statement was dropped from one version of the title page of the first edition and from the title pages of subsequent editions of the *Inquiry*; see Wolfgang Leidhold's textual note to Francis Hutcheson, *An Inquiry into the Original of Our Ideas of Beauty and Virtue in Two Treatises* (2008), pp. xxiii–xxviii. For the specific passages in which Hutcheson applies rudimentary mathematical reasoning in order to 'find a universal Canon to compute the Morality of any Actions, with all their Circumstances, when we judge of the Actions done by our selves, or by others', see Hutcheson, *Inquiry*, pp. 128–35, 190. Hutcheson deleted his use of mathematical equations in the fourth edition of the *Inquiry*, which appeared in 1738. On the contexts for Hutcheson's moral calculus, see the Introduction, pp. cii–ciii.

33/17 Reid refers to the *vis viva* controversy. Newton and his followers asserted that the correct measure of the force of a moving body was mv, whereas Leibniz and his camp of natural philosophers affirmed that the proper measure was mv^2. On the controversy and Reid's response to it, see the Introduction, pp. xxiv–xxv, ciii–civ, cvii–cxi.

33/30 For Reid's most detailed treatment of the distinctiveness of mathematical evidence, see Reid, 'A System of Logic', pp. 49–52. In his logic lectures he classified the different kinds of evidence found in the sciences and argued that the certainty of mathematics was due to the fact that mathematical reasoning dealt with 'relations we observe between Quantities compared together' (p. 49). And because he maintained that the evidence of mathematics was distinctive, he criticised John Locke's suggestion that morality was capable of being a demonstrative science with the same degree of evidence as algebra or geometry. Locke asserted that 'the *Ideas* of Quantity are not those alone that are capable of Demonstration and Knowledge'; John Locke, *An Essay concerning Human Understanding* (1690), p. 549 (IV.iii.18). Reid agreed, but insisted that the nature of the evidence we have for propositions in mathematics, morals and natural philosophy is different; compare Reid, *On Logic, Rhetoric and the Fine Arts*, pp. 145–9, 186, and Reid, *Intellectual Powers*, pp. 40–1, 99–100, 121, 228–33, 255–6, 439–40, 456, 494, 546, 547–55. On the distinctive evidence of mathematics, compare Colin Maclaurin, *A Treatise of Fluxions. In Two Books* (1742), vol. I, pp. 51–2.

33/34 Reid's comment reflects the fact that following the revival of algebra by Girolamo Cardano and François Viète in sixteenth-century

Europe and the subsequent blending of arithmetic and algebra, mathematicians were obliged to reconcile traditional conceptions of number inherited from Aristotle and Euclid with new algebraic objects such as surds and infinite series. Newton himself acknowledged the difficulties in conceptualising the latter and compared infinite series to surd roots; see the translation of Newton's 'De analysi per aequationes numero terminorum infinitas' in John Stewart, *Sir Isaac Newton's Two Treatises of the Quadrature of Curves and Analysis by Equations of an Infinite Number of Terms, Explained* (1745), p. 340. The *Oxford English Dictionary* defines an adfected equation as one which includes 'different powers of an unknown quantity' and cites the use of the term applied to the equation '$ax^2 - Bx + A = 0$' in George Campbell, 'A Method for Determining the Number of Impossible Roots in Adfected Æquations', p. 515. Reid may well have known this paper since it was published in the *Philosophical Transactions* for 1727–28.

5/I/22: Essay Concerning the Object of Mathematicks

34/3 Hutcheson, *An Inquiry into the Original of Our Ideas of Beauty and Virtue in Two Treatises*; see above, editorial note 33/2.

34/10 Aristotle, *Categories*, 5a.38-b.10, in Aristotle, *The Complete Works of Aristotle: The Revised Oxford Translation* (1984); compare Reid, 'A System of Logic', pp. 7–8, and Reid, *On Logic, Rhetoric and the Fine Arts*, p. 150.

34/18 Inertia and the impressed, motive and accelerative forces.

34/18 Colin Maclaurin was later to provide such measures for fluxions; see the Introduction, p. xxvii, and Maclaurin, *Treatise of Fluxions*, vol. I, pp. 51–9.

34/23 Reid refers to definitions 1–8 in Book I of Newton, *The Principia*, pp. 403–8; compare Reid's comments on Newton's definitions in AUL, MS 2131/5/II/8, fol. 1r.

34/34 The key figure in defining a quantitative measure of velocity during the first half of the seventeenth century was Galileo; see the mathematical treatment of local motion in the Third Day of Galileo Galilei, *Two New Sciences: Including Centers of Gravity and Force of Percussion*, trans. Stillman Drake (1974), esp. pp. 147–53, and, for analysis, J. L. Heilbron, *Galileo*, pp. 126–42, 336–40. Reid later paid tribute to Galileo in *Intellectual Powers*, pp. 13, 102.

35/4 The *Oxford English Dictionary* defines the adjective 'sopite' as 'put to rest or sleep; settled'. The verb is defined in the first instance as 'to put or lull to sleep; to render drowsy, dull, or inactive', and secondarily as 'to put an end to, to settle' or 'to pass over or suppress'.

35/6 Compare Reid, *Intellectual Powers*, pp. 18, 457.

35/25 See definition 8, Book I of Newton, *The Principia*, p. 407.

36/7 Compare Leibniz's comments in §99 of his 'Fifth Paper' in *A Collection of Papers, Which Passed between the Late Learned Mr Leibnitz, and Dr Clarke, in the Years 1715 and 1716* (1717), pp. 251–3.

36/9 That is, 'the Adversary' need not accept the relationship between force, mass and velocity stipulated by Newton's definition of the measure of the quantity of motion in definition 2, Book I of *The Principia*, p. 404. It is precisely this relationship, as it had earlier been expressed in the writings of Descartes, that Gottfried Wilhelm Leibniz questioned in 'A Brief Demonstration of a Notable Error of Descartes and Others concerning a Natural Law, According to Which God Is Said Always to Conserve the Same Quantity of Motion; a Law Which They also Misuse in Mechanics', pp. 296–8, wherein Leibniz distinguished between different measures of the quantity of motion and of the force of motion. Even though Reid adopted Newton's definition because it was simple and conformed to the use of the word 'force' in ordinary language (p. 56, ll. 5–7), he recognised that these considerations were not in themselves sufficient to compel the acceptance of Newton's definition. Consequently, he admitted that it was open to Leibniz and his followers to propose an alternative definition of the measure of the force of moving bodies. In this sense, Reid agreed with the view initially expressed by W. J. 's Gravesande in 1729 and later echoed by J. T. Desaguliers in 1744 that the *vis viva* controversy was, at least in part, as Desaguliers subsequently put it, 'only a Dispute about Words'; J. T. Desaguliers, *A Course of Experimental Philosophy* (1734–44), vol. II, p. vi; see also vol. II, pp. 49–50. On 's Gravesande, see Thomas L. Hankins, 'Eighteenth-century Attempts to Resolve the *Vis viva* Controversy', pp. 287–8.

37/3 Leibniz, 'A Brief Demonstration of a Notable Error of Descartes and Others concerning a Natural Law', pp. 297–8.

37/20 Reid had earlier made a similar point regarding arithmetical and geometrical reasoning; see AUL, MS 2131/5/II/1, fol. 1r, for a note dated 19 July 1729, which partly deals with the mathematical analysis of planetary motion.

37/23 Compare the definition given above, p. 34/11–13.

3061/7: An Essay on Quantity, October 1748

38/6 Reid's epigraph is a paraphrase from Aristotle, *Categories*, 5a.38-b.10.

38/24 Compare Reid's definition of mathematics with those found in Jacques Ozanam, *Cursus mathematicus: Or, a Compleat Course of the Mathematicks* (1712), vol. I, p. 1, and Willem Jacob 's Gravesande, *The Elements of Universal Mathematics, or Algebra: To Which Is Added, a Specimen of a Commentary on Sir Isaac Newton's Universal Arithmetic* (1728), pp. 1–2. Reid's definition of mathematics should also be read in the context of Query 25 of George Berkeley's *The Analyst: Or, A Discourse Addressed to an Infidel Mathematician* (1734), in *The Works of George Berkeley, Bishop of Cloyne*, vol. IV, p. 98: 'Whether the finding out proper expressions or notations of quantity be not the most general character and tendency of the mathematics? And arithmetical operation that which limits and defines their use?' See also editorial note 32/3, above.

38/27 In this version of 'An Essay on Quantity' Reid does not name Francis Hutcheson as one of his targets.

39/18 Compare Reid's definition of algebra with those found in 's Gravesande, *Elements of Universal Mathematics*, pp. 1–2, and Joseph Raphson, *A Mathematical Dictionary: Or, a Compendious Explication of All Mathematical Terms, Abridg'd from Monsieur Ozanam, and Others* (1702), s.v. 'Algebra'.

39/25 Aristotle, *Categories*, 5a.38-b.1: 'Only [number, language, lines, surfaces, bodies, time and place] are called quantities strictly, all the others derivatively; for it is to these we look when we call the others quantities'.

40/9 See above, editorial note 34/34.

40/24 On probability, compare above, p. 32/29, and Reid, *Intellectual Powers*, pp. 559–60. As we have seen in the Introduction, pp. xxx–xxxii, xlvi–xlvii, little evidence survives regarding the works on mathematics that Reid was reading before his move to Glasgow in 1764. An extensive undated set of notes taken from Abraham de Moivre's *The Doctrine of Chances: Or, a Method of Calculating the*

Probability of Events in Play (1718) suggests that Reid may have known de Moivre's work at the time he was writing this manuscript; see 'Of our reasoning concerning Chance', AUL, MS 2131/5/II/49, fols 3r–6v. A second, enlarged edition of *The Doctrine of Chances* appeared in 1738 and a third in 1756. Reid's notes give no indication of which edition he had read.

40/26 Inertia and impressed force.

40/39 Definitions 1–8 in Book I of Newton, *The Principia*, pp. 403–8. Compare Reid's comments in AUL, MS 2131/5/II/8, fol. 1r.

41/15 Definitions 5–8 in Book I of Newton, *The Principia*, pp. 405–8. In these definitions Newton distinguished between three different types of centripetal force: motive, accelerative and absolute.

41/26 Book I of *The Principia* is devoted to the mathematical analysis of the motions of bodies, including the laws governing planetary motion, while Book II investigates mathematically and experimentally the principles of fluid mechanics; for a detailed discussion of the contents and the structure of Newton's argument in Books I and II, see I. Bernard Cohen's editorial introduction in Newton, *The Principia*, pp. 128–94.

42/19 Reid's comments register the debate in Britain initiated by Isaac Barrow and John Wallis over the respective merits of geometry and algebra that is delineated in Pycior, *Symbols, Impossible Numbers and Geometric Entanglements*. In opposition to Wallis' claims regarding the precedence of algebra over geometry, Barrow contended that geometry was fundamental to the mathematical sciences, stating that 'the Whole of Mathematics [is] in some sort contained and circumscribed within the Bounds of Geometry'. According to Barrow, arithmetic was rooted in geometry, and algebra was no more than 'a Part or Species of *Logic*, or a certain Manner of using Reason in the Solution of Questions'. Consequently, when Barrow argued for the value of the study of the mathematical sciences on the ground that it was only in mathematics that examples of demonstrative reasoning could be found, he indicated that Euclidean geometry served as the model for rigorous logical thought; Isaac Barrow, *The Usefulness of Mathematical Learning Explained and Demonstrated: Being Mathematical Lectures Read in the Publick Schools at the University of Cambridge* (1734), pp. 28, 29, 52–3. Later, George Berkeley echoed Barrow in *The Analyst* when he wrote that 'Geometry is an excellent Logic'. Berkeley affirmed that 'when the definitions are clear; when the postulata cannot be refused, nor the axioms denied; [and] when from the distinct contemplation and comparison of figures, their properties are derived, by a perpetual well-connected chain of consequences', as in geometry 'there is acquired an habit of reasoning, close and exact and methodical; which habit strengthens and sharpens the mind'; Berkeley, *The Analyst*, pp. 65–6. For Reid, on the other hand, algebra served as the foundation of the mathematical sciences. He was willing to grant that geometrical reasoning had a greater degree of evidence than that found in algebra, because 'in Geometry a Figure is used to represent other figures', whereas in algebra quantities are represented by 'certain arbitrary Symbols'. Nevertheless, he insisted that 'the Elements of Algebra seem in the order of Nature to go before those of Geometry' because 'Algebra considers Quantity in General' whereas 'Geometry treats particularly of those Quantitys which have figure'; 'A View of Algebra in the Order in which it is to be taught', AUL, MS 2131/7/I/1, fol. 1r–v. And in arguing for the value of the study of the mathematical sciences, he followed Locke in emphasising that it was mathematics in general (including algebra), rather than simply geometry, that could be used to 'exercise and strengthen the reasoning powers'; compare Reid, 'A Brief Account of Aristotle's Logic. With Remarks', in Reid, *On Logic, Rhetoric and the Fine Arts*, pp. 142–3, with John Locke, 'Of the Conduct of the Understanding', in John Locke, *Posthumous Works of Mr. John Locke* (1706), pp. 29–35. Reid's characterisation of algebra as 'involving an Artificial kind of Operation' may have been inspired by Berkeley's view of algebra as an 'artificial sort of language' in which the permutations of signs are governed by various 'logistic operations'; George Berkeley, *Alciphron, or the Minute Philosopher* (1732), in *The Works of George Berkeley*, vol. III, p. 307. For discussions of Berkeley's conception of algebra as a science of signs see Pycior, *Symbols, Impossible Numbers and Geometric Entanglements*, pp. 212–31, and Douglas M. Jesseph, *Berkeley's Philosophy of Mathematics*, pp. 114–17. As well as registering this general debate, Reid's juggling of the rival claims of algebra and geometry also reflects the unstable nature of Newton's mathematical legacy; on this point, see especially Niccolò Guicciardini, 'Dot-Age: Newton's Mathematical Legacy in the Eighteenth Century', pp. 224–32, and Pycior, *Symbols, Impossible Numbers and Geometric Entanglements*, pp. 200–8. While the mathematical style of

The Principia was modelled on the rigour of classical geometry, Newton's *Arithmetica universalis* (1707) served as the starting point for much of the work done on algebra (and, ironically, geometry) by British mathematicians during the course of the eighteenth century. Reid can thus be interpreted as endeavouring to negotiate a workable compromise between these two strands of Newton's work.

42/26 See editorial note 33/30.

42/36 Compare p. 32, l. 29, and p. 40, ll. 20–1.

43/3 Francis Hutcheson, *An Inquiry into the Original of Our Ideas of Beauty and Virtue in Two Treatises*; see editorial note 33/2.

43/17 For the meaning of 'sopite', see editorial note 35/4.

43/19 Compare Reid, *Intellectual Powers*, pp. 18, 457.

44/7 See definition 2 in Book I of Newton, *The Principia*, p. 404.

44/28 Reid has in mind demonstration experiments involving levers and balances. See, for example, the experiment described in Desaguliers, *A Course of Experimental Philosophy*, vol. I, pp. 46–7.

44/34 Because Reid believed that the axiomatic structure of mathematics and natural philosophy served as a model for at least some branches of the human sciences, he paid careful attention to matters of definition, as can be seen, for example, in his work on Euclid's *Elements*. His most extensive discussion of the issues associated with defining basic terms in the sciences is found in Reid, *Intellectual Powers*, pp. 17–20, 362–4, 373–4, 381–5, 457–9. Reid also continued to appeal to common usage as providing a standard of meaning; see Reid, *Intellectual Powers*, p. 303.

45/3 For experimental demonstrations that purported to prove that mv was the correct measure of the force of moving bodies see, for example, J. T. Desaguliers, 'Animadversions upon Some Experiments Relating to the Force of Moving Bodies; with Two New Experiments on the Same Subject'; Desaguliers, *A Course of Experimental Philosophy*, vol. I, pp. 393–9; James Jurin, 'An Inquiry into the Measure of the Force of Bodies in Motion: With a Proposal of an *Experimentum crucis*, to Decide the Controversy About It'.

45/16 Apart from Leibniz, the most notable proponents of the view that the correct measure of the force of a moving body is mv^2 were Jakob Hermann, Giovanni Poleni, Willem Jacob 's Gravesande, Jean Bernoulli and Pieter van Musschenbroek; see, *inter alia*, Jakob Hermann, *Phoronomia, sive de viribus et motibus corporum solidorum et fluidorum libri duo* (1716); Giovanni Poleni, *De castellis per quae derivantur fluviorum aquae habentibus latera convergentia liber: Quo etiam continentur nova experimenta ad aquas fluentes, & ad percussionis vires pertinentia* (1718); Willem Jacob 's Gravesande, 'Essai d'une nouvelle théorie du choc des corps fondée sur l'expérience'; Jean Bernoulli, *Discours sur les loix de la communication du movement* (1727); and Pieter van Musschenbroek, *The Elements of Natural Philosophy* (1744), vol. I, pp. 75–88. On the controversy in Paris surrounding Bernoulli's *Discours*, see also Mary Terrall, '*Vis viva* Revisited', esp. pp. 191–5 and the references cited there.

46/3 Hermann, Poleni, 's Gravesande and van Musschenbroek all described experiments which they claimed demonstrated that mv^2 was the correct measure of the force of a moving body. In Britain, the most widely discussed experimental demonstrations of *vis viva* were those of Poleni, 's Gravesande and van Musschenbroek. For the critical response to their experiments by English natural philosophers see: Henry Pemberton, 'A Letter to Dr Mead, Coll. Med. Lond. & Soc. Reg. S. concerning an Experiment, Whereby It Has Been Attempted to Shew the Falsity of the Common Opinion, in Relation to the Force of Bodies in Motion'; J. T. Desaguliers, 'An Account of Some Experiments Made to Prove, That the Force of Moving Bodies Is Proportionable to Their Velocities: (Or Rather That the *Momentum* of Moving Bodies Is to be Found by Multiplying the Masses into the Velocities) in Answer to Such Who Have Sometime Ago Affirm'd, That That Force is Proportionable to the Square of the Velocity, and to Those Who Still Defend the Same Opinion'; J. T. Desaguliers, 'Animadversions upon Some Experiments Relating to the Force of Moving Bodies; with Two New Experiments on the Same Subject'; John Eames, 'A Remark upon the New Opinion Relating to the Forces of Moving Bodies, in the Case of the Collision of Non-elastic Bodies'; John Eames, 'Remarks upon Some Experiments in Hydraulics, Which Seem to Prove, That the Forces of Equal Moving Bodies Are as the Squares of Their Velocities'. A helpful overview of this debate is found in Carolyn Iltis, 'The Leibnizian–Newtonian Debates: Natural Philosophy and Social Psychology'.

46/10 Leibniz, 'A Brief Demonstration of a Notable Error of Descartes and Others concerning a Natural Law', pp. 297–8.

46/29 See especially the experimental data discussed in Book III of Pierre Louis Moreau de Maupertuis, *The Figure of the Earth, Determined from Observations Made by Order of the French King, at the Polar*

Circle (1738), pp. 199–232. As noted in the Introduction, p. cxiv, Reid read this work in November 1739. He may also have read the important paper alluded to by Maupertuis: George Graham, Colin Campbell and James Bradley, 'An Account of Some Observations Made in London … and at Black-River in Jamaica … concerning the Going of a Clock; in Order to Determine the Difference between the Lengths of Isochronal Pendulums in Those Places'. The starting point for all of these experiments was Newton's discussion of the weights of bodies at the pole and at the equator in proposition 20, Book III, of Newton, *The Principia*, pp. 826–32.

47/33 For this distinction, see definition 8, Book I, of Newton, *The Principia*, pp. 407–8.

48/7 See, for example, the work attributed to Euclid, *Sectio canonis*, and Ptolemy's *Harmonics*, both in *Greek Musical Writings*, ed. Andrew Barker, vol. II, pp. 200–1, 285–6, 289, 298–301. Reid would have been familiar with the text of the *Sectio canonis* in David Gregory's edition of Euclid's writings; Euclid, *ΕΥΚΛΕΙΔΟΥ ΤΑ ΣΩΖΟΜΕΝΑ*, pp. 547–56. Gregory's edition also included the text of a work incorrectly attributed to Euclid, the *Introduction to Harmonics*, which is now thought to be written by Cleonides; Euclid, *ΕΥΚΛΕΙΔΟΥ ΤΑ ΣΩΖΟΜΕΝΑ*, pp. 533–43. It should be noted that in both the *Sectio canonis* and the *Harmonics*, the interval of the octave is said to be a duple ratio, that is, 2:1, not Reid's 1:2. In addition to reading these texts and perhaps others by ancient writers on music, Reid may have learned about the science of harmonics in classical antiquity through Alexander Malcolm's *A Treatise of Musick, Speculative, Practical and Historical*. First published in Edinburgh in 1721, Malcolm's *Treatise* was republished in London in 1730. Another work of Malcolm's to appear that year, *A New System of Arithmetick, Theorical and Practical*, was dedicated to the Aberdeen Town Council and on the title page Malcolm was styled 'Teacher of the Mathematicks at *Aberdeen*'. It is thus likely that Reid knew of Malcolm's *Treatise* and that he may have met Malcolm himself. For further evidence of Reid's interest in harmonics and musical theory, see AUL, MSS 2131/6/V/16, 6/V/16a, 6/V/17, 7/III/4, 7/III/5, 7/V/25 and 8/I/16, fols 2r–v; the first three manuscripts listed relate to his reading of Robert Smith's *Harmonics, or the Philosophy of Musical Sounds* (1749), while the last manuscript cited contains a reference to Charles Avison's *An Essay on Musical Expression* (1752).

48/11 See the discussion between Sagredo and Salviati in the First Day of Galileo's *Two New Sciences*, pp. 100–2.

50/6 This 'fact' comes from Desaguliers, *A Course of Experimental Philosophy*, vol. II, pp. v, 50; Desaguliers provides an extensive review of the *vis viva* debate on pp. 49–94 of the long-delayed second volume.

50/10 Desaguliers, *A Course of Experimental Philosophy*, vol. II, pp. vi, 49–50.

An Essay on Quantity, November 1748

50/14 Henry Miles was a noted Dissenting minister and man of science in England, who had been elected a Fellow of the Royal Society of London in 1743. Miles was said by his biographer and fellow Dissenter, Philip Furneaux, to have been given an honorary doctor of divinity degree by one of the Aberdeen colleges in 1744; see Philip Furneaux, *A Sermon Occasioned by the Death of the Reverend Henry Miles, D.D. and F.R.S.* (1763), p. 34. Miles' name does not, however, appear on the lists of honorary degree recipients of either college. It may be that Reid was in contact with English Nonconformists such as Miles through David Fordyce, who corresponded with a number of prominent Dissenters he had befriended during his sojourn in England in the period 1737–40, including Miles, Miles' associate John Canton and Philip Doddridge. Martin Folkes succeeded Sir Hans Sloane as President of the Royal Society in 1741 and resigned from his post in November 1753 due to ill-health. According to Dugald Stewart, Reid had met Folkes in London in 1736; Dugald Stewart, *Account of the Life and Writings of Thomas Reid, D.D. F.R.S.Edin. Late Professor of Moral Philosophy in the University of Glasgow* (1803), p. 13.

50/27 An allusion to Francis Hutcheson; compare above, p. 43, ll. 1–11.

51/12 Compare Reid, 'A System of Logic', pp. 49–50, and Reid, *On Logic, Rhetoric and the Fine Arts*, p. 178.

51/26 Aristotle, *Categories*, 5a.38-b.1: 'Only [number, language, lines, surfaces, bodies, time and place] are called quantities strictly, all the others derivatively; for it is to these we look when we call the others quantities'.

52/12 See editorial note 34/34.

52/26 As mentioned in editorial note 40/24, it is unclear which works on the calculus of chances Reid had read by the autumn of 1748. One likely candidate is Abraham de Moivre's *The Doctrine of Chances*.

52/28 Inertia and impressed force.

53/2 Definitions 1–8 in Book I of Newton, *The Principia*, pp. 403–8; compare above, editorial note 40/39.

53/17 Definitions 5–8 in Book I of Newton, *The Principia*, pp. 405–8. Newton here distinguished between three different types of centripetal force: motive, accelerative and absolute.

53/27 Book I of Newton's *Principia* is devoted to the mathematical analysis of the motions of bodies, including the laws governing planetary motion, while Book II investigates mathematically and experimentally the principles of fluid mechanics; for a detailed discussion of the contents and the structure of Newton's argument in Books I and II see I. Bernard Cohen's editorial introduction in Newton, *The Principia*, pp. 128–94.

54/19 See editorial note 42/19.

54/24 On the distinctiveness of mathematical evidence, compare Reid, 'A System of Logic', pp. 49–52; Reid, *On Logic, Rhetoric and the Fine Arts*, pp. 145–9, 186; and Reid, *Intellectual Powers*, pp. 40–1, 99–100, 121, 228–33, 255–6, 439–40, 456, 494, 546, 547–55.

54/37 Hutcheson, *An Inquiry into the Original of Our Ideas of Beauty and Virtue in Two Treatises*, pp. 128–35, 190; see editorial note 33/2.

55/14 Compare Reid, *Intellectual Powers*, pp. 18, 457.

56/1 Definition 2, Book I, of Newton, *The Principia*, p. 404. For Newton's followers who endeavoured to demonstrate experimentally Newton's measure of the force of moving bodies see, *inter alia*: Desaguliers, 'Animadversions upon Some Experiments Relating to the Force of Moving Bodies'; Desaguliers, *A Course of Experimental Philosophy*, vol. I, pp. 393–9; Jurin, 'An Inquiry into the Measure of the Force of Bodies in Motion'.

56/22 As mentioned in editorial note 44/28, Reid had in mind demonstration experiments involving levers and balances; see, for example, the experiment described in Desaguliers, *A Course of Experimental Philosophy*, vol. I, pp. 46–7.

56/39 Reid recognised that there were no logically compelling reasons to adopt Newton's definition of the measure of the force of moving bodies, even though Newton's definition had the virtues of simplicity, clarity and conformity to the common use of the term 'force'.

57/8 See Leibniz, 'A Brief Demonstration of A Notable Error of Descartes and Others concerning a Natural Law', pp. 296–8.

57/31 See, *inter alia*, Hermann, *Phoronomia*; Poleni, *De castellis per quae derivantur fluviorum aquae habentibus latera convergentia liber*; 's Gravesande, 'Essai d'une nouvelle théorie du choc des corps fondée sur l'expérience'; Bernoulli, *Discours sur les loix de la communication du movement*; and van Musschenbroek, *The Elements of Natural Philosophy*, vol. I, pp. 75–88.

57/38 Leibniz, 'A Brief Demonstration of a Notable Error of Descartes and Others concerning a Natural Law', pp. 296–7.

58/17 Maupertuis, *The Figure of the Earth*, pp. 199–232; Graham, Campbell and Bradley, 'An Account of Some Observations Made in London … and at Black-River in Jamaica'; Newton, *The Principia*, pp. 826–32.

59/15 Newton distinguishes between the accelerative quantity of centripetal force and the motive quantity of centripetal force in definitions 7 and 8, Book I, of *The Principia*, pp. 407–8.

Part Three: Astronomy

7/VIII/10: Observations concerning Whiston's Astronomical Tables

60/3 William Whiston, *Prælectiones astronomicæ … Quibus accedunt tabulæ plurimæ astronomicæ Flamstedianæ correctæ, Halleianæ, Cassinianæ et Streetianæ* (1707).

60/6 Whiston, *Prælectiones astronomicæ*, p. 332. In the English version of Whiston's table the title is translated as 'The Mean Motion of the Perihelion of the Earth from the Vernal Equinox'; William Whiston, 'A Collection of Astronomical Tables…', in William Whiston, *Astronomical Lectures, Read in the Publick Schools at Cambridge*, second edition (1728), p. 6.

60/8 The 'Apogeum', apogæum or apogee of the Sun is the point in the Sun's orbit at which it is at its greatest distance from the Earth.

60/12 Whiston, *Prælectiones astronomicæ*, pp. 328–31. For Newton's calculation of the motion of the Sun's apogee, see 'Sir Isaac Newton's Theory of the Moon', in David Gregory, *The Elements of Astronomy, Physical and Geometrical* (1715), vol. II, p. 564. Newton's 'Theory of the Moon' first appeared in the Latin edition of Gregory's *Elements* published in 1702 and was then translated as Isaac Newton, *A New and Most Accurate Theory of the Moon's Motion; Whereby All*

Her Irregularities May Be Solved, and Her Place Truly Calculated to Two Minutes (1702).

60/24 Whiston, *Prælectiones astronomicæ*, pp. 118–23.

61/8 Whiston, *Prælectiones astronomicæ*, pp. 142–3.

61/27 In astronomy, the nodes are the points of intersection between the orbital path of a planet and the ecliptic. A syzygy occurs when a planet is in opposition or conjunction with the Sun. In a lunar eclipse the Moon is in opposition to the Sun, and is eclipsed when the Earth moves between them such that the three bodies are aligned in a right line. In a solar eclipse, the Moon is in conjunction with the Sun and eclipses the Sun by interposing between the Sun and the Earth so that the positions of the three bodies are in a right line.

7/II/22: A Degree of Latitude

62/2 The lengths for a degree of latitude in France and at the Arctic Circle are likely taken from Maupertuis, *The Figure of the Earth*, pp. 7–8, 96–7, 162. The figure of 57,060 toises for the length of a degree of latitude in France was first established by Jean Picard and later confirmed by Jacques Cassini; see Jean Picard, *Mesure de la Terre* (1671), and Jacques Cassini, *Traité de la grandeur et de la figure de la Terre* (1723). Cassini's result was also announced in 1718 in the Académie royale des sciences and later published in the Académie's *Mémoires*. In Reid's period it was commonly accepted that a toise was equivalent to six French feet, and that a French foot was equal to 1.068 English feet. A toise was thus equal to 6.408 English feet. These standardised measures are noted in Chambers, *Cyclopædia*, s.v. 'Measure', vol. II, p. 516. But compare the figures given for French and English feet given in David Gregory, *A Treatise of Practical Geometry* (1745), p. 6; for Reid's reading notes from this work see AUL, MS 2131/3/I/12, esp. fol. 1r.

62/14 In the late 1730s Reid was both conversing and corresponding with John Stewart regarding the geometrical analysis of the figure of the Earth; see Reid to John Stewart, [1739], in Reid, *Correspondence*, pp. 4–5. The figures of 3,255,854 toise for the Earth's semi-axis and 3,278,858 toise for the equatorial semi-diameter in this manuscript are the same as those given in Reid's letter to Stewart.

2/I/1: Solar Eclipse, July 1748

63/14 Reid's observations are of the annular solar eclipse that occurred on 14 July 1748; for modern data on this eclipse see http://astro.ukho.gov.uk/eclipse/0331748. In Britain, the full annular eclipse was observable only in Scotland. Slightly different times for the eclipse than those given in the manuscript are found in the report of Reid's findings in James Douglas, fourteenth earl of Morton, Pierre Charles Le Monnier and James Short, 'An Eclipse of the Sun, July 14. 1748', p. 593. The modern values for the latitude and longitude of New Machar are 57.2670° N by 2.1920° W. Those for Aberdeen and Edinburgh are 57.1526° N by 2.1100° W, and 55.9531° N by 3.1889° W respectively. The values for the City of London are 51.5072° N by 0.1275° W.

7/II/20: The greatest Aberration of the fixed Stars

63/16 James Bradley, 'A Letter from the Reverend Mr James Bradley Savillian Professor of Astronomy at Oxford, and F.R.S. to Dr Edmond Halley Astronom. Reg. &c. Giving an Account of a New Discovered Motion of the Fix'd Stars', p. 653.

63/25 Bradley's figure was also 16' 26"; see Bradley, 'An Account of a New Discovered Motion of the Fix'd Stars', p. 655.

7/VIII/3: Of the Aberration of the fixed Stars

64/2 Thomas Simpson, *Essays on Several Curious and Useful Subjects, in Speculative and Mix'd Mathematicks* (1740), pp. vi, 10, 12–14. As well as being a physician with a practice in London, John Bevis was a dedicated astronomer who constructed his own observatory in Stoke Newington in 1738. He later saw through the press the astronomical tables of his close associate, Edmond Halley, and published numerous papers of his own in the *Philosophical Transactions*.

64/6 A star is in quadrature with the Sun when the star, as viewed from the Earth, is located at an angle of 90° in relation to the location of the Sun.

2/I/7: The Transit of Venus

65/11 Folios 1r–4r of this manuscript presumably served as the basis for the discourse Reid read to four of his fellow members of the

Aberdeen Philosophical Society on 14 July 1761; see H. Lewis Ulman (ed.), *The Minutes of the Aberdeen Philosophical Society, 1758–1773* (1990), p. 103. In response to Robert Traill's question 'What are the proper Methods of determining the Suns Parallax by the Transit of Venus over his Disk in 1761?', the Society had already discussed on 12 April 1758 how best to observe the transit that had been predicted by Edmond Halley; Ulman, *Minutes of the Aberdeen Philosophical Society*, pp. 83, 189.

65/15 On the basis of observations made in August and September 1756 and 1757 Reid calculated that the latitude of the Canonists Manse adjacent to King's College to be 57° 10' 40"; AUL, MS 2131/7/III/12, fol. 2v. In the *Inquiry* Reid mentions that 'in May 1761' he was 'occupied in making an exact meridian, in order to observe the transit of Venus'; Reid, *Inquiry*, p. 131. His revised figure for the latitude of King's College is presumably the result of the observations he made in the month before the transit occurred.

65/27 Other than Reid, the members of the Wise Club known to have an interest in astronomy were the Humanist at King's, Thomas Gordon, the Marischal Professor of Natural Philosophy, George Skene, John Stewart and Robert Traill. Reid's friends the physicians John Gregory and David Skene may also have been involved in the observation of the transit.

65/33 Observers elsewhere in Britain complained about the poor weather at the time of the transit; see, for example, Nathaniel Bliss, 'Observations on the Transit of Venus over the Sun, on the 6th of June 1761: In a Letter to the Right Honourable George earl of Macclesfield, President of the Royal Society', pp. 173–5.

66/5 Reid's reference to 'Halley's Tables' implies that he was taking his data from Edmond Halley, *Astronomical Tables with Precepts both in English and Latin for Computing the Places of the Sun, Moon, Planets and Comets* (1752). However, we have seen in the Introduction (p. xlvii) that in 1761 he took extensive notes from the Latin edition of Halley's tables published in 1749. In what follows I cite the Latin edition which, like the 1752 version, is unpaginated.

66/9 Compare the tables 'Logarithmi distantiarum Solis a Terra' and 'Logarithmi distantiarum Veneris a Sole' in Edmond Halley, *Tabulæ astronomicæ. Accedunt de usu tabularum præcepta* (1749). Reid's values do not match any of those given in Halley's tables.

66/26 Reid uses the figure at the foot of the right column in the 'Tabula latitudinaria Veneris', in Halley, *Tabulæ astronomicæ.*

66/38 Reid presumably derived his value for the semi-diameter of the Sun from the 'Tabula motuum horariorum, diametrorum et paralaxium Solis et Lunæ, in eclipsibus', in Halley, *Tabulæ astronomicæ.*

68/33 Compare the figure of 3° 23' 20" given above, p. 66, l. 26. Reid presumably based his calculation of the location of the node of Venus partly on the table 'Medius motus Veneris ab æquinoctio', in Halley, *Tabulæ astronomicæ.* This table includes Halley's data for the motion of the nodes of Venus over the centuries.

69/4 Reid probably took his figures from Anon., 'Observations of the Transit of Venus over the Sun's Disc, June 6. 1761', pp. 429–30, which reproduces observational data first published in Stephen Bolton et al., 'Observations of the Transit of Venus over the Sun'.

69/13 A brief summary of the observations of the transit of Venus made in Paris by Benedict Ferner is found in Anon., 'Observations of the Transit of Venus over the Sun's Disc, June 6. 1761', p. 429; see also Benedict Ferner, 'An Account of the Observations on the Same Transit [of Venus] Made in and near Paris'.

69/17 Anon., 'Observations of the Transit of Venus over the Sun's Disc, June 6. 1761', p. 429.

70/37 The modern value for the longitude of Aberdeen is 2.1100° west of Greenwich. Reid's computations thus involved an error of 26'.

70/38 James Ferguson, *A Plain Method of Determining the Parallax of Venus, by Her Transit over the Sun: And from Thence, by Analogy, the Parallax and Distance of the Sun, and of All the Rest of the Planets*, second edition (1761), pp. 32, 42.

70/39 Ferguson, *A Plain Method of Determining the Parallax of Venus*, p. 33.

71/17 Volume 52 of the *Philosophical Transactions*, covering the years 1761–62, was issued in two parts, the first appearing in late 1762 and the second in the summer of 1763. Summaries of the volume were included in successive issues of the *Gentleman's Magazine* from December 1762 to March 1763 and from August to December 1763. The summary in the February issue listed the various reports on the transit and stated that 'the reader is referred to the originals'; Anon., 'An Epitome of the *Philosophical Transactions*, vol. [LII]', pp. 67–8. Reid would thus have been able to read these reports only at some point in early 1763. Before then, he would have had to rely on accounts of the observations of the transit published in newspapers and the periodical press.

71/25 Based on his observations of the eclipses of Jupiter's moon Io, the Danish astronomer Ole Römer concluded that the speed of light was finite and he informed the Académie des sciences of his discovery in November 1676; see Anon., 'A Demonstration concerning the Motion of Light, Communicated from Paris, in the *Journal des scavans*, and Here Made English'. Römer's discovery served as the starting point for James Bradley's explanation of the aberration of the fixed stars; Bradley, 'An Account of a New Discovered Motion of the Fix'd Stars'. The speed of light was factored into the interpretation of the data on Io included in Halley's astronomical tables; see the note 'Viri Reverendi D. Jacobi Bradley, in has suas satellitum tabulas notæ', in Halley, *Tabulæ astronomicæ*. Reid read this note, as well as a note on the tables of the motions of the moons of Saturn, with care in 1761; see AUL, MS 2131/3/I/21, fol. 3r–v.

71/38 On the basis of his data, Halley predicted that the 1769 transit would begin at 11 o'clock on the night of 3 June 1769. Although Halley calculated the length of time it would take Venus to pass across the Sun's disc in 1761, he did not estimate the duration of the transit in 1769; see Edmond Halley, 'Methodus singularis quâ Solis parallaxis sive distantia à Terra, ope Veneris intra Solem conspiciendæ, tuto determinari poterit', p. 464.

73/3 On the time of the 1769 transit, compare Halley, 'Methodus singularis quâ Solis parallaxis sive distantia à Terra, ope Veneris intra Solem conspiciendæ, tuto determinari poterit', p. 464; Ferguson, *A Plain Method of Determining the Parallax of Venus*, p. 44. Reid's timing was likely taken from Guillaume-Joseph-Hyacinthe-Jean-Baptiste Le Gentil de la Galaisière, 'Observation de la conjonction inférieure de Vénus avec le Soleil, arrivée le 31 Octobre 1751, faite à l'Observatoire royal de Paris; avec des remarques sur les deux conjonctions écliptiques de cette planète avec le Soleil, qui doivent arriver en 1761 & 1769', p. 33, where Le Gentil calculates that the beginning of the transit in 1769 would occur at $7^h\ 47'\ 00''$ in the evening. Le Gentil's figure also appeared in a summary of his paper published in January 1759; Anon., 'A Summary of the Last Volume of Memoires of the Royal Academy at Paris', p. 24.

73/15 In Britain, the most prominent critic of Halley's method of determining the distance between the Earth and the Sun was the Edinburgh

Professor of Mathematics Matthew Stewart; see Matthew Stewart, *Tracts, Physical and Mathematical* (1761), pp. v–vi, and Matthew Stewart, *The Distance of the Sun from the Earth Determined, by the Theory of Gravity* (1763), pp. v–vi. Stewart rejected Halley's method, and maintained that the distance between the Earth and the Sun could be accurately determined only by using the observed motion of the Moon's nodes or apogee to calculate the gravitational attraction of the Sun on the Moon and the Earth. Once the value of the attractive force had been ascertained, the distance between the Sun and the Earth could be calculated on the basis of Newton's inverse square law. Reid's comments here show that he was not persuaded by Stewart's argument. Halley's method was also criticised by the French astronomer Nicolas-Louis de La Caille; see X.Y.Z., 'Differences in the Observations of the Transit of Venus', pp. 317–18.

73/28 For the time of internal contact, see Bliss, 'Observations on the Transit of Venus over the Sun', pp. 175–6. For the time of the middle of the transit, see Nathaniel Bliss, 'A Second Letter to the Right Hon. the earl of Macclesfield, President of the Royal Society, concerning the Transit of Venus over the Sun, on the 6th of June 1761', p. 249.

73/36 On the basis of a painstaking comparison of the observations of the 1761 transit made across the globe, James Short established that the parallax of the Sun was 8"56; James Short, 'Second Paper concerning the Parallax of the Sun Determined from the Observations of the Late Transit of Venus, in Which This Subject Is Treated of More at Length, and the Quantity of the Parallax More Fully Ascertained', p. 340. Short's figure was also used in James Ferguson, *Astronomy Explained upon Sir Isaac Newton's Principles, and Made Easy to Those Who Have Not Studied Mathematics*, third edition (1764), p. 352.

7/III/14: Proposition 26, Book III, of Newton's *Principia*

74/2 Propositions 22 and 25–35 in Book III of the *Principia* contain Newton's theory of the Moon, which incorporates his investigation of the perturbations in the motions of the Moon caused by the gravitational force of the Sun. Proposition 26 deals with the calculation of the effect the attraction of the Sun has on the orbital motion of the Moon. The proposition reads: 'To find the hourly increase of the area that the moon, by a radius drawn to the earth, describes in a circular orbit'; Newton, *The Principia*, p. 840. In what follows, Reid refers to the diagrams accompanying propositions 25 and 26 in the second and subsequent Latin editions of the *Principia*:

Proposition 25:

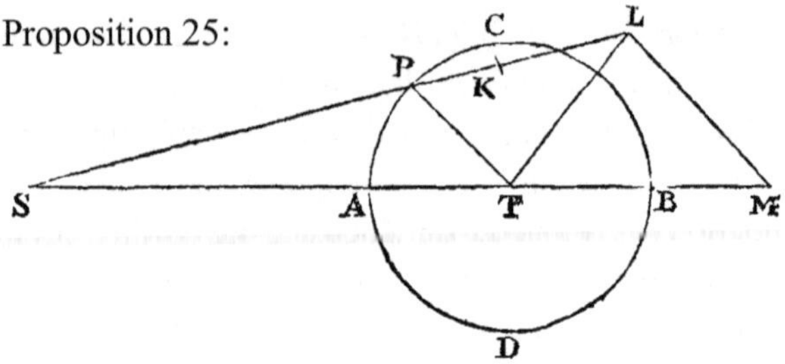

Proposition 26:

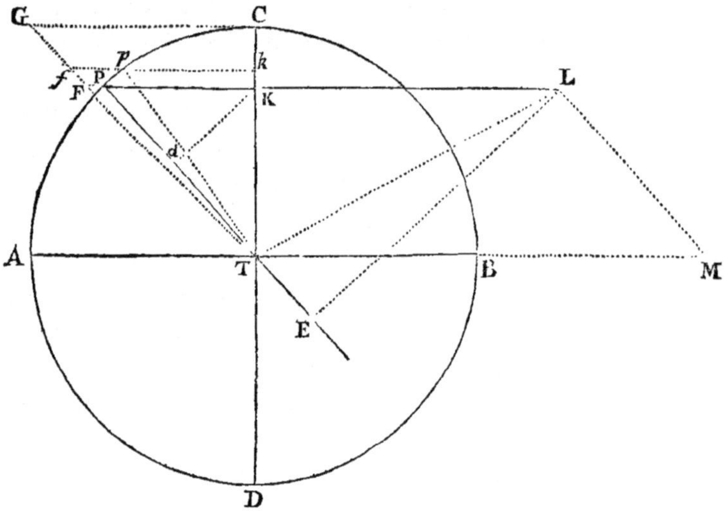

74/21 Reid's Latin may be translated as follows:

Scholium to Proposition 26 Book 3 Principia of Newton

see the figure by the Author

Let S be the accelerative force of the Earth towards the Sun, or of the Moon in its own mean distance from the Sun; let T be the accelerative force of gravity of the Moon towards the Earth. And let L indicate the force LT by which the motion of the Moon around the Earth is disturbed by the force of the Sun. Let the force L or LT be resolved into the two forces LE and ET of which the first acting in the direction of the orbit can be called orbital and should be indicated by O.

The second, which can be called central, is that which we indicate with the letter C.

Since $PT : TK :: PL(=3PK) : LE$ will be the straight line $LE = 3PK \times TK/PT$; the straight line $PE = 3PK^2/PT$ and the straight line $ET = PT - 3PK^2/PT$ and ET will be positive or negative just as PT is greater or less than the straight line PE.

Indeed, the forces S, O, C are proportional to the straight lines ST (i.e., the distance of the Sun from the Earth), LE and TE. Therefore $C : S :: PT - 3PK^2/PT : ST$. And $C = PT^2 - 3PK^2 \times S/PT \times ST$.

Let the periodic time of the Earth around the Sun be P, the Moon around the Earth π and since $C : S :: PT^2 - 3PK^2 \times PT$ is indeed $S : T :: ST/P^2 : PT/\pi^2$ and it will be from the equality of the ratios $C : T :: PT^2 - 3PK^2 \times ST/P^2 : ST \times PT \times PT/\pi^2 : PT^2 - 3PK^2 \times \pi^2 : PT^2 \times P^2 : ⅓PT^2 - PK^2 \times \pi^2 : ⅓PT \times P^2$.

The length of the segment ET at quadratures is = +PT, at syzgies is −2PT, therefore the mean quantity is −½PT. The magnitude of the straight line LE is greatest in octants, that is, it is 3/2PT. Its mean is therefore ¾PT.

74/24 Compare Stewart, *Tracts, Physical and Mathematical*, pp. 301–7, with Newton, *The Principia*, pp. 840–3.

74/26 Stewart, *Tracts, Physical and Mathematical*, pp. 308–9.

74/29 Stewart, *Tracts, Physical and Mathematical*, pp. 309–13.

74/31 Stewart, *Tracts, Physical and Mathematical*, pp. 301–7.

75/3 Stewart, *Tracts, Physical and Mathematical*, pp. 347–9.

75/8 Stewart, *Tracts, Physical and Mathematical*, p. 349; for proposition 16, see Stewart, *Tracts, Physical and Mathematical*, pp. 334–7.

75/12 Proposition 66:

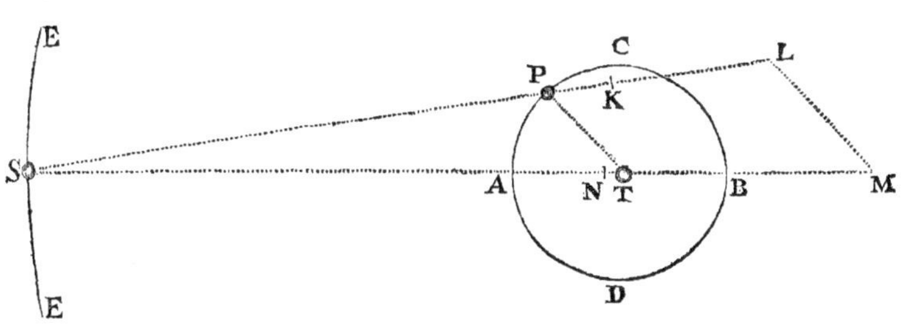

In Book I of the *Principia*, proposition 66 and its twenty-two corollaries deal with what in the eighteenth century came to be known as 'the three-body problem'. Proposition 66 states: 'Let three bodies – whose forces decrease as the squares of the distances – attract one another, and let the accelerative attractions of any two toward the third be to each other inversely as the squares of the distances, and let the two lesser ones revolve about the greatest. Then I say that if that greatest body is moved by these attractions, the inner body [of the two revolving bodies] will describe about the innermost and greatest body, by radii drawn to it, areas more nearly proportional to the times and a figure more closely approaching the shape of an ellipse (having its focus in the meeting point of the radii) than would be the case if that greatest body were not attracted by the smaller ones and were at rest, or if it were much less or much more attracted and were acted on either much less or much more'; Newton, *The Principia*, p. 570. Corollary 14 is the more important of the two that Reid comments on. The corollary reads: 'When body S is extremely far away, the forces NM and ML are very nearly as the force SK and the ratio of PT to ST jointly (that is, if both the distance PT and the absolute force of body S are given, as ST^3 inversely), and those forces NM and ML are the causes of all the errors and effects that have been dealt with in the preceding corollaries; hence it is manifest that all these effects – if the system of bodies T and P stays the same and only the distance ST and the absolute force of body S are changed – are very nearly in a ratio compounded of the direct ratio of the absolute force of body S and the inverse ratio of the cube of the distance ST. Accordingly, if the system of bodies T and P revolves about the distant body S, those forces NM and ML and their effects will … be inversely as the square of the periodic time. And hence also, if the magnitude of body S is proportional to its absolute force, those forces NM and ML and their effects will be directly as the cube of the apparent diameter of the distant body S when looked at from body T, and conversely. For these ratios are the same as the above-mentioned compounded ratio'. For corollaries 14 and 15 see Newton, *The Principia*, pp. 579–80, as well as I. Bernard Cohen's editorial commentary on the significance of corollary 14 on pp. 355–9.

76/20 For proposition 66, corollary 12 see Newton, *The Principia*, p. 579. Reid refers to the retrograde motion of the nodes of the Moon.

77/25 See the figure in editorial note 75/12.

77/32 See the figures reproduced in editorial note 74/2.

79/3 For Stewart's determination of the mean disturbing force of the Sun, see Stewart, *The Distance of the Sun from the Earth Determined*, pp. 43–8. Stewart's calculations were later attacked in [John Dawson], *Four Propositions, &c. Shewing, not only, that the Distance of the*

Sun, as Attempted to be Determined from the Theory of Gravity, by a Late Author, Is, upon His Own Principles, Erroneous; but also, that It Is More than Probable This Capital Question Can Never Be Satisfactorily Answered by Any Calculus of the Kind (1769); John Dawson, 'Observations on Dr Stewart's Conclusions, &c. on the Sun's Distance from the Earth'; John Landen, *Animadversions on Dr Stewart's Computation of the Sun's Distance from the Earth* (1771).

79/34 Stewart, *The Distance of the Sun from the Earth Determined*, p. 48.

79/35 That is, according to Reid's corollary 9, P is the periodical time of the Earth around the Sun and π is the periodical time of the Moon around the Earth. Stewart states that 'the duplicate ratio of the periodic time of the moon round the earth to the periodic time of the earth round the sun is as 1 to 178.725', which agrees with the figure in proposition 25, Book I, of Newton's *Principia*; compare Stewart, *The Distance of the Sun from the Earth Determined*, p. 77, with Newton, *The Principia*, p. 840.

79/36 'Mr Trail' is most likely William Trail, rather than Reid's professorial colleague and associate from the north-east Robert Traill. The latter was briefly the Glasgow Professor of Oriental Languages before succeeding William Leechman as the Professor of Divinity in 1761. William Trail matriculated at Glasgow in 1763 and took his MA in 1766. Reid was one of the examiners who chose Trail to become the Professor of Mathematics at Marischal College, Aberdeen, in August 1766. While he was a student at Glasgow, Trail was befriended by Robert Simson, whose life he later chronicled in his *Account of the Life and Writings of Robert Simson, M.D. Late Professor of Mathematics in the University of Glasgow* (1812). He was also part of a circle of aspiring young men of science that included John Robison and William Irvine; see John Robison to James Watt, [October 1800], in Eric Robinson and Douglas McKie (eds), *Partners in Science: Letters of James Watt and Joseph Black*, pp. 359–60. See also the Introduction, p. clxxxii. The value for the mean distance of the Moon from the Earth is taken from the *Principia*, Book III, propositions 4 and 25; Newton, *The Principia*, pp. 803–5, 840.

80/30 Reid probably copied these two canons from a published work. I have been unable to identify his source.

80/35 Reid's note speaks to the sharp disagreement between French and British astronomers over the values for basic astronomical phenomena such as the parallax of the Sun. For example, whereas James

Short argued that the observations of the transit of Venus in 1761 showed that the parallax of the Sun was 8"56 and the Savilian Professor of Astronomy at Oxford, Thomas Hornsby, calculated that the mean parallax of the Sun was 9"732, the French astronomer Alexandre-Guy Pingré insisted that the true value was over 10" and eventually concluded that 10"1 was correct. Pingré's colleague, Nicolas Louis de La Caille, had earlier maintained that the Sun's parallax was between 9"94 and 10"2 on the basis of observations obtained before the transit took place; Thomas Hornsby, 'A Discourse on the Parallax of the Sun'; Nicolas Louis de La Caille, 'Mémoire sur la parallaxe du Soleil, qui résulte de la comparison des observations simultanées de Mars & de Vénus, faites en l'année 1751 en Europe & au cap de Bonne-espérance', pp. 96–7; Alexandre-Guy Pingré, 'Mémoire sur quelques observations du passage de Vénus. Faites le 6 de Juin 1761, au-delà de l'équateur; & sur les secours qu'on peut en tirer pour la détermination de la parallaxe du Soleil'; Alexandre-Guy Pingré, 'Nouvelle recherche sur la détermination de la parallaxe du Soleil par le passage de Vénus du 6 Juin 1761'; Short, 'Second Paper concerning the Parallax of the Sun Determined from the Observations of the Late Transit of Venus'. However, after the transit of Venus in June 1769 there was a convergence among the French and British astronomers over the figure for the mean parallax of the Sun. Hornsby arrived at 8"78, while Joseph-Jérôme de Lalande concluded that the value lay somewhere between 8"55 and 8"63 after having first suggested a range between 8" to 10" and Pingré eventually settled on 8"80; Thomas Hornsby, 'The Quantity of the Sun's Parallax, as Deduced from the Observations of the Transit of Venus, on June 3, 1769', p. 579; Joseph-Jérôme de Lalande, 'Mémoire sur la parallaxe du Soleil, déduite des observations faites dans la mer du Sud, dans le royaume d'Astracan, & à la Chine', p. 798; Joseph-Jérôme de Lalande, 'Mémoire sur la parallaxe du Soleil, qui résulte du passage de Vénus, observé en 1769', p. 14; Alexandre-Guy Pingré, 'Mémoire sur la parallaxe du Soleil, déduite des meilleures observations de la durée du passage de Vénus sur son disque le 3 Juin 1769', p. 419. I have been unable to identify Reid's source(s) for his numerical values.

7/III/1: Of the Centripetal Forces

81/19 Inherent force or inertia.

81/27 A term derived from the Latin, literally meaning an arrow. As Reid indicates, in mathematics it refers to the versed sine of an arch.

82/8 Proposition 4, Book I, of the *Principia* reads: 'The centripetal forces of bodies that describe different circles with uniform motion tend toward the centres of those circles and are to one another as the squares of the arcs described in the same time divided by the radii of the circles'; Newton, *The Principia*, p. 449. There follow nine corollaries in which Newton stated various mathematical relationships fundamental to his analysis of planetary motion. Newton then added a scholium in which he indicated that the centripetal force of gravity obeys the inverse square law.

7/VIII/8: Glasgow College September 1769

82/9 When Reid wrote this set of notes, he and his family lived in one of the houses adjoining what was known as the New (later Professors) Court at the College. These houses occupied the north side of the College buildings on the High Street; see David Murray, *Memories of the Old College of Glasgow: Some Chapters in the History of the University*, pp. 372–3.

82/11 This house was presumably located on the New Vennel, which bounded the University to the north, although it could also have been on the High Street to the west.

82/13 Reid watched the comet now known as Comet C/1769 P1 (Messier). It was first observed by Charles Messier at the Observatoire de la Marine in Paris on 8 August 1769 and remained visible in the skies over Europe until early December of that year. The comet was also tracked by the Astronomer Royal, Nevil Maskelyne, at Greenwich, as well as other observers scattered across Europe, China and the South Pacific. Maskelyne began to record data on the comet on 25 August, while the general public were alerted to appearance of the comet in the heavens by the mathematical practitioner Samuel Dunn in a letter published in *Lloyd's Evening Post and British Chronicle* at the beginning of September; see Charles Messier, 'Mémoire contenant les observations de la Xe. comète observée à Paris, de l'Observatoire de la Marine pendant les mois d'Août, Septembre, Octobre, Novembre, jusqu'au 1er. Décembre 1769'; Nevil Maskelyne, *Astronomical Observations Made at the Royal*

Observatory at Greenwich, in the Years 1765, 1766, 1767, 1768 and 1769 (1774), pp. 65–71; Samuel Dunn, 'Letter to the Editor', *Lloyd's Evening Post and British Chronicle*, 1–4 September 1769, p. 221; David Seargent, *The Greatest Comets in History: Broom Stars and Celestial Scimitars*, pp. 121–4.

82/17 α Orion is the star Betelgeuse in the constellation Orion. α Ceti is the star Menkar in the constellation Cetus (the Whale).

82/31 Procyon is the brightest star in the constellation Canis Minor and is itself known as the Little Dog Star. When Reid recorded his observations on 10 September, the comet was at its closest to the Earth.

83/6 The thermometer that Reid evidently kept in his room was probably made by his colleague Alexander Wilson.

3/I/20: Extracts from Pemberton

83/7 Henry Pemberton, *A View of Sir Isaac Newton's Philosophy* (1728). Pemberton had worked closely with Newton in preparing the third edition of the *Principia*, published in 1726, and stated that he had made 'an english translation of [the] *Principia*, which I have had some time by me'; Pemberton, *A View of Sir Isaac Newton's Philosophy*, sig. a2v. His translation never appeared. Reid's notes most likely date from the period 12 to 15 March 1784, when he borrowed Pemberton's *View* from Glasgow University Library; 'Professors Receipt Book, 1770–[1789]', GUL, Spec Coll MS Lib (uncatalogued), entries for 1784. After returning the Pemberton, he then borrowed Colin Maclaurin's *An Account of Sir Isaac Newton's Philosophical Discoveries, in Four Books* (1748). It is not clear why he was interested in (re?)reading these popular expositions of Newton's natural philosophy at this juncture.

83/20 Reid summarises Pemberton. *A View of Sir Isaac Newton's Philosophy*, pp. 89–90; in the original Reid's figure of '69' reads '690'.

83/29 See especially Pemberton, *A View of Sir Isaac Newton's Philosophy*, pp. 124–34. The perihelion is the point in a planet's orbit when it is closest to the Sun, whereas the aphelion is the point at which the planet is furthest from the Sun.

83/32 Inherent force or inertia.

84/18 Reid summarises the basics of Newton's analysis of planetary motion in the two preceding paragraphs. Compare proposition 1, Book I, of Newton, *The Principia*, pp. 444–5.

84/25 Pemberton praises Newton's method of prime and ultimate ratios on page 129. According to Pemberton, Newton had 'introduced [this method] into geometry, thereby greatly inriching that science'; Pemberton, *A View of Sir Isaac Newton's Philosophy*, p. 129.

84/25 Reid added this remark on Archimedes and Euclid; see textual note 84/25. His addition indicates that he believed that Pemberton's assessment of the originality of Newton's method of prime and ultimate ratios required qualification. Reid presumably has in mind the method of exhaustion as it was used by Euclid in, for example, proposition 2, Book XII of the *Elements*, and by Archimedes in the *Measurement of a Circle*; see Euclid, *The Thirteen Books of Euclid's Elements*, trans. T. L. Heath, vol. III, pp. 371–4, and Archimedes, *Measurement of a Circle*, in *The Works of Archimedes*, pp. 91–8. Compare Colin Maclaurin, *A Treatise of Fluxions*, vol. I, pp. 2–50, where Maclaurin traces Newton's method of fluxions back to its origins in the work of Euclid and especially Archimedes.

84/27 Pemberton, *A View of Sir Isaac Newton's Philosophy*, pp. 143–60. Newton discusses the resistance of fluids at length in Book II of the *Principia*, partly in order to undermine Descartes' theory of vortices.

84/28 Pemberton, *A View of Sir Isaac Newton's Philosophy*, pp. 283–96.

3/I/19: Facts from Bailly

84/29 Jean Sylvain Bailly's *Histoire de l'astronomie moderne depuis la fondation de l'école d'Alexandrie, jusqu'a l'époque de M.D.CC. XXX* was first published in three volumes between 1779 and 1782. My references are to the 1785 edition of the work, which is available online. Reid's notes are written in the late form of his handwriting characteristic of the 1780s and 1790s.

85/8 Bailly, *Histoire de l'astronomie moderne*, vol. III, p. 137.

85/12 Isaac Newton, 'Tractatus de quadratura curvarum', in Newton, *Opticks: Or, a Treatise of the Reflexions, Refractions, Inflexions and Colours of Light* (1704); Roger Cotes, *Harmonia mensurarum, sive analysis & synthesis per rationum & angulorum mensuras promotæ: Accedunt alia opuscula mathematica* (1722). On Cotes' *Harmonia mensurarum*, see especially Niccolò Guicciardini, *The Development of Newtonian Calculus in Britain, 1700–1800*, pp. 28–31. For Reid's knowledge of the *Harmonia mensurarum*, see AUL, MSS 2131/3/III/2, fol. 1r–v, and 2131/5/I/15.

85/13 Reid's comment is revealing, for it suggests that his knowledge of the work done by mathematicians on the continent during the course of the eighteenth century was limited; compare Niccolò Guicciardini, 'Thomas Reid's Mathematical Manuscripts: A Survey', p. 73.

86/2 Bailly, *Histoire de l'astronomie moderne*, vol. III, pp. 141–54. The controversy over the solution to the three-body problem surveyed by Bailly was sparked by the Académie des sciences' essay prize competition in 1748, which was won by Leonhard Euler for the paper he then published as *Piece qui a remporté le prix de l'Académie royale des sciences en M.DCC.XLVIII. Sur les inégalités du mouvement de Saturne & de Jupiter* (1749). Alexis-Claude Clairaut and Jean D'Alembert were both members of the prize committee and worked independently to solve the three-body problem after the topic for the 1748 essay competition had been decided in the spring of 1746. Having read Euler's submission, which questioned the validity of the inverse square law, Clairaut presented a paper to the Académie in November 1747 which likewise challenged the truth of Newton's law in relation to the motion of the apse of the Moon. The debate between Clairault and Buffon that ensued can be followed in the *Histoire de l'Académie royale des sciences. Année M.DCCXLV* (1749). The papers by the main disputants were rushed into print in this volume of the *Histoire*, even though it was ostensibly devoted to the proceedings of the Académie in 1745. D'Alembert later outlined his solution to the three-body problem in his *Recherches sur differens points importans du systême du monde* (1754–56). Clairaut and Euler subsequently accepted that the irregularities in the Moon's motions could in fact be reconciled with the inverse square law; see Leonhard Euler, 'Extract of a Letter from Professor Euler, of Berlin, to the Rev. Mr Caspar Wetstein, Chaplain to Her Royal Highness the Princess Dowager of Wales'.

86/10 Bailly, *Histoire de l'astronomie moderne*, vol. III, pp. 155–6; Alexis-Claude Clairaut, *Tables de la Lune, calculées suivant la théorie de la gravitation universelle* (1754); Tobias Mayer, 'Novae tabulae motuum Solis et Lunae'. From 1751 until his death in 1762 Mayer was a professor at the Georg-August Academy in Göttingen. Shortly after taking up his chair he initiated a correspondence with Leonhard Euler in which they discussed, *inter alia*, the motions of the Moon. Bailly also refers to the French astronomer Louis Robert Joseph Cornelier Lémery.

86/15 Bailly, *Histoire de l'astronomie moderne*, vol. III, pp. 156.

86/19 Bailly, *Histoire de l'astronomie moderne*, vol. III, pp. 156–8; Jean

Le Rond D'Alembert, *Recherches sur la précession des equinoxes, et sur la nutation de l'axe de la Terre dans le systême Newtonien* (1749).

86/21 Bailly, *Histoire de l'astronomie moderne*, vol. III, pp. 162.

86/32 Bailly, *Histoire de l'astronomie moderne*, vol. III, pp. 162–3; Bailly cites Pierre-Simon Laplace, 'Suite des recherches sur plusieurs points du système du monde'.

86/39 Bailly, *Histoire de l'astronomie moderne*, vol. III, pp. 168–9; Euler, *Piece qui a remporté le prix de l'Académie royale des sciences en M.DCC.XLVIII* and Leonhard Euler, *Recherches sur les irrégularités du mouvement de Jupiter et de Saturne* (1752).

87/8 Bailly, *Histoire de l'astronomie moderne*, vol. III, pp. 171–4.

87/9 Bailly, *Histoire de l'astronomie moderne*, vol. III, pp. 170; Nicolas-Louis de La Caille, *Tabulæ solares* (1758).

87/11 Bailly, *Histoire de l'astronomie moderne*, vol. III, pp. 193; Alexis Claude Clairaut, *Théorie du mouvement des comètes, dans laquelle on a égard aux altérations que leurs orbites éprouvent par l'action de planètes* (1760). See also Curtis Wilson, 'Clairaut's Calculation of the Eighteenth-Century Return of Halley's Comet'.

87/13 Bailly, *Histoire de l'astronomie moderne*, vol. III, pp. 198–200; Bailly refers to Achille-Pierre Dionus du Séjour, 'Nouvelles méthodes analytiques pour calculer les éclipses de Soleil, les occultations des étoiles fixes et des planètes par la Lune; et en général pour réduire les observations de cet astre, faites à la surface de la Terre, a lieu vu du centre'.

87/14 Bailly, *Histoire de l'astronomie moderne*, vol. III, pp. 208; Paolo Frisi, *De gravitate universali corporum libri tres* (1768).

87/18 Bailly, *Histoire de l'astronomie moderne*, vol. III, pp. 102, note b; in this note Bailly cites Patrick Murdoch, 'Of the Moon's Distance and Parallax'; Patrick Murdoch, 'An Essay on the Connexion between the Parallaxes of the Sun and Moon; Their Densities; and Their Disturbing Forces on the Ocean'; and Samuel Horsley, 'A Computation of the Distance of the Sun from the Earth'.

87/21 Bailly, *Histoire de l'astronomie moderne*, vol. III, pp. 175–87, 334 note a. Reid has registered the references to the following works by Bailly: 'Deuxiéme mémoire sur la théorie des satellites de Jupiter'; *Essai sur la théorie des satellites de Jupiter, suivi des tables de leurs mouvemens, déduits du principe de la gravitation universelle* (1766); 'Mémoire sur la cause de la variation de l'inclinaison de l'orbite du second satellite de Jupiter'; 'Mémoire sur les inégalités de la lumière des satellites de Jupiter, sur la mesure de leurs diamètres; & sur un moyen, aussi simple que commode, de rendre les observations comparables, en remédiant à la difference des vues & des lunettes'; 'Mémoire sur le mouvement des nœuds, et sur la variation de l'inclinaison des satellites de Jupiter'; 'Premier mémoire sur la théorie des satellites de Jupiter'; 'Troisiéme mémoire sur la théorie des satellites de Jupiter'.

Part Four: Optics

7/II/5: October 13 1770

88/3 Patrick Wilson was the son of the Glasgow Professor of Practical Astronomy and Observer in the University, Alexander Wilson, whom he succeeded in 1786. The younger Wilson had matriculated at the University of Glasgow in 1756 and, following his father's appointment to the newly established regius chair in 1760, assisted the older man in his activities as an astronomer, type founder and instrument maker. Patrick Wilson seems to have first given the phenomenon of the aberration of light serious consideration in the late 1760s. In 1784 John Robison recalled that while he held the lectureship in chemistry at Glasgow (1766–69), 'about the year 1767 or 1768, Mr Wilson entertained an opinion that the aberration of the fixed stars indicated the proportion between the orbital v[e]locity of the earth, and the velocity of light in the vitreous humour of the eye'; [John Robison], 'Mr Wilson on the Solar System', pp. 28–9. Shortly after discussing his ideas with Reid in October 1770, Patrick Wilson was permitted to read to the Glasgow Literary Society 'a Dissertation concerning the Velocity of Light shewing that this as deduced from Dr. Bradleys Theory of the Aberration of the fixed Stars is erroneous and a new Principle which affects that pointed out and examined' on 23 November 1770; 'Laws of the Literary Society in Glasgow College', p. 36. A summary of Wilson's discourse was shortly thereafter published in the *London Chronicle*; see X., 'To the Printer of the London Chronicle', *London Chronicle*, 1–4 December 1770, p. 532. A copy of his discourse was then probably sent by Alexander Wilson to his friend the Astronomer Royal, Nevil Maskelyne, who discussed the younger Wilson's ideas with Richard Price and other men of science in London. Maskelyne was apparently highly critical of his analysis of the aberration of light, although both Wilson the younger and Reid thought that Maskelyne's criticisms were misplaced; see Reid to [Richard Price], [1772–73], in Reid, *Correspondence*, p. 63. Wilson did, however, subsequently abandon his initial theory of aberration; see the editorial note to the 1809 abridged edition of his paper 'An Experiment Proposed for Determining, by the Aberration of the Fixed Stars, whether the Rays of Light, in Pervading Different Media, Change Their Velocity According to the Law Which Results from Sir Isaac Newton's Ideas concerning the Cause of Refraction; and for Ascertaining Their Velocity in Every Medium Whose Refractive Density Is Known', pp. 192–5. His revised theory likely formed the basis for the discourse he delivered to the Glasgow Literary Society on 31 January 1777, 'Of the Refraction and Refrangibility of Light'; 'Laws of the Literary Society in Glasgow College', pp. 45–6. He also outlined aspects of his new theory in a paper read to the Royal Society in London on 24 January 1782 and later published in the *Philosophical Transactions*, 'An Experiment Proposed for Determining, by the Aberration of the Fixed Stars, whether the Rays of Light, in Pervading Different Media, Change Their Velocity According to the Law Which Results from Sir Isaac Newton's Ideas concerning the Cause of Refraction; and for Ascertaining Their Velocity in Every Medium Whose Refractive Density Is Known'.

88/9 For expositions of the 'common theory of Aberration', see: Bradley, 'An Account of a New Discovered Motion of the Fix'd Stars'; Maclaurin, *An Account of Sir Isaac Newton's Philosophical Discoveries*, pp. 264–6; Simpson, *Essays on Several Curious and Useful Subjects*, pp. 1–20; Robert Smith, *A Compleat System of Opticks in Four Books, viz. a Popular, a Mathematical, a Mechanical and a Philosophical Treatise* (1738), vol. II, pp. 449–55; Roger Long, *Astronomy, in Five Books* (1742), vol. I, pp. 303–6; Thomas Rutherforth, *A System of Natural Philosophy, Being a Course of Lectures in Mechanics, Optics, Hydrostatics and Astronomy; Which Are Read in St Johns College Cambridge* (1748), vol. II, pp. 941–53.

88/15 Patrick Wilson was not the first commentator to draw a connection between the aberration of light and the optics of the eye. His father's late friend, Thomas Melvill, had made a similar suggestion in 1753 to the Astronomer Royal, James Bradley. In a letter to Bradley that dealt primarily with Melvill's hypothesis that the differently coloured rays of light have different velocities, Melvill mentioned that he had also been thinking about Bradley's theory of the aberration of light and had realised that 'it is not the proportion of the velocity of light in the celestial space, or in the air, but in the humours of the eye to that of the earth which is determined by your observations'; Melvill to James Bradley, 2 June 1753, in Stephen Peter Rigaud (ed.), *Miscellaneous Works and Correspondence of the Rev. James Bradley, D.D. F.R.S. Astronomer Royal, Savilian Professor of Astronomy in the University of Oxford, &c.* (1832), p. 485.

88/16 The vitreous humour is located between the lens and the retina in the eye. Robert Smith likened the vitreous humour to 'the white of an egg' in his description of the anatomy of the eye; Smith, *A Compleat System of Opticks*, vol. I, p. 26. See also William Porterfield, *A Treatise on the Eye, the Manner and Phænomena of Vision* (1759), vol. I, pp. 240–6.

88/23 According to Newton's theory of refraction, when a ray of light is refracted moving from air to water the ratio of the sine of incidence to the sine of refraction is 4:3 and the ratio of the velocities of the ray in the different media is 3:4; see especially propositions 94–8, Book I of Newton, *The Principia*, pp. 622–9; Sir Isaac Newton, *Opticks: Or, a Treatise of the Reflections, Refractions, Inflections and Colours of Light* (1730), pp. 6, 130–1; and Geoffrey Cantor, *Optics after Newton: Theories of Light in Britain and Ireland, 1704–1840*, pp. 27–8. Reid probably took his figure for the ratio between the velocity of light in the vitreous humour and the air from James Jurin, 'An Essay upon Distinct and Indistinct Vision', p. 135, although Jurin qualifies his figure on p. 141. Wilson derived his figure from the writings of Francis Hauksbee and Henry Pemberton, whom Jurin also cites; see X. 'To the Printer of the London Chronicle', p. 532. On the refractive power of the vitreous humour, see also Porterfield, *A Treatise on the Eye*, vol. I, pp. 245–6, 282–3.

88/31 Reid elaborates on this suggestion below, pp. 105–6.

5/II/55: All Visible objects are seen by Rays of Light

89/14 See, for example, Newton, *Opticks*, pp. 18–19.

90/12 As mentioned in editorial note 71/25, Ole Römer discovered the progressive motion of light on the basis of his observations of Jupiter's satellite Io. Observational astronomers made careful studies of the motions of the moons of Jupiter in the late seventeenth and eighteenth centuries, partly because their data were used for navigational purposes but also because the consideration of the orbital motions of Jupiter, Saturn and their moons was central to the development of perturbation theory. In Glasgow, Alexander Wilson had made observations of Io in the late 1760s; see Alexander Wilson, 'Eclipses of Jupiter's First Satellite, with an Eighteen Inch Reflector of Mr Short's. Observed by Dr Wilson at the Glasgow Observatory'. Given that Reid maintained that it took approximately eight minutes for light emitted by the Sun to reach the Earth, his statement that 'Jupiter & his Satellites give intelligence to us of their Motions sometimes in 33' sometimes not in less than 48" implies that, at a minimum, the distance between the Earth and Jupiter is 4.125 times greater than the distance between the Earth and the Sun. The maximum distance is 6 times greater. His numbers are roughly compatible with eighteenth-century estimates; see, for example, William Whiston's figures for the distances between the Sun and the planets as tabulated in Whiston, *Astronomical Lectures, Read in the Publick Schools at Cambridge*, p. 85.

90/35 Reid's conception of absolute space is taken from Newton; see especially the scholium to the definitions prefacing Book I in Newton, *The Principia*, pp. 408–15.

92/11 Reid's wording implies that this manuscript served as the basis for an oral presentation; compare the comment in his discourse on Euclid delivered to the Glasgow Literary Society above, p. 28, ll. 7–9. If so, this manuscript was likely read as a discourse or an occasional paper at a meeting of the Society. As we have seen in editorial note 15/17, there is a lengthy gap in the Society's minutes from 5 November 1779 until 7 November 1794. In addition, there are shorter periods for which minutes of the Society's meetings do not survive. In the 1770s, there are gaps in the minutes for February through May 1771, 29 November 1771 to late January 1773 and 11 February 1773 to May 1776; see 'Laws of the Literary Society in Glasgow College', pp. 38–40. Reid's manuscript probably dates to one of the earlier periods not covered by the extant records of the Literary Society, and before Patrick Wilson revamped his theory of the aberration of light.

94/2 That is, the Sun is assumed to be stationary.

94/13 See, for example, Bradley, 'An Account of a New Discovered Motion of the Fix'd Stars', pp. 646–8.

94/20 Bradley, 'An Account of a New Discovered Motion of the Fix'd Stars', p. 653; and above, p. 63, ll. 15–25.

94/35 Reid here follows Thomas Melvill in suggesting that the rays of the different prismatic colours of light have different velocities. Melvill, however, had hypothesised that the rays of red have the greatest velocity; Thomas Melvill, 'A Letter from Mr T. Melvil to the Rev. James Bradley, D.D. F.R.S. with a Discourse concerning the Cause of the Different Refrangibility of the Rays of Light', p. 262. See also Reid's undated note on the appearance of the fixed stars transcribed below, p. 123, ll. 7–13, in which he takes up Melvill's suggestion. It may be that Reid knew of Pieter van Musschenbroek's claim that the red rays of light have a lower velocity than violet rays; see Joseph Priestley, *The History and Present State of Discoveries Relating to Vision, Light and Colours* (1772), p. 404, and Pieter van Musschenbroek, *Introductio ad philosophiam naturalem* (1762), vol. II, esp. pp. 724–6.

94/39 Reid's figure is taken from Bradley, 'An Account of a New Discovered Motion of the Fix'd Stars', pp. 653, 655. Bradley calculated that light propagated from the Sun takes 8'12" to 8'13" to reach the Earth.

95/5 According to Newton, the velocity of light varies with the density of the medium through which it travels. He supposed that in glass, the velocity is greater than it is in a less dense medium such as air. Hence Newton's doctrine implies that light travels more slowly in a vacuum than it does in the air. What is not clear from Reid's remarks is whether he discounted the etherial medium pervading all of space postulated by Newton in Queries 17–24 of Newton's *Opticks*. These queries were added to the second English edition of the *Opticks*, published in 1718; see Newton, *Opticks*, pp. 347–54.

95/13 Reid considers in his third case the implications of Patrick Wilson's initial account of the aberration of light, whereas in the first two cases he discussed standard topics dealt with by previous writers on aberration.

95/24 Euclid was the first to analyse vision geometrically in terms of a cone of rays and his idea was subsequently reformulated by a succession of writers on optics; see David C. Lindberg, *Theories of Vision from Al-Kindi to Kepler*, pp. 12–13. The double cone of rays explains why the image of an object is inverted on the retina. Reid credited Johannes Kepler with the discovery of this phenomenon, although it was known before Kepler published his *Ad Vitellionem paralipomena, quibus astronomiæ pars optica traditur* (1604). In *Ad Vitellionem paralipomena*, Kepler endeavoured to explain how we see objects in their proper orientation, even though our retinal images of them are inverted and laterally reversed; see Reid, *Inquiry*, pp. 114–15, and also p. 126, where he mentions the 'inward Cone' in the eye. For a discussion of Kepler's 'discovery' in context, see Lindberg, *Theories of Vision from Al-Kindi to Kepler*, esp. ch. 9, and A. Mark Smith, *From Sight to Light: The Passage from Ancient to Modern Optics*, pp. 353–72.

96/5 Compare Smith, *A Compleat System of Opticks*, vol. I, p. 13.

96/15 Compare Reid's definition of the axis of the eye with that found in Ephraim Chambers' *Cyclopædia*: '[The] Optick Axis, or Visual Axis, is a Ray passing through the Centre of the Eye; or it is that Ray which proceeding out of the Middle of the luminous Cone, falls perpendicularly on the crystalline Humour, and consequently passes through the Centre of the Eye'; Chambers, *Cyclopædia*, s.v. 'Axis', vol. I, p. 183; see also Smith, *A Compleat System of Opticks*, vol. I, p. 46.

96/19 Reid had previously discussed the axis of the eye and the centre of the retina in much the same terms in the sections of the *Inquiry* devoted to single vision and the phenomenon of squinting; see Reid, *Inquiry*, pp. 132–67.

96/37 Jurin, 'An Essay upon Distinct and Indistinct Vision', pp. 135, 141.

97/22 On this 'Law of our Nature' see Reid, *Inquiry*, pp. 120–31, and Porterfield, *A Treatise on the Eye*, vol. II. pp. 293–304.

97/34 Compare Reid, *Inquiry*, pp. 120, 123, 130; Smith, *A Compleat System of Opticks*, vol. I, p. 27.

98/7 Reid recapitulates the argument of the *Inquiry*, pp. 121–2, 125–6, 130, 132–3.

98/26 Reid, *Inquiry*, pp. 127–9. In the *Inquiry* Reid notes that he derived his experiments from the writings of the Edinburgh physician William Porterfield, who was in turn indebted to the seventeenth-century Jesuit natural philosopher Christoph Scheiner. See William Porterfield, 'An Essay concerning the Motions of Our Eyes: Of Their External Motions'; William Porterfield, 'An Essay concerning the Motions of Our Eyes: Of Their Internal Motions'; William Porterfield, *A Treatise on the Eye*; Christoph Scheiner, *Oculus hoc est: Fundamentum opticum* (1619).

99/13 Jurin, 'An Essay upon Distinct and Indistinct Vision', pp. 135, 141.

99/22 Proposition 4 registers Reid's adoption of Patrick Wilson's early theory of the aberration of light, for whereas James Bradley maintained that aberration was a function of the velocity of light as it travels between the fixed stars and the Earth, Wilson asserted that aberration is related to the velocity of light in the eye. Reid thus reformulated the mathematics of Bradley's analysis of aberration in terms of the basics of the Wilson's theory; see above, p. 88, ll. 1–26.

100/10 Compare above, p. 88, ll. 18–2.

7/II/15: Axiom

102/4 Reid states some of the fundamental elements of Newton's theory of optics in this postulate. Although Newton indicated that he believed that light consisted of material particles in his first published papers on light and colours, in the *Principia* he adopted a stance of nescience regarding the nature of light. In the scholium to proposition 96, Book I, he stated that while he acknowledged 'the analogy that exists between the propagation of rays of light and the motion of bodies' he denied that he was 'arguing at all about the nature of the rays (that is, whether they are bodies or not), but only determining the trajectories of bodies, which are very similar to the trajectories of rays'; Newton, *The Principia*, p. 626. Similarly, in what became Query 29 in fourth edition of the *Opticks*, published in 1730, he suggested that 'the Rays of Light [are] very small Bodies emitted from shining Substances' but in the body of the text he spoke simply of 'rays' of light and did not speculate on the physical nature of these rays; Newton, *Opticks*, p. 370. Reid thus adopts Newton's posture of neutrality regarding the question of the particulate or wave-like nature of light. Moreover, according to propositions 94–96 in Book I of the *Principia*, rays of light are reflected and refracted by a force acting perpendicularly to the surface of reflecting and refracting media. And, as Reid says, the force is taken to be 'equal at equal distances from the Surface'; Newton, *The Principia*, pp. 622–6. Compare Reid's 'Law of Motion respecting Light' enunciated in his lectures on optics at King's College, Aberdeen: 'The rays of Light when they approach very near the surface of bodies, are attracted to, or repelled from these bodies in a direction perpendicular to the Surface; and this attracting or repelling force is equal at equal distances from the surface of the same body, it is greater in more dense bodies and less in those that are more rare'; Anon., 'Natural Philosophy 1758', AUL, MS K 160,

p. 258. Lastly, the analogy between the behaviour of bodies and rays of light drawn by Newton in the scholium to proposition 96, Book I of the *Principia* points to the fact that he maintained that the behaviour of light was governed by the three laws of motion stated at the beginning of Book I. For a brief but lucid discussion of Newton's optical ideas (as opposed to those of his followers), see Cantor, *Optics after Newton*, pp. 26–31.

2/I/3: Of the Path of a Ray of Light

104/6 This manuscript almost certainly has its origins in conversations between Reid and Patrick Wilson. Wilson also discussed the theoretical issues addressed by Reid in this manuscript with John Robison. Robison investigated these issues in a paper which was read to the Royal Society of Edinburgh in April 1788 and published the Society's *Transactions* in 1790; see John Robison, 'On the Motion of Light, as Affected by Refracting and Reflecting Substances, Which Are Also in Motion'. On Robison's paper, see: Cantor, *Optics after Newton*, pp. 75–6, 212–16; Kurt Møller Pedersen, 'Roger Joseph Boscovich and John Robison on Terrestrial Aberration'; Kurt Møller Pedersen, 'Water-filled Telescopes and the Pre-history of Fresnel's Ether Dragging', pp. 522–9.

104/12 The 'noble Discovery of the Progressive Motion and Velocity of the Rays of Light' was announced by Ole Römer in 1677; see Anon., 'A Demonstration concerning the Motion of Light, Communicated from Paris, in the *Journal des scavans*, and Here Made English'.

104/16 Bradley, 'An Account of a New Discovered Motion of the Fix'd Stars'.

104/21 The link that Reid draws here between 'the progressive Motion of Light' and 'the *Phœnomena* of Vision' reflects the lingering influence of Patrick Wilson's initial conception of aberration; see editorial note 88/3.

104/26 The general case involves the observation of the fixed stars.

104/30 See propositions 94–96, Book I in Newton, *The Principia*, pp. 622–6; compare Reid's 'Law of Motion respecting Light' in his King's College lectures on optics quoted above in editorial note 102/4.

106/3 The inequality of the angles of incidence and reflection that Reid posits in this case contradicts the law of reflection, which states that the angle of incidence is equal to the angle of reflection; for statements of the law, see Newton, *Opticks*, p. 5, axiom II, and Smith, *A Compleat System of Opticks*, vol. I, p. 3.

106/6 Compare above, p. 88, ll. 4–11; Bradley, 'An Account of a New Discovered Motion of the Fix'd Stars', p. 653; Maclaurin, *An Account of Sir Isaac Newton's Philosophical Discoveries*, p. 265; and Charles Hutton, *A Mathematical and Philosophical Dictionary* (1795–96), s.v. 'aberration', vol. I, p. 7.

106/37 Compare the second case considered by Reid in his earlier manuscript on the aberration of light, p. 90, l. 31 to p. 95, l. 5.

107/35 Both Patrick Wilson and John Robison came to the same conclusion regarding terrestrial objects seen through a telescope; see the editorial note to the 1809 abridged edition of his paper, 'An Experiment Proposed for Determining, by the Aberration of the Fixed Stars, whether the Rays of Light, in Pervading Different Media, Change Their Velocity', p. 193, and Robison, 'On the Motion of Light', pp. 87–8, 91–2.

110/13 Reid's second figure appears to have been lost.

7/II/11: The Sine of Incidence to the Sine of Refraction

111/19 Newton, *Opticks*, p. 131.

111/21 Newton, *Opticks*, pp. 128–9.

111/24 Reid's figures differ significantly from those in Newton, *Opticks*, p. 131; his numbers are either based on his own calculations or taken from an unidentified source.

113/3 Water prisms consisted of an outer glass shell filled with water. They feature in various experiments described in Newton, *Opticks*, pp. 31, 72, 78, 116.

7/III/13: Indistinct Vision

115/10 Pages 1–4 of this manuscript no longer survive.

115/18 Reid's wording implies that this manuscript may have been a commentary on the work of one of his friends or even the draft of a letter to one of his close scientific associates outside of Glasgow. The most likely recipient of a set of comments or a letter was John Robison.

117/3 For Reid's earlier experiments using pin holes in paper cards to illustrate the laws governing distinct and indistinct vision, see Reid, *Inquiry*, pp. 127–9, where he notes that his experiments were based on those described in Porterfield, *A Treatise on the Eye*, vol. I, pp. 130–1, 392–4, and vol. II, pp. 293–6.

117/26 Reid's experience of double vision may have been related to the injury to his right eye that occurred in May 1761; see Reid, *Inquiry*, p. 131.

117/29 Reid refers to a missing part of his text. James Jurin described a similar experiment in Jurin, 'An Essay upon Distinct and Indistinct Vision', p. 156, and a variant on pp. 163–4.

118/26 Reid alludes to Virgil's phrase 'hoc opus, hic labor est', which translated literally means 'this is the task, this the toil' but which can also be translated as 'there's the rub'; Virgil, *The Aeneid*, Book VI, line 129, in *Eclogues, Georgics and Aeneid and the Minor Poems*.

118/31 Newton outlined his problematic theory of 'Fits of easy Reflexion' and 'Fits of easy Transmission' in propositions 13–20, Book II, Part iii of his *Opticks*, pp. 281–8; see Cantor, *Optics after Newton*, pp. 83–6. James Jurin took up Newton's theory to explain indistinct vision in Jurin, 'An Essay upon Distinct and Indistinct Vision', pp. 157–8, 160–8.

118/39 See especially William Herschel, 'Account of Some Observations Tending to Investigate the Construction of the Heavens' and 'On the Parallax of the Fixed Stars'; these papers were published in 1785 and 1782, respectively. Herschel's 1782 paper explicitly addressed issues in optical theory and the question of indistinct vision. Reid's claim that the investigation of indistinct vision would lead to progress in the science of optics registers Herschel's statement that: 'Too much has hitherto been taken for granted in optics: every natural philosopher is ready enough to allow the necessity of making experiments, and tracing out the steps of nature; why this method should not be more pursued in the art of seeing does not appear. Theories are only to be used when proper data are assigned; but the data are carefully to be re-examined, when new improvements may widely alter the result of former experiments. Thus, we are told, that we gain nothing by magnifying *too much*. I grant it; but shall never believe I magnify too much till by experience I find, that I can see better with a lower power'; Herschel, 'On the Parallax of the Fixed Stars', pp. 91–2. See also William Herschel, 'Catalogue of Double Stars', 'Catalogue of Double Stars … Communicated by Dr Watson, Jun.' and 'On the Construction of the Heavens'.

119/27 See the figures in Herschel, 'Account of Some Observations Tending to Investigate the Construction of the Heavens' and 'On the Parallax of the Fixed Stars'.

7/VIII/2: The aberration of the Axis of a Telescope

119/29 The use of telescopes filled with water to investigate the aberration of light was discussed by Ruđer Josip Bošković, John Robison and Patrick Wilson; see Ruđer Josip Bošković, 'De modo determinandi discrimen velocitatis, quam habet lumen, dum percurrit diversa media, per duo telescopia dioptrica, alterum commune, alterum novi cujusdam generis', in *Opera pertinentia ad opticam, et astronomiam*, vol. II, esp. pp. 297–314; Ruđer Josip Bošković, 'De calculanda aberratione astrorum orta e propagatione luminis successiva', in *Opera pertinentia ad opticam et astronomiam*, vol. V, pp. 417–37; Robison, 'On the Motion of Light', p. 84; Wilson, 'An Experiment Proposed for Determining, by the Aberration of the Fixed Stars, whether the Rays of Light, in Pervading Different Media, Change Their Velocity'. Bošković seems to have been the first of the three to hit on the idea of using a water telescope, for he proposed an experiment on the aberration of light that employed both ordinary and water telescopes in a letter to the French astronomer Joseph-Jérôme Lefrançois de Lalande written in 1766; see Joseph-Jérôme Lefrançois de Lalande, *Astronomie*, second edition (1771–81), vol. IV, pp. 687–8. Bošković may have been inspired by Newton's description of a water telescope in Newton, *Opticks*, pp. 101–2, 110. According to John Robison, Patrick Wilson hit on the idea of using a water telescope to detect the motion of the centre of the solar system in the mid-1770s; Robison, 'Mr Wilson on the Solar System', p. 29. Moreover, in an undated letter to John Robison commenting on Robison's paper 'On the Motion of Light', Reid noted that 'Some years ago I had some intercourse with our Friend P Wilson about the Water Telescope' and said that they had concluded that 'according to the optical Principles of Newton, it can neither give any aberration to terrestrial Objects, nor any other aberration to celestial than the common telescope gives'; Reid to Robison, [April 1788–90], in Reid, *Correspondence*, p. 198. Reid and Wilson thus agreed with one of the most important claims made in Robison's 1788 paper; compare Reid's first proposition in this manuscript and his remarks in his letter to Robison with Robison, 'On the Motion of Light', p. 92.

120/34 Compare Wilson, 'An Experiment Proposed for Determining, by the Aberration of the Fixed Stars, whether the Rays of Light, in Pervading Different Media, Change Their Velocity', p. 60, and Robison, 'On the Motion of Light', p. 94.

3/III/11: To find the aberration of the Axis of a Water telescope

122/10 According to Newton, when a ray of light is refracted moving from water to air the ratio of the velocities of the ray in the different media is 4:3; see editorial note 88/23.

7/VIII/9: Rays of Light

123/13 Copernicus' heliocentric theory implied that there is a shift in the apparent position of the fixed stars due to annual motion of the Earth. In the seventeenth and eighteenth centuries observers repeatedly failed to discover the parallax of the fixed stars, and James Bradley affirmed that 'it must be granted, that the Parallax of the fixt Stars is much smaller, than hath been hitherto supposed by those who have pretended to deduce it from their Observations. I believe, that I may venture to say, that … it does not amount to 2". I am of Opinion, that if it were 1", I should have perceived it.… There appearing therefore after all, no sensible Parallax in the fixt Stars, the Anti-Copernicans have still room on that Account, to object against the Motion of the Earth'; Bradley, 'An Account of a New Discovered Motion of the Fix'd Stars', pp. 659–60. Reid was evidently looking for evidence to confirm heliocentrism and in this brief note he considers the implications of the suggestion initially made by Thomas Melvill in 1753 that the 'differently-colour'd rays' of light 'are projected with different velocities from [a] luminous body; the red with the greatest, violet with the least, and the intermediate colours with intermediate degrees of velocity'; Melvill, 'A Letter from Mr T. Melvil to the Rev. James Bradley, D.D. F.R.S.', p. 262. On eighteenth-century attempts to observe the parallax of the fixed stars, see Mari Williams, 'James Bradley and the Eighteenth-Century "Gap" in Attempts to Measure Annual Stellar Parallax'.

Part Five: Electricity

6/V/11: Electricity

124/1 Reid's teaching cycle was such that he lectured on natural philosophy during the 1754–55 academic session at King's College, Aberdeen. His lecture notes on fols 1r–2r of this manuscript were probably written in this period. For a brief set of notes covering the same introductory material see AUL, MS 2131/8/V/3, fol. 2v.

124/3 While Reid's definition of electricity as 'a Power or Virtue in Bodies, excited by Friction' is firmly rooted in the conception of electricity current in the seventeenth century, he adds a Newtonian element when he speaks of powers of attraction *and* repulsion; see Niels H. de V. Heathcote, 'The Early Meaning of Electricity: Some *Pseudodoxia Epidemica*'. Propositions 1–13 draw heavily on J. T. Desaguliers' writings on electricity and, in particular, Desaguliers' *A Dissertation concerning Electricity* (1742). On Desaguliers' *Dissertation*, see J. L. Heilbron, *Electricity in the Seventeenth and Eighteenth Centuries: A Study of Early Modern Physics*, pp. 293–4. Heilbron writes that Desagulier's *Dissertation* 'offers an inconsistent yet briefly influential theory', p. 293.

124/8 Reid's discussion of the differences between electric and non-electric bodies is indebted to Desaguliers, *Dissertation*, pp. 2–3; J. T. Desaguliers, 'Some Further Observations concerning Electricity'; J. T. Desaguliers, 'Some Things concerning Electricity'; J. T. Desaguliers, 'Some Thoughts and Experiments concerning Electricity'.

124/28 Reid's distinction between vitreous and resinous forms of electricity was initially drawn the French electrician Charles François de Cisternay Dufay; see Dufay, 'A Letter from Mons. Du Fay, F.R.S. and of the Royal Academy of Sciences at Paris, to his Grace Charles Duke of Richmond and Lenox, concerning Electricity', pp. 263–5. Desaguliers also employed Dufay's distinction; see Desaguliers, *Dissertation*, p. 39, and J. T. Desaguliers, 'Some Conjectures concerning Electricity, and the Rise of Vapours', p. 140.

125/3 On the effect of moist or cold and dry air, see J. T. Desaguliers, 'An Account of Some Electrical Experiments Made before the Royal Society on Thursday the 16th February 1737–38', p. 204, and Desaguliers, 'Some Thoughts and Experiments concerning Electricity', p. 192.

125/7 Reid's 'Conjectures concerning Electricity' address salient features of Benjamin Franklin's theory of electricity. Three editions of Franklin's text were available to him in 1754–55: Benjamin Franklin, *Experiments and Observations on Electricity, Made at Philadelphia in America* (1751); Benjamin Franklin, *Experiments and Observations on Electricity, Made at Philadelphia in America* (1751–53), which included a new section added in 1753 entitled 'Supplemental Experiments and Observations on Electricity, Part II'; and Benjamin Franklin, *New Experiments and Observations on Electricity. Made at Philadelphia in America* (1754). A copy of the 1754 edition of Franklin's text survives in Aberdeen University Library, which suggests that Reid used this edition.

125/10 For Franklin's conjecture that the 'electrical fire' is a subtle fluid diffused through material bodies, see Franklin, *New Experiments and Observations on Electricity*, esp. p. 51. The notion of there being an equilibrium in the fluid features prominently in the discussion of the Leyden jar at the beginning of the text; Franklin, *New Experiments and Observations on Electricity*, pp. 1–9.

125/17 Franklin, *New Experiments and Observations on Electricity*, pp. 10–12, 57, 62, 117.

125/19 On the two electricities, see Franklin, *New Experiments and Observations on Electricity*, esp. pp. 2–3, 15. For Franklin's concept of a 'natural quantity' of the electrical fluid in a body, see Franklin, *New Experiments and Observations on Electricity*, pp. 117–19.

125/29 On the equilibrium of the electrical fluid being disturbed by friction, see Franklin, *New Experiments and Observations on Electricity*, pp. 15, 75–8. On the manifestations of the restoration of the equilibrium of the electrical fluid, see Franklin, *New Experiments and Observations on Electricity*, pp. 6, 8–9, 14–15, 20–1, 44, 117, 119–20.

125/34 Franklin, *New Experiments and Observations on Electricity*, pp. 23–33, 70–9, 83–5. On the theoretical significance of Franklin's description of the electrical properties of the glass in a Leyden jar, see Heilbron, *Electricity in the Seventeenth and Eighteenth Centuries*, pp. 330–4.

125/36 Franklin, *New Experiments and Observations on Electricity*, p. 51.

126/14 Reid's query appears to have been prompted by the experiments using glass and sulphur globes described by Franklin and his associate Ebenezer Kinnersley in Franklin, *New Experiments and Observations on Electricity*, pp. 99–106. Franklin interpreted the experimental data in terms of the behaviour of positive and negative electricities, whereas Reid's query is framed in terms of Dufay's distinction between vitreous and resinous electricities.

126/18 Franklin, *New Experiments and Observations on Electricity*, pp. 28, 48, 49, 90–2.

126/21 Franklin, *New Experiments and Observations on Electricity*, pp. 36–49, 111–29.

126/23 Franklin, *New Experiments and Observations on Electricity*, pp. 25–6; Heilbron, *Electricity in the Seventeenth and Eighteenth Centuries*, p. 337.

126/29 Both Dufay and Franklin maintained that there are two forms of electricity, in Dufay's case vitreous and resinous, in Franklin's positive and negative. Reid here blends together their ideas. His reference to 'Male' and 'Female' electricities suggests that he also knew of the ideas of the German electrical theorist G. M. Bose, who speculated that there were two kinds of electrical 'fire', male and female. Reid may have learned of Bose's work through a detailed report on research that had been carried out in Germany on electricity that appeared in the *Gentleman's Magazine* in April 1745; Anon., 'An Historical Account of the Wonderful Discoveries, Made in Germany, &c. concerning Electricity', esp. p. 195. This report also mentioned Dufay's distinction between vitreous and resinous electricities (p. 193).

126/34 Reid thought of his piece of sulphur as an electrical body; see proposition 2, p. 124, ll. 9–10. The sulphur was, however, behaving in an anomalous fashion because it should not have been giving shocks to Reid and his friends in wet weather; compare what he states in propositions 11–12 above, p. 125, ll. 1–3; Desaguliers, 'An Account of Some Electrical Experiments Made before the Royal Society on Thursday the 16th of February 1737–38', p. 204, and Desaguliers, *Dissertation*, p. 6.

126/35 The physician David Skene was one of Reid's closest associates in Aberdeen, despite the twenty-year age gap between them. Beginning in 1753, Skene shared a medical practice in Aberdeen with his father Andrew, who was also a close friend of Reid. The younger Skene was a skilled man-midwife, having studied under the Scot William Smellie in London, and, like his father, was involved in the care of the sick poor at the Aberdeen Infirmary. Along with Reid, he was later a member of the Gordon's Mill Farming Club and the Aberdeen Philosophical Society.

127/5 Reid's silk stockings were in his view electrical bodies; see proposition 2, p. 124, ll. 9–10. The electrical effect of the stockings was increased by the cold weather conditions; compare Desaguliers, 'An Account of Some Electrical Experiments Made before the Royal Society on Thursday the 16th of February 1737–38', p. 204. Silk was first identified as an electric in Stephen Gray, 'An Account of Some New Electrical Experiments', published in the 1720–21 volume of *Philosophical Transactions*. Reid's comments on the behaviour of his silk and worsted stockings are similar to the observations later made by the Scottish electrician Robert Symmer; see especially Robert Symmer, 'New Experiments and Observations concerning Electricity', esp. pp. 340–70, 390–3.

127/7 Reid's experiments involving potatoes described in his notes from January and February 1758 speak not only to his interest in agricultural improvement but also to the fact that potatoes were increasingly a feature of the diet of the lower and middling ranks of Scottish society during the course of the eighteenth century. As T. C. Smout has pointed out, potatoes had been cultivated as a food crop for the Scottish nobility since the seventeenth century; T. C. Smout, *A History of the Scottish People, 1560–1830*, pp. 251–2.

6/V/20: Mr Epinus

127/18 Reid's summary of F. U. T. Aepinus' account of the electrical properties of the tourmaline crystal were almost certainly taken from Aepinus' paper published in 1758, 'Mémoire concernant quelques nouvelles expériences électriques remarquables'. On p. 106 Aepinus mentions the Dutch name 'Aschentrekcer' noted by Reid. But this detail is not recorded in the digest of Aepinus' essay that appeared in the supplement to the *Gentleman's Magazine* for 1758; see Anon., 'An Account of the Surprizing Electrical Qualities of a Precious Stone, Found in the Island of Ceylon'.

127/22 Aepinus, 'Mémoire concernant quelques nouvelles expériences électriques remarquables', p. 106.

127/26 Aepinus, 'Mémoire concernant quelques nouvelles expériences électriques remarquables', p. 106.

127/31 Aepinus, 'Mémoire concernant quelques nouvelles expériences électriques remarquables', pp. 111–12.

128/3 Aepinus, 'Mémoire concernant quelques nouvelles expériences électriques remarquables', pp. 113–15.

128/5 Aepinus, 'Mémoire concernant quelques nouvelles expériences électriques remarquables', pp. 115–16.

128/6 Giovanni Carafa, Duca di Noja, *Lettre du Duc de Noya Carafa sur la tourmaline, a Monsieur de Buffon* (1759). The authorship of this work is uncertain. R. W. Home suggests that even though Giovanni Carafa, the Duca di Noja, was an aristocratic man of science from Naples who was elected a Fellow of the Royal Society in 1759, much of the *Lettre* was written by the French natural historian Michel Adanson, who was an associate of the noted French electrician Jean Antoine Nollet; see Home's editorial introduction to F. U. T. Aepinus, *An Essay on the Theory of Electricity and Magnetism* (1759), p. 94.

128/20 Carafa, *Lettre du Duc de Noya Carafa sur la tourmaline*, esp. pp. 5–6, 33–4.

Part Six: Chemistry

7/III/6: Of Heat

129/22 Compare Reid's discussion of heat in an undated manuscript titled 'Of Secondary Qualities', AUL, MS 2131/2/III/3. See also AUL, MSS 2131/6/I/5; 2131/7/II/4, fol. 2v; 2131/7/II/18; Reid, *Inquiry*, pp. 42–3, 54–5; and Reid, *Intellectual Powers*, pp. 200–11. Joseph Black made a similar distinction between heat as a property of bodies and heat as a sensation; Joseph Black, *Lectures on the Elements of Chemistry, Delivered in the University of Edinburgh, by the Late Joseph Black, M.D.* (1803), vol. I, pp. 22–3. Reid's formulation of the distinction in the *Inquiry* was later used in Adair Crawford, *Experiments and Observations on Animal Heat, and the Inflammation of Combustible Bodies* (1779), p. 1. For an earlier example of the distinction, see Chambers, *Cyclopædia*, s.v. 'Heat', vol. I, pp. 223–4.

130/9 One possible source for at least some of the material in this paragraph is the entry on heat in Chambers, *Cyclopædia*, s.v. 'Heat', vol. I, pp. 223–8; see also Chambers, *Cyclopædia*, s.v. 'Potential', vol. II, p. 851. In herbals and pharmacopoeias of the early eighteenth century, the efficacy of nettles as a herbal remedy was attributed to the fact that they were 'hot' and 'dry'; see, for example, Nicholas

Culpeper, *Pharmacopœia Londinensis: Or, the London Dispensatory Further Adorned by the Studies and Collections of the Fellows Now Living, of the Said College* (1718), pp. 56–7, 84–6. Given Reid's competence as a natural historian and his interest in medicine he would have been familiar with this medical lore.

130/31 On the expansion of bodies as a measure of heat and the use of thermometers in chemistry, see for example: Black, *Lectures on the Elements of Chemistry*, vol. I, pp. xxxix–xl, 36, 50–105; Herman Boerhaave, *Elements of Chemistry: Being the Annual Lectures of Herman Boerhaave, M.D. Formerly Professor of Chemistry and Botany, and at Present Professor of Physick in the University of Leyden* (1735), vol. I, p. 85; Edmond Halley, 'An Account of Several Experiments Made to Examine the Nature of the Expansion and Contraction of Fluids by Heat and Cold, in Order to Ascertain the Divisions of the Thermometer, and to Make That Instrument, in All Places, without Adjusting by a Standard'; George Martine, *Essays Medical and Philosophical* (1740), esp. pp. 288–9; van Musschenbroek, *The Elements of Natural Philosophy*, vol. II, pp. 2, 9–15; and Brook Taylor, 'An Account of an Experiment, Made to Ascertain the Proportion of the Expansion of the Liquor in the Thermometer, with Regard to the Degrees of Heat'. Given that Reid had studied notes taken from the chemistry lectures of William Cullen (see the Introduction, pp. clxxviii–clxxx), he would also have known Cullen's discussion of this topic; see Arthur L. Donovan, *Philosophical Chemistry in the Scottish Enlightenment: The Doctrines and Discoveries of William Cullen and Joseph Black*, pp. 132–5.

131/29 Compare William Irvine in William Irvine and William Irvine, Jr, *Essays, Chiefly on Chemical Subjects* (1805), p. 156; Black, *Lectures on the Elements of Chemistry*, vol. I, p. 87; and Martine, *Essays Medical and Philosophical*, pp. 288–9. This point was later discussed in Jean-André Deluc, *Récherches sur les modifications de l'atmosphere* (1772) and, following Deluc, Adair Crawford, *Experiments and Observations on Animal Heat, and the Inflammation of Combustible Bodies; Being an Attempt to Resolve These Phenomena into a General Law of Nature* (1788), pp. 18–63.

132/15 Reid attended Black's Glasgow lectures on chemistry in the winter of 1765–66; see Reid to David Skene, 20 December [1765], in Reid, *Correspondence*, p. 44. It may be that the material on pages 4–8 of the manuscript consists of notes he took at Black's lectures or of summaries made afterwards of what he had heard and the experiments that he had witnessed.

132/16 A Florence flask is a long-necked flask made of thin glass that Black routinely employed in his teaching and research; see Robert G. W. Anderson, *The Playfair Collection and the Teaching of Chemistry at the University of Edinburgh, 1713–1858*, pp. 137 (fig. 50), 144.

132/26 Compare the description of this experiment with that in Reid to David Skene, 23 March 1766, in Reid, *Correspondence*, p. 49; see also Black, *Lectures on the Elements of Chemistry*, vol. I, pp. 120–2.

132/29 Compare Reid to David Skene, 23 March 1766, in Reid, *Correspondence*, p. 49; Black, *Lectures on the Elements of Chemistry*, vol. I, pp. 122–5. Black maintained that his experiments on the change of state of ice to water confirmed his theory of latent heat and refuted the idea that heat is a form of motion, as well as Boerhaave's theory of fire. He later wrote of the first two experiments described by Reid that 'these two experiments, and the reasoning which accompanies them, were read by me in the Philosophical Club, or Society of Professors and others in the University of Glasgow, in the year 1762, and have been described and explained in my lectures, there, and in Edinburgh, every year since'. In his preface to his edition of Black's lectures, John Robison stated that Black announced his discovery of latent heat to the Glasgow Literary Society in a paper delivered on 23 April 1762; Black, *Lectures on the Elements of Chemistry*, vol. I, p. xxxviii, 125 note.

132/35 On the equilibrium of heat, see Black, *Lectures on the Elements of Chemistry*, vol. I, pp. 25, 76–84; compare Martine, *Essays Medical and Philosophical*, p. 233, who cites Boerhaave, *Elements of Chemistry*, vol. I, pp. 152, 169, and Pieter van Musschenbroek, *Essai de physique* (1739–51), vol. I, pp. 475–6. See also van Musschenbroek, *The Elements of Natural Philosophy*, vol. II, pp. 23–5.

133/4 Compare Reid to David Skene, 23 March 1766, in Reid, *Correspondence*, p. 49; Black, *Lectures on the Elements of Chemistry*, vol. I, pp. 110, for the reference to van Musschenbroek. The experiment described by Reid does not appear in the published version of Black's lectures, but variants are found in Black, *Lectures on the Elements of Chemistry*, vol. I, pp. 127–30.

133/10 Compare Reid to David Skene, 23 March 1766, in Reid, *Correspondence*, p. 48; Black, *Lectures on the Elements of Chemistry*, vol. I, pp. 137, 139. Black notes that these experiments were performed by

William Irvine. The *Oxford English Dictionary* defines spermaceti as 'a white waxy substance obtained from an organ in the head of the sperm whale [which] was formerly used in candles and ointments'.

133/17 Black, *Lectures on the Elements of Chemistry*, vol. I, p. 138.

133/20 For Black on the freezing of mercury and on fluidity, see Black, *Lectures on the Elements of Chemistry*, vol. I, pp. 68–76, 106–44.

133/33 Reid's use of rudimentary mathematical formulae to calculate the temperature of the mixture of different quantities of two fluids reflects the emphasis placed on quantification by Black's circle of chemists in Glasgow. According to John Robison, William Irvine 'had a true mathematical taste, and delighted in reducing every thing to measure by means of Equations. He hunted for opportunities of doing this, and was very quick sighted in discovering the means of procedure'; John Robison to James Watt, [October 1800], in Robinson and McKie, *Partners in Science*, pp. 359–60. Both Robison and Black later praised Irvine's mathematical abilities; Black, *Lectures on the Elements of Chemistry*, vol. I, pp. xliv, 171. For evidence of Irvine's quantitative bent, see Irvine and Irvine, Jr, *Essays, Chiefly on Chemical Subjects*, pp. 157–8.

133/36 Black, *Lectures on the Elements of Chemistry*, vol. I, pp. 131–2.

134/2 Black, *Lectures on the Elements of Chemistry*, vol. I, pp. 132–3.

134/7 Descriptions of experiments 9, 10, 14, 15 and 16 are not found in Black, *Lectures on the Elements of Chemistry*. On experiment 9, compare Black, *Lectures on the Elements of Chemistry*, vol. I, pp. 118–19, where he discusses the example of melting ice in ice houses, and Thomas Cochrane, *Notes from Doctor Black's Lectures on Chemistry, 1767/8* (1966), p. 13.

134/8 Black regarded vaporisation as one of the five 'general effects of heat'; see the section of his lectures headed 'Of Vapour and Vaporisation' in Black, *Lectures on the Elements of Chemistry*, vol. I, pp. 35, 144–202.

134/17 Black, *Lectures on the Elements of Chemistry*, vol. I, pp. 148–9.

134/24 Black, *Lectures on the Elements of Chemistry*, vol. I, p. 158.

134/29 A description of this experiment appears in Anon., *An Enquiry into the General Effects of Heat with Observations on the Theories of Mixture* (1770), pp. 48–9, although some of the numerical values differ from those recorded by Reid.

134/30 This experiment was probably related to those described in Black, *Lectures on the Elements of Chemistry*, vol. I, pp. 157–8.

134/35 Black, *Lectures on the Elements of Chemistry*, vol. I, p. 159; compare Anon., *An Enquiry into the General Effects of Heat with Observations on the Theories of Mixture*, pp. 50–1.

135/3 Black, *Lectures on the Elements of Chemistry*, vol. I, pp. 149–50, 160–1; compare Crawford, *Experiments and Observations on Animal Heat* (1788), p. 77, and Anon., *An Enquiry into the General Effects of Heat with Observations on the Theories of Mixture*, pp. 51–2. Black's experiments were originally performed by James Watt. 'Papins Digester' was an early form of a pressure cooker, invented in 1679 by Denis Papin; see Denis Papin, *A New Digester or Engine for Softning Bones, Containing the Description of Its Make and Use in These Particulars: Viz. Cookery, Voyages at Sea, Confectionary, Making of Drinks, Chymistry and Dying* (1681).

135/16 Black, *Lectures on the Elements of Chemistry*, vol. I, pp. 161–4.

135/21 For Black on evaporation, see Black, *Lectures on the Elements of Chemistry*, vol. I, pp. xiv, 202–16.

135/22 Compare 'Of Inflammation' in Black, *Lectures on the Elements of Chemistry*, vol. I, pp. 227–42. The lectures Black gave on combustion late in life were very different from those he had earlier given in Glasgow.

135/27 Black, *Lectures on the Elements of Chemistry*, vol. I, pp. 48–9. In the summer of 1765 Reid wrote to David Skene: 'I have had but very imperfect hints of Dr Blacks Theory of Fire. He has a Strong Apprehension that the Phlogistick principle is so far from adding to the weight of Bodies by being Joyned to them that it diminishes, it. And on the contrary by taking the Ph‹l›ogistick from any body you make it heavier. He brings many Experiments to prove this; The Calcination of Metals and the Decomposition of Sulphur, you will easily guess to be among the Number, but he is very modest and cautious in his conclusions and wants to have them amply confirmed before he asserts them positively. I am told that Blacks Theory is not known at Edinburgh'; Reid to David Skene, 13 July 1765, in Reid, *Correspondence*, p. 39. On Black's theory of phlogiston, see especially Carleton E. Perrin, 'Joseph Black and the Absolute Levity of Phlogiston'.

135/29 Given that the second edition of Adair Crawford's *Experiments and Observations on Animal Heat* was published in 1788 and that in this edition Crawford mentions that 'it has been observed, by my learned friend Dr. Reid, in a very accurate communication, with which he favoured me on this subject, that "until the ratio between one temperature and another be ascertained by experiment and induction, we ought to consider temperature as a measure which admits of degrees, but not of ratios…"', it may be that the last three paragraphs of this manuscript were prompted by Reid's reading of a draft of, or a set of the proofs for, Crawford's revised text; see Crawford, *Experiments and Observations on Animal Heat* (1788), p. 6.

135/33 The relationships between temperature, relative heat and absolute heat were first defined in William Irvine's lectures on chemistry; see especially the fragment included in Irvine and Irvine, Jr, *Essays, Chiefly on Chemical Subjects*, pp. 153–8. Irvine's ideas were developed in print by Adair Crawford; see Crawford, *Experiments and Observations on Animal Heat* (1779), pp. 1–3, and Crawford, *Experiments and Observations on Animal Heat* (1788), pp. 1–6. In the first edition of his *Experiments and Observations on Animal Heat*, Crawford stated: '*absolute* heat expresses that *power* or *element*, which, when it is present to a *certain* degree, excites in all animals the *sensation* of heat; and sensible heat expresses the same power, considered as *relative* to the *effects* which it produces' (p. 2). Crawford also identified absolute heat with 'the element of fire' (p. 3). Sensible or relative heat, on the other hand, was the quantity of heat of a body that was manifest to our senses and measured by temperature. In the second edition of *Experiments and Observations on Animal Heat*, Crawford reformulated the distinction and contrasted 'absolute' and 'relative' heat: '*absolute* heat expresses, in the abstract, that *power* or *element*, which, when it is present to a *certain* degree, excites in all animals the sensation of heat; and *relative* heat expresses the same power, considered as having a relation to the effects by which it is known and measured'. He also distinguished between three types of relative heat: (i) the sensations of heat felt by animals (which he termed 'sensible heat'); (ii) temperature as measured by thermometers; and (iii) the varying heat capacities of different substances (which he called 'comparative heat'); Crawford, *Experiments and Observations on Animal Heat* (1788), pp. 2–3, 8. Reid's wording follows the second edition of Crawford's text.

136/2 For a similar formulation, see the fragment by William Irvine published in Irvine and Irvine, Jr, *Essays, Chiefly on Chemical Subjects*, pp. 153–4.

136/24 Compare William Irvine in Irvine and Irvine, Jr, *Essays, Chiefly on Chemical Subjects*, pp. 154–8.

7/III/11: The Quantity of Heat

136/26 It is unclear what the 'discovery' that prompted this note was. Reid's comments presumably relate to the research on heat capacity and absolute heat carried out by his colleague William Irvine, who served as the University of Glasgow's Lecturer in Materia Medica (1766–87) and Lecturer in Chemistry (1769–87); compare the account of Irvine's researches given by his son in Irvine and Irvine, Jr, *Essays, Chiefly on Chemical Subjects*, pp. 51–63, 115–23.

137/31 This supposition was rejected by Joseph Black and William Irvine; see Black, *Lectures on the Elements of Chemistry*, vol. I, pp. 79–83, and Irvine, 'On Heat Produced by Mixture', in Irvine and Irvine, Jr, *Essays, Chiefly on Chemical Subjects*, pp. 187–8. Adair Crawford, however, affirmed that the quantity of absolute heat in a body was proportional to the quantity of matter; see Adair Crawford, *Experiments and Observations on Animal Heat* (1779), pp. 5, 10–16, and Crawford, *Experiments and Observations on Animal Heat* (1788), pp. 17–18, 85–93.

138/3 William Irvine.

138/21 On the equilibrium of heat, see editorial note 132/35.

138/33 Reid refers to his erstwhile colleague Joseph Black.

2/I/4: Of the Chemical Elements of Bodies

139/11 The chemical composition of water was established in the early 1780s by James Watt, Henry Cavendish and Antoine-Laurent Lavoisier. See James Watt, 'Sequel to the Thoughts on the Constituent Parts of Water and Dephlogisticated Air'; James Watt, 'Thoughts on the Constituent Parts of Water and of Dephlogisticated Air; with an Account of Some Experiments on That Subject'; Henry Cavendish, 'Experiments on Air'; and Anon., 'Extrait d'un mémoire lu par M. Lavoisier, à la séance publique de l'Académie royale des sciences du 12 Novembre, sur la nature de l'eau, & sur des expériences qui paroissent prouver que cette substance n'est point un élément proprement dit, mais qu'elle est susceptible de décomposition & de recomposition'.

139/25 Compare Lavoisier: 'All that can be said upon the number and nature of elements is, in my opinion, confined to discussions entirely of a metaphysical nature. The subject only furnishes us with indefinite problems, which may be solved in a thousand different ways, not one of which, in all probability, is consistent with nature. I shall therefore only add upon this subject, that if, by the term *elements*, we mean to express those simple and indivisible atoms of which matter is composed, it is extremely probable we know nothing at all about them; but, if we apply the term *elements*, or *principles of bodies*, to express our idea of the last point which analysis is capable of reaching, we must admit, as elements, all the substances into which we are capable, by any means, to reduce bodies by decomposition. Not that we are entitled to affirm, that these substances we consider as simple may not be compounded of two, or even of a greater number of principles; but, since these principles cannot be separated, or rather since we have not hitherto discovered the means of separating them, they act with regard to us as simple substances, and we ought never to suppose them compounded until experiment and observation has proved them to be so'; Antoine-Laurent Lavoisier, *Elements of Chemistry, in a New Systematic Order, Containing all the Modern Discoveries* (1790), p. xxiv.

139/30 On the relationship between light and caloric, compare Lavoisier, *Elements of Chemistry*, p. 6. Significantly, Lavoisier does not claim here that light and caloric 'may according to circumstances be converted one into the other' as Reid has written. Rather, he states that it 'is indisputable, that, in a system where only decided facts are admissible, and where we avoid, as far as possible, to suppose any thing to be that is not really known to exist, we ought provisionally to distinguish, by distinct terms, such things as are known to produce different effects. We therefore distinguish light from caloric; though we do not therefore deny that these have certain qualities in common, and that, in certain circumstances, they combine with other bodies almost in the same manner, and produce, in part, the same effects'. Reid's gloss on this passage indicates that he was reading Lavoisier through the lens of the phlogiston theory. Phlogiston theorists typically saw light and heat as modifications of a basic subtle or ethereal fluid which they identified as phlogiston. Black, for example, is recorded as having stated in his lectures that 'From what appears to our Senses we would say that this principle [i.e. the 'principle of inflammability' or phlogiston] is Heat and Light, or some modification of Matter upon which Heat and Light depends'; quoted in David B. Wilson, *Seeking Nature's Logic: Natural Philosophy in the Scottish Enlightenment*, p. 148. Black's student Patrick Dugud Leslie asserted in a similar vein that 'phlogiston is fire and light, or a certain subtile elastick fluid, upon the modifications of which the phænomena of heat and light immediately depend' and cited Black's lectures as the source for his ideas; Patrick Dugud Leslie, *A Philosophical Inquiry into the Cause of Animal Heat: With Incidental Observations on Several Physiological and Chymical Questions, Connected with the Subject* (1778), p. 104. Like Black, Leslie also said that phlogiston was an imponderable fluid (p. 119).

139/32 On light as 'the instrument of Vision', compare Reid, *Inquiry*, p. 77. Reid's phrase 'enlivens the whole Face of Nature' echoes Stephen Hales' characterisation of 'elastick air'; see Stephen Hales, *Vegetable Staticks: Or, an Account of Some Statical Experiments on the Sap in Vegetables: Being an Essay towards a Natural History of Vegetation* (1727), p. 178. Hales cited various passages in the Queries to Newton's *Opticks* in order to justify the comparison he drew between the operations of light and those of elastic air. George Berkeley drew on Hales in his description of the function of 'pure æther, or light, or fire' in the economy of nature in *Siris: A Chain of Philosophical Reflexions and Inquiries* (1747), p. 82. Hales was later quoted in David Macbride, *Experimental Essays on the Following Subjects: I. On the Fermentation of Alimentary Mixtures. II. On the Nature and Properties of Fixed Air. III. On the Respective Powers, and Manner of Acting, of the Different Kinds of Antiseptics. IV. On the Scurvy; with a Proposal for Trying New Methods to Prevent or Cure the Same, at Sea. V. On the Dissolvent Power of Quick-lime* (1764), p. 30. Compare also Lavoisier, *Elements of Chemistry*, p. 184.

139/36 Reid's wording recalls the full title of Newton's *Opticks*: *Or, a Treatise of the Reflections, Refractions, Inflections and Colours of Light.*

139/37 Newton explained basic optical phenomena such as refraction in terms of inherent and unchanging properties of the 'rays' of light; see, for example, Isaac Newton, 'A Letter of Mr Isaac Newton, Professor of the Mathematicks in the University of Cambridge; Containing His New Theory about Light and Colors', p. 3081, and

Query 29 in Newton, *Opticks*, pp. 370–4. See also Reid, *Inquiry*, p. 77, on the original and unchanging properties of rays of light.

140/5 Compare Reid, *Inquiry*, p. 77.

140/13 During the course of the eighteenth century a succession of chemists, including Étienne François Geoffroy, Wilhelm Homberg, Lavoisier, Pierre Joseph Macquer and Joseph Priestley, had employed lenses and mirrors for the purposes of chemical analysis; see W. A. Smeaton, 'Some Large Burning Lenses and Their Use by Eighteenth-Century French and British Chemists', and Lavoisier, *Elements of Chemistry*, pp. 58–9.

140/19 On heat as uncombined or 'free' caloric, see Lavoisier, *Elements of Chemistry*, pp. 19–21.

140/36 See Lavoisier, *Elements of Chemistry*, pp. 185–6, who was indebted to the work on the interaction of light and vegetables carried out by Jan Ingenhousz, *Experiments upon Vegetables, Discovering Their Great Power of Purifying the Common Air in the Sun-shine, and of Injuring It in the Shade and at Night* (1779).

141/2 Reid's comment that heat is a material substance registers his close reading of Lavoisier as well as his familiarity with the theories of fire advanced earlier in the eighteenth century by Herman Boerhaave, W. J. 's Gravesande and Pieter van Musschenbroek, and the speculations of William Cullen and Joseph Black; Boerhaave, *Elements of Chemistry*, vol. I, pp. 78–247; W. J. 's Gravesande, *Mathematical Elements of Natural Philosophy, Confirm'd by Experiments; or, an Introduction to Sir Isaac Newton's Philosophy* (1721–26), vol. II, pp. 1–22; van Musschenbroek, *The Elements of Natural Philosophy*, vol. II, pp. 1–49; John R. R. Christie, 'Ether and the Science of Chemistry: 1740–1790'; Donovan, *Philosophical Chemistry in the Scottish Enlightenment*, pp. 131–43; Perrin, 'Joseph Black and the Absolute Levity of Phlogiston'. Reid's claim that heat was weightless owed something to Lavoisier, but Black had earlier maintained that phlogiston was an imponderable fluid and Cullen had flirted with the idea that fire does not gravitate; on Black and Cullen, see Perrin, 'Joseph Black and the Absolute Levity of Phlogiston', esp. pp. 112–24. As for Reid's comments on light, the claim that light is an imponderable substance follows from his earlier comment that caloric and light can be 'converted one into the other' (p. 139, l. 29). Yet Reid had previously assumed that rays of light obeyed Newton's laws of motion and were governed by the law of gravitation. The question of whether light was ponderable or imponderable was, however, discussed in the closing decades of the eighteenth century; see Cantor, *Optics after Newton*, pp. 62–4.

141/10 On the double meaning of the word 'heat', see Lavoisier, *Elements of Chemistry*, p. 20, and editorial note 129/22.

141/13 Lavoisier, *Elements of Chemistry*, pp. 1–3.

141/16 Lavoisier, *Elements of Chemistry*, pp. 5–6. Reid's discussion of caloric in this section of the manuscript indicates that he saw Lavoisier as putting forward a *chemical* theory of heat and was thus following the lead of Black, Irvine and Crawford; on this point, see Donovan, *Philosophical Chemistry in the Scottish Enlightenment*, pp. 222–31; Jan Golinski, '"Fit Instruments": Thermometers in Eighteenth-Century Chemistry', pp. 196–200.

141/26 Lavoisier, *Elements of Chemistry*, pp. 6–8, 15–17.

141/32 Lavoisier, *Elements of Chemistry*, p. 20.

142/8 See editorial note 139/30 and, for a summary of Lavoisier's theory of combustion, Lavoisier, *Elements of Chemistry*, pp. 414–18.

142/32 Reid here translates the distinction between absolute and relative heat initially drawn by William Irvine and Adair Crawford into the language of Lavoisier's caloric theory; compare Crawford, *Experiments and Observations on Animal Heat* (1788), pp. 2–5. The translation process was relatively straightforward because Lavoisier learned of the work of Joseph Black and Adair Crawford at a crucial stage in the development of his theory of caloric; see Robert J. Morris, 'Lavoisier and the Caloric Theory', pp. 9–12.

142/37 Lavoisier, *Elements of Chemistry*, p. 19; for Crawford's table of the comparative heats or heat capacities of different substances, see Crawford, *Experiments and Observations on Animal Heat* (1788), pp. 489–91.

143/8 In this paragraph Reid uses the terminology of the caloric theory to reformulate the explanations of changes of state advanced by Joseph Black, William Irvine and Adair Crawford; compare Lavoisier, *Elements of Chemistry*, pp. 7–8, 29.

143/14 On fixed caloric, see Lavoisier's comments in Lavoisier, *Elements of Chemistry*, p. 183; see also below, p. 151, ll. 1–7.

143/34 Lavoisier, *Elements of Chemistry*, pp. 36–7, 38, 46–7, 51–2, 89, 94, 414.

144/6 Lavoisier, *Elements of Chemistry*, pp. 165–7.

144/12 Lavoisier, *Elements of Chemistry*, pp. 61–2, 78–81.

144/14 See the 'Table of the binary Combinations of Oxygen with simple Substances' in Lavoisier, *Elements of Chemistry*, facing p. 185.

144/18 See the 'Table of the binary Combinations of Oxygen with simple Substances' in Lavoisier, *Elements of Chemistry*, facing p. 185.

144/19 Lavoisier, *Elements of Chemistry*, pp. 63–4, 66, 69, and the 'Table of the binary Combinations of Oxygen with simple Substances' facing p. 185.

144/21 Lavoisier, *Elements of Chemistry*, pp. 79–80.

144/23 See the 'Table of the binary Combinations of Oxygen with simple Substances' in Lavoisier, *Elements of Chemistry*, facing p. 185.

144/26 See the 'Table of the binary Combinations of Oxygen with simple Substances' in Lavoisier, *Elements of Chemistry*, facing p. 185.

144/27 Lavoisier, *Elements of Chemistry*, pp. 117, 179, 190.

144/28 Lavoisier, *Elements of Chemistry*, pp. 121, 179, 190.

145/5 Lavoisier, *Elements of Chemistry*, pp. 194–7. According to Lavoisier, 'azotic gas … composes nearly two thirds of the atmosphere', not 'near three fourths' as Reid has it; Lavoisier, *Elements of Chemistry*, p. 195.

145/10 Lavoisier, *Elements of Chemistry*, p. 198.

145/13 Lavoisier, *Elements of Chemistry*, pp. 123–5.

145/18 Lavoisier, *Elements of Chemistry*, pp. 209–10.

145/22 Lavoisier, *Elements of Chemistry*, pp. 151–6.

145/23 Lavoisier, *Elements of Chemistry*, pp. 202–9.

145/27 Lavoisier, *Elements of Chemistry*, pp. 110, 194, 198, 204, 207.

145/29 Lavoisier, *Elements of Chemistry*, pp. 159–60, 175–6.

145/34 Lavoisier, *Elements of Chemistry*, pp. 157–8, 177.

146/5 Lavoisier, *Elements of Chemistry*, pp. 56–7, 61.

146/8 Lavoisier, *Elements of Chemistry*, pp. 60, 97, 99, 100.

146/11 Lavoisier, *Elements of Chemistry*, pp. 63, 97, 98.

146/19 Lavoisier, *Elements of Chemistry*, pp. 75–6, and the 'Table of the binary Combinations of Oxygen with simple Substances', facing p. 185, which refers to 'Smoaking nitrous acid'.

146/21 The experiments Reid refers to, discussed on pages 87 and 89 of the *Elements of Chemistry*, were related to Lavoisier's investigation of the chemical composition of water. Reid has calculated the proportion of oxygen to hydrogen per 100 units of water on the basis of Lavoisier's experimental results recorded on page 87.

146/23 Lavoisier, *Elements of Chemistry*, pp. 88–9.

146/27 Lavoisier, *Elements of Chemistry*, pp. 104–5.

146/29	Lavoisier, *Elements of Chemistry*, p. 112.
147/7	Lavoisier, *Elements of Chemistry*, pp. 123–5.
147/9	Lavoisier, *Elements of Chemistry*, p. 132.
147/16	Lavoisier, *Elements of Chemistry*, p. 126.
147/21	Lavoisier, *Elements of Chemistry*, pp. 126–7.
147/25	Lavoisier, *Elements of Chemistry*, p. 127.
147/38	Lavoisier, *Elements of Chemistry*, pp. 127–8. Dippel's oil is distilled from bone. It was first produced by the German Pietist theologian, physician and alchemist Johann Conrad Dippel; on Dippel see Karl Hufbauer, *The Formation of the German Chemical Community (1720–1795)*, pp. 169–70.
148/10	Reid has based his calculations of the proportions between the chemical constituents of dry yeast, dry acetous acid, carbonic acid and dry alcohol on the data tabulated in Lavoisier, *Elements of Chemistry*, p. 136. Lavoisier had elsewhere stated that there are seventy-two parts of oxygen and twenty-eight parts of charcoal (carbon) per hundred of carbonic acid; Lavoisier, *Elements of Chemistry*, pp. 63–4, 86.
148/12	Lavoisier, *Elements of Chemistry*, p. 139.
148/16	Lavoisier, *Elements of Chemistry*, pp. 109–110.
148/24	Lavoisier, *Elements of Chemistry*, pp. 110–11.
148/30	Lavoisier, *Elements of Chemistry*, p. 111.
148/34	Lavoisier, *Elements of Chemistry*, p. 111.
148/37	Lavoisier, *Elements of Chemistry*, pp. 111–12.
149/4	Lavoisier, *Elements of Chemistry*, p. 112.
149/8	Lavoisier, *Elements of Chemistry*, pp. 112–13.
149/12	Lavoisier, *Elements of Chemistry*, p. 162.
149/19	Lavoisier, *Elements of Chemistry*, p. 161.
149/26	Lavoisier, *Elements of Chemistry*, p. 162.
149/30	Lavoisier, *Elements of Chemistry*, p. 163.
149/34	Lavoisier, *Elements of Chemistry*, p. 163.
149/37	Lavoisier, *Elements of Chemistry*, pp. 163–4.
150/6	Lavoisier, *Elements of Chemistry*, p. 164.
150/17	Lavoisier, *Elements of Chemistry*, pp. 165–70.
150/19	Lavoisier, *Elements of Chemistry*, pp. 169–70 and 170 note.
150/23	Lavoisier, *Elements of Chemistry*, pp. 179, 182.
150/31	Lavoisier, *Elements of Chemistry*, pp. 180–1.
150/34	See the 'Table of compound oxydable and acidifiable bases' in Lavoisier, *Elements of Chemistry*, p. 179.
150/37	Lavoisier, *Elements of Chemistry*, p. 181.

151/3 Reid paraphrases Lavoisier. The passage to which Reid refers reads: 'We know, in general, that all bodies in nature are imbued, surrounded, and penetrated in every way with caloric, which fills up every interval left between their particles; that, in certain cases, caloric becomes fixed in bodies, so as to constitute a part even of their solid substance, though it more frequently acts upon them with a repulsive force, from which, or from its accumulation in bodies to a greater or lesser degree, the transformation of solids into fluids, and of fluids to aeriform elasticity, is entirely owing.'; Lavoisier, *Elements of Chemistry*, pp. 182–3.

151/4 Compare above, p. 143, ll. 9–20.

151/17 Lavoisier, *Elements of Chemistry*, pp. 177, 187.

151/21 Lavoisier, *Elements of Chemistry*, pp. 186–7.

151/24 Lavoisier, *Elements of Chemistry*, p. 186.

151/31 Lavoisier, *Elements of Chemistry*, p. 186.

151/37 Lavoisier, *Elements of Chemistry*, pp. 187–8.

152/10 Lavoisier, *Elements of Chemistry*, pp. 188–9.

152/13 Lavoisier, *Elements of Chemistry*, p. 189.

152/16 Lavoisier, *Elements of Chemistry*, p. 195; see also editorial note 145/5.

152/37 Reid's speculation regarding the beneficial role of azote in the physiology of animals runs directly counter to Lavoisier's characterisation of the properties of oxygen and azote; compare Lavoisier, *Elements of Chemistry*, pp. 46, 51–2 and esp. 183–4, where Lavoisier links together oxygen and light and emphasises the beneficial effects of the latter on vegetables, animals and humankind.

153/5 Lavoisier, *Elements of Chemistry*, pp. 156–7, 165, 177, 194, 195–7.

153/9 Lavoisier, *Elements of Chemistry*, pp. 155–6.

153/14 Reid here addresses a point made in Lavoisier, *Elements of Chemistry*, p. 156.

153/18 Lavoisier, *Elements of Chemistry*, p. 216.

153/21 Lavoisier, *Elements of Chemistry*, p. 220.

153/23 Lavoisier, *Elements of Chemistry*, p. 223.

153/25 Lavoisier, *Elements of Chemistry*, p. 223.

153/27 Lavoisier, *Elements of Chemistry*, p. 226.

Textual Notes

Part One: Euclidean Geometry

5/II/47: Observations on the Elements of Euclid

3/15	perhaps *inserted.*
4/19	as also in Proposition 19 Book 2 *added.*
6/7	this *inserted.*
6/10	together *inserted.*
7/24	conceive] consider
7/25	circumstances] conditions
8/21–4	Dr D. Gregory … by and by. *added from the left margin.*
8/26	at Pavia *added.*
9/7–8	Definition of parallel lines and *inserted.*
9/11	aid of the disputed *inserted.*
9/12–14	Is it not possible … with demonstrative Evidence] I believe that Mathematicians will easily grant, that when the properties of parallel lines have been demonstrated, The 11 Axiom may be deduced from them with demonstrative Evidence
9/16	May we … deduce] May I therefore presume to attempt to deduce
9/24	right] streight
9/31–2	than the 11 Axiom of Euclid *added.*

3/I/13: Simson's Euclid

11/10	since] as
11/16	right *inserted.*
12/10–14	If a right line … be a right line *written in the right margin.*
12/13–14	that is … from it *inserted.*

5/I/1: Proposition 6

13/3	meeting Dm in r *inserted.*
15/7	of parallel lines *inserted.*
15/8	it is said *inserted.*
15/16	*Reid has here cancelled the following passage:*

Dr Simson thinks that we must take it for an Axiom that such lines do neither approach Nor recede from one another but I apprehend it may be strictly demonstrated from the following Lemma

If two right angled Triangles have a Side & the Hypothenuse of one equal to a Side & Hypothenuse of the other, the remaining side will be equal to the remaining side & the remaining Angles equal to those that are subtended by the equal Sides. This is assumed in Trigonometry. It ought to have a place in the Elements of Geometry, and is easily demonstrated from Propositions 16 & 18 Lib 1.

This being assumed let the first proposition about parallel lines be Parallel Lines are in every point equally distant and if one be cut at right angles by a right line all will be so

3061/11: The Elements of Euclid

19/4 shall throw] will throw
19/27 his Definition] the Definition
20/13 1 Book 11 *inserted.*
20/33–7 for as a … of a right line. *inserted from the right margin.*
21/6–7 whose very words I revere too much to change them] being unwilling to change his very words
21/25 supposed] assumed
21/35 falling upon a right line *inserted.*
21/36 Now *inserted.*
22/7–23/13 Corollary 4 … will be superceded *inserted. Reid's insertion replaced the following cancelled passage*:

I shall add a fourth Corollary from the Definition of a right line, by which I apprehend the properties of parallel lines may be demonstrated without any dependance upon the eleventh Axiom. or any other Axiom.

Corollary 4 If a line be in every point at the same distance from a right line and in the same plane, it must be a right line For if it be brought in contact with the right line in its extreme points it will touch it in every point, that is will coincide with it, & therefore be a right line by the Definition.

It is evident that a crooked line in the same plane with a right line cannot in every point have the same distance from the right line from this it necessarily follows that a line which in every point has the same distance from a right line in the same plane is not crooked, that is, it is a right line, & if applied to the right line must coincide with it.

It may be observed that the way in which we draw a right line in practice shews that the truth of this corollary is obvious to all Men. Being possessed of a streight Ruler, we direct the right line to be drawn so as that it may be in every point at an equal distance from the right line of the ruler, concluding that a line thus directed by a right line so as to be in every point at an equal distance from it, must be a right line.

We may observe by the way that the Axiom of Archimedes, That a right line is the shortest of all that can be drawn between two points, of which he makes so great use in his beautiful demonstrations, appears also to be the universal belief of Mankind from | [11] another way of drawing a right line commonly practised, that is by stretching a flexible line from one point to another.

23/31 this Axiom] it

23/36 which is lost *inserted.*

24/23–4 works in his own Language that is extant;] works that are extant, in his own Language;

24/35 bequeathed] left

25/3 David *inserted.*

25/33 so that We may justly say] We may therefore say

26/22 1st *inserted.*

27/1–2 neither thinks the 11th Axiom selfevident, nor pretends strictly to demonstrate it,] neither pretends strictly to demonstrate the 11 Axiom,

27/7 *A note here in the right margin reads* See Dr Simson's later thoughts page 22

27/20 might] may

27/23 *Reid has cancelled the following passage*:

I am not indeed able to do this without assuming some Axiom different from those in the Elements. That which appears most for the purpose is this. A Cur‹v›e line or a crooked line in the same plane with a right line, cannot in all its points have the same distance from the right line. A Curve line is distinguished from a crooked line by this, that a crooked line may be composed of right lines put together at angles whereas a curve line cannot be so composed. The curve & the crooked taken together are adequately opposed to the right line, so that a line that is neither Curve nor crooked must be one right line. This being understood, the Axiom is, That a line which is either curve or crooked, being in the same plane with a right line cannot have all its points at the same distance from the right line

This Axiom, surely is not more true than Euclids; but whether its Evidence be more suited to the Apprehension of young Mathematicians, must be left to the Judgment of those who attend to it. Taking it then for granted my first ‹I›nference from it is this A line which is in every point at the same distance from a right line and in the same plane must be a right line. For by the Axiom if it were either curve or crooked, it could not be at the same distance. Being therefore neither curve nor crooked, it must be a right line.

This proposition indeed appears evident to all Men from the common way in which we draw a right line in practice. Being possessed of a streight Ruler we direct the line to be drawn by the right line of the Ruler, so as to be at an equal distance from it in every point concluding that a line thus directed by a right line so as to be at an equal distance from it in every point, must be a right line.

We may observe by the way that the Axiom of Archimedes, of which he makes so great Use in his elegant Demonstrations, That a right line is the shortest of all that can be drawn between two points, appears to be the Universal belief of Mankind, from another way of drawing a right line commonly practised, that is, by stretching a flexible line between two points.

But to proceed

27/24–5 Corollary 4 from the Definition of right Lines *inserted. I have omitted Reid's page reference* pag 20, 21.

28/9–16 In the same way … serves for both *inserted from the right margin.*

28/16 axiom *inserted.*

28/18 of the kind *inserted.*

28/28 if it be a *inserted.*

28/31 if any] or no

28/35–30/17 *Inserted.*

29/5 His last Opinion onely ought to be imputed to him, & *inserted.*

29/10–11 & ought not to be placed among the Axioms *inserted.*

29/14 *Reid has here cancelled the following passage*: Those who think Dr Simsons Axiom has all the Evidence required in a Mathematical Axiom, or even that it is more evident than that which is assumed by Euclid must approve of the pains he has taken to demonstrate the last by the aid of the first.

29/28 determine] judge

29/32 be] appear

29/34 be] appear

29/35 was not so] did not appear] required the aid of reasoning.

30/7 very *inserted.*

30/10–12 but that in the confidence … without proof rather] but that he chose to assume it, in the confidence … of no doubt, rather

30/18 *At the top right of the page Reid has written*: This was not read at the Meeting

31/22 *Reid has here written*: (N. B. This Problem with its Corollaries was not read in the Society)

31/22 *Reid has written on the last blank folio*: If the Author had thought this Discourse worthy to be shewn to Mathematicians, he would have transcribed it fair, & put in their proper places the parts that are disjoyned; but as Dr Hope informs him that Mr. Playfair desires to see it, he will easily perceive that it is not worth that trouble

Part Two: 'An Essay on Quantity'

5/I/20: Object of Mathematicks

32/8 Velocity is measured by time & space] Velocity is measured by time & space & if we should consider abstractly from these measures of it we will hardly be able to demonstrate any thing about it.

32/9 resistance *inserted.*

32/18–20 But other mathematical quantitys … without] But other quantitys cannot be conceived immediately nor any ratio perceived between them but far less any ratio between them perceived without

32/23 mult‹i›plied or divided *added.*

33/4 already *added.*

33/6 *Reid has here written* See f. *His addition has subsequently been lost.*

33/7 mathematical *inserted.*

33/10 or as some Quantity Measured by these *added.*

33/11 an Improper quantity] a quantity

33/12 mathematical *inserted.*

33/25–34 3 Th‹o›se quantitys … Converging Series's *added.*
33/29 Quantities] Magnitudes
33/30 lines or Numbers] Number

5/I/22: Essay Concerning the Object of Mathematicks

This manuscript has been heavily worked over and contains numerous deletions, especially on folios 1 verso and 2 recto. The manuscript is made up of two sheets of paper which differ in size. Folio 3 recto is written on a sheet which is larger than that used in folios 1 recto to 2 verso and the material on this folio was probably written at a different date. The state and contents of the manuscript are similar to that of AUL, MS 2131/2/I/1, which contains disjointed fragments of the 'Essay' which can be pieced together to produce a more or less complete draft.

34/1 Essay *added.*
34/2–3 occasioned by … Compound Ratios *added.*
34/9 observed is either] observed in his Categories is either
34/13 Solids by Solids *added.*
34/18 Ratios Fluxions *added.*
34/24–7 any two … Mathematical Reasoning] Betwixt any two portions of Space or Duration, or Numbers we can immediately perceive a Ratio as that one is Double Treble or Half the other
34/27–8 But other Quantitys … without] But betwixt other Quantitys no such Ratio can be conceived without
34/29–30 And 'tis onely … Mathematical Quantitys. *added.*
34/35 Suppose I should take it in my head] But if any Mathematician should take it in his head
34/36 *After this sentence Reid has cancelled the following*: I do not think it is possible to confute him either by mathematical Reasoning or Experiments, Because he differs from me not in a Mathematical proposition but onely in a Mathematical Definition
35/1 *At the top of folio 1v Reid has cancelled the following lengthy passage*:

I differ from other Mathematicians not in a Mathematical proposition but in a Mathematical Definition. For a just Definition of an improper Quantity must include the Measure of it. My Notion will lead me to no false conclusion for tho my conclusions may differ in words from those of others yet in reality are the Same. Thus if I say upon my principle that the velocity of a body falling by its own gravity is as the Square of the time in which it acquires that velocity; I mean the very same thing with those who say that the velocity is Simply as the time. The consequence of this is that such a Notion cannot be refuted neither by Mathematical Reasoning nor by Experience. For being a Definition it can neither be true nor false Is there no preference then due to the common Notion of velocity besides it being Common? I think there is first because it is common for a Common Definition that is clear ought not to be rejected for one that hath nothing to recommend it but its Singularity. But there is another reason why the common Notion ought to be preferred & that is because it is more simple For why introduce a duplicate Ratio into a Definition when the simple one will do equally well. The Simple one ought to be chosen for the very Simplicity of it if you leave that you may take the duplicate the subduplicate or any other Ratio as well as the duplicate. Were I to define a foot by the Number of inches it containes I would much rather say it contained twelve inches than that it contained 36/3 of an inch yet both amount to the same thing.

35/1 I think … som‹e› light] It is easy to apply all this

35/5–6 is not compatible] cannot consist

35/6 This very thing has often made me Suspect that the Subject of this debate cannot be a Mathematical proposition. And I believe the Disputants on both sides on a little reflection must acknowledge that their arguments do not even in their own breasts breed that kind of Conviction which arises from Mathematical Demonstrations or Experimental Proofs. But from the principles above explained I conceive we may justly infer that this debate is onely about a Definition and that the Notions so earnestly contended for and against are Neither true nor False
The Vulgar agree with *cancelled.*

35/17 *Reid has written above the line at the end of this sentence* See i.

35/17 For Nothing can be the object of Mathematical Reasoning but what can be doubled or trebled or halfed by Corolary first. This cannot be done with regard to force till the measure of it be ascertained. Till this is done we fight in the dark, we fight about a vague indeterminate Word. And when Such a Measure is aggreed upon and such a Definition given It does then indeed become a Mathe‹mati›cal Quantity and a Subject of Mathematical Reasoning. But when this Definition is given the Controversy is at an end and no Person can be at a loss how to determine it. *cancelled.*

35/18 *Reid has written* m *in the margin beside the start of this paragraph.*

35/25 while the quantity of Matter is the same. *added. Reid has written at the end of the sentence* See g.

35/25 But if you lay down this Notion that the Force of a Moving Body is as its Velocity, As a proposition capable of proof or Demonstration you cannot take one step in the proof of it till you define what you mean by the force of a moving body by laying down some measure of it. If you should give this definition of it that the force of a Body is as the time in which that force may be generated by a uniform power such as that of Gravity. Here first, you take it for granted that the force of Gravity is uniform so as to generate a force proportional to the time which is a proposition neither plainer nor more evident than that which is to be proved by it. Reason cannot assure us that the power of Gravity is always the Same Experience indeed shews that Gravity generates a velocity proportional to the time. But if we infer from this that it generates a force proportional to the time we take for granted the thing to be proved that the force is as the Velocity. *cancelled.*

35/26 Secondly You take] But Secondly If I should yield to you that Gravity is always the Same you will find it difficult to prove from this that it must generate a force proportional to the time. ‹?› You take

35/27 conspiring] concurring

35/30 Experiments] Experience

35/31 Suppose] If

36/17 *Reid has written* h *in the margin.*

36/17 Sup‹p›ose] If

37/1 a given force] an impulse

37/8 measured by the length] proportional to the Space

3061/7: An Essay on Quantity, October 1748

38/6 N.B. This was published in the London Philosophical Transactions, pt. 45 & is republished in the 8vo Edition of Dr Reids works Edinburgh 1819. *added in a different hand.*

39/2 must have] behoved to have

43/14 Controversy about the *inserted.*

47/34–48/34 The Intervalls of Musical Notes … is true or false. *inserted. The inserted material is written on the recto and verso of a separate sheet of paper.*

48/10 at least *inserted.*

50/1–10 P.S. … well understood. *added in the form of handwriting that dates from late in Reid's life. Compare, for example, AUL, MSS 3061/6 and 3061/11, which were written in the 1790s.*

An Essay on Quantity, November 1748

50/11 'An Essay on *Quantity*; occasioned by reading a Treatise, in which *Simple* and *Compound Ratio's* are applied to *Virtue* and *Merit*, by the Rev. Mr. *Reid*; communicated in a Letter from the Rev. *Henry Miles* D.D. & F.R.S. to *Martin Folkes* Esq; Pr.R.S.'. *Philosophical Transactions* 45 (1748): 505–20. *This volume of the Philosophical Transactions was published in 1750.*

As L. L. Laudan has pointed out the widely used text of 'An Essay on Quantity' *published in Sir William Hamilton's edition of Reid's works is corrupt and is based on an unreliable version of the text that appeared in an early nineteenth-century abridgement of the* Philosophical Transactions. *See L. L. Laudan, 'The* Vis viva *Controversy, a Post-mortem', p. 140, note 42, and Charles Hutton, George Shaw and Richard Pearson,* The Philosophical Transactions of the Royal Society of London, from their Commencement, in 1665, to the Year 1800; Abridged, with Notes and Biographic Illustrations, *vol. IX, pp. 559–66. Editions of Reid's works published before Hamilton's also included this corrupt version of Reid's original text; see, for example, Thomas Reid,* Essays on the Powers of the Human Mind; to which are Prefixed, An Essay on Quantity and An Analysis of Aristotle's Logic *(1812), vol. I, pp. iii–xviii.*

Part Three: Astronomy

7/VIII/10: Observations concerning Whiston's Astronomical Tables

60/13–14 for the apparent time *inserted.*
60/14 absolute *inserted.*
61/7 Whiston is right & I was in the mistake *added.*
61/9 or of the nearest approximation of the centers *inserted.*

2/I/1: Solar Eclipse, July 1748

Reid has recorded his observations of the eclipse of the sun that occurred on 14 July 1748 on fol. 2r of a manuscript that is otherwise taken up with a heavily worked over, and apparently incomplete, draft of 'An Essay on Quantity'.

2/I/7: The Transit of Venus

65/13–14 As to … which make it] Its Longitude by the Maps
65/26–7 and particularly … this Society *inserted.*
66/7 from the Sun *inserted.*
67/11 segment *inserted.*
67/21 horizontal *added.*
67/29–30 So that … 5 hours 8'59" *inserted.*
71/27 as she appears in the heavens *inserted.*
71/33–8 but if we … of our Longitude] which ought to be considered in calculating the next transit] which indeed is so small a matter that it may be owing to an error in the Observation or to either of the other two causes mentioned, and we cannot pretend to say to which it is most probably owing till we have access to compare our Observation with those that have been made by others.
72/1 differs from] agrees with
72/16 alteration] correction
72/19 side *inserted.*
72/30 by halleys tables *inserted.*
73/2 begins] happens
73/4 forwarding] protracting
73/6 accelerated] protracted
73/10 & *inserted.*
73/16–17 that is … Micrometer *added.*

7/III/14: Proposition 26, Book III, of Newton's *Principia*

74/26 By proposition 8 Tract 4 *inserted from the left margin.*
74/31 first *inserted.*
75/23–6 The force MN … the Force LM *inserted from the left margin.*
75/27 accelerative *inserted.*
75/35 both *inserted.*
77/27 4 Let the force PL (see figure of Proposition 26 Liber 3 Principia) be resolved into the two forces PE & EL. In the triangle PEL the angle PEL is a right Angle the angle EPL = ∠ STP + ∠ TSP. *cancelled.*
77/35 path] orbit

79/18 let us consider whether *inserted.*

79/23 *Reid has cancelled the following material on folios 9r to 9v:*

11 By Article 9 $A = \pi^2 / P^2 \times ½ + 2\delta^2 / D^2 + 3\delta^4 / D^4$ &c therefore $2P^2A = \pi^2 + 4\delta^2\pi^2 / D^2 + 6\delta^4\pi^2 / D^4$ and neglecting the terms after the second and substituting the value of A in place of it we should have $D = 2\delta\pi D / 2\delta\pi$ That is D = D which makes us no wiser |

12 Construction. With the center T and Radius TP describe the circle PQR. From the Center T, in the right line TN take TM such

Take TY:TQ::as the periodical time of the Earth to the periodical time of the moon; and take TM to TY in the Subduplicate ratio of twice the mean disturbing central force of the Sun upon the moon to the accelerative force of the moon towards the Earth. With the center T and Radius TQ describe the circle PQR & from M draw a tangent to the circle MP. draw TP through the point of contact P &

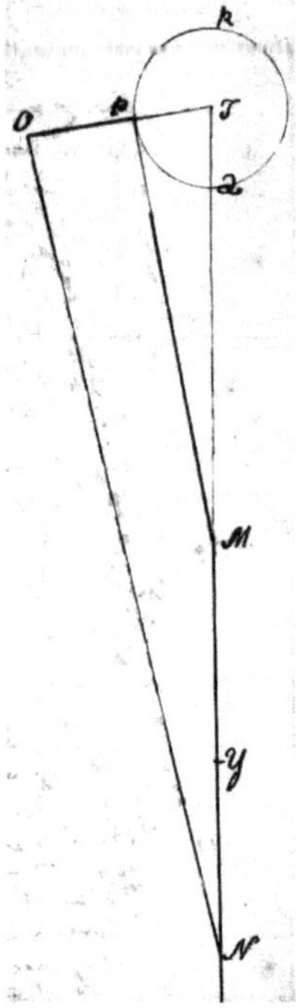

7/III/1: Of the Centripetal Forces

This manuscript consists of two very different sheets of paper. Reid has folded one sheet in half twice to produce the standard quire that he typically used. Part of the top fold of the quire remains uncut. I have transcribed the material written on this part of the manuscript but have omitted a note on a theorem of Matthew Stewart dealing with the quadrature of the circle which is written on a much smaller single sheet (measuring 105 mm high × 83 mm wide) tucked into the inner side fold of the quire.

3/I/20: Extracts from Pemberton

83/8 page 90 *written in the left margin.*

83/21 130 & c *written in the left margin.*
84/19 129 *written in the right margin.*
84/25 Did not … the first *added.*

Part Four: Optics

7/II/5: October 13 1770

88/4 annual *inserted.*
88/23–4 of the Motions of the Heavenly Bodies *inserted.*
89/1–6 N.B. As … Telescope onely. *written in the left margin.*

5/II/55: All Visible objects are seen by Rays of Light

89/12 last *inserted.*
89/12–13 their deviations from a right line] their Reflexions or Refractions
89/16–17 discern to a … of Objects] discern a small fraction of a minute in the direction of the Heavenly Bodies
89/26–9 4 When … the Eye. *inserted from the left margin.*
90/5–6 which really changes its place *inserted.*
90/7–9 The Sun …set out. *inserted from the right margin to replace the cancelled passage*: Thus the Sun must allwise appear more backward in the Eccliptick than he really is, by the space which he moves in 7 or 8 Minutes of time.
90/17 If they … of Saturn *inserted from the right margin.*
90/34 & absolute *inserted.*
90/35 fixed & *inserted.*
92/37 having its Axis in the real direction of the Ray *inserted.*
93/1 axis of the *inserted.*
93/4 Tangent]Subtense
93/11 *Reid's diagram is drawn in the upper left margin of the page.*
93/36–94/2 But it is …relative onely. *inserted from the right margin to replace the cancelled passage*: The first is always nearly the case of the Sun, both Errors with regard to the Sun's motion are all ways nearly the Same being contrary they can very little affect his apparent Motion. When any of the Planets or Comets are progressive the two Errors are contrary & one in part compensates the other but when retrograde they concur & produce an Error equal to their Sum
94/5 apparent *inserted.*
94/29 & therefore needs not be regarded. *added.*

95/5 *Reid has cancelled the following material on pages 7 through 9*:

Corollary 8. If a Telescope should be filled with Water having an object and Eye Glass fitted to give distinct Vision, the aberration in this telescope must correspond to the velocity of the Rays in Water and consequently upon the Newtonian hypothesis be one fourth part less than in a common telescope. And if a Telescope could be made of one piece of Solid Glass the Aberration by the same Hypothesis would be reduced a third part.

Case 3

We have hitherto considered the aberration of the Axis of a Telescope from the true place of the Star, which is required to make the ray move along that Axis. And it appears I think from what has been said that this aberration will be different according as the telescope is a Vacuum, or filled with Air, with Water, or with Glass. If the Velocities of Light be different in these different Media It remains to be considered how far the Structure of the Eye affects the aberration of a Star. In order to this it will be necessary to premise some Definitions & Propositions concerning the Eye that have not been commonly attended to by writers in Opticks.

Definition 1. The Axis of the Eye is that right line in which the Rays falling upon the cornea and upon all the Humours at Right Angles & therefore pass on to the Retina without any Refraction. In an Eye truly centered the Axis will pass through the Center of the Pupil. & this is probably the case in all perfect Eyes

In perfectly distinct Vision an Object placed in the Axis of the Eye is seen in the direction of that Axis. But it is to be observed that this object is seen by a pencil of Rays which fall on all parts of the pupil & which are collected again at that point of the Retina where it is cut by the Axis. And although the Rays which compose this pencil have various directions before they fall upon the Eye when the Object is Near, & much more when they pass through it, yet the Object is seen by them onely in one direction, to wit, in the direction of the Axis of the Eye‹.› | [8]

Definition 2. When a pencil of Rays proceeds from any point without the Axis but within the field of Vision, this pencil will fall upon all parts of the Pupil; but, in distinct vision will be collected in one point of the Retina at some distance from the Axis of the Eye. Of the various Rays of which this pencil is composed there will be one which either undergoes no Refraction or by contrary refractions retains its first direction, so as to pass through the vitreous humor in a line parallel to that in which it fell upon the Cornea. We shall beg leave to call the Line in which this line passes through the Vitreous Humor a Diameter of the Eye. And here it is to be observed that the Object is seen by the whole pencil or by any part of it in the direction of this Diameter, although the Rays which compose the pencil come to the Eye in various directions when the object is near, and all of them excepting the Ray which passes along the Diameter have other directions in passing through the Eye.

Every one who understands the principles of Opticks will perceive that the principles laid down in explaining these two definitions necessarily result from conceiving the Eye as an Optical Instrument formed for giving us true Vision, & particularly that they are necessary in order to our perceiving truly the Angular distance of objects which fall at once within the field of Vision Those who would see these principles confirmed by particular Experiments may consult Reids Inquiry Ch 6 §12. Where it is shewn to be a Law of Nature that we see an Object in the direction of the Right line which passes from the Image of the Object upon the Retina through the Optical Center of the Eye; or which is the same, in the direction of that Diameter of the Eye which proceeds from the Image upon the Retina

Definition 3 Let us call that Point of the Axis in which a Diameter of the Eye meets it, the Optical Center of the Eye. If all the Diameters meet the Axis and one another in one & the same Point this point will accurately ‹*word or phrase omitted*› the Center of the Eye whatever Diameters we take. But as we use this term (the Center of the Eye) onely as a technical term to denote the point where two diameters meet forming an Angle which measures the angular Distance of two visible points, it is of no importance whether the Diameters all meet in one point or not. We desire to be understood to mean | [9] by the Center of the Eye the Vertex of an Angle whose two sides are diameters of the Eye and whose Subtense is a line upon the Retina joyning the pictures of two visible points. This angle is the measure of the angular Distance at which the two visible Points appear and the intersection of the two diameters which form this Angle is what I call the Center of the Eye with relation to these two diameters. If two other Diameters meet in another point that other point is to be accounted the center of the eye with regard to them. It is Demonstrable that two Diameters of the Eye which are drawn to two points seen at the Same time must meet, otherwise they would not be in the same plain. Now it is Known from Experience that any two Objects seen at the Same time are all ways seen in one plain, to wit, in the plain passing through the Objects and the Center of the Eye.

It is moreover certain from the Structure of the Eye that all its Diameters meet one another & meet the Axis if not accurately, yet very nearly in one Point & that that point is very near the center of the Cornea. For Dr Jurin has found that the refracting power of the Cornea and of all the Humors of the Eye is very nearly the same. So that almost the whole refraction of the rays is performed at their entring the cornea. Hence it follows the Rays which fall at Right angles upon the cornea will have no Sensible refraction, & therefore will be Diameters of the Eye and will all meet at or very near the center of the Cornea, which therefore we may without sensible Error take for the Optical Center of the Eye.

Definition. The Deviation of the Axis of the Eye from the true place of a Star, which is necessary to make the Rays from that Star move along the Axis of the Eye, I shall call the Aberration of the Eye. And the Deviation of the Axis of the Telescope from the true place of a Star which is necessary to make the Rays from that Star to pass along the Axis of the Telescope, I shall call the aberration of the Telescope.

96/5–6 The Axis of the two cones] This Axis

96/33 *The manuscript has been damaged and a semi-circular fragment of the page has fallen away. I have supplied the missing letters or words.*

98/22 both of distinct & indistinct vision *added.*

100/11 very nearly *added.*

100/18 *Reid's diagram is in the upper right margin of the page.*

100/37 another] two other

101/1 The Ray] 1 The Ray

101/6 2 The cross hairs of the telescope having the same Motion with the Earth and the Eye will have no aberration. but be seen in their true direction from the Eye *cancelled.*

101/12–14 at Most … Axis of the Eye *written in the lower right margin.*

101/14 But a Telescope filled with Water should lessen the aberration about one fourth part. *written in the lower right margin and cancelled.*

7/II/15: Axiom

101/16 either *inserted.*
101/29 real *inserted.*
102/16 real Velocity] Velocity
102/16 real velocity] velocity
102/34 the Radius diminished by *inserted.*
103/34 unper‹c›eptibly *inserted.*

2/I/3: Of the Path of a Ray of Light

104/7 December 4 1781 *written in the left margin.*
104/30 onely *inserted.*
105/3–4 Problem … Motion. *inserted from the right margin.*
105/4 the Surface of *inserted.*
105/6–7 excepting … Cases *added.*
105/20 that] the same
107/5 the position *inserted.*
108/26 the path of *inserted.*
108/27 which goes along its Axis *inserted.*
109/3 pellucid *inserted.*
109/6 upon the Ray *inserted.*
109/21–2 Path & Velocity and the Path & Velocity of] Path and the Path of
110/5 at most *inserted.*
110/13 *The second figure does not appear in the manuscript.*
110/24 at most *inserted.*

7/II/11: The Sine of Incidence to the Sine of Refraction

114/19 *There follows an unpaginated folio containing various calculations and the note*: As it is now found that of different Refracting Media, one Scatters the different colloured rays more, another less; May we not hope to find some Medium that refracts without separating them at all Sir Isaac Newton finding in all his Experiments, that the Ratio of the difference between the Sines of Incidence and of Refraction of the Most Refrangible rays, to the difference of the Sine of incidence to that of refraction of the least refrangible was always as 27 to 28. concluded too rashly that this Ratio holds in all Refractions. He conjectured that the disposition of one ray to be more refracted than another depended solely on different inherent qualities of the differently coloured rays.

7/III/13: Indistinct Vision

115/11–12 the surface … spherical Surface] their Cornea was not perfectly spherical
115/20 falling upon] passing through
115/21 the Eye, produced] the Eye, that is, the Pupil, produced
115/24 possible *inserted.*
115/25 of them *inserted.*
115/34 known] common
115/34 the Rays passing through *inserted.*
116/9–12 The half of … which form it. *inserted from the left margin.*
116/39 Objects] Images
117/1 contrary order to that of the Spaces] same order as the Spaces
117/2 that form them *inserted.*
117/3–4 in strong white Paper *inserted.*
117/5 all *inserted.*
118/9 in their passage *inserted.*
118/35 particularly *inserted.*
119/16 In a good Light *inserted.*

7/VIII/2: The aberration of the Axis of a Telescope

119/35 the Axis of *inserted.*
121/5 absolute *inserted.*

3/III/11: To find the aberration of the Axis of a Water telescope

122/14 in water] in a water Telescope

Part Five: Electricity

6/V/11: Electricity

124/3 other Bodies] other light Bodies at considerable distances as of several Inches
124/12 to any Considerable Degree *inserted.*
124/22–3 Some times … weak Animals *inserted from the right margin.*
124/25 Non Electricks] bodies
125/4–5 13 The greatest … one another. *added from the right margin.*
125/9–10 & so … Equilibrium *added.*
125/15 *I have renumbered Reid's conjectures because his numerical sequence is muddled.*

125/36 Glass] Electric
126/9–18 *Some of Reid's wording is difficult to read or illegible because the manuscript is stained and the ink is badly faded.*
126/30 *The dates on folio 2 verso are all written in the left margin.*

Part Six: Chemistry

7/III/6: Of Heat

129/3 or State *inserted.*
129/10–11 and which … of it *inserted.*
129/27 when applied *inserted.*
130/5–6 more slowly and more permanently] in a different manner & more Slowly
130/18 any] the
130/23 or temperature *inserted.*
131/19 equal *inserted.*
132/24 by calculation *insertion.*
132/25 necessary] found necessary
135/29 December 1787 *written in the left margin.*
136/6 Resolution … Bodies mixed] Decomposition
136/24 that the Relative … Quicksilver *written in the left margin.*

7/III/11: The Quantity of Heat

138/6 The Quantitys … are equal *added.*
138/23 *Reid has here cancelled the following*:

like a fluid that is elastick and uniform which can never be at rest untill all parts of it be equally compressed.

Now it is evident that if all heat were of this volatile Nature, it would like water that is agitated come to rest after some time & remain at rest untill some new cause destroyed its æquilibrium. And when it is at rest from what has been shewn it becomes imperceptible or latent. For it is onely by passing from one body to another that it becomes sensible

Thus if our body were allways of the same temperature and all other bodies of the temperature with it we should never have felt either heat or cold. Thermometers would never have risen nor fallen. We should have had no Name for heat nor any suspicion that there was any such thing in Nature. It would have been qu‹i›escent and altogether latent, every body retaining its own portion without giving to or receiving from any other.

As it appears from what has been said that there may be heat in bodies which is latent or insensible so there may be heat which is fixed and not volatile, as we have hitherto supposed it to be. I would call that fixed heat which adheres to its Subject while the texture of that Subject continues. so as not to pass into other contiguous bodies even when they have a lower temperature. It is evident that such heat must be latent & continue so, as long as it is fixed to one Subject.

That there is a quantity of heat thus fixed in certain changes of the texture of certain Bodies, which they retain while that texture continues and which again becomes volatile when that texture is destroyed has been shewn by Dr Black by a beautifull Induction from Experiments. Thus in water of the temperature of 32 there is 140° of more heat than in ice of the Same temperature, and all this 140° it must lose before it be wholly turned to Ice, and after all the ‹?› is no colder by all this loss of heat. That this heat is not annihilate‹d› but onely become fixed & thereby latent appears from this that the Ice of 32 requires 140 degrees of additional heat to bring it wholly to water of the temperature 32 | [2v]

This noble Discovery leads us to consider heat as an Element which enters into the Composition of Bodies. It leads us to judge that the heat or cold that is perceived by f‹er›mintations Solutions and other operations which alter the texture of Bodies does not infer a creation of heat which did not formerly exist in the bodies but onely the bringing forth some secret Store they contained or the locking up some quantity of heat which was formerly within the Notice of our Senses. A great part of the heat which belongs to this Systeme which we inhabit is bound up as it were in chains and reserved perhaps for some grand purpose in some future revolution

2/I/4: Of the Chemical Elements of Bodies

139/24–5 hence in … a compound] hence be found to be a compound, in a more advanced State of the Chemical Art.

140/6–7 in opaque … greater] in Bodies greater

140/12–13 the composition and decomposition of bodies] the decomposition of compound bodies

140/13–19	Query … uncombined *added.* Effect … uncombined *written in the right margin.*
140/30–1	Query … in Vacuo *added.*
140/35	or even pure Air or Oxygen gas *added.*
140/35–6	The Rays … oxigen gas *inserted from the right margin.*
142/6–7	give them … totally, &] give a free passage, &
142/7–8	All inflammation … Vital Air *added.*
142/34–7	The whole … one another. *inserted from the right margin.*
143/9–20	Query. May … in the body? *inserted from the left margin.*
143/27	inflammation] combustion
143/32	more than a fourth] near one third
144/2	names have been given to] there have been found
144/3–4	there being … acidifiable *inserted.*
144/31	near three fourths] about two thirds
146/4–5	; which attract Water very strongly *added.*
146/7–8	One pound of Hydrogen melted 295 pounds 9 oz 3½ gros Ice *added.*
146/13	is not miscible with Water *inserted.*
147/2	Absolute heat in one pound, at Temperature 32, of Oxygen 66,⅓. Of Carbonic Acid 21. Of Water 12⅓ Of the Nitric Acid 58,72 *written in the right margin.*
148/1	In 100 parts of dry yeast there is 1,074 of azote 14,296 of hydrogen of oxygen 75,666 34,873 of charcoal. oxygen 72,677 hydrogen 12,967 *cancelled.*
149/14	Acids to form *inserted.*
152/15	near *inserted.*
152/21	lost not onely] not lost onely

Bibliography

Manuscript Sources

Aberdeen University Library (AUL)

Anon. 'Natural Philosophy 1758'. MS K 160.
Gregory, David. 'Notae in Isaaci Newtoni Principia Philosophiæ'. MS 465.
Gregory, John. 'Reflexions on the Principles of Agriculture'. MS 2206/7/18.
'Minute Book of the Farming Club at Gordon's Mill 1758'. MS 49.
Reid, Thomas. Birkwood Collection. MSS 2131/1–8.
— 'The Chemical History of Salts'. MS 2343.
— 'Letters and Papers Relating to the Life and Work of Prof. Thomas Reid'. MSS 2814/1/1–80.
— 'Original Papers of Thomas Reid, 1710–1796'. MSS 3061/1–26.
— 'A Short System of Book-keeping for the Farmer drawn up at the desire of the Farmers Club at Gordon Milne Aberdeen A° 1761'. MS 2341.
Skene, David. 'Correspondence, 1751–1770'. MS 38.

Birmingham Archives and Heritage, Library of Birmingham

James Watt Papers.

British Library

'Memoirs of the Society for the Incouragement of Learning taken from the Register of their Meetings, and Minute Books of the Committee'. Add MS 6185.

Edinburgh University Library

Reid, Thomas. 'A System of Logic, Taught at Aberdeen 1763'. MS Dk.3.2.

Glasgow University Archive Services

'Minutes of Meetings of the Faculty, 1784–1789'. GUA MS 26693.
'Minutes of Senate Meetings, 1771–1787'. SEN 1/1/1.
'Minutes of Senate Meetings, 1787–1802'. SEN 1/1/2.
'Minutes of the University Meetings, 1763–1768'. GUA MS 26643.

Glasgow University Library (GUL)

Erskine, David Steuart, eleventh earl of Buchan. 'Concerning John Anderson Professor of Natural & experimental Philosophy in the University of Glasgow'. MS Murray 502/76.
— 'Extracts from the Diaries and Letter-books of the 11th earl of Buchan. No. 9. The Story of His Life, 1764'. MS Murray 502/65.
'Laws of the Literary Society in Glasgow College'. MS Murray 505.
'Professors Receipt Book, 1765–1770'. Spec Coll MS Lib (uncatalogued).
'Professors Receipt Book, 1770–[1789]'. Spec Coll MS Lib (uncatalogued).
'Rules &c Relating to Dr R Simsons Collection of Books 1768'. MS Simson Ea5-b.1.

Mitchell Library, Glasgow

Baird, George. 'Notes from the Lectures of Dr Thomas Reid, 1779–80'. 8 vols. MS 104929–37.

National Library of Scotland

Home, Henry, Lord Kames. 'Cause of Gravity Read in the Philosophical Society February 1 1781'. MS Acc 10,073/4.

Royal College of Physicians of Edinburgh

Carmichael, David. 'Dr William Cullen's Chemical Lectures 1757. & 1758. &c'. MS CUL/2/2/7.
— 'Notes from William Cullen's Chemistry Lectures'. MS CUL/2/2/3–4.

Royal Society of London

John Canton Papers.
'Journal Book (copy)', xv (1735–36).

University of Strathclyde Archives and Special Collections

John Anderson Papers.

Newspapers and Journals

Aberdeen Magazine
Acta eruditorum
Bibliothèque des sciences et des beaux-arts

Commentarii Academiae scientiarum imperialis Petropolitanae
Commentarii Societatis regiae scientiarum Gottingensis
Commercium litterarium ad rei medicae et scientiae naturalis incrementum institutum quo quicquid novissime observatum agitatum scriptum vel peractum est succincte dilucideque exponitur
Edinburgh Journal of Science
Gentleman's Magazine
Histoire de l'Académie royale des sciences
Histoire de l'Académie royale des science et belles lettres
Le Journal littéraire de la Haye
Lloyd's Evening Post and British Chronicle
London Chronicle
Medical Commentaries
Monthly Review
Nova acta eruditorum
Nouveaux mémoires de l'Académie royale des sciences et belles-lettres
Observations sur la physique, sur l'histoire naturelle et sur les arts
Philosophical Transactions
Scots Magazine
Transactions of the Royal Society of Edinburgh

Primary Sources

Aepinus, F. U. T. *An Essay on the Theory of Electricity and Magnetism* [1759]. Translated by P. J. Connor and edited by R. W. Home. Princeton, NJ, 1979.

— 'Mémoire concernant quelques nouvelles expériences électriques remarquables'. In *Histoire de l'Academie royale des sciences et belles lettres. Année MDCCLVI* (pp. 105–21). Berlin, 1758.

Aguilón, François de. *Opticorum libri sex: Philosophis iuxta ac mathematicis utiles.* Antwerp, 1613.

Algarotti, Francesco. *Il Newtonianismo per le dame ovvero dialoghi sopra la luce e i colori.* [Venice], 1737.

— *The Philosophy of Sir Isaac Newton Explained, in Six Dialogues, on Light and Colours, between a Lady and the Author.* Glasgow, 1765.

— *Sir Isaac Newton's Philosophy Explain'd for the Use of the Ladies.* 2 vols. London, 1739.

Allardyce, Alexander (ed.). *Scotland and Scotsmen in the Eighteenth Century: From the MSS. of John Ramsay, Esq. of Ochtertyre.* 2 vols. Edinburgh and London, 1888.

Alston, Charles. *A Dissertation on Quick-lime and Lime-water.* Edinburgh, 1752.

Anderson, John. *A Compend of Experimental Philosophy; Containing Propositions Proved by a Course of Experiments in Natural Philosophy, and the General Heads of Lectures Which Accompany Them.* Glasgow, 1760.

Anon. *Abstract of Some Statutes and Orders of King's College in Old Aberdeen. M.DCC.LIII. With Additions M.DCC.LIV.* [Aberdeen], [1754].

— 'An Account of the Life and Writings of Dr John Gregory'. In John Gregory, *The Works of the Late John Gregory, M.D.* (vol. I, pp. 1–85). 4 vols. Edinburgh, 1788.

— 'An Account of the Surprizing Electrical Qualities of a Precious Stone, Found in the Island of Ceylon'. *Gentleman's Magazine* 28 (1758): 617–19.

— 'Affairs in Scotland'. *Scots Magazine* 19 (1757): 431.

— *An Authentic Narrative of the Dissensions and Debates in the Royal Society.* London, 1784.

— 'Biographical Account of Dr William Irvine'. *Medical Commentaries*, decade second 2 (1788): 455–61.

— *The Case of Patronage Stated, According to the Laws, Civil and Ecclesiastical, of the Realm of Scotland.* Glasgow, 1782.

— 'A Demonstration concerning the Motion of Light, Communicated from Paris, in the *Journal des scavans*, and Here Made English'. *Philosophical Transactions* 12 (1677–78): 893–4.

— *An Enquiry into the General Effects of Heat with Observations on the Theories of Mixture.* London, 1770.

— 'An Epitome of the *Philosophical Transactions*, vol. [LII]'. *Gentleman's Magazine* 33 (1763): 67–9.

— *Excerpta quædam e Newtoni Principiis philosophiæ naturalis, cum notis variorum.* Cambridge, 1765.

— 'Extrait d'un mémoire lu par M. Lavoisier, à la séance publique de l'Académie royale des sciences du 12 Novembre, sur la nature de l'eau, & sur des expériences qui paroissent prouver que cette substance n'est point un élément proprement dit, mais qu'elle est susceptible de décomposition & de recomposition'. *Observations sur la physique, sur l'histoire naturelle et sur les arts* 23 (1783): 452–5.

— 'Ferguson's Method of Determining the Parallax of Venus, &c.'. *Monthly Review* 24 (1761): 169–80.

— 'An Historical Account of the Wonderful Discoveries, Made in Germany, &c. concerning Electricity'. *Gentleman's Magazine* 15 (1745): 193–7.

— 'Observations of the Transit of Venus over the Sun's Disc, June 6. 1761'. *Aberdeen Magazine* 1 (1761): 429–30.

— Review of *Histoire et mémoires de l'Académie royale des sciences, inscriptions et belles-lettres de Toulouse*, vols II and III. *Monthly Review* 80 (1790): 489–502.

— 'A Short Account of the University of Edinburgh, the Present Professors in It, and the Several Parts of Learning Taught by Them'. *Scots Magazine* 3 (1741): 371–4.
— 'Steward's [*sic*] Method of Finding the Sun's Distance from the Earth'. *Monthly Review* 30 (1764): 51–6.
— 'Stewart's *Tracts, Physical and Mathematical*'. *Monthly Review* 25 (1761): 458–60.
— 'A Summary of the Last Volume of Memoires of the Royal Academy at Paris'. *Gentleman's Magazine* 29 (1759): 23–5.
Archimedes. *Measurement of a Circle*. In *The Works of Archimedes* (pp. 91–8). Edited by T. L. Heath. Cambridge, 1897.
Aristotle. *Categories*. In *The Complete Works of Aristotle: The Revised Oxford Translation* (vol. 1, pp. 3–24). Edited by Jonathan Barnes. 2 vols. Princeton, NJ, 1984.
Astrophilus. 'Transit of Venus over the Sun'. *Gentleman's Magazine* 31 (1761): 77–9.
Avison, Charles. *An Essay on Musical Expression*. London, 1752.
Bailly, Jean Sylvain. 'Deuxiéme mémoire sur la théorie des satellites de Jupiter'. In *Histoire de l'Académie royale des sciences. Année M.DCCLXIII* (pp. 172–89). Paris, 1766.
— *Essai sur la théorie des satellites de Jupiter, suivi des tables de leurs mouvemens, déduits du principe de la gravitation universelle*. Paris, 1766.
— *Histoire de l'astronomie moderne depuis la fondation de l'école d'Alexandrie, jusqu'a l'époque de M.D.CC.XXX*. New edition. 3 vols. Paris, 1785.
— 'Mémoire sur la cause de la variation de l'inclinaison de l'orbite du second satellite de Jupiter'. In *Histoire de l'Académie royale des sciences. Année M.DCCLXV* (pp. 499–504). Paris, 1768.
— 'Mémoire sur les inégalités de la lumière des satellites de Jupiter, sur la mesure de leurs diamètres; & sur un moyen, aussi simple que commode, de rendre les observations comparables, en remédiant à la difference des vues & des lunettes'. In *Histoire de l'Académie royale des sciences. Année M.DCCLXXI* (pp. 580–667). Paris, 1774.
— 'Mémoire sur le mouvement des nœuds, et sur la variation de l'inclinaison des satellites de Jupiter'. In *Histoire de l'Académie royale des sciences. Année M.DCCLXVI* (pp. 346–64). Paris, 1769.
— 'Premier mémoire sur la théorie des satellites de Jupiter'. In *Histoire de l'Académie royale des sciences. Année M.DCCLXIII* (pp. 121–36). Paris, 1766.
— 'Troisiéme mémoire sur la théorie des satellites de Jupiter'. In *Histoire de l'Académie royale des sciences. Année M.DCCLXIII* (pp. 377–84). Paris, 1766.

Barrow, Isaac. *Euclide's Elements; the Whole Fifteen Books Compendiously Demonstrated.* London, 1705.

— *The Usefulness of Mathematical Learning Explained and Demonstrated: Being Mathematical Lectures Read in the Publick Schools at the University of Cambridge.* London, 1734.

Berkeley, George. *Alciphron, or the Minute Philosopher* [1732]. In *The Works of George Berkeley, Bishop of Cloyne* (vol. III, pp. 21–329). Edited by A. A. Luce and T. E. Jessop. 9 vols. London, 1948–57.

— *The Analyst: Or a Discourse Addressed to an Infidel Mathematician* [1734]. In *The Works of George Berkeley, Bishop of Cloyne* (vol. IV, pp. 63–102). Edited by A. A. Luce and T. E. Jessop. 9 vols. London, 1948–57.

— *Arithmetica absque algebra aut Euclide demonstrata* [1707]. In *The Works of George Berkeley, Bishop of Cloyne* (vol. IV, pp. 163–230). Edited by A. A. Luce and T. E. Jessop. 9 vols. London, 1948–57.

— *An Essay Towards a New Theory of Vision.* Dublin, 1709.

— *Siris: A Chain of Philosophical Reflexions and Inquiries* [1747]. In *The Works of George Berkeley, Bishop of Cloyne* (vol. V, pp. 25–164). Edited by A. A. Luce and T. E. Jessop. 9 vols. London, 1948–57.

Bernoulli, Jean. *Discours sur les loix de la communication du movement.* Paris, 1727.

Bernoulli II, Jean. 'Essai d'une nouvelle méthode de déterminer la diminution séculaire de l'obliquité de l'écliptique, au moyen de l'Étoile polaire'. In *Nouveaux mémoires de l'Académie royale des sciences et belles-lettres. Année M.DCCLXXIX* (pp. 211–42). Berlin, 1781.

Birch, Thomas. *The Life of the Honourable Robert Boyle.* London, 1741.

Black, Joseph. *The Correspondence of Joseph Black.* Edited by Robert G. W. Anderson and Jean Jones. 2 vols. Farnham and Burlington, VT, 2012.

— 'Experiments upon Magnesia alba, Quick-lime and Some Other Alkaline Substances'. In *Essays and Observations, Physical and Literary* (vol. II, pp. 157–225). 3 vols. Edinburgh, 1754–71.

— *Lectures on the Elements of Chemistry, Delivered in the University of Edinburgh, by the Late Joseph Black, M.D.* Edited by John Robison. 2 vols. Edinburgh, 1803.

Blair, Robert. 'Experiments and Observations on the Unequal Refrangibility of Light'. *Transactions of the Royal Society of Edinburgh* 3 (1794): 3–76.

Bliss, Nathaniel. 'Observations on the Transit of Venus over the Sun, on the 6th June 1761: In a Letter to the Right Honourable George earl of Macclesfield, President of the Royal Society'. *Philosophical Transactions* 52 (1761–62): 173–7.

— 'A Second Letter to the Right Hon. the earl of Macclesfield, President of the Royal Society, concerning the Transit of Venus over the Sun, on the 6th June 1761'. *Philosophical Transactions* 52 (1761–62): 232–50.

Boerhaave, Herman. *Elementa chemiae*. 2 vols. Leiden, 1732.

— *Elements of Chemistry: Being the Annual Lectures of Herman Boerhaave, M.D. Formerly Professor of Chemistry and Botany, and at Present, Professor of Physick in the University of Leyden*. Translated by Timothy Dallowe. 2 vols. London, 1735.

Bolton, Stephen, et al. 'Observations of the Transit of Venus over the Sun'. *Gentleman's Magazine* 31 (1761): 251–2.

Bošković, Ruđer Josip. 'De calculanda aberratione astrorum orta e propagatione luminis successiva'. In *Opera pertinentia ad opticam et astronomiam* (vol. V, pp. 417–37). 5 vols. Bassano, 1785.

— *Dissertationis de lumine pars prima*. Rome, 1748.

— *De lumine*. Rome, 1749.

— 'De modo determinandi discrimen velocitatis, quam habet lumen, dum percurrit diversa media, per duo telescopia dioptrica, alterum commune, alterum novi cujusdam generis'. In *Opera pertinentia ad opticam et astronomiam* (vol. II, pp. 248–314). 5 vols. Bassano, 1785.

— *De viribus vivis dissertatio*. Rome, 1745.

Bouguer, Pierre. *La figure de la Terre, déterminée par les observations de messieurs Bouguer, & de la Condamine, de l'Académie royale des sciences, envoyés par ordre du Roy au Pérou, pour observer aux environs de l'equateur. Avec une relation abregée de le voyage, qui contient la description du pays dans lequel les opérations ont été faites*. Paris, 1749.

Boyle, Robert. *Essays of the Strange Subtilty, Determinate Nature [and] Great Efficacy of Effluviums*. London, 1673.

— *The Works of the Honourable Robert Boyle*. 5 vols. London, 1744.

Bradley, James. 'Directions for Using the Common Micrometer, Taken from a Paper in the late Dr Bradley's Hand-writing'. *Philosophical Transactions* 62 (1772): 46–53.

— 'A Letter from the Reverend Mr James Bradley Savilian Professor of Astronomy at Oxford, and F.R.S. to Dr Edmond Halley Astronom. Reg. &c. Giving an Account of a New Discovered Motion of the Fix'd Stars'. *Philosophical Transactions* 35 (1727–28): 637–61.

— 'A Letter to the Right Honourable George Earl of Macclesfield, concerning an Apparent Motion Observed in Some of the Fixed Stars'. *Philosophical Transactions* 45 (1748): 1–43.

— *Miscellaneous Works and Correspondence of the Rev. James Bradley, D.D.*

F.R.S. Astronomer Royal, Savilian Professor of Astronomy in the University of Oxford, &c. Edited by Stephen Peter Rigaud. Oxford, 1832.

Burnett, James, Lord Monboddo. *Antient Metaphysics: Or, the Science of Universals.* 6 vols. Edinburgh, 1779–99.

Campbell, Archibald. *An Enquiry into the Original of Moral Virtue.* Edinburgh, 1733.

Campbell, George. 'A Method for Determining the Number of Impossible Roots in Adfected Æquations'. *Philosophical Transactions* 35 (1727–28): 515–31.

Canton, John. 'Electrical Properties of the Tour-malin'. *Gentleman's Magazine* 29 (1759): 424–5.

Carafa, Giovanni, Duca di Noja. *Lettre du Duc de Noya Carafa sur la tourmaline, a Monsieur de Buffon.* Paris, 1759.

Cassini, Jacques. *Traité de la grandeur et de la figure de la Terre.* Amsterdam, 1723.

Cavallo, Tiberius. 'An Account of Some Thermometrical Experiments; Containing, I. Experiments Relating to the Cold Produced by the Evaporation of Various Fluids, with a Method of Purifying Ether. II. Experiments Relating to the Expansion of Mercury. III. Description of a Thermometrical Barometer'. *Philosophical Transactions* 71 (1781): 509–25.

Cavendish, Henry. 'An Account of a New Eudiometer'. *Philosophical Transactions* 73 (1783): 106–35.

— 'An Account of Some Attempts to Imitate the Effects of the Torpedo by Electricity'. *Philosophical Transactions* 66 (1776): 196–225.

— 'Experiments on Air'. *Philosophical Transactions* 74 (1784): 119–53.

Chambers, Ephraim. *Cyclopædia: Or, an Universal Dictionary of Arts and Sciences.* 2 vols. London, 1728.

Clairaut, Alexis Claude. *Tables de la Lune, calculées suivant la théorie de la gravitation universelle.* Paris, 1754.

— *Théorie du mouvement des comètes, dans laquelle on a égard aux altérations que leurs orbites éprouvent par l'action des planètes.* Paris, 1760.

Clarke, Samuel. *A Collection of Papers, Which Passed between the Late Learned Mr Leibnitz, and Dr Clarke, in the Years 1715 and 1716.* London, 1717.

— 'A Letter from the Rev. Dr Samuel Clarke to Mr Benjamin Hoadly, F.R.S. Occasion'd by the Present Controversy Among Mathematicians, concerning the Proportion of Velocity and Force in Bodies in Motion'. *Philosophical Transactions* 35 (1727–28): 381–8.

Clavius, Christopher. *Euclidis elementorum libri XV.* Second edition. Rome, 1589.

[Cleghorn, Robert]. *Sketch of the Character of the Late Thomas Reid, D.D. Professor of Moral Philosophy in the University of Glasgow; with Observations on the Danger of Political Innovation, from a Discourse Delivered on*

28th. Nov. 1794 by Dr Reid, Before the Literary Society in Glasgow College. Glasgow, 1796.

Cochrane, Thomas. *Notes from Doctor Black's Lectures on Chemistry, 1767/8.* Edited by Douglas McKie. Wilmslow, 1966.

Collier, Arthur. *Clavis universalis: Or, a New Inquiry After Truth. Being a Demonstration of the Non-existence, or Impossibility, of an External World.* London, 1713.

Collins, John. *Commercium epistolicum D. Johannis Collins, et aliorum, de analysi promota, jussu Societatis Regiæ in lucem editum.* London, 1722.

Cooper, Anthony Ashley, third earl of Shaftesbury. *Characteristicks of Men, Manners, Opinions, Times* [1711]. Edited by Douglas J. Den Uyl. 3 vols. Indianapolis, 2001.

Cotes, Roger. *Harmonia mensurarum, sive analysis & synthesis per rationum & angulorum mensuras promotæ: Accedunt alia opuscula mathematica.* Cambridge, 1722.

Cramer, Johann Andreas. *Elementa artis docimasticae.* Leiden, 1739.

— *Elements of the Art of Assaying Metals.* London, 1741.

Crawford, Adair. *Experiments and Observations on Animal Heat, and the Inflammation of Combustible Bodies.* London, 1779.

— *Experiments and Observations on Animal Heat, and the Inflammation of Combustible Bodies; Being an Attempt to Resolve These Phenomena into a General Law of Nature.* Second edition. London, 1788.

Cronstedt, Axel Fredrik. *An Essay Towards a System of Mineralogy.* Translated by Gustav von Engstrom. Revised by Emanuel Mendes da Costa. London, 1770.

Cullen, William. 'Of the Cold Produced by Evaporating Fluids, and of Some Other Means of Producing Cold'. In *Essays and Observations, Physical and Literary* (vol. II, pp. 145–56). 3 vols. Edinburgh, 1754–71.

Culpeper, Nicholas. *Pharmacopœia Londinensis: Or, the London Dispensatory Further Adorned by the Studies and Collections of the Fellows Now Living, of the Said College.* London, 1718.

D'Alembert, Jean Le Rond. *Recherches sur differens points importans du systême du monde.* 3 vols. Paris, 1754–56.

— *Recherches sur la précession des equinoxes, et sur la nutation de l'axe de la Terre, dans le systême Newtonien.* Paris, 1749.

Dawson, John. *Four Propositions, &c. Shewing, not only, that the Distance of the Sun, as Attempted to be Determined from the Theory of Gravity, by a Late Author, Is, upon His Own Principles, Erroneous; but also, that It Is More Probable This Capital Question Can Never be Satisfactorily Answered by Any Calculus of the Kind.* Newcastle upon Tyne, 1769.

— ‘Observations on Dr Stewart’s Conclusions, &c. on the Sun’s Distance from the Earth’. *Gentleman’s Magazine* 40 (1770): 452–3.

Dechales, Claude François Milliet. *The Elements of Euclid, Explained and Demonstrated in a New and Most Easy Method.* Translated by Reeve Williams. London, 1703.

Degravers, Peter. *A Complete Physico-medical and Chirurgical Treatise on the Human Eye: And a Demonstration of Natural Vision.* London, 1780.

Deluc, Jean-André. *Récherches sur les modifications de l’atmosphere.* Geneva, 1772.

De Moivre, Abraham. *The Doctrine of Chances: Or, a Method of Calculating the Probability of Events in Play.* London, 1718.

Desaguliers, J. T. ‘An Account of Some Experiments Made to Prove, That the Force of Moving Bodies is Proportionable to Their Velocities: (Or Rather That the *Momentum* of Moving Bodies is to be Found by Multiplying the Masses into the Velocities) in Answer to Such Who Have Sometime Ago Affirm’d, That That Force is Proportionable to the Square of the Velocity, and to Those Who Still Defend the Same Opinion’. *Philosophical Transactions* 32 (1722–23): 269–79.

— ‘An Account of Some Electrical Experiments Made before the Royal Society on Thursday the 16th February 1737–38’. *Philosophical Transactions* 41 (1739–41): 200–8.

— ‘Animadversions upon Some Experiments Relating to the Force of Moving Bodies; with Two New Experiments on the Same Subject’. *Philosophical Transactions* 32 (1722–23): 285–90.

— *A Course of Experimental Philosophy.* 2 vols. London, 1734–44.

— *A Dissertation concerning Electricity.* London, 1742.

— ‘Optical Experiments Made in the Beginning of August 1728, before the President and Several Members of the Royal Society, and Other Gentlemen of Several Nations, upon Occasion of Signior Rizzetti’s Opticks, with an Account of the Said Book’. *Philosophical Transactions* 35 (1727–28): 596–629.

— ‘Some Conjectures concerning Electricity, and the Rise of Vapours’. *Philosophical Transactions* 42 (1742–43): 140–3.

— ‘Some Further Observations concerning Electricity’. *Philosophical Transactions* 42 (1742–43): 14–18.

— ‘Some Things concerning Electricity’. *Philosophical Transactions* 41 (1739–41): 634–7.

— ‘Some Thoughts and Experiments concerning Electricity’. *Philosophical Transactions* 41 (1739–41): 186–93.

Dodson, James. *The Mathematical Repository.* 3 vols. London, 1748–55.

Dollond, John. 'An Account of Some Experiments concerning the Different Refrangibility of Light'. *Philosophical Transactions* 50 (1758): 733–43.

Dollond, Peter. 'A Letter from Mr Peter Dollond, to Nevil Maskelyne, F.R.S. and Astronomer Royal; Describing Some Additions and Alterations Made to Hadley's Quadrant, to Render It More Serviceable at Sea'. *Philosophical Transactions* 62 (1772): 95–8.

Douglas, James, fourteenth earl of Morton, Pierre Charles Le Monnier and James Short. 'An Eclipse of the Sun, July 14. 1748'. *Philosophical Transactions* 45 (1748): 582–97.

Dufay, Charles François de Cisternay. 'A Letter from Mons. Du Fay, F.R.S. and of the Royal Academy of Sciences at Paris, to His Grace Charles Duke of Richmond and Lenox, concerning Electricity'. *Philosophical Transactions* 38 (1733–34): 258–66.

Dunn, Samuel. 'Letter to the Editor'. *Lloyd's Evening Post and British Chronicle*, 1–4 September 1769, p. 221.

Du Séjour, Achille-Pierre Dionus. 'Nouvelles méthodes analytiques pour calculer les éclipses de Soleil, &c.'. In *Histoire de l'Académie royale des sciences. Année M.DCCLXXII. Première partie* (pp. 161–296). Paris, 1775.

— 'Nouvelles méthodes analytiques pour calculer les éclipses de Soleil, les occultations des étoiles fixes et des planètes par la Lune; et en général pour réduire les observations de cet astre, faites à la surface de la Terre, au lieu vu du centre'. In *Histoire de l'Académie royale des sciences. Année M.DCCLXVII* (pp. 137–232). Paris, 1770.

Eames, John. 'A Remark upon the New Opinion Relating to the Forces of Moving Bodies, in the Case of the Collision of Non-elastic Bodies'. *Philosophical Transactions* 34 (1726–27): 183–7.

— 'Remarks upon Some Experiments in Hydraulics, Which Seem to Prove, That the Forces of Equal Moving Bodies Are as the Squares of Their Velocities'. *Philosophical Transactions* 35 (1727–28): 343–7.

Elliott, John. *Philosophical Observations on the Senses of Vision and Hearing; to Which Are Added, a Treatise on Harmonic Sounds and an Essay on Combustion and Animal Heat*. London, 1780.

Emerson, William. *The Principles of Mechanics; Explaining and Demonstrating the General Laws of Motion, the Laws of Gravity, Motion of Descending Bodies, Projectiles, Mechanic Powers, Pendulums, Centers of Gravity, &c. Strength and Stress of Timber, Hydrostatics and Construction of Machines*. London, 1754.

Erskine, David Stuart, earl of Buchan, and Walter Minto. *An Account of the Life, Writings, and Inventions of John Napier, of Merchiston*. Perth, 1787.

Euclid. ΕΥΚΛΕΙΔΟΥ ΤΑ ΣΩΖΟΜΕΝΑ. *Euclidis quae supersunt omnia*. Edited by David Gregory. Oxford, 1703.

— *Sectio canonis*. In *Greek Musical Writings* (vol. II, pp. 191–207). Edited by Andrew Barker. 2 vols. Cambridge, 1984–89.

— *The Thirteen Books of Euclid's Elements*. Translated by Thomas L. Heath. Second edition. 3 vols. Cambridge, 2014.

Euler, Johann Albrecht. 'A Deduction of the Quantity of the Sun's Parallax from the Comparison of the Several Observations of the Late Transit of Venus, Made in Europe, with Those Made in George Island in the South Seas'. *Philosophical Transactions* 62 (1772): 69–76.

Euler, Leonhard. 'Extract of a Letter from Professor Euler, of Berlin, to the Rev. Mr Caspar Wetstein, Chaplain to Her Royal Highness the Princess Dowager of Wales'. *Philosophical Transactions* 47 (1751–52): 263–4.

— *Mechanica sive motus scientia analytice exposita*. 2 vols. St Petersburg, 1736.

— *Opuscula varii argumenti*. Berlin, 1746.

— *Piece qui a remporté le prix de l'Académie royale des sciences en M.DCC. XLVIII. Sur les inégalités du mouvement de Saturne & de Jupiter*. Paris, 1749.

— *Recherches sur les irrégularités du mouvement de Jupiter et de Saturne*. [Place of publication not stated], 1752.

— 'Recherches sur le mouvement de rotation des corps celestes'. In *Histoire de l'Académie royale des sciences et belles-lettres. Année M.DCCLIX* (pp. 265–309). Berlin, 1766.

Ferguson, James. *Astronomy Explained upon Sir Isaac Newton's Principles, and Made Easy to Those Who Have Not Studied Mathematics*. Third edition. London, 1764.

— *A Plain Method of Determining the Parallax of Venus by Her Transit over the Sun: And from Thence, by Analogy, the Parallax and Distance of the Sun, and of All the Rest of the Planets*. Second edition. London, 1761.

Ferner, Benedict. 'An Account of the Observations on the Same Transit [of Venus] Made in and Near Paris'. *Philosophical Transactions* 52 (1761–62): 221–5.

Fisher, T. 'Transit of Venus over the Sun in 1761'. *Gentleman's Magazine* 28 (1758): 367–8.

Fontana, Felice. 'Account of the Airs Extracted from Different Kinds of Waters; with Thoughts on the Salubrity of Air at Different Places'. *Philosophical Transactions* 69 (1779): 432–53.

Fontenelle, Bernard Le Bovier de. *Histoire du renouvellement de l'Académie royale des sciences en M.DC.XCIX. Et les éloges historiques de tous les académiciens morts depuis ce renouvellement: Avec un discours préliminaire sur l'utilité des mathématiques & de la physique*. Paris, 1708.

— *Oeuvres*. New enlarged edition. 12 vols. Amsterdam, 1764.

Fourcroy, Antoine François, comte de. *Elements of Natural History, and of Chemistry.* Translated by William Nicholson. 4 vols. London, 1788.

Franklin, Benjamin. *Experiments and Observations on Electricity, Made at Philadelphia in America.* London, 1751.

— *Experiments and Observations on Electricity, Made at Philadelphia in America* [with] *Supplemental Experiments and Observations on Electricity, Part II.* London, 1751–53.

— *Experiments and Observations on Electricity ... to Which Are Added, Letters and Papers on Philosophical Subjects.* London, 1769.

— *New Experiments and Observations on Electricity.* London, 1754.

— *The Papers of Benjamin Franklin.* Edited by Leonard W. Labaree et al. 41 vols, continuing. New Haven, CT, 1959–.

Freind, John. *Chymical Lectures: In Which Almost All the Operations of Chymistry Are Reduced to Their True Principles, and the Laws of Nature.* London, 1712.

Frisi, Paolo. *De gravitate universali corporum libri tres.* Milan, 1768.

— *De motu diurno Terrae dissertatio.* Pisa, 1756.

Furneaux, Philip. *A Sermon Occasioned by the Death of the Reverend Henry Miles, D.D. and F.R.S.* London, 1763.

Galilei, Galileo. *The Assayer* [1623]. In *Discoveries and Opinions of Galileo* (pp. 229–80). Translated by Stillman Drake. New York, NY, 1957.

— *Two New Sciences: Including Centers of Gravity and Force of Percussion.* Translated by Stillman Drake. Madison, WI, and London, 1974.

Galt, John. *The Last of the Lairds: Or, the Life and Opinions of Malachi Mailings Esq. of Auldbiggings* [1826]. Edited by Ian A. Gordon. Edinburgh and London, 1976.

Geoffroy, Etienne-François. 'Des differents rapports observés en chimie entre differentes substances'. In *Histoire de l'Académie royale des sciences. Année M.DCCXVIII* (pp. 202–12). Paris, 1741.

Gerard, Alexander. *Plan of Education in the Marischal College and University of Aberdeen, with the Reasons of It.* Aberdeen, 1755.

Gordon, Alexander. *Itinerarium septentrionale: Or, a Journey thro' Most of the Counties of Scotland, and Those in the North of England.* London, 1726.

Graham, George, Colin Campbell and James Bradley. 'An Account of Some Observations Made in London ... and at Black-River in Jamaica ... concerning the Going of a Clock; in Order to Determine the Difference between the Lengths of Isochronal Pendulums in Those Places'. *Philosophical Transactions* 38 (1733–34): 302–14.

Gray, Stephen. 'An Account of Some New Electrical Experiments'. *Philosophical Transactions* 31 (1720–21): 104–7.

Gregory, David. *The Elements of Astronomy, Physical and Geometrical.* 2 vols. London, 1715.

— *A Treatise of Practical Geometry.* Edinburgh, 1745.

Gregory, James. *Optica promota, seu abdita radiorum reflexorum & refractorum mysteria, geometrice enucleata; cui subnectitur appendix, subtilissimorum astronomiæ problematωn resolutionem exhibens.* London, 1663.

Gregory, John. *Lectures on the Duties and Qualifications of a Physician.* New edition. London, 1772.

— *The Works of the Late John Gregory, M.D.* 4 vols. Edinburgh, 1788.

Grimaldi, Francesco. *Physico-mathesis de lumine, coloribus et iride, alijsque sequenti pagina indicatis.* Bologna, 1665.

Hales, Stephen. *Vegetable Staticks: Or, an Account of Some Statical Experiments on the Sap in Vegetables: Being an Essay towards a Natural History of Vegetation* [1727]. London, 1969.

Halley, Edmond. 'An Account of Several Experiments Made to Examine the Nature of the Expansion and Contraction of Fluids by Heat and Cold, in Order to Ascertain the Divisions of the Thermometer, and to Make That Instrument, in all Places, without Adjusting by a Standard'. *Philosophical Transactions* 17 (1693): 650–6.

— *Astronomical Tables with Precepts both in English and Latin for Computing the Places of the Sun, Moon, Planets and Comets.* London, 1752.

— 'Dissertation on the Method of Finding the Sun's Parallax and Distance from the Earth, by the Transit of Venus over the Sun's Disc, June the 6th, 1761'. In James Ferguson, *Astronomy Explained upon Sir Isaac Newton's Principles, and Made Easy to Those Who Have Not Studied Mathematics* (pp. 316–27). Third edition. London, 1764.

— 'Methodus singularis quâ Solis parallaxis sive distantia à Terra, ope Veneris intra Solem conspiciendæ, tuto determinari poterit'. *Philosophical Transactions* 29 (1714–16): 454–64.

— *A Synopsis of the Astronomy of Comets.* London, 1705.

— *Tabulæ astronomicæ. Accedunt de usu tabularum præcepta.* London, 1749.

— 'De visibili conjunctione inferiorum planetarum cum Sole, dissertatio astronomica'. *Philosophical Transactions* 16 (1686–92): 511–22.

Harris, John. *Lexicon technicum: Or, an Universal English Dictionary of Arts and Sciences: Explaining not only the Terms of Art, but the Arts Themselves.* Second edition. 2 vols. London, 1708.

Harrison, John. *The Principles of Mr Harrison's Time-keeper, with Plates of the Same.* London, 1767.

— *Remarks on a Pamphlet Lately Published by the Rev. Mr Maskelyne, under the Authority of the Board of Longitude.* London, 1767.

Helsham, Richard. *A Course of Lectures in Natural Philosophy.* London, 1739.

Hermann, Jacob. *Phoronomia, sive de viribus et motibus corporum solidorum et fluidorum libri duo.* Amsterdam, 1716.

Herschel, William. 'Account of Some Observations Tending to Investigate the Construction of the Heavens'. *Philosophical Transactions* 74 (1784): 437–51.

— 'Catalogue of Double Stars'. *Philosophical Transactions* 75 (1785): 40–126.

— 'Catalogue of Double Stars … Communicated by Dr Watson, Jun.' *Philosophical Transactions* 72 (1782): 112–62.

— 'On the Construction of the Heavens'. *Philosophical Transactions* 75 (1785): 213–66.

— 'On the Parallax of the Fixed Stars'. *Philosophical Transactions* 72 (1782): 82–111.

Hill, Sir John. *A Review of the Works of the Royal Society of London; Containing Animadversions on such of the Papers as Deserve Particular Observation.* Second edition. London, 1780.

Home, Francis. *Experiments on Bleaching.* Edinburgh, 1756.

Home, Henry, Lord Kames. *Essays on the Principles of Morality and Natural Religion.* Edinburgh, 1751.

— 'Of the Laws of Motion'. In *Essays and Observations, Physical and Literary* (vol. I, pp. 1–69). 3 vols. Edinburgh, 1754–71.

Hooke, Robert. *An Attempt to Prove the Motion of the Earth from Observations.* London, 1674.

— *The Posthumous Works of Robert Hooke, M.D. S.R.S. Geom. Prof. Gresh. &c. Containing His Cutlerian Lectures and Other Discourses, Read at the Meetings of the Illustrious Royal Society.* London, 1705.

Hornsby, Thomas. 'A Discourse on the Parallax of the Sun'. *Philosophical Transactions* 53 (1763): 467–95.

— 'The Quantity of the Sun's Parallax, as Deduced from the Observations of the Transit of Venus, on June 3, 1769'. *Philosophical Transactions* 61 (1771): 574–9.

Horsley, Samuel. 'A Computation of the Distance of the Sun from the Earth'. *Philosophical Transactions* 57 (1767): 179–85.

— 'On the Computation of the Sun's Distance from the Earth, by the Theory of Gravity'. *Philosophical Transactions* 59 (1769): 153–4.

Hume, David. *The Letters of David Hume.* Edited by J. Y. T. Greig. 2 vols. Oxford, 1932.

— *A Treatise of Human Nature* [1739–40]. Edited by David Fate Norton and Mary J. Norton. Oxford, 2000.

Hunter, John. 'An Account of the Gymnotus electricus'. *Philosophical Transactions* 65 (1775): 395–407.

Hutcheson, Francis. *An Inquiry into the Original of Our Ideas of Beauty and Virtue in Two Treatises* [1725]. Edited by Wolfgang Leidhold. Revised edition. Indianapolis, IN, 2008.

Hutton, Charles. *A Mathematical and Philosophical Dictionary: Containing an Explanation of the Terms, and an Account of the Several Subjects, Comprized under the Heads Mathematics, Astronomy and Philosophy both Natural and Experimental: With an Historical Account of the Rise, Progress and Present State of These Sciences: Also Memoirs of the Lives and Writings of the Most Eminent Authors, both Ancient and Modern, Who by Their Discoveries or Improvements Have Contributed to the Advancement of Them*. 2 vols. London, 1795–96.

— *Mathematical Tables: Containing Common, Hyperbolic and Logistic Logarithms*. London, 1785.

— *A Treatise on Mensuration, Both in Theory and Practice*. Newcastle upon Tyne, 1770.

Ingenhousz, Jan. 'On the Degree of Salubrity of the Common Air at Sea, Compared with That of the Sea-shore, and That of Places Far Removed from the Sea'. *Philosophical Transactions* 70 (1780): 354–77.

— *Experiments upon Vegetables, Discovering Their Great Power of Purifying the Common Air in the Sun-shine, and of Injuring It in the Shade and at Night*. London, 1779.

Irvine, William, and William Irvine, Jr. *Essays, Chiefly on Chemical Subjects*. London, 1805.

Jurin, James. 'An Essay upon Distinct and Indistinct Vision'. In Robert Smith, *A Compleat System of Opticks in Four Books, viz. a Popular, a Mathematical, a Mechanical and a Philosophical Treatise* (vol. II, pp. 115–71). 2 vols. Cambridge, 1738.

— 'An Inquiry into the Measure of the Force of Bodies in Motion: With a Proposal of an *Experimentum crucis*, to Decide the Controversy About It'. *Philosophical Transactions* 43 (1744–45): 423–40.

— *A Reply to Mr Robins's Remarks on the Essay upon Distinct and Indistinct Vision Published at the End of the Dr Smith's Compleat System of Opticks*. London, 1739.

Keill, James. *The Anatomy of the Humane Body Abridg'd: Or, a Short and Full View of All the Parts of the Body*. Second edition. London, 1703.

Keill, John. *Euclid's Elements of Geometry*. Translated and revised by Samuel Cunn. London, 1723.

— *An Introduction to Natural Philosophy: Or, Philosophical Lectures Read in the University of Oxford, Anno Dom. 1700*. London, 1720.

Kepler, Johannes. *Ad Vitellionem paralipomena, quibus astronomiæ pars optica traditur.* Frankfurt, 1604.

Kirwan, Richard. 'Conclusion of the Experiments and Observations concerning the Attractive Powers of the Mineral Acids'. *Philosophical Transactions* 73 (1783): 15–84.

— 'Continuation of the Experiments and Observations on the Specific Gravities and Attractive Powers of Various Saline Substances'. *Philosophical Transactions* 72 (1782): 179–236.

— *Elements of Mineralogy.* London, 1784.

— 'Experiments and Observations on the Specific Gravities and Attractive Powers of Various Saline Substances'. *Philosophical Transactions* 71 (1781): 7–41.

La Caille, Nicolas-Louis de. 'An Account of the Astronomical, Geographical and Physical Observations Made at the Cape of Good Hope, in 1751, 1752 and 1753, by Order of the French King, by the Abbe de la Caille'. *Gentleman's Magazine* 25 (1755): 511–13.

— *Avis aux astronomes, à l'occasion des observations qu'il va faire par ordre du Roi dans l'hémisphère austral.* Paris, 1750.

— *The Elements of Astronomy, Deduced from Observations; and Demonstrated upon the Mathematical Principles of the Newtonian Philosophy: With Practical Rules Whereby the Principal Phenomena Are Determined.* Translated by John Robertson. London, 1750.

— 'Mémoire sur la parallaxe du Soleil, qui résulte de la comparison des observations simultanées de Mars & de Vénus, faites en l'année 1751 en Europe & au cap de Bonne-espérance'. In *Histoire de l'Académie royale des sciences. Année M.DCCLX* (pp. 73–97). Paris, 1766.

— *Tabulæ solares.* Paris, 1758.

Lagrange, Joseph-Louis. 'Théorie de la libration de la Lune, & des autres phénomenes qui dépendent de la figure non sphérique de cette planète'. In *Nouveaux mémoires de l'Académie royale des sciences et belles-lettres. Année M.DCCLXXX* (pp. 203–309). Berlin, 1782.

Lalande, Joseph-Jérôme Lefrançois de. *Astronomie.* Second edition. 4 vols. Paris, 1771–81.

— 'Mémoire sur les équations séculaires, et sur les moyens mouvemens du Soleil, de la Lune, de Saturne, de Jupiter & de Mars, avec les observations de Tycho-Brahé, faites sur Mars en 1593, tirées des manuscrits de cet auteur'. In *Histoire de l'Académie royale des sciences. Année M.DCCLVII* (pp. 411–70). Paris, 1762.

— 'Mémoire sur le flux & le reflux de la mer, & spécialement sur les marées des équinoxes'. In *Histoire de l'Académie royale des sciences. Année M.DCCLXX. Première partie* (pp. 297–324). Paris, 1775.

— 'Mémoire sur la parallaxe du Soleil, déduite des observations faites dans la mer du Sud, dans le royaume d'Astracan, & à la Chine'. In *Histoire de l'Académie royale des sciences. Année M.DCCLXXI* (pp. 776–98). Paris, 1774.

— 'Mémoire sur la parallaxe du Soleil, qui résulte du passage de Vénus, observé en 1769'. In *Histoire de l'Académie royale des sciences. Année M.DCCLXX* (pp. 9–14). Paris, 1773.

Lambert, Johann Heinrich. 'Sur les irrégularités du mouvement de Saturne'. In *Nouveaux mémoires de l'Académie royale des sciences et belles-lettres. Année M.DCCLXXIX* (pp. 244–92). Berlin, 1781.

Landen, John. *Animadversions on Dr Stewart's Computation of the Sun's Distance from the Earth*. London, 1771.

Laplace, Pierre-Simon. 'Mémoire sur les solutions particulières des équations différentielles, & sur les inégalités séculaires des planètes'. In *Histoire de l'Académie royale des sciences. Année M.DCCLXXII. Première partie* (pp. 343–72). Paris, 1775.

— 'Suite des recherches sur plusieurs points du système du monde'. In *Histoire de l'Académie royale des sciences. Année M.DCCLXXVI* (pp. 177–267, 525–52). Paris, 1779.

Lavoisier, Antoine-Laurent. *Elements of Chemistry, in a New Systematic Order, Containing all the Modern Discoveries*. Translated by Robert Kerr. Edinburgh, 1790.

— 'Mémoire sur l'usage de l'esprit-de-vin dans l'analyse des eaux minérales'. In *Histoire de l'Académie royale des science. Année M.DCCLXXII. Seconde partie* (pp. 555–63). Paris, 1776.

— 'Premier mémoire sur la destruction du diamant par le feu'. In *Histoire de l'Académie royale des science. Année M.DCCLXXII. Seconde partie* (pp. 564–91). Paris, 1776.

— 'Second mémoire sur la destruction du diamant, &c.'. In *Histoire de l'Académie royale des science. Année M.DCCLXXII. Seconde partie* (pp. 591–616). Paris, 1776.

Le Gentil de la Galaisière, Guillaume-Joseph-Hyacinthe-Jean-Baptiste. 'Observation de la conjonction inférieure de Vénus avec la Soleil, arrivée le 31 Octobre 1751, faite à l'Observatoire royal de Paris; avec des remarques sur les deux conjonctions écliptiques de cette planète avec le Soleil, qui doivent arriver en 1761 & 1769'. In *Histoire de l'Académie royale des sciences. Année M.DCCLIII* (pp. 27–34). Paris, 1757.

Leibniz, Gottfried Wilhelm. 'A Brief Demonstration of a Notable Error of Descartes and Others concerning a Natural Law, According to Which God Is Said Always to Conserve the Same Quantity of Motion; a Law Which They also Misuse in Mechanics'. In *Philosophical Papers and Letters* (pp. 296–8). Edited and translated by Leroy E. Loemaker. Second edition. Dordrecht, 1969.

Le Roy, David. *A Succinct Account of the Attempts of Mess. Harrison and Le Roy, for Finding the Longitude at Sea, and of the Proofs Made of Their Works.* London, 1768.

Leslie, Patrick Dugud. *A Philosophical Inquiry into the Cause of Animal Heat: With Incidental Observations on Several Phisiological and Chymical Questions, Connected with the Subject.* London, 1778.

L'Hôpital, Guillaume François Antoine, Marquis de. *Analyse des infiniment petits, pour l'intelligence des lignes courbes.* Paris, 1696.

— 'Solution d'un probleme physico-mathématique'. In *Histoire de l'Académie royale des sciences. Année M.DCC.* (pp. 9–21). Second revised and enlarged edition. Paris, 1761.

Locke, John. 'Of the Conduct of the Understanding'. In *Posthumous Works of Mr John Locke* (pp. 3–137). London, 1706.

— *An Essay concerning Human Understanding* [1690]. Edited by Peter H. Nidditch. Oxford, 1975.

Long, Roger. *Astronomy, in Five Books.* 2 vols. Cambridge, 1742.

Macbride, David. *Experimental Essays on the Following Subjects: I. On the Fermentation of Alimentary Mixtures. II. On the Nature and Properties of Fixed Air. III. On the Respective Powers, and Manner of Acting, of the Different Kinds of Antiseptics. IV. On the Scurvy; with a Proposal for Trying New Methods to Prevent or Cure the Same, at Sea. V. On the Dissolvent Power of Quick-lime.* London, 1764.

Machin, John. 'The Laws of the Moon's Motion According to Gravity'. In Sir Isaac Newton, *The Mathematical Principles of Natural Philosophy* (vol. II, pp. 1–71). Translated by Andrew Motte. 2 vols. London, 1729.

Maclaurin, Colin. *An Account of Sir Isaac Newton's Philosophical Discoveries, in Four Books.* London, 1748.

— *The Collected Letters of Colin Maclaurin.* Edited by Stella Mills. Nantwich, 1982.

— *Geometria organica: Sive descriptio linearum curvarum universalis.* London, 1720.

— 'An Observation of the Eclipse of the Sun, on Feb. 18. 1737. Made at Edinburgh'. *Philosophical Transactions* 40 (1737–38): 177–95.

— *A Treatise of Algebra, in Three Parts.* London, 1748.

— *A Treatise of Fluxions. In Two Books.* 2 vols. Edinburgh, 1742.

Malcolm, Alexander. *A New System of Arithmetick, Theorical and Practical.* London, 1730.

— *A Treatise of Musick, Speculative, Practical and Historical.* London, 1730.

Martin, Benjamin. *Philosophia Britannica: Or, a New and Comprehensive System of the Newtonian Philosophy, Astronomy and Geography.* 2 vols. Reading, 1747.

— *Venus in the Sun: Being an Explication of the Rationale of that Great Phœnomenon; of the Several Methods Used by Astronomers for Computing the Quantity and Phases Thereof; and of the Manner of Applying a Transit of Venus over the Solar Disk, for the Discovery of the Parallax of the Sun; Settling the Theory of That Planet's Motion, and Ascertaining the Dimensions of the Solar System.* London, 1761.

Martine, George. *Essays Medical and Philosophical.* London, 1740.

— *Essays and Observations on the Construction and Graduation of Thermometers, and on the Heating and Cooling in Bodies.* Second edition. Edinburgh, 1772.

Maskelyne, Nevil. *An Account of the Going of Mr John Harrison's Watch, at the Royal Observatory, from May 6th, 1766, to March 4th, 1767.* London, 1767.

— 'An Account of Observations Made on the Mountain Schehallien for Finding Its Attraction'. *Philosophical Transactions* 65 (1775): 500–42.

— *Astronomical Observations Made at the Royal Observatory at Greenwich, from the Year MDCCLXV to the Year MDCCLXXIV.* London, 1776.

— *Astronomical Observations Made at the Royal Observatory at Greenwich, in the Years 1765, 1766, 1767, 1768 and 1769.* London, 1774.

— *The Nautical Almanac and Astronomical Ephemeris, for the Year 1767.* London, 1766.

— 'Remarks on the Hadley's Quadrant, Tending Principally to Remove the Difficulties Which Have Hitherto Attended the Use of the Back-observation, and to Obviate the Errors That Might Arise from a Want of Parallelism in the Two Surfaces of the Index-glass'. *Philosophical Transactions* 62 (1772): 99–122.

Maupertuis, Pierre-Louis Moreau de. *La figure de la Terre déterminée par les observations ... faites par ordre du Roy au cercle polaire.* Paris, 1738.

— *The Figure of the Earth, Determined from Observations Made by Order of the French King, at the Polar Circle.* London, 1738.

Mayer, Tobias. 'Novae tabulae motuum Solis et Lunae'. *Commentarii Societatis regiae scientiarum Gottingensis* 2 (1753): 383–430.

Melvill, Thomas. 'A Letter from Mr T. Melvil [*sic*] to the Rev. James Bradley, D.D. F.R.S. with a Discourse concerning the Cause of the Different Refrangibility of the Rays of Light'. *Philosophical Transactions* 48 (1753): 261–70.

— 'Observations on Light and Colours'. In *Essays and Observations, Physical and Literary* (vol. II, pp. 12–90). 3 vols. Edinburgh, 1754–71.

Messier, Charles. 'Mémoire contenant les observations de la Xe. comète observée à Paris, de l'Observatoire de la Marine pendant les mois d'Août, Septembre,

Octobre, Novembre, jusqu'au 1er. Décembre 1769'. In *Histoire de l'Académie royale des sciences. Année M.DCCLXXV* (pp. 392–444). Paris, 1778.

— 'Mémoire contenant les observations de la première comète qui a paru en 1760, & qui est la cinquante-unième dont l'orbite ait été calculée; observée de l'Observatoire de la Marine à Paris, depuis le 8 Janvier jusqu'au 30 du même mois'. In *Histoire de l'Académie royale des sciences. Année M.DCCLXXII. Première partie* (pp. 333–42). Paris, 1775.

Montucla, Jean Étienne. *Histoire des mathématiques, dans laquelle on rend compte de leurs progrès depuis leur origine jusq'à nos jours; où l'on expose le tableau & le développement des principales découvertes, les contestations qu'elles ont fait naître, & les principaux traits de la vie des mathématiciens le plus célebres.* 2 vols. Paris, 1758.

Murdoch, Patrick. 'An Account of the Life and Writings of the Author'. In Colin Maclaurin, *An Account of Sir Isaac Newton's Philosophical Discoveries, in Four Books* (pp. i–xx). London, 1748.

— 'An Essay on the Connexion between the Parallaxes of the Sun and Moon; Their Densities; and Their Disturbing Forces on the Ocean'. *Philosophical Transactions* 58 (1768): 24–33.

— 'Of the Moon's Distance and Parallax'. *Philosophical Transactions* 54 (1764): 29–37.

Musschenbroek, Pieter van. *The Elements of Natural Philosophy.* Translated by John Colson. 2 vols. London, 1744.

— *Essai de physique.* 2 vols. Leiden, 1739–51.

— *Introductio ad philosophiam naturalem.* 2 vols. Leiden, 1762.

Newton, Isaac. 'De analysi per æquationes numero terminorum infinitas'. In *Analysis per quantitatum series, fluxiones, ac differentias: Cum enumeratione linearum tertii ordinis* (pp. 1–21). London, 1711.

— *Analysis per quantitatum series, fluxiones, ac differentias: Cum enumeratione linearum tertii ordinis.* London, 1711.

— *Arithmetica universalis; sive de compositione et resolutione arithmetica. Cum commentario Johannis Castillionei.* 2 vols. Amsterdam, 1761.

— *Arithmetica universalis; sive de compositione et resolutione arithmetica liber.* Cambridge, 1707.

— *The Correspondence of Isaac Newton.* Edited by H. W. Turnbull, A. Rupert Hall, Laura Tilling and J. F. Scott. 7 vols. Cambridge, 1959–77.

— *Isaac Newton's Papers and Letters on Natural Philosophy and Related Documents.* Edited by I. Bernard Cohen. Cambridge, 1958.

— 'A Letter of Mr Isaac Newton, Professor of Mathematicks in the University of Cambridge; Containing His New Theory about Light and Colors'. *Philosophical Transactions* 6 (1671): 3075–87.

— 'Methodus differentialis'. In *Analysis per quantitatum series, fluxiones, ac differentias: Cum enumeratione linearum tertii ordinis* (pp. 93–101). London, 1711.

— *A New and Most Accurate Theory of the Moon's Motion; Whereby All Her Irregularities May Be Solved, and Her Place Truly Calculated to Two Minutes.* London, 1702.

— *Opera quæ exstant omnia.* Edited by Samuel Horsley. 5 vols. London, 1779–85.

— *Opticks: Or, a Treatise of the Reflections, Refractions, Inflections and Colours of Light.* Fourth edition. London, 1730. Reprinted New York, NY, 1952.

— *Opuscula mathematica, philosophica et philologica.* Edited by Johann Castillon. 3 vols. Lausanne and Geneva, 1744.

— *Philosophiæ naturalis principia mathematica.* London, 1687.

— *Principes mathématiques de la philosophie naturelle.* Translated by the Marquise du Châtelet. 2 vols. Paris, 1759.

— *The Principia: Mathematical Principles of Natural Philosophy* [1687]. Translated by I. Bernard Cohen, Anne Whitman and Julia Budenz. Berkeley, CA, Los Angeles, CA, and London, 1999.

— 'Tractatus de quadratura curvarum'. In *Opticks: Or, a Treatise of the Reflexions, Refractions, Inflexions and Colours of Light* (pp. 170–211). London, 1704.

— *Universal Arithmetick: Or, a Treatise of Arithmetical Composition and Resolution. To Which Is Added, Dr Halley's Method of Finding the Roots of Æquations Arithmetically.* Translated by Joseph Raphson and Samuel Cunn. London, 1720.

Nieuwentijdt, Bernard. *The Religious Philosopher: Or, the Right Use of Contemplating the Works of the Creator* [1715]. Translated by John Chamberlayne. 3 vols. London, 1718–19.

Oliver, Thomas. *De sophismatum præstigiis cavendis admonitio....* Cambridge, 1604.

Ozanam, Jacques. *Cursus mathematicus: Or, a Compleat Course of the Mathematicks.* 5 vols. London, 1712.

Papin, Denis. *A New Digester or Engine for Softning Bones, Containing the Description of Its Make and Use in These Particulars: Viz. Cookery, Voyages at Sea, Confectionary, Making of Drinks, Chymistry and Dying.* London, 1681.

Pemberton, Henry. 'Geometrical Solutions of Three Celebrated Astronomical Problems'. *Philosophical Transactions* 62 (1772): 434–46.

— 'A Letter to Dr Mead, Coll. Med. Lond. & Soc. Reg. S. concerning an Experiment, Whereby It Has Been Attempted to Shew the Falsity of the

Common Opinion, in Relation to the Force of Bodies in Motion'. *Philosophical Transactions* 32 (1722–23): 57–68.
— *A View of Sir Isaac Newton's Philosophy.* London, 1728.
Picard, Jean. *Mesure de la terre.* Paris, 1671.
Pingré, Alexandre-Guy. 'Mémoire sur la parallaxe du Soleil, déduite des meilleures observations de la durée du passage de Vénus sur son disque le 3 Juin 1769'. In *Histoire de l'Académie royale des sciences. Année M.DCCLXXII. Première partie* (pp. 398–420). Paris, 1775.
— 'Mémoire sur quelques observations du passage de Vénus. Faites le 6 de Juin 1761, au-delà de l'équateur; & sur les secours qu'on peut en tirer pour la détermination de la parallaxe du Soleil'. In *Histoire de l'Académie royale des sciences. Année M.DCCLXIII* (pp. 354–7). Paris, 1766.
— 'Nouvelle recherche sur la détermination de la parallaxe du Soleil par le passage de Vénus du 6 juin 1761'. In *Histoire de l'Académie royale des sciences. Année M.DCCLXV* (pp. 1–34). Paris, 1768.
Playfair, John. 'Biographical Account of James Hutton, M.D.' In *The Works of John Playfair, Esq.* (vol. IV, pp. 31–118). 4 vols. Edinburgh, 1822.
— *Elements of Geometry; Containing the First Six Books of Euclid, with Two Books on the Geometry of Solids.* Edinburgh, 1795.
Poleni, Giovanni. *De castellis per quae derivantur fluviorum aquae habentibus latera convergentia liber: Quo etiam continentur nova experimenta ad aquas fluentes, & ad percussionis vires pertinentia.* Padua, 1718.
Pope, Alexander. *Poetical Works.* Edited by Herbert Davis with an introduction by Pat Rogers. Oxford and New York, NY, 1985.
Porterfield, William. 'An Essay concerning the Motions of Our Eyes: Of Their External Motions'. In *Medical Essays and Observations, Published by a Society in Edinburgh* (vol. III, pp. 133–214). Third edition. 6 vols. Edinburgh, 1747.
— 'An Essay concerning the Motions of Our Eyes: Of Their Internal Motions'. In *Medical Essays and Observations, Published by a Society in Edinburgh* (vol. IV, pp. 103–236). Third edition. 6 vols. Edinburgh, 1747.
— *A Treatise on the Eye, the Manner and Phænomena of Vision.* 2 vols. Edinburgh, 1759.
Price, Richard. 'A Letter from Richard Price, D.D. F.R.S. to Benjamin Franklin, LL.D. F.R.S. on the Effect of the Aberration of Light on the Time of a Transit of Venus over the Sun'. *Philosophical Transactions* 60 (1770): 536–40.
Priestley, Joseph. *Disquisitions Relating to Matter and Spirit.* Second edition. 2 vols. Birmingham, 1782.
— *Experiments and Observations on Different Kinds of Air.* 3 vols. London, 1774–77.

— 'Experiments Relating to Phlogiston, and the Seeming Conversion of Water into Air'. *Philosophical Transactions* 73 (1783): 398–434.

— *The History and Present State of Discoveries Relating to Vision, Light and Colours*. London, 1772.

— *The History and Present State of Electricity, with Original Experiments*. London, 1767.

— *Institutes of Natural and Revealed Religion*. Second edition. 2 vols. Birmingham, 1782.

— 'Observations on Different Kinds of Air'. *Philosophical Transactions* 62 (1772): 147–264.

Pringle, Sir John. *A Discourse on the Torpedo, Delivered at the Anniversary Meeting of the Royal Society, November 30 1774*. London, 1775.

Proclus. *A Commentary on the First Book of Euclid's Elements*. Translated by Glenn R. Morrow. Princeton, NJ, 1970.

Ptolemy. *Harmonics*. In *Greek Musical Writings* (vol. II, pp. 275–391). Edited by Andrew Barker. 2 vols. Cambridge, 1984–89.

Raphson, Joseph. *Historia fluxionum, sive tractatus originem & progressum peregregiæ istius methodi brevissimo compendio (et quasi synopticè) exhibens*. London, 1715.

— *A Mathematical Dictionary: Or, a Compendious Explication of All Mathematical Terms, Abridg'd from Monsieur Ozanam, and Others*. London, 1702.

Reid, Thomas. *The Correspondence of Thomas Reid*. Edited by Paul Wood. Edinburgh, 2002.

— 'An Essay on Quantity; Occasioned by Reading a Treatise, in Which Simple and Compound Ratio's Are Applied to Virtue and Merit'. *Philosophical Transactions* 45 (1748): 505–20.

— 'An Essay on Quantity: Occasioned by Reading a Treatise, in Which Simple and Compound Ratios Are Applied to Virtue and Merit'. In Charles Hutton, George Shaw and Richard Pearson, *The Philosophical Transactions of the Royal Society of London, from their Commencement, in 1665, to the Year 1800; abridged, with Notes and Biographic Illustrations* (vol. IX, pp. 559–66). 18 vols. London, 1809.

— *Essays on the Active Powers of Man* [1788]. Edited by Knud Haakonssen and James A. Harris. Edinburgh, 2010.

— *Essays on the Intellectual Powers of Man* [1785]. Edited by Derek R. Brookes and Knud Haakonssen. Edinburgh, 2002.

— *An Inquiry into the Human Mind on the Principles of Common Sense* [1764]. Edited by Derek R. Brookes. Edinburgh, 1997.

— *The Philosophical Orations of Thomas Reid Delivered at Graduation Ceremonies in King's College, Aberdeen, 1753, 1756, 1759, 1762*; reproduced in Reid, *Thomas Reid and the University.*

— 'Some Farther Particulars of the Family of the Gregorys and Andersons, Communicated by Dr Thomas Reid, Professor of Moral Philosophy in the University of Glasgow, a Nephew of the Late Dr David Gregory Savilian Professor at Oxford'. In Charles Hutton, *A Mathematical and Philosophical Dictionary: Containing an Explanation of the Terms, and an Account of the several Subjects, Comprized under the Heads Mathematics, Astronomy and Philosophy both Natural and Experimental: With an Historical Account of the Rise, Progress and Present State of These Sciences: Also Memoirs of the Lives and Writings of the Most Eminent Authors, Both Ancient and Modern, Who by Their Discoveries or Improvements Have Contributed to the Advancement of Them* (vol. I, pp. 555–8). 2 vols. London, 1795–96.

— *Thomas Reid on the Animate Creation: Papers Relating to the Life Sciences.* Edited by Paul Wood. Edinburgh, 1995.

— *Thomas Reid on Logic, Rhetoric and the Fine Arts.* Edited by Alexander Broadie. Edinburgh, 2005.

— *Thomas Reid on Society and Politics: Papers and Lectures.* Edited by Knud Haakonssen and Paul Wood. Edinburgh, 2015.

— *Thomas Reid and the University.* Edited by Alexander Broadie and Paul Wood. Edinburgh, forthcoming.

Robins, Benjamin. *A Full Confutation of Dr Jurin's Reply to the Remarks on his Essay upon Distinct and Indistinct Vision.* London, 1740.

— *Remarks on Mr Euler's Treatise of Motion, Dr Smith's Compleat System of Opticks and Dr Jurin's Essay upon Distinct and Indistinct Vision.* London, 1739.

Robinson, Bryan. *A Dissertation on the Æther of Sir Isaac Newton.* Dublin, 1743.

Robinson, Eric, and Douglas McKie (eds). *Partners in Science: Letters of James Watt and Joseph Black.* London, 1970.

Robison, John. 'Impulsion'. In *Supplement to the Third Edition of the Encyclopaedia Britannica* (vol. I, pp. 782–810). Second edition. 2 vols. Edinburgh, 1803.

— 'Mr Wilson on the Solar System'. *Transactions of the Royal Society of Edinburgh* 1 (1788): 28–30.

— 'On the Motion of Light, as Affected by Refracting and Reflecting Substances, Which are Also in Motion'. *Transactions of the Royal Society of Edinburgh* 2 (1790): 83–111.

— *A System of Mechanical Philosophy.* 4 vols. Edinburgh, 1822.

Rowning, John. *A Compendious System of Natural Philosophy*. 2 vols. London, 1753.

Rutherforth, Thomas. *A System of Natural Philosophy, Being a Course of Lectures in Mechanics, Optics, Hydrostatics and Astronomy; Which are Read in St Johns College Cambridge*. 2 vols. Cambridge, 1748.

Saccheri, Girolamo. *Euclid Vindicated from Every Blemish* [1733]. Edited by Vincenzo De Risi. Translated by George Bruce Halstead and Linda Allegri. Heidelberg, 2014.

— *Euclides ab omni nævo vindicatus: Sive conatus geometricus quo stabiliuntur prima ipsa universæ geometriæ principia*. Milan, 1733.

— *Logica demonstrativa*. Turin, 1697.

— *Neo-statica*. Milan, 1708.

Savile, Sir Henry. *Praelectiones tresdecim in principium Elementorum Euclidis*. Oxford, 1621.

Scarburgh, Edmund. *The English Euclide, Being the First Six Elements of Geometry, Translated Out of the Greek, with Annotations and Useful Supplements*. Oxford, 1705.

Scheiner, Christoph. *Oculus hoc est: Fundamentum opticum*. Innsbruck, 1619.

's Gravesande, Willem Jacob. *The Elements of Universal Mathematics, or Algebra: To Which Is Added, a Specimen of a Commentary on Sir Isaac Newton's Universal Arithmetic*. London, 1728.

— 'Essai d'une nouvelle théorie du choc des corps fondée sur l'expérience'. *Le Journal littéraire de la Haye* 12 (1722): 1–54, 190–7.

— *An Explanation of the Newtonian Philosophy, in Lectures Read to the Youth of the University of Leyden*. London, 1735.

— *Mathematical Elements of Natural Philosophy Confirm'd by Experiments, or an Introduction to Sir Isaac Newton's Philosophy*. Translated by J. T. Desaguliers. Second edition. 2 vols. London, 1721–26.

— *Matheoseos universalis elementa*. Leiden, 1727.

Shaw, Peter. *Philosophical Principles of Universal Chemistry: Or, the Foundation of a Scientifical Manner of Inquiring into and Preparing the Natural and Artificial Bodies for the Uses of Life: Both in the Smaller Way of Experiment, and the Larger Way of Business ... Drawn from the Collegium Jenense of Dr George Ernest Stahl*. London, 1730.

Short, James. 'Second Paper concerning the Parallax of the Sun Determined from the Observations of the Late Transit of Venus, in Which This Subject is Treated of More at Length, and the Quantity of the Parallax More Fully Ascertained'. *Philosophical Transactions* 53 (1763): 300–45.

Simpson, Thomas. *Essays on Several Curious and Useful Subjects, in Speculative and Mix'd Mathematicks*. London, 1740.

— *A Treatise of Algebra; Wherein the Fundamental Principles Are Fully and Clearly Demonstrated, and Applied to the Solution of a Great Variety of Problems*. London, 1745.

Simson, Robert. *The Elements of Euclid, viz. the First Six Books, Together with the Eleventh and Twelfth*. Glasgow, 1756.

— *The Elements of Euclid, viz. the First Six Books, Together with the Eleventh and Twelfth*. Second edition. Glasgow, 1762.

— *The Elements of Euclid, viz. the First Six Books, Together with the Eleventh and Twelfth*. Fourth edition. Edinburgh, 1772.

— *The Elements of Euclid, viz. the First Six Books, Together with the Eleventh and Twelfth*. Fifth edition. Edinburgh, 1775.

— *Euclidis elementorum libri priores sex, item undecimus et duodecimus, ex versione Latina Federici Commandini*. Glasgow, 1756.

— *Opera quaedam reliqua*. Glasgow, 1776.

— *Sectionum conicarum libri V.* Edinburgh, 1735.

— *Sectionum conicarum libri quinque*. Second edition. Edinburgh, 1750.

Smith, Robert. *A Compleat System of Opticks in Four Books, viz. a Popular, a Mathematical, a Mechanical and a Philosophical Treatise*. 2 vols. Cambridge, 1738.

— *Harmonics, or the Philosophy of Musical Sounds*. Cambridge, 1749.

Stahl, Georg Ernst. *Experimenta, observationes, animadversiones, CCC numero, chymicae et physicae*. Berlin, 1731.

Stay, Benedetto. *Philosophiae recentioris a Benedicto Stay ... versibus traditae libri X ... cum adnotationibus, et supplementis P. Rogerii Josephi Boscovich S.J.* 2 vols. Rome, 1755–60.

Stewart, Dugald. *Account of the Life and Writings of Thomas Reid, D.D. F.R.S. Edin. Late Professor of Moral Philosophy in the University of Glasgow*. Edinburgh, 1803.

Stewart, John. *Sir Isaac Newton's Two Treatises of the Quadrature of Curves, and Analysis by Equations of an Infinite Number of Terms, Explained*. London, 1745.

Stewart, John, M.D. 'Some Remarks on the Laws of Motion, and the Inertia of Matter'. In *Essays and Observations, Physical and Literary* (vol. I, pp. 70–140). 3 vols. Edinburgh, 1754–71.

Stewart, Matthew. *The Distance of the Sun from the Earth Determined, by the Theory of Gravity*. Edinburgh, 1763.

— 'A Solution of Kepler's Problem'. In *Essays and Observations, Physical and Literary* (vol. II, pp. 105–44). 3 vols. Edinburgh, 1754–71.

— *Tracts, Physical and Mathematical*. Edinburgh, 1761.

Stirling, James. *Methodus differentialis: Sive tractatus de summatione et interpolatione serierum infinitarum*. London, 1730.

Stone, Edmund. *Euclid's Elements. Volume 2*. London, 1731.

— *Euclid's Elements of Geometry, the First Six, the Eleventh and Twelfth Books; Translated into English, from Dr Gregory's Edition, with Notes and Additions*. London, 1752.

Symmer, Robert. 'New Experiments and Observations concerning Electricity'. *Philosophical Transactions* 51 (1759–60): 340–93.

Taylor, Brook. 'An Account of an Experiment, Made to Ascertain the Proportion of the Expansion of the Liquor in the Thermometer, with Regard to the Degrees of Heat'. *Philosophical Transactions* 32 (1722–23): 291.

Trail, William. *Account of the Life and Writings of Robert Simson, M.D. Late Professor of Mathematics in the University of Glasgow*. Bath, 1812.

Turnbull, George. 'On the Association of Natural Science with Moral Philosophy'. In *Education for Life: Correspondence and Writings on Religion and Practical Philosophy* (pp. 45–57). Edited by M. A. Stewart and Paul Wood. Indianapolis, IN, 2014.

— *Observations upon Liberal Education, in All Its Branches* [1742]. Edited by Terrence O. Moore, Jr. Indianapolis, IN, 2003.

— *The Principles of Moral and Christian Philosophy* [1740]. Edited by Alexander Broadie. 2 vols. Indianapolis, IN, 2005.

Ulman, H. Lewis (ed.). *The Minutes of the Aberdeen Philosophical Society, 1758–1773*. Aberdeen, 1990.

Van Schooten, Frans. *De organica conicarum sectionum in plano descriptione, tractatus*. Leiden, 1646.

Vilant, Nicolas. 'An Easy Method of Extracting the Cubic Roots of Binomials, whether Possible or Impossible'. *Scots Magazine* 21 (1759): 121–3, 189–90.

Virgil. *Eclogues, Georgics and Aeneid and the Minor Poems*. Translated by H. Rushton Fairclough. Revised edition. 2 vols. Cambridge, MA, and London, 1986.

Wallis, John. *Mathesis universalis; seu opus arithmeticum* [1657]. In *Opera mathematica* (vol. I, pp. 11–228). 3 vols. Oxford, 1693–99.

— *Opera mathematica*. 3 vols. Oxford, 1693–99.

— *A Treatise of Algebra, Both Historical and Practical*. London, 1685.

Watt, James. 'Sequel to the Thoughts on the Constituent Parts of Water and Dephlogisticated Air'. *Philosophical Transactions* 74 (1784): 354–7.

— 'Thoughts on the Constituent Parts of Water and of Dephlogisticated Air; with an Account of Some Experiments on That Subject'. *Philosophical Transactions* 74 (1784): 329–53.

Wells, William Charles. *An Essay upon Single Vision with Two Eyes: Together with Experiments and Observations on Several Other Subjects in Optics*. London, 1792.

West, William. *Mathematics*. Revised by John Rowe. London, 1762.

Whiston, William. *Astronomical Lectures, Read in the Publick Schools at Cambridge*. Second edition. London, 1728.

— *Prælectiones astronomicæ ... Quibus accedunt tabulæ plurimæ astronomicæ Flamstedianæ correctæ, Halleianæ, Cassinianæ et Streetianæ*. Cambridge, 1707.

Whytt, Robert. 'Of the Various Strength of Different Lime-waters'. In *Essays and Observations, Physical and Literary* (vol. I, pp. 372–85). 3 vols. Edinburgh, 1754–71.

Wilson, Alexander. 'An Account of the Remarkable Cold Observed at Glasgow, in the Month of January, 1768'. *Philosophical Transactions* 61 (1771): 326–31.

— 'Eclipses of Jupiter's First Satellite, with an Eighteenth Inch Reflector of Mr Short's. Observed by Dr Wilson at the Glasgow Observatory'. *Philosophical Transactions* 59 (1769): 402–3.

— 'Observations on the Solar Spots'. *Philosophical Transactions* 64 (1774): 1–30.

— 'Observations of the Transit of Venus over the Sun, Contained in a Letter to the Reverend Nevil Maskeleyne, Astronomer Royal'. *Philosophical Transactions* 59 (1769): 333–8.

— *Thoughts on General Gravitation, and Views Thence Arising as to the State of the Universe*. London, 1777.

Wilson, Patrick. 'An Account of a Most Extraordinary Degree of Cold at Glasgow in January Last; Together with Some New Experiments and Observations on the Comparative Temperature of Hoar-Frost and the Air Near to It, Made at the Macfarlane Observatory Belonging to the College'. *Philosophical Transactions* 70 (1780): 451–73.

— 'Biographical Account of Alexander Wilson, M.D. Late Professor of Practical Astronomy in Glasgow'. *Edinburgh Journal of Science* 10 (1829): 1–17.

— 'An Experiment Proposed for Determining, by the Aberration of the Fixed Stars, whether the Rays of Light, in Pervading Different Media, Change Their Velocity According to the Law Which Results from Sir Isaac Newton's Ideas concerning the Cause of Refraction; and for Ascertaining Their Velocity in Every Medium Whose Refractive Density Is Known'. *Philosophical Transactions* 72 (1782): 58–70.

— 'An Experiment Proposed for Determining, by the Aberration of the Fixed Stars, whether the Rays of Light, in Pervading Different Media, Change Their

Velocity According to the Law Which Results from Sir Isaac Newton's Ideas concerning the Cause of Refraction; and for Ascertaining Their Velocity in Every Medium Whose Refractive Density Is Known'. In *The Philosophical Transactions of the Royal Society of London, from their Commencement, in 1665, to the Year 1800; Abridged, with Notes and Biographic Illustrations* (vol. XV, pp. 184–95). Charles Hutton, George Shaw and Richard Pearson. 18 vols. London, 1809.

X. 'To the Printer of the London Chronicle'. *London Chronicle*, 1–4 December 1770, p. 532.

X.Y.Z. 'Differences in the Observations of the Transit of Venus'. *Gentleman's Magazine* 31 (1761): 317–19.

Secondary Sources

Addison, W. Innes. *The Matriculation Albums of the University of Glasgow, from 1728 to 1858*. Glasgow, 1913.

— *A Roll of the Graduates of the University of Glasgow, from 31st December 1727 to 31st December 1897*. Glasgow, 1898.

Anderson, Peter John. *Fasti Academiae Mariscallanae Aberdonensis: Selections from the Records of the Marischal College and University, 1593–1860*. 3 vols. Aberdeen, 1889–98.

— *Officers and Graduates of University and King's College Aberdeen, 1495–1860*. Aberdeen, 1893.

Anderson, Robert G. W. 'Boerhaave to Black: The Evolution of Chemistry Teaching'. *Ambix* 53 (2006): 237–54.

— *The Playfair Collection and the Teaching of Chemistry at the University of Edinburgh, 1713–1858*. Edinburgh, 1978.

— 'Thomas Charles Hope and the Limiting Legacy of Joseph Black'. In *Cradle of Chemistry: The Early Years of Chemistry at the University of Edinburgh* (pp. 147–62). Edited by Robert G. W. Anderson. Edinburgh, 2015.

Bechler, Zev. '"A Less Agreeable Matter": The Disagreeable Case of Newton and Achromatic Refraction'. *British Journal for the History of Science* 8 (1975): 101–26.

Bennett, Jim. 'The Travels and Trials of Mr Harrison's Timekeeper'. In *Instruments, Travel and Science: Itineraries of Precision from the Seventeenth to the Twentieth Century* (pp. 75–95). Edited by Marie-Noëlle Bourguet, Christian Licoppe and Heinz Otto Sibum. London, 2002.

Biagioli, Mario. 'The Social Status of Italian Mathematicians, 1450–1600'. *History of Science* 27 (1989): 41–95.

Boantza, Victor. 'The Phlogistic Role of Heat in the Chemical Revolution and the Origins of Kirwan's "Ingenious Modifications ... into the Theory of Phlogiston'. *Annals of Science* 65 (2008): 309–38.

Bortolotto, Andréa. 'Johann Andreas Cramer and Chemical Mineral Assay in the Eighteenth Century'. *Ambix* 62 (2015): 312–32.

Bryden, D. J. 'James Watt, Merchant: The Glasgow Years, 1754–1774'. In *Perceptions of Great Engineers: Fact and Fantasy* (pp. 9–21). Edited by Denis Smith. London, 1994.

Callergård, Robert. *An Essay on Thomas Reid's Philosophy of Science*. Stockholm, 2006.

Campbell Fraser, Alexander. *Thomas Reid*. Edinburgh and London, [1898].

Cantor, Geoffrey. 'Berkeley, Reid and the Mathematization of Mid-Eighteenth-Century Optics'. *Journal of the History of Ideas* 38 (1977): 429–48.

— 'Henry Brougham and the Scottish Methodological Tradition'. *Studies in History and Philosophy of Science* 2 (1971): 69–89.

— *Optics after Newton: Theories of Light in Britain and Ireland, 1704–1840*. Manchester, 1983.

Christie, John R. R. 'Ether and the Science of Chemistry: 1740–1790'. In *Conceptions of Ether: Studies in the History of Ether Theories, 1740–1900* (pp. 85–110). Edited by G. N. Cantor and M. J. S. Hodge. Cambridge, 1981.

— 'The Origins and Development of the Scottish Scientific Community, 1680–1760'. *History of Science* 12 (1974): 122–41.

Clarke, David. *Reflections on the Astronomy of Glasgow: A Story of Some Five Hundred Years*. Edinburgh, 2013.

Clericuzio, Antonio. '"Sooty Empiricks" and Natural Philosophers: The Status of Chemistry in the Seventeenth Century'. *Science in Context* 23 (2010): 329–50.

Cohen, I. Bernard. *Franklin and Newton: An Inquiry into Speculative Newtonian Experimental Science and Franklin's Work in Electricity as an Example Thereof*. Philadelphia, PA, 1956.

Cook, Alan. *Edmond Halley: Charting the Heavens and the Seas*. Oxford, 1998.

Coutts, James. *A History of the University of Glasgow: From Its Foundation in 1451 to 1909*. Glasgow, 1909.

Crosland, Maurice P. 'The Use of Diagrams as Chemical "Equations" in the Lecture Notes of William Cullen and Joseph Black'. *Annals of Science* 15 (1959): 75–90.

Daniels, Norman. *Thomas Reid's 'Inquiry': The Geometry of Visibles and the Case for Realism*. Stanford, CA, 1989.

Darrigol, Olivier. *A History of Optics: From Greek Antiquity to the Nineteenth Century*. Oxford, 2012.

Daston, Lorraine. *Classical Probability in the Enlightenment.* Princeton, NJ, 1988.

Davie, George Elder. *The Democratic Intellect: Scotland and Her Universities in the Nineteenth Century.* Third edition. Edinburgh, 2013.

Dea, Shannon. 'Thomas Reid's Rigourised Anti-hypotheticalism'. *Journal of Scottish Philosophy* 3 (2005): 123–38.

Dobbin, Leonard. 'A Cullen Chemical Manuscript of 1753'. *Annals of Science* 1 (1936): 138–56.

Donovan, Arthur L. *Philosophical Chemistry in the Scottish Enlightenment: The Doctrines and Discoveries of William Cullen and Joseph Black.* Edinburgh, 1975.

— 'Scottish Responses to the New Chemistry of Lavoisier'. In *Studies in Eighteenth-Century Culture* (vol. IX, pp. 237–49). Edited by Roseann Runte. Madison, WI, and London, 1979.

— 'William Cullen and the Research Tradition of Eighteenth-Century Scottish Chemistry'. In *The Origins and Nature of the Scottish Enlightenment* (pp. 98–114). Edited by R. H. Campbell and Andrew S. Skinner. Edinburgh, 1982.

Duncan, Alexander. *Memorials of the Faculty of Physicians and Surgeons of Glasgow, 1599–1850: With a Sketch of the Rise and Progress of the Glasgow Medical School and of the Medical Profession in the West of Scotland.* Glasgow, 1896.

Duncan, Alistair M. 'The Functions of Affinity Tables and Lavoisier's List of Elements'. *Ambix* 17 (1970): 28–42.

Eddy, Matthew D. 'The "Doctrine of Salts" and Rev. John Walker's Analysis of a Scottish Spa (1749–1761)'. *Ambix* 48 (2001): 137–60.

— 'The Interactive Notebook: How Students Learned to Keep Notes During the Scottish Enlightenment'. *Book History* 19 (2016): 86–131.

Eisenstaedt, Jean. 'Light and Relativity, a Previously Unknown Eighteenth-Century Manuscript by Robert Blair (1748–1828)'. *Annals of Science* 62 (2005): 347–76.

— 'L'optique balistique Newtonienne à l'épreuve des satellites de Jupiter'. *Archive for History of Exact Sciences* 50 (1996): 117–56.

Emerson, Roger L. *Academic Patronage in the Scottish Enlightenment: Glasgow, Edinburgh and St Andrews Universities.* Edinburgh, 2008.

— *Neglected Scots: Eighteenth Century Glaswegians and Women.* Edinburgh, 2015.

— 'The Philosophical Society of Edinburgh, 1737–1747'. *British Journal for the History of Science* 12 (1979): 154–91.

— 'The Philosophical Society of Edinburgh, 1748–1768'. *British Journal for the History of Science* 14 (1981): 133–76.

— *Professors, Patronage and Politics: The Aberdeen Universities in the Eighteenth Century*. Aberdeen, 1992.

— and Paul Wood. 'Science and Enlightenment in Glasgow, 1690–1802'. In *Science and Medicine in the Scottish Enlightenment* (pp. 79–142). Edited by Charles W. J. Withers and Paul Wood. East Linton, 2002.

Eyles, V. A. 'The Evolution of a Chemist: Sir James Hall, Bt., F.R.S., P.R.S.E., of Dunglass, Haddingtonshire, (1761–1832), and His Relations with Joseph Black, Antoine Lavoisier and Other Scientists of the Period'. *Annals of Science* 19 (1963): 153–82.

Falkenstein, Lorne. 'Reid and Smith on Vision'. *Journal of Scottish Philosophy* 2 (2004): 103–18.

Feingold, Mordechai. 'A Jesuit among Protestants: Boscovich in England *c*. 1745–1820'. In *R. J. Boscovich: His Life and Scientific Work* (pp. 511–26). Edited by Piers Bursill-Hall. Rome, 1993.

Forbes, Eric Gray. 'Tobias Mayer (1723–62): A Case of Forgotten Genius'. *British Journal for the History of Science* 5 (1970): 1–20.

Fraser, Kevin J. 'John Hill and the Royal Society in the Eighteenth Century'. *Notes and Records of the Royal Society of London* 48 (1994): 43–67.

Gascoigne, John. *Cambridge in the Age of the Enlightenment: Science, Religion and Politics from the Restoration to the French Revolution*. Cambridge, 1989.

— *Joseph Banks and the English Enlightenment: Useful Knowledge and Polite Culture*. Cambridge, 1994.

Gaskell, Philip. *A Bibliography of the Foulis Press*. Second edition. Winchester, 1986.

Gavine, David Myles. 'Astronomy in Scotland, 1745–1900'. PhD thesis, Open University, 1981.

Glass, I. S. *Nicolas-Louis de La Caille: Astronomer and Geodesist*. Oxford, 2013.

Golinski, Jan. 'Chemistry'. In *The Cambridge History of Science, Volume 4: Eighteenth-Century Science* (pp. 375–96). Edited by Roy Porter. Cambridge, 2003.

— '"Fit Instruments": Thermometers in Eighteenth-Century Chemistry'. In *Instruments and Experimentation in the History of Chemistry* (pp. 185–210). Edited by Frederic L. Holmes and Trevor H. Levere. Cambridge, MA, 2000.

— *Science as Public Culture: Chemistry and Enlightenment in Britain, 1760–1820*. Cambridge, 1992.

Grabiner, Judith V. 'Maclaurin and Newton: The Newtonian Style and the Authority of Mathematics'. In *Science and Medicine in the Scottish Enlightenment* (pp. 143–71). Edited by Charles W. J. Withers and Paul Wood. East Linton, 2002.

— 'Was Newton's Calculus a Dead End? The Continental Influence of Maclaurin's *Treatise of Fluxions*'. *American Mathematical Monthly* 104 (1997): 393–410.

Grandi, Giovanni B. 'Thomas Reid's Geometry of Visibles and the Parallel Postulate'. *Studies in History and Philosophy of Science* 36 (2005): 79–103.

Grant, Robert. *History of Physical Astronomy, from the Earliest Ages to the Middle of the Nineteenth Century.* London, 1852.

Greenberg, John L. *The Problem of the Earth's Shape from Newton to Clairaut: The Rise of Mathematical Science in Eighteenth-Century Paris and the Fall of 'Normal' Science.* Cambridge, 1995.

Guerlac, Henry. 'Joseph Black's Work on Heat'. In *Joseph Black, 1728–1799: A Commemorative Symposium* (pp. 13–22). Edited by A. D. C. Simpson. Edinburgh, 1982.

Guicciardini, Niccolò. *The Development of Newtonian Calculus in Britain, 1700–1800.* Cambridge, 1989.

— 'Dot-Age: Newton's Mathematical Legacy in the Eighteenth Century'. *Early Science and Medicine* 9 (2004): 218–56.

— 'Editing Newton in Geneva and Rome: The Annotated Edition of the *Principia* by Calandrini, Le Seur and Jacquier'. *Annals of Science* 72 (2015): 337–80.

— *Reading the Principia: The Debate on Newton's Mathematical Methods for Natural Philosophy from 1687 to 1736.* Cambridge, 1999.

— 'Thomas Reid's Mathematical Manuscripts: A Survey'. *Reid Studies* 4 (2001): 71–86.

Hall, A. Rupert. 'Beyond the Fringe: Diffraction as Seen by Grimaldi, Fabri, Hooke and Newton'. *Notes and Records of the Royal Society of London* 44 (1990): 13–23.

Hankins, Thomas L. 'Eighteenth-Century Attempts to Resolve the *Vis viva* Controversy'. *Isis* 56 (1965): 281–97.

— *Science and the Enlightenment.* Cambridge, 1985.

Harris, Bob. *The Scottish People and the French Revolution.* London, 2008.

Heathcote, Niels H. de V. 'The Early Meaning of Electricity: Some *Pseudodoxia Epidemica*'. *Annals of Science* 23 (1967): 261–75.

Heilbron, J. L. *Electricity in the Seventeenth and Eighteenth Centuries: A Study of Early Modern Physics.* Berkeley, CA, Los Angeles, CA, and London, 1979.

— *Galileo.* Oxford, 2010.

— 'A Mathematicians' Mutiny, With Morals'. In *World Changes: Thomas Kuhn and the Nature of Science* (pp. 81–129). Edited by Paul Horwich. Cambridge, MA, and London, 1993.

Heimann, P. M. 'Ether and Imponderables'. In *Conceptions of Ether: Studies in the History of Ether Theories, 1740–1900* (pp. 61–83). Edited by G. N. Cantor and M. J. S. Hodge. Cambridge, 1981.

Henry, John. *The Scientific Revolution and the Origins of Modern Science.* Second edition. Basingstoke and New York, 2002.

Hill, Katherine. '"Juglers or Schollers?": Negotiating the Role of a Mathematical Practitioner'. *British Journal for the History of Science* 31 (1998): 253–74.

Hills, Richard L. *James Watt, Volume I: His Time in Scotland, 1736–1774*. Ashbourne, 2002.

Homann, Frederick A., S.J. 'Christopher Clavius and the Renaissance of Euclidean Geometry'. *Archivum historicum Societatis Iesu* 52 (1983): 233–46.

Home, Roderick W. 'Franklin's Electrical Atmospheres'. *British Journal for the History of Science* 6 (1972): 131–51.

Howell, Wilbur Samuel. *Eighteenth-Century British Logic and Rhetoric*. Princeton, NJ, 1971.

Howse, Derek. *Nevil Maskelyne: The Seaman's Astronomer*. Cambridge, 1989.

Hufbauer, Karl. *The Formation of the German Chemical Community (1720–1795)*. Berkeley, CA, Los Angeles, CA, and London, 1982.

Humphries, Walter R. 'The Philosopher, the Farmer and Commercial Education'. *Scottish Educational Journal* 14 and 21 May 1937, 619–20, 661–3.

Iliffe, Rob. '"Aplatisseur du monde et de Cassini": Maupertuis, Precision Measurement and the Shape of the Earth in the 1730s'. *History of Science* 31 (1993): 335–75.

Iltis, Carolyn. 'The Leibnizian–Newtonian Debates: Natural Philosophy and Social Psychology'. *British Journal for the History of Science* 6 (1973): 343–77.

Jesseph, Douglas M. *Berkeley's Philosophy of Mathematics*. Chicago, IL, and London, 1993.

Johnston, Stephen. 'Mathematical Practitioners and Instruments in Elizabethan England'. *Annals of Science* 48 (1991): 319–44.

Kuhn, Thomas S. 'Mathematical Versus Experimental Traditions in the Development of Physical Science'. In Thomas S. Kuhn, *The Essential Tension: Selected Studies in Scientific Tradition and Change* (pp. 31–65). Chicago, IL, and London, 1977.

Laudan, L. L. 'The Medium and Its Message: A Study of Some Philosophical Controversies About Ether'. In *Conceptions of Ether: Studies in the History of Ether Theories, 1740–1900* (pp. 157–85). Edited by G. N. Cantor and M. J. S. Hodge. Cambridge, 1981.

— 'Thomas Reid and the Newtonian Turn of British Methodological Thought'. In *The Methodological Heritage of Newton* (pp. 103–31). Edited by Robert E. Butts and John W. Davis. [Toronto], 1970.

— 'The *Vis viva* Controversy, a Post-mortem'. *Isis* 59 (1968): 131–43.

Lenman, B. P., and J. B. Kenworthy. 'Dr David Skene, Linnaeus and the Applied Geology of the Scottish Enlightenment'. *Aberdeen University Review* 47 (1977–78): 32–44.

— 'Dr David Skene and the Applied Geology of the Scottish Enlightenment II: Skene's Study of Contemporary Coalmining Practice'. *Aberdeen University Review* 47 (1977–78): 231–7.

Lindberg, David C. *Theories of Vision from al-Kindi to Kepler*. Chicago, IL, 1976.

Lubbock, Constance A. *The Herschel Chronicle: The Life-story of William Herschel and His Sister Caroline Herschel*. Cambridge, 1933.

Macdonald, D. C. *Birthright in Land*. London, 1891.

Melhado, Evan M. 'Mineralogy and the Autonomy of Chemistry Around 1800'. In *Lychnos 1990: Annual of the Swedish History of Science Society* (pp. 229–62). Uppsala, 1990.

— 'Oxygen, Phlogiston and Caloric: The Case of Guyton'. *Historical Studies in the Physical Sciences* 13 (1983): 311–34.

Mendelsohn. Everett. *Heat and Life: The Development of the Theory of Animal Heat*. Cambridge, MA, 1964.

Millburn, John R., and Henry C. King. *Wheelwright of the Heavens: The Life and Work of James Ferguson, FRS*. London, 1988.

Morris, Robert J. 'Lavoisier and the Caloric Theory'. *British Journal for the History of Science* 6 (1972): 1–38.

Murray, David. *Memories of the Old College of Glasgow: Some Chapters in the History of the University*. Glasgow, 1927.

Nolan, J. Bennet. *Benjamin Franklin in Scotland and Ireland, 1759 and 1771*. Philadelphia, PA, 1956.

Nyden, Tammy. 'Living Force at Leiden: De Volder, 's Gravesande and the Reception of Newtonianism'. In *Newton and Empiricism* (pp. 207–22). Edited by Zvi Biener and Eric Schliesser. Oxford and New York, NY, 2014.

Olson, Richard. 'The Reception of Boscovich's Ideas in Scotland'. *Isis* 60 (1969): 91–103.

— *Scottish Philosophy and British Physics, 1750–1880: A Study in the Foundations of the Victorian Scientific Style*. Princeton, NJ, 1975.

— 'Scottish Philosophy and Mathematics, 1750–1830'. *Journal of the History of Ideas* 32 (1971): 29–44.

Pagli, Paolo. 'Two Unnoticed Editions of Girolamo Saccheri's *Logica demonstrativa*'. *History and Philosophy of Logic* 30 (2009): 331–40.

Pannekoek, Anton. *A History of Astronomy*. London, 1961.

Pedersen, Kurt Møller. 'Roger Joseph Boscovich and John Robison on Terrestrial Aberration'. *Centaurus* 24 (1980): 335–45.

— 'Water-filled Telescopes and the Pre-history of Fresnel's Ether Dragging'. *Archive for History of Exact Sciences* 54 (2000): 499–564.

Perrin, Carleton E. 'Joseph Black and the Absolute Levity of Phlogiston'. *Annals of Science* 40 (1983): 109–37.

— 'A Reluctant Catalyst: Joseph Black and the Edinburgh Reception of Lavoisier's Chemistry'. *Ambix* 29 (1982): 141–76.

Ponting, Betty. 'Mathematics at Aberdeen: Developments, Characters and Events, 1717–1860'. *Aberdeen University Review* 48 (1979–80): 162–76.

Porter, Theodore M. 'The Promotion of Mining and the Advancement of Science: The Chemical Revolution of Mineralogy'. *Annals of Science* 38 (1981): 543–70.

Proverbio, Edoardo (ed.). *Catalogo delle opere a stampa di Ruggiero Giuseppe Boscovich (1711–1787)*. Rome, 2007.

Powers, John C. 'Measuring Fire: Herman Boerhaave and the Introduction of Thermometry into Chemistry'. *Osiris* 29 (2014): 158–77.

Pycior, Helena M. *Symbols, Impossible Numbers and Geometric Entanglements: British Algebra through the Commentaries on Newton's* Universal Arithmetick. Cambridge, 1997.

Rait, Robert Sangster. *The Universities of Aberdeen: A History*. Aberdeen, 1895.

Richards, Robert J. 'Christian Wolff's Prolegomena to Empirical and Rational Psychology: Translation and Commentary'. *Proceedings of the American Philosophical Society* 124 (1980): 227–39.

Ross, Ian Simpson. *Lord Kames and the Scotland of His Day*. Oxford, 1972.

Schofield, Robert E. *Mechanism and Materialism: British Natural Philosophy in an Age of Reason*. Princeton, NJ, 1970.

Seargent, David. *The Greatest Comets in History: Broom Stars and Celestial Scimitars*. New York, NY, 2009.

Shapiro, Barbara J. *Probability and Certainty in Seventeenth-Century England: A Study of the Relationships between Natural Science, Religion, History, Law and Literature*. Princeton, NJ, 1983.

Shepherd, Christine M. 'Newtonianism in Scottish Universities in the Seventeenth Century'. In *The Origins and Nature of the Scottish Enlightenment* (pp. 65–85). Edited by R. H. Campbell and Andrew S. Skinner. Edinburgh, 1982.

Sher, Richard B. *Church and University in the Scottish Enlightenment: The Moderate Literati of Edinburgh*. Edinburgh, 1985.

Siegfried, Robert. *From Elements to Atoms: A History of Chemical Composition*. Transactions of the American Philosophical Society, vol. 92, part 4. Philadelphia, PA, 2002.

Siegfried, Robert, and Betty Jo Dobbs. 'Composition, a Neglected Aspect of the Chemical Revolution'. *Annals of Science* 24 (1968): 275–93.

Smeaton, W. A. *Fourcroy: Chemist and Revolutionary, 1755–1809*. Cambridge, 1962.

— 'Some Large Burning Lenses and Their Use by Eighteenth-Century French and British Chemists'. *Annals of Science* 44 (1987): 265–76.

Smith, A. Mark. *From Sight to Light: The Passage from Ancient to Modern Optics*. Chicago, IL, and London, 2015.

Smith, J. H. *The Gordon's Mill Farming Club, 1758–1764*. Edinburgh and London, 1962.

Smout, T. C. *A History of the Scottish People, 1560–1830*. Glasgow, 1979.

Sorrenson, Richard. 'Dollond & Son's Pursuit of Achromaticity, 1758–1789'. *History of Science* 39 (2001): 31–55.

Stewart, Larry. *The Rise of Public Science: Rhetoric, Technology and Natural Philosophy in Newtonian Britain, 1660–1750*. Cambridge, 1992.

Taylor, E. G. R. *The Mathematical Practitioners of Hanoverian England, 1714–1840*. Cambridge, 1966.

— *The Mathematical Practitioners of Tudor and Stuart England, 1485–1714*. Cambridge, 1970.

Taylor, Georgette. 'Marking Out a Disciplinary Common Ground: The Role of Chemical Pedagogy in Establishing the Doctrine of Affinity at the Heart of British Chemistry'. *Annals of Science* 65 (2008): 465–86.

Terrall, Mary. *The Man Who Flattened the Earth: Maupertuis and the Sciences in the Enlightenment*. Chicago, IL, and London, 2002.

— '*Vis viva* Revisited'. *History of Science* 42 (2004): 189–209.

Thackray, Arnold. *Atoms and Powers: An Essay on Newtonian Matter Theory and the Development of Chemistry*. Cambridge, MA, and London, 1970.

Van Helden, Albert. *Measuring the Universe: Cosmic Dimensions from Aristarchus to Halley*. Chicago, IL, and London, 1985.

Van Leeuwen, Henry G. *The Problem of Certainty in English Thought, 1630–1690*. Second edition. The Hague, 1970.

Walker, W. Cameron. 'Animal Electricity before Galvani'. *Annals of Science* 2 (1937): 84–113.

Wallis, Ruth. 'John Bevis, M.D., F.R.S. (1695–1771): Astronomer Loyal'. *Notes and Records of the Royal Society of London* 36 (1982): 211–25.

Walters, Alice N. 'Ephemeral Events: English Broadsides of Early Eighteenth-Century Solar Eclipses'. *History of Science* 37 (1999): 1–43.

Williams, Mari. 'James Bradley and the Eighteenth Century "Gap" in Attempts to Measure Annual Stellar Parallax'. *Notes and Records of the Royal Society of London* 37 (1982–83): 83–100.

Wilson, Curtis. 'Astronomy and Cosmology'. In *The Cambridge History of Science, Volume 4: Eighteenth-Century Science* (pp. 328–53). Edited by Roy Porter. Cambridge, 2003.

— 'Clairaut's Calculation of the Eighteenth-Century Return of Halley's Comet'. *Journal for the History of Astronomy* 24 (1993): 1–15.

Wilson, David B. *Seeking Nature's Logic: Natural Philosophy in the Scottish Enlightenment*. University Park, PA, 2009.

Wood, Paul. *The Aberdeen Enlightenment: The Arts Curriculum in the Eighteenth Century*. Aberdeen, 1993.

— '"The Fittest Man in the Kingdom": Thomas Reid and the Glasgow Chair of Moral Philosophy'. *Hume Studies* 23 (1997): 277–313.

— 'The Hagiography of Common Sense: Dugald Stewart's *Account of the Life and Writings of Thomas Reid*'. In *Philosophy, Its History and Historiography* (pp. 305–22). Edited by A. J. Holland. Dordrecht, 1985.

— '"Jolly Jack Phosphorous" in the Venice of the North; or, Who Was John Anderson?' In *The Glasgow Enlightenment* (pp. 111–32). Edited by Andrew Hook and Richard B. Sher. East Linton, 1995.

— *The Life of Thomas Reid: The Culture of the Mind*. Edinburgh, forthcoming.

— 'Reid, Parallel Lines and the Geometry of Visibles'. *Reid Studies* 2 (1998): 27–41.

— 'Science, Philosophy and the Mind'. In *The Cambridge History of Science, Volume IV: Eighteenth-Century Science* (pp. 800–24). Edited by Roy Porter. Cambridge, 2003.

— 'Science in the Scottish Enlightenment'. In *The Cambridge Companion to the Scottish Enlightenment* (pp. 94–116). Edited by Alexander Broadie. Cambridge, 2003.

— 'Science, the Universities and the Public Sphere in Eighteenth-Century Scotland'. *History of Universities* 13 (1994): 99–135.

— 'Thomas Reid, Natural Philosopher: A Study of Science and Philosophy in the Scottish Enlightenment'. PhD thesis, University of Leeds, 1984.

— 'A Virtuoso Reader: Thomas Reid and the Practices of Reading in Eighteenth-Century Scotland'. *Journal of Scottish Thought* 4 (2011): 33–74.

— 'Who Was Thomas Reid?' *Reid Studies* 5 (2001): 35–51.

Woolf, Harry. *The Transits of Venus: A Study of Eighteenth-Century Science*. Princeton, NJ, 1959.

Wulf, Andrea. *Chasing Venus: The Race to Measure the Heavens*. New York, NY, 2012.

Index of Persons and Titles

General Index